Springer Proceedings in Mathematics & Statistics

Volume 277

Springer Proceedings in Mathematics & Statistics

This book series features volumes composed of selected contributions from workshops and conferences in all areas of current research in mathematics and statistics, including operation research and optimization. In addition to an overall evaluation of the interest, scientific quality, and timeliness of each proposal at the hands of the publisher, individual contributions are all refereed to the high quality standards of leading journals in the field. Thus, this series provides the research community with well-edited, authoritative reports on developments in the most exciting areas of mathematical and statistical research today.

More information about this series at http://www.springer.com/series/10533

Jörg Feldvoss · Lauren Grimley ·
Drew Lewis · Andrei Pavelescu ·
Cornelius Pillen
Editors

Advances in Algebra

SRAC 2017, Mobile, Alabama, USA, March 17–19

Springer

Editors
Jörg Feldvoss
Department of Mathematics and Statistics
University of South Alabama
Mobile, AL, USA

Lauren Grimley
Department of Mathematics
Spring Hill College
Mobile, AL, USA

Drew Lewis
Department of Mathematics and Statistics
University of South Alabama
Mobile, AL, USA

Andrei Pavelescu
Department of Mathematics and Statistics
University of South Alabama
Mobile, AL, USA

Cornelius Pillen
Department of Mathematics and Statistics
University of South Alabama
Mobile, AL, USA

ISSN 2194-1009 ISSN 2194-1017 (electronic)
Springer Proceedings in Mathematics & Statistics
ISBN 978-3-030-11520-3 ISBN 978-3-030-11521-0 (eBook)
https://doi.org/10.1007/978-3-030-11521-0

Library of Congress Control Number: 2018967445

Mathematics Subject Classification (2010): 13F20, 14M17, 14M25, 14M27, 15A42, 16D10, 16D20, 16T05, 16Z05, 17B35, 20C20, 20D10, 20D15, 20G05, 20G10, 20N20, 52B20, 81R10

This Springer imprint is published by the registered company Springer Nature Switzerland AG
The registered company address is: Gewerbestrasse 11, 6330 Cham, Switzerland

To Henry Heatherly and Temple Fay

Preface

The Southern Regional Algebra Conference (SRAC) is an annual weekend conference that has been in existence since 1988. It serves an important role for the research community in algebra in the Gulf Coast Region. The conference series helped to build a network of mathematicians that fosters collaboration and increases the visibility of research programs in the region at the national and international level.

SRAC 2017 was held March 17–19, 2017, on the campus of the University of South Alabama. A total of 89 participants from 37 different institutions attended. At the center of the program were two well-received plenary talks on *Computing with Matrix and Basic Algebras* by Jon Carlson and *Basic Algorithms in the Representation Theory of Groups and Algebras* by Klaus Lux, two leading experts on computational methods in algebra and representation theory. In addition, there were 48 shorter talks, organized in two parallel sessions, including 9 presentations by graduate students. With the help of NSF support, the number of participants increased significantly and, as a result, a broad range of topics was discussed at the conference. The presentations covered both theoretical as well as computational aspects in ring theory, group theory, Lie theory, commutative algebra, algebraic geometry, linear algebra, quantum groups, and other research areas in algebra. Conference abstracts, the program, and a list of participants can be found on the conference webpage at `http://www.southalabama.edu/colleges/artsandsci/mathstat/srac/index.html`.

The Southern Regional Algebra Conference has its origins at the University of Louisiana at Lafayette when, in 1970, algebraists such as Prof. Laszlo Fuchs from Tulane University and Prof. Gilbert Baumslag from Rice University, together with other algebraists from the region, primarily in Louisiana and Texas, first met in Lafayette for the Lloyd Roeling Mathematics Conferences. Henry Heatherly, who arrived at Lafayette in 1968, is one of the founding fathers of the conference series.

The name of the conference has at times varied. Referred to as the "Weekend Algebra Meeting" among other variations, it is now known as the Southern Regional Algebra Conference. Although the conference location has remained primarily in the southeastern United States, the event has attracted algebraists from across the country and indeed the world. Over the years, SRAC has included participating algebraists from Africa, Asia, Europe, and South America.

The Southern Regional Algebra Conference has a rich tradition of embracing new mathematics from across all areas in algebra in an atmosphere of collegial discourse and collaboration. SRAC has always been well known for its friendly and supportive atmosphere that attracts many graduate students and researchers from different institutions, ranging from small undergraduate colleges to large comprehensive research institutions. The conference format is simple, which might account in part for its popularity. There has never been a registration fee and the host institution provides for conference dinners and picnics. The University of Southern Mississippi hosted the meeting regularly from the late 80s until the beginning of the new Millennium. These meetings included a crawfish boil, as well as volleyball and fierce soccer games at Temple "Buzz" Fay's house in Hattiesburg.

The papers in this volume reflect the broad spectrum of topics covered at the conference. They include work on algebraic groups, combinatorial commutative algebra, computational methods for representations of groups and algebras, group theory, Hopf-Galois theory, hypergroups, Lie superalgebras, matrix analysis, ring theory, spherical and algebraic spaces, and tropical algebraic geometry.

We would like to thank the editorial staff at Springer for their initial interest and their help in producing this volume. A special thanks goes to the anonymous referees for their insightful reports. Finally, we would like to take this opportunity to express our appreciation to the participants for making the conference such a success, to the speakers for their interesting talks, and, especially, to the authors of this volume for their contributions.

Mobile, AL, USA
July 2018

Jörg Feldvoss
Lauren Grimley
Drew Lewis
Andrei Pavelescu
Cornelius Pillen

The following is a list of SRAC's host institutions:

1988 University of Southern Mississippi
1989 University of Houston
1990 University of Louisiana, Lafayette
1992 University of Southern Mississippi
1993 Baylor University
1994 University of Louisiana, Lafayette
1995 University of Southern Mississippi
1996 Texas A&M University
1996 University of Louisiana, Lafayette
1998 University of Southern Mississippi
1999 University of Louisiana, Lafayette
2000 University of Louisiana, Lafayette (Special Session, AMS Sectional Meeting)
2001 University of Southern Mississippi
2002 Loyola University, New Orleans
2003 Florida Atlantic University
2003 University of Louisiana, Lafayette
2004 Southeastern Louisiana University
2006 Auburn University
2007 University of Louisiana, Lafayette
2008 University of Colorado, Colorado Springs
2009 University of South Alabama
2010 Auburn University, Montgomery
2010 University of Louisiana, Lafayette
2012 Clayton State University and Georgia College and State University
2013 Southeastern Louisiana University
2014 Auburn University, Montgomery
2015 University of Louisiana, Lafayette
2016 Auburn University
2017 University of South Alabama
2018 Auburn University, Montgomery
2019 University of Louisiana, Lafayette

Southern Regional Algebra Conference, University of South Alabama, Mobile, Alabama, March 17–19, 2017

Acknowledgements

The organizers of SRAC 2017 are grateful for the financial support received from the National Science Foundation and the University of South Alabama. Many thanks to the College of Arts and Sciences and the Department of Mathematics and Statistics at the University of South Alabama for their invaluable help in hosting this event.

Contents

Contributors

Ulrich Albrecht Department of Mathematics and Statistics, Auburn University, Auburn, AL, USA

Arindam Banerjee Ramakrishna Mission Vivekananda University, Belur, West Bengal, India

Selvi Kara Beyarslan Department of Mathematics and Statistics, University of South Alabama, Mobile, AL, USA

Gary F. Birkenmeier Department of Mathematics, University of Louisiana at Lafayette, Lafayette, LA, USA

Mahir Bilen Can Mathematics Department, Tulane University, New Orleans, LA, USA

Jon F. Carlson Department of Mathematics, University of Georgia, Athens, GA, USA

Tianran Chen Department of Mathematics and Computer Science, Auburn University Montgomery, Montgomery, AL, USA

Robert Davis Department of Mathematics, Michigan State University, East Lansing, MI, USA

Christopher M. Drupieski Department of Mathematical Sciences, DePaul University, Chicago, IL, USA

Huajun Huang Department of Mathematics and Statistics, Auburn University, Auburn, AL, USA

Hà Huy Tài Department of Mathematics, Tulane University, New Orleans, LA, USA

Michael Joyce Mathematics Department, Tulane University, New Orleans, LA, USA

Yeliz Kara Department of Mathematics, Bursa Uludağ University, Bursa, Turkey

Alan Koch Department of Mathematics, Agnes Scott College, Decatur, GA, USA

Timothy Kohl Department of Mathematics and Statistics, Boston University, Boston, MA, USA

Jonathan R. Kujawa Department of Mathematics, University of Oklahoma, Norman, OK, USA

Mark L. Lewis Department of Mathematical Sciences, Kent State University, Kent, OH, USA

Klaus Lux Department of Mathematics, University of Arizona, Tucson, AZ, USA

Bradley McQuaig Department of Mathematics and Statistics, Auburn University, Auburn, AL, USA

Nham V. Ngo University of North Georgia–Gainesville, Oakwood, GA, USA

Elena Poletaeva School of Mathematical and Statistical Sciences, University of Texas Rio Grande Valley, Edinburg, TX, USA

Matthew F. Ragland Auburn University at Montgomery, Montgomery, USA

Alexander Ryba Department of Computer Science, Queens College/CUNY, Queens, NY, USA

Leonard L. Scott Department of Mathematics, University of Virginia, Charlottesville, VA, USA

Adnan Tercan Department of Mathematics, Hacettepe University, Ankara, Turkey

Paul J. Truman School of Computing and Mathematics, Keele University, Staffordshire, UK

Robert Underwood Department of Mathematics and Computer Science, Auburn University at Montgomery, Montgomery, AL, USA

Ethan C. Zell Department of Mathematics, University of Virginia, Charlottesville, VA, USA

Paul-Hermann Zieschang School of Mathematical and Statistical Sciences, University of Texas Rio Grande Valley, Edinburg, TX, USA

Hattori-Torsion-Freeness and Endomorphism Rings

Ulrich Albrecht and Bradley McQuaig

Abstract This paper introduces the notions of K^r-faithfulness and quasi-flatness. They are used to discuss non-singularity and Hattori-torsion-freeness in the context of endomorphism rings. Several additional examples are given.

Keywords Endomorphism ring · Non-singular · Torsion-free

1 Introduction

Every right R-module A can be viewed as an E-R-bimodule where $E = End_R(A)$ denotes the R-endomorphism ring of A. There is an adjoint pair (H_A, T_A) of functors between the category M_R of right R-modules and the category M_E of right E-modules defined by $H_A(M) = \mathrm{Hom}_R(A, M)$ and $T_A(N) = N \otimes_E A$ for all $M \in M_R$ and all $N \in M_E$. These functors induce natural maps $\theta_M : T_A H_A(M) \to M$ and $\Phi_N : N \to H_A T_A(N)$ defined by $\theta_M(\alpha \otimes a) = \alpha(a)$ and $[\Phi_N(x)](a) = x \otimes a$ for all $\alpha \in H_A(M)$, $x \in N$ and $a \in A$. An R-module M is *A-solvable* if θ_M is an isomorphism, it is *(finitely) A-generated* if it is an epimorphic image of $\oplus_I A$ for some index-set I (with $|I| < \omega$). It is easy to see that M is A-generated if and only if $S_A(M) = M$ where $S_A(M) = im(\theta_M)$.

An important class of A-solvable modules was introduced by Arnold, Lady, and Murley in [10, 11]. A right R-module is *(finitely) A-projective* if it is isomorphic to a direct summand of $\oplus_I A$ for some (finite) index-set I. While finitely A-projective modules are always A-solvable, this holds for all A-projective modules if A is *self-small*, i.e., if the natural map $H_A(\oplus_I A) \to \Pi_I E$ actually maps into $\oplus_I E$ for all index-sets I [11].

U. Albrecht (✉) · B. McQuaig
Department of Mathematics and Statistics, Auburn University,
Auburn, AL 36849, USA
e-mail: albreuf@auburn.edu

B. McQuaig
e-mail: bsm0012@auburn.edu

J. Feldvoss et al. (eds.), *Advances in Algebra*, Springer Proceedings
in Mathematics & Statistics 277, https://doi.org/10.1007/978-3-030-11521-0_1

It is the main goal of this paper to investigate the behavior of torsion-free E-module with respect to the functors H_A and T_A. However, our discussion encounters the immediate difficulty that the notion of torsion-freeness, which was originally defined for modules over integral domains, can be extended in several nonequivalent ways to noncommutative rings [6]. In the following, we mostly focus on Hattori's notion of torsion-freeness, and call a right R-module M *(H-)torsion-free* if $\mathrm{Tor}_1^R(M, R/Rr) = 0$ for all $r \in R$. In particular, M is H-torsion-free if and only if $xr = 0$ yields $x \in M\ ann_\ell(r)$ for all $x \in M$ and $r \in R$ [20]. On the other hand, Goodearl [19] and Stenstrom [22] called a right R-module M *non-singular (singular)* provided $Z(M) = 0$ $(Z(M) = M)$. Here, $Z(M) = \{x \in M \,|xI = 0$ for some essential right ideal I of $R\}$ is the *singular submodule of* M. Moreover, R is *a right non-singular ring* if $Z_r(R) = Z(R_R) = 0$. An important class of rings in the discussion of torsion-free and non-singular modules are the *p.p.-rings*, i.e., the rings for which all its principal right ideals are projective. Moreover, if R is a right non-singular ring, its injective hull $Q^r = Inj(R_R)$ of R_R carries a natural ring-structure induced by R. This ring is called the maximal right ring of quotients of R. We want to emphasize that non-singular modules need not be H-torsion-free and vice versa [6].

2 *H*-Torsion-Freeness and P.P-Rings

We want to remind the reader that an R-module A is *flat over its endomorphism ring* if the functor T_A is exact. Ulmer showed in [23] that a right R-module A is flat as a module over its endomorphism ring if and only if ker α is A-generated for all $\alpha \in \mathrm{Hom}_R(A^n, A)$ and all $n < \omega$. In particular, A-generated submodules of A-solvable modules are A-generated in this case [3]. Moreover, the module A is faithful as an E-module if $T_A(M) \neq 0$ for all nonzero finitely generated right E-modules M. A faithfully flat module is both faithful and flat as an E-module, and has the additional property that it is fully faithful, i.e., $T_A(M) \neq 0$ for all nonzero right E-modules M. Abelian groups which are faithfully flat over their endomorphism ring were discussed in detail in [1, 3], and the basic results of these papers carry over to modules over an arbitrary ring R.

We begin our discussion by giving a similar description of the fact that A is H-torsion-free as an E-module.

Lemma 1 *Let A be a right R-module.*

(a) A is H-torsion-free as a left E-module if and only if ker α is A-generated for all $\alpha \in E$.
(b) If A is H-torsion-free as a left E-module, then every regular endomorphism of A is one-to-one.
(c) A is flat as an E-module if and only if A^n is H-torsion-free as an $E_R(A^n)$-module for all $n < \omega$.

Proof (a) Suppose that A is H-torsion-free. If $a \in \ker\ \alpha$ for some $\alpha \in E$, then $a \in ann_r(\alpha)A$ yields $a = \beta_1 a_1 + \cdots + \beta_n a_n \in \beta_1(A) + \cdots + \beta_n(A)$ with $\beta_i \in ann_r(\alpha)$ and $a_i \in A$ for $i = 1, \ldots, n$. Since $\alpha\beta_i = 0$, we obtain $\beta(A) \subseteq S_A(\ker\ \alpha)$. Conversely, suppose $ker\ \alpha$ is A-generated for all $\alpha \in E$. If $\alpha a = 0$, then $a = \beta_1(a_1) + \cdots + \beta_n(a_n)$ for $\beta_1, \ldots, \beta_n \in H_A(\ker\ \alpha)$ and $a_1, \ldots, a_n \in A$. Since $ann_r(\alpha) = H_A(\ker\ \alpha)$, we obtain $a \in ann_r(\alpha)A$.

(b) For a regular endomorphism c of E, we consider the commutative diagram

$$\begin{array}{ccccc} 0 = \mathrm{Tor}_1^E(E/cE, A) & \longrightarrow & T_A(E) & \xrightarrow{T_A(c1_E)} & T_A(E) \\ & & \cong \big\downarrow \theta_A & & \cong \big\downarrow \theta_A \\ & & A & \xrightarrow{c1_E} & A \end{array}$$

which yields that c is one-to-one.

(c) Suppose that A^n is H-torsion-free as a module over its endomorphism ring. Since every map $\alpha : A^n \to A$ can be viewed as an endomorphism of A^n, we obtain $S_A(\ker\ \alpha) = S_{A^n}(\ker\ \alpha) = \ker\ \alpha$ by (a). The converse follows directly from Ulmer's Theorem and (a).

Our next results describe the R-modules whose endomorphism ring is a right p.p.-ring:

Theorem 1 *The following conditions are equivalent for a right R-module A:*

(a) A is H-torsion-free as a left E-module, and E is a right p.p.-ring.
(b) $\ker\ \alpha$ *is a direct summand of A for all* $\alpha \in E$.

Proof *(a)* $\Rightarrow$ *(b)*: For $\alpha \in E$, consider the sequence $0 \to U \to A \xrightarrow{\alpha} A$ in which $U = \ker\ \alpha$ is A-generated by Lemma 1. It induces the exact sequence

$$0 \to H_A(U) \to E \xrightarrow{H_A(\alpha)} E$$

where $M = im\ H_A(\alpha) = \alpha E$ is projective since E is a right p.p.-ring. Therefore, the monomorphism $0 \to H_A(U) \to H_A(A)$ splits. This yields the splitting of the top-row of the diagram

$$\begin{array}{ccccc} 0 & \longrightarrow & T_A H_A(U) & \longrightarrow & T_A H_A(A) \\ & & \theta_U \big\downarrow & & \theta_A \big\downarrow \wr \\ 0 & \longrightarrow & U & \longrightarrow & A \end{array}$$

in which θ_U is onto because U is A-generated. Since a simple diagram chase shows that it also is one-to-one, the bottom-row has to split too.

(b) $\Rightarrow$ *(a)*: Clearly, A is a H-torsion-free as an E-module by Lemma 1. For $\alpha \in E$, let $I = \alpha E$, and consider the commutative diagram

$$\begin{array}{ccccc} & & H_AT_A(I) & \longrightarrow & H_AT_A(E) \\ & & \uparrow{\scriptstyle \phi_I} & & \wr\uparrow{\scriptstyle \phi_E} \\ 0 & \longrightarrow & I & \longrightarrow & E \end{array}$$

which yields that ϕ_I is a monomorphism. The epimorphism $\pi : E \to I$ of right E-modules defined by $\pi(1_E) = \alpha$ induces the sequence

$$0 \to \ker\ T_A(\pi) \to T_A(E) \stackrel{T_A(\pi)}{\longrightarrow} T_A(I) \to 0.$$

Since A is H-torsion-free as an E-module, the sequence

$$0 = \mathrm{Tor}_1^R(E/I, A) \to T_A(I) \to T_A(E)$$

is exact. Thus, $T_A(\pi)$ can be viewed as an endomorphism of A whose kernel splits by (b). Consequently, the top-row in the diagram

$$\begin{array}{ccccc} H_AT_A(E) & \stackrel{H_AT_A(\pi)}{\longrightarrow} & H_AT_A(I) & & \\ \wr\uparrow{\scriptstyle \phi_E} & & \uparrow{\scriptstyle \phi_I} & & \\ E & \stackrel{\pi}{\longrightarrow} & I & \longrightarrow & 0, \end{array}$$

splits. If $\delta : H_AT_A(I) \to H_AT_A(E)$ satisfies $H_AT_A(\pi)\delta = 1$, then

$$\begin{aligned} \phi_I\pi[H_AT_A(\phi_E)]^{-1}\delta\phi_I &= H_AT_A(\pi)H_AT_A(\phi_E)[H_AT_A(\phi_E)]^{-1}\delta\phi_I \\ &= \phi_I H_AT_A(\pi)\delta\phi_I = \phi_I 1_I. \end{aligned}$$

Since ϕ_I is one-to-one, $[H_AT_A(\phi_E)]^{-1}\delta\phi_I$ is a splitting map for π; and E is a right p.p.-ring.

A ring R is *right semi-hereditary* if all finitely generated right ideals of R are projective.

Corollary 1 *The following conditions are equivalent for a right R-module A:*

(*a*) *A is flat as a left E-module, and E is right semi-hereditary.*
(*b*) ker α *is a direct summand of* A^n *for all* $\alpha \in E_R(A^n)$ *and all* $n < \omega$.
(*c*) *If* P_1 *and* P_2 *are finitely A-projective, then* ker α *is a direct summand of* P_1 *for all* $\alpha \in Hom_R(P_1, P_2)$.

Proof (*a*) $\Rightarrow$ (*c*): We select finitely A-projective modules Q_1 and Q_2 such that

$$P_1 \oplus Q_1 \cong P_2 \oplus Q_2 \cong A^n$$

for some $n < \omega$. Then α induces an endomorphism α' of A^n with the property *ker* $\alpha' = Q_1 \oplus \ker\ \alpha$. By Part (d) of Lemma 1, A^n is H-torsion-free as a module over its endomorphism ring. Moreover, $E_R(A^n)$ is a right p.p.-ring since E is right semi-hereditary [13, Theorem 8.17]. By Theorem 1, ker α' is a direct summand of $P_1 \oplus Q_1$, from which we obtain that ker α is a direct summand of P_1.

It remains to show $(b) \Rightarrow (a)$. For this, observe that A^n is H-torsion-free as a module over its endomorphism ring, and that $E_R(A^n)$ is a right p.p.-ring by Theorem 1. Because of Part (c) of Lemma 1 and [13, Theorem 8.17], (a) holds.

We want to remind the reader that A has the *finite Baer-splitting property* if every epimorphism $A^n \to P$ such that P is A-projective of finite A-rank splits. The Baer-splitting property was considered first by Baer in 1937 in [12], and plays an important role in the study of torsion-free modules over integral domains [18].

Corollary 2 ([21]) *If A is a right R-module which is flat as an E-module and has a right semi-hereditary endomorphism ring, then A has the finite Baer-splitting property.*

The ring R of algebraic integers is a Prüfer domain which is not hereditary. Faticoni showed in [16] that there is an Abelian group A with R as its endomorphism ring, which is not flat over as a module over its endomorphism ring. Moreover, since R is an integral domain, A is H-torsion-free as an E-module. On the other hand, we obtain from [4]:

Example 1 Let R be a right semi-hereditary, right strongly non-singular ring without an infinite set of orthogonal idempotents. If A is a right essential two-sided R-submodule of Q^r, then A is flat as a module over its endomorphism ring and $E_R(A)$ is right semi-hereditary.

3 Quasi-Flatness and H-Torsion-Freeness

The right R-module $K^r = Q^r/R$ plays a central role in the study of torsion-free and non-singular modules over right non-singular rings. If A is a non-singular R-module, then every endomorphism of A extends uniquely to an endomorphism of its injective hull $Inj(A)$. We can therefore view $Inj(A)$ as an E-R-bimodule, and call A *quasi-flat* if $Inj(A)$ is flat as a left E-module. The notion of quasi-flatness was introduced in [8] in the case $R = \mathbb{Z}$, and most of its homological properties carry over from that case. In particular, there exist quasi-flat torsion-free Abelian groups which are not E-flat.

Proposition 1 *Let A be a non-singular right module over a non-singular ring R.*

(a) If A is quasi-flat, then $Tor_1^R(M, A)$ *is a singular right R-module for all right E-modules M.*
(b) If R has finite Goldie dimension, and $Tor_1^R(M, A)$ *is a singular right R-module for all right E-modules M, then A is quasi-flat.*
(c) If R has finite Goldie dimension and A is flat as an E-module, then A is quasi-flat.

Proof (a) We consider the induced sequence

$$0 = \mathrm{Tor}_2^R(M, Inj(A)) \to \mathrm{Tor}_2^R(M, Inj(A)/A)$$

$$\to \mathrm{Tor}_1^R(M, A) \to \mathrm{Tor}_1^R(M, Inj(A)) = 0.$$

Since $Inj(A)/A$ is singular, it therefore suffices to show that $\mathrm{Tor}_2^R(M, N)$ is singular whenever N is an E-R-bimodule which is singular as an R-module. For this, we consider exact sequences $0 \to U \to F \to M \to 0$ and $0 \to V \to P \to U \to 0$ with F and P projective. They induce the exact sequences

$$0 = \mathrm{Tor}_2^R(F, N) \to \mathrm{Tor}_2^R(M, N) \to \mathrm{Tor}_1^R(U, N) \to \mathrm{Tor}_1^R(F, N) = 0$$

and

$$0 = \mathrm{Tor}_1^R(P, N) \to \mathrm{Tor}_1^R(U, N) \to V \otimes_E N$$

of right R-modules. Since $V \otimes_E N$ is singular as a right R-module, a) follows.

(b) Consider the exact sequence

$$\mathrm{Tor}_1^R(M, A) \to \mathrm{Tor}_1^R(M, Inj(A)) \to \mathrm{Tor}_1^R(M, Inj(A)/A).$$

Arguing as in (a) we obtain that $\mathrm{Tor}_1^R(M, Inj(A)/A)$ is singular. By (b), $\mathrm{Tor}_1^R(M, A)$ is singular, from which we get that $\mathrm{Tor}_1^R(M, Inj(A))$ is singular. On the other hand, the latter is also a right Q^r-module. By [22], Q^r is a semi-simple Artinian ring whenever R has finite right Goldie dimension. Hence, all right Q^r-modules are non-singular as R-modules. This is only possible if $\mathrm{Tor}_1^R(M, Inj(A)) = 0$.

(c) For a right E-module M, we have an exact sequence

$$0 = \mathrm{Tor}_1^R(M, A) \to \mathrm{Tor}_1^R(M, Inj(A)) \to \mathrm{Tor}_1^R(M, Inj(A)/A).$$

Since $Inj(A)$ is a module over the semi-simple ring Q^r, it is non-singular as a right R-module. On the other hand, a projective resolution $0 \to U \to F \to M \to 0$ of M induces the exact sequence

$$0 = \mathrm{Tor}_1^R(F, Inj(A)/A) \to \mathrm{Tor}_1^R(M, Inj(A)/A) \to U \otimes_E Inj(A)/A$$

from which we get that $\mathrm{Tor}_1^R(M, Inj(A)/A)$ is singular as an R-module. Thus, $\mathrm{Tor}_1^R(M, Inj(A)) = 0$.

A right R-module A is *faithful* over its endomorphism ring if $IA \neq A$ for all proper right ideals of E. It is *fully faithful* if $M \otimes_E A \neq 0$ for all nonzero right E-modules M. If A is faithfully flat over E, then it is fully faithful. Finally, a non-singular right R-module A with a right non-singular endomorphism ring is K^r*-faithful* if $T_A(M) \neq 0$ for all nonzero submodules M_E of K^r. While an R-module which is faithfully flat over its endomorphism ring is K^r-faithful, the converse may fail:

Example 2 There exist a torsion-free Abelian group A of finite rank such that

(i) E is a p.p.-ring.
(ii) A is E-flat (and hence K^r-faithful).
(iii) A is not faithful.

Proof The Abelian group $A = \mathbb{Z} \oplus \mathbb{Q}$ is flat as an E-module by Ulmer's Theorem, but not faithful as a module over its endomorphism ring since there exist a non-splitting sequence $P_1 \to P_2 \to 0$ with P_1 and P_2 A-projective ([1, 10]). Moreover, $E(A) \cong \begin{pmatrix} \mathbb{Z} & 0 \\ \mathbb{Q} & \mathbb{Q} \end{pmatrix}$ which is a right and left p.p.-ring [6].

We now give an example of a right R-module which is not K^r-faithful although its endomorphism ring is a p.p.-ring. We want to remind the reader that the symbol J_p denotes the p-adic integers, e.g., see [17], which is the completion of $\mathbb{Z}$ in the p-adic topology with basis $\{p^n\mathbb{Z} | n < \omega\}$.

Example 3 The Abelian group $A = \mathbb{Z}(p^\infty)$ has J_p its endomorphism ring. Since $F = \mathbb{Q} \otimes_{\mathbb{Z}} J_p$ is its field of quotients, the additive group of every submodule M of F/J_p is a p-group. Thus, $M \otimes_{J_p} A = 0$, and A is not K^r-faithful. Moreover, there exists an endomorphism of A whose kernel is not a direct summand of A, although the endomorphism ring of A is a principal ideal domain. Thus, the two conditions in part (a) of Theorem 1 are independent.

Although all submodules of a direct sum of copies of K^r are singular, not all singular modules can be embedded into such a direct sum [5].

Theorem 2 *Let R be a right non-singular ring R. If A is a non-singular quasi-flat right R-module with a right non-singular endomorphism ring, then*

(a) A is K^r-faithful and $T_A(M) \neq 0$ for all nonzero finitely presented right R-modules M with projective dimension at most 1.
(b) If E also has finite right Goldie dimension, then $T_A(M) \neq 0$ for all nonzero right R-modules M with projective dimension at most 1.

Proof Suppose that M is a submodule of Q^r containing E with $T_A(M/E) = 0$, and let $\iota : E \to M$ be the inclusion map. The quasi-flatness of A yields an exact sequence

$$\mathrm{Tor}_1^R(M/E, A) \to T_A(E) \to T_A(M) \to 0$$

in which $\mathrm{Tor}_1^R(M/E, A)$ is singular by Proposition 1. Since $T_A(E) \cong A$ is non-singular, the map $T_A(E) \to T_A(M) \to 0$ is an isomorphism. Applying the functor H_A gives the top-row in the commutative diagram

$$\begin{array}{ccccccc} 0 & \longrightarrow & H_A T_A(E) & \xrightarrow{H_A T_A(\iota)} & H_A T_A(M) & \longrightarrow & 0 \\ & & \wr \uparrow \Phi_E & & \uparrow \Phi_M & & \\ 0 & \longrightarrow & E & \xrightarrow{\iota} & M. & & \end{array}$$

Since $H_A T_A(\iota)\Phi_E = \Phi_M \iota$ is an isomorphism, ι splits. Because E_R is essential in M, this is only possible if $M = E$.

Now, consider a nonzero right E-module M of projective dimension at most 1 such that $T_A(M) = 0$. Select a free E-module $F \cong \oplus_J E$ and a projective module P containing F such that $P/F \cong M$. If M is finitely presented, then we can choose J to be finite, and $\oplus_J Q^r$ is the injective hull of F. If E has finite Goldie dimension, this remains true even if J is infinite [22].

If M is not singular, then $Z(M/Z(M)) \neq 0$ because E is right non-singular. If $0 \neq x \in M/Z(M)$, then $ann_r(x)$ is not an essential right ideal of E. Thus, we can find a nonzero right ideal I of E and a monomorphism $\alpha : I \to M/Z(M)$. Since A is quasi-flat, the kernel of the evaluation map $T_A(I) \to IA$ is singular as an R-module. Because $IA \neq 0$ is non-singular, $T_A(I)$ cannot be singular. However, $T_A(M) = 0$ yields $T_A(M/Z(M)) = 0$, from which we get the exact sequence

$$\mathrm{Tor}_1^R(coker\ \alpha, A) \to T_A(I) \to T_A(M/Z(M)) = 0$$

which yields that $T_A(I)$ is singular, a contradiction. Therefore, M is singular. Because $M \cong P/F$ and P is non-singular, F is an essential submodule of P. In view of $Inj(F) = \oplus_I Q^r$, we obtain $M \cong P/F \subseteq [\oplus_I Q^r]/F \cong \oplus_I K^r$. Since $M \neq 0$, we can find a nonzero image N of M in Q^r/E. But $T_A(N) = 0$ contradicts the K^r-faithfulness of A.

An A-generated group G *A-torsion-free* [2] if every finitely A-generated subgroup of G is isomorphic to a subgroup of an A-projective group. In particular, every A-torsion-free group is A-solvable. Moreover, we showed in [2] that the kernel of every map $A^n \to G$ splits whenever G is A-torsion-free and E is a right semi-hereditary right strongly non-singular ring without an infinite family of orthogonal idempotents.

We now turn to the related question which modules have the property that the kernel of every homomorphism $A \to M$ splits whenever $H_A(M)$ is H-torsion-free. We want to emphasize that an A-generated module need not be A-solvable if $H_A(M)$ is H-torsion-free. For instance, if A is a torsion-free group A of finite rank with a semi-prime p.p. endomorphism ring, then $H_A(G)$ is Hattori-torsion-free for all A-generated torsion-free groups G, although such a G need not be A-solvable.

Theorem 3 *Let A a quasi-flat non-singular right R-module which is H-torsion-free as an E-module. If E is a p.p-ring without an infinite family of orthogonal idempotents, then the following conditions are equivalent for an A-generated R-module M:*

(a) ker ϕ *is direct summand of A for all* $\phi \in H_A(M)$.
(b) $\phi(A)$ *is isomorphic to an A-projective submodule of A for all* $\phi \in H_A(M)$.
(c) $H_A(M)$ *is H-torsion-free, and* $\phi(A)$ *is A-solvable for all* $\phi \in H_A(M)$.

Proof Observe that A is K^r-faithful by Theorem 2, and consider a right E-module M.

$(a) \Rightarrow (c)$: Since finitely A-projective modules are A-solvable, it remains to show that $H_A(M)$ is torsion-free. Suppose that $\phi \in H_A(M)$ and $\alpha \in E$ with $\phi\alpha = 0$. By (a), ker ϕ is a direct summand of A, and can be written as $e(A)$ for some idempotent e of E. Hence, $\phi = \phi e + \phi(1-e) = \phi(1-e)$. On the other hand, $\phi\alpha = 0$ yields $\alpha(A) \subseteq e(A)$. Thus, $1 - e \in ann_\ell(\alpha)$ and $\phi \in H_A(M)ann_\ell(\alpha)$, and $H_A(M)$ is torsion-free.

$(c) \Rightarrow (b)$: Suppose $\phi \in H_A(M)$, and observe that ϕE is a projective E-module by [6] because $H_A(M)$ is H-torsion-free. In particular, $H_A(M)$ is a non-singular E-module. For $U = \ker\ \phi$, we obtain the induced exact sequence

$$0 \to H_A(U) \to H_A(A) \overset{H_A(\phi)}{\longrightarrow} H_A(\phi(A))$$

in which $\phi(A)$ is A-solvable by (c). The embedding $\phi E \subseteq H_A(\phi(A))$ induces the top-row of the commutative diagram

$$\begin{array}{ccccccc} T_A(\phi E) & \longrightarrow & T_A H_A(\phi(A)) & \longrightarrow & T_A(H_A(\phi(A))/\phi E) & \longrightarrow & 0 \\ \downarrow \theta & & \wr \downarrow \theta_{\phi(A)} & & & & \\ \phi(A) & \overset{1}{\longrightarrow} & \phi(A) & & & & \end{array}$$

in which the evaluation map θ is onto. Therefore, $T_A(H_A(\phi(A))/\phi E) = 0$.

If $H_A(\phi(A))/\phi E$ were not singular, then it would contain an element x whose right annihilator is not essential in E. Hence, we could find a nonzero submodule of xE which is isomorphic to a right ideal I and a monomorphism $\alpha : I \to H_A(\phi(A))/\phi E$. Since A is quasi-flat, this would yield that $T_A(I)$ is singular as in the proof of Theorem 2. On the other hand, it cannot be singular since it has IA as a nonzero non-singular image. Since $H_A(M)$ is non-singular, $H_A(\phi(A))$ is an essential extension of ϕE. But the latter is isomorphic to a direct summand of E since it is projective, say $\phi E \oplus X \cong E$. Thus, $H_A(\phi(A)) \oplus X$ is an essential extension of E, and hence can be embedded into Q^r. Consequently, $H_A(\phi(A))/\phi E$ is isomorphic to a submodule of K^r with $T_A(H_A(\phi(A))/\phi E) = 0$, from which we get $H_A(\phi(A)) = \phi E$ is isomorphic to a direct summand of E. Hence, $\phi(A) \cong T_A H_A(\phi(A))$ is isomorphic to a direct summand of A.

$(b) \Rightarrow (a)$: We consider an exact sequence $0 \to U \to A \overset{\pi}{\to} P \to 0$ in which P is an A-projective submodule of M. By (b), P is isomorphic to a submodule of A, and we can view π as an endomorphism of A. Theorem 1 yields that U is a direct summand of A.

We want to remind the reader that $X^* = \{\alpha \in E | \alpha(x) = 0\}$ for any subset X of a right R-module A and $Y^* = \{a \in A | \alpha(Y) = 0\}$ for any subset Y of E [11].

Corollary 3 *Let A a non-singular right R-module which is flat as an E-module and whose endomorphism ring does not contain an infinite family of orthogonal idempotents.*

(a) E is a p.p.-ring if and only if the three conditions in Theorem 3 are equivalent for all right E-modules M.

(b) A is self-small if E is a p.p.-ring.

Proof (a) The proof of Theorem 3 remains valid if one assumes that A is E-flat instead of H-torsion-free and quasi-flat as an E-module. Therefore, it suffices to show that E is a p.p-ring if the three conditions are equivalent. Because A is E-flat, $\phi(A)$ is A-solvable for all $\phi \in E$. Since $E = H_A(A)$ is torsion-free, we obtain that ker ϕ is a direct summand of A. Now apply Theorem 1.

(b) To see that A is self-small if E is a right p.p.-ring, consider a subset X of E. We show that there are $x_0, \ldots, x_n \in X$ such that $X^* = \{x_0, \ldots, x_n\}^*$. If this is not the case, then we obtain a subset $\{x_n | n < \omega\}$ of E such that

$$\{x_0, \ldots, x_{n+1}\}^* \neq \{x_0, \ldots, x_n\}^*$$

for all n. Once we have shown that $U_n = \{x_0, \ldots, x_n\}^*$ is a direct summand of A, then E has an infinite family of orthogonal idempotents since a p.p.-ring without an infinite family of orthogonal idempotents satisfies the ascending and descending chain condition for right annihilators [13, Lemma 8.4]. We consider the exact sequence

$$0 \to U_n \to A \xrightarrow{\beta} A^n$$

in which U_n is A-generated since A is E-flat. Thus,

$$0 \to H_A(U_n) \to H_A(A) \xrightarrow{H_A(\beta)} H_A(A^n)$$

is an exact sequence of right E-modules in which *im* $H_A(\beta)$ is projective by [6]. Thus $H_A(U_n)$ is a direct summand of $H_A(A)$. As before, this yields that U_n is a direct summand of A in view of the commutative diagram

$$\begin{array}{ccccc} 0 & \longrightarrow & T_A H_A(U_n) & \longrightarrow & T_A H_A(A) \\ & & \downarrow \theta_{U_n} & & \wr\downarrow \theta_A \\ 0 & \longrightarrow & U_n & \longrightarrow & A \end{array}$$

in which θ_{U_n} is onto since U_n is A-generated.

The next result shows that the equivalences in Part (a) may fail if there are infinite families of orthogonal idempotents. For further results relating non-singularity and H-torsion-freeness, we refer the reader to [6].

Example 4 Let F be a field, and $R = F^I$ for some infinite index-set I. Then, R is a commutative p.p.-ring, and $I = \oplus_I F$ is an essential ideal of R such that $Rr \cap I = Ir$ for all $r \in R$. Thus, $M = R/I$ is a H-torsion-free singular R-module. Setting $A = R$ yields that $H_A(M) \cong M$ is H-torsion-free and A-solvable. Thus, A and M satisfy Condition (iii) in Part (a) of Theorem 3. However, ker ϕ is essential in $A = R$ for all $\phi \in H_A(M)$. Hence, Condition (i) in Part (a) fails.

Moreover, an A-projective module P which is an image of A need not always be isomorphic to a submodule of A:

Example 5 Suppose that A_1 and A_2 be two subgroups of $\mathbb{Q}$ of incomparable type, and let $A_3 = A_1 \cap A_2$ and $A_4 = A_1 + A_2$. The group $A = A_1 \oplus A_2 \oplus A_3 \oplus A_4$ is flat as an E-module by Ulmer's Theorem. However, $A_4 \oplus A_4$ is an A-projective image of A which cannot be embedded into A.

4 Non-singular Endomorphism Rings

We begin our discussion with a result similar to Lemma 1 which describes when A is non-singular as a left E-module.

Lemma 2 *A right R-module A is non-singular as a left E-module if and only if $I^{**} = E$ for all essential left ideals of E.*

Proof Suppose $J^{**} = E$ for all essential left ideals J of E. If $0 \neq a \in A$ such that $Ia = 0$ for some left ideal I of E, then $1_A \notin I^{**}$, and hence I is not essential. Conversely, assume that A is non-singular. If $I^{**} \neq E$, then $I^* \neq 0$. But every nonzero $x \in I^*$ satisfies $Ix = 0$, a contradiction.

An A-generated submodule U of A is *A-essential* if there is no nonzero A-generated subgroup V of A with $U \cap V = 0$. If M is a right R-module, then

$$R_A(M) = \cap \{\ker \phi \mid \phi \in \mathrm{Hom}_R(M, A)\}$$

is the *A-radical of M*. A sequence of R-modules is *A-balanced* if A is projective with respect to it. Every sequence $0 \to U \to G \to H \to 0$ of Abelian groups in which G is A-generated and H is A-solvable is A-balanced if A is faithfully flat as an E-module [3].

Proposition 2 *Let A be a right R-module which is faithfully flat as an E-module.*

(a) A right ideal I is essential if and only if IA is A-essential.
(b) E is right non-singular if $R_A(A/U) = A/U$ for all A-essential submodules U of A.

Proof (a) Suppose that $IA \cap V \neq 0$ for all nonzero A-generated submodules of A. If $I \cap \alpha E = 0$ for some nonzero $\alpha \in E$, then $\alpha(A) \cap IA \neq 0$. Since A is flat as an E-module, $\alpha(A)$, IA and $\alpha(A) + IA$ are A-solvable as A-generated submodules of A by the remarks preceding Lemma 1. Moreover, the faithful flatness of A guarantees that the sequence $0 \to \alpha(A) \cap IA \to \alpha(A) \oplus IA \to \alpha(A) + IA \to 0$ is A-balanced. But then, $0 \neq \alpha(A) \cap IA$ is A-generated because A is E-flat [3]. We pick a nonzero $\beta \in H_A(\alpha(A)) \cap H_A(IA)$. Using the faithful flatness of A as a left E-module once more, we obtain $J = H_A(JA)$ for all right ideals J of E. To see this consider the commutative diagram

$$\begin{array}{ccccccc} T_A(J) & \longrightarrow & T_AH_A(JA) & \longrightarrow & T_A(H_A(JA)/J) & \longrightarrow & 0 \\ \downarrow \theta & & \wr \downarrow \theta_{JA} & & & & \\ JA & \xrightarrow{1_{JA}} & JA & & & & \end{array}$$

in which the evaluation map θ is an isomorphism since A is E-flat. Thus the first map in the top-sequence is an isomorphism, and $T_A(H_A(JA)/J) = 0$ from which we get $H_A(JA)/J)) = 0$. Hence, $\beta \in \alpha E \cap I$, contradiction.
Conversely, suppose that I is essential. If V is a nonzero A-generated subgroup of A with $IA \cap V = 0$. Since I is essential, we can find $0 \neq \alpha \in I \cap H_A(V)$. Then, $\alpha(A) \subseteq IA \cap V = 0$, a contradiction.

(b) Suppose that E is non-singular. If there is an A-essential submodule U of A for which we can find a nonzero $\alpha : A/U \rightarrow A$, then we can view α as an endomorphism of A with $\alpha(U) = 0$. Since $U = H_A(U)A$, we obtain that $H_A(U)$ is an essential right ideal of E by (a). But $\alpha H_A(U) = 0$ yields $\alpha = 0$. Conversely, suppose that we can find $0 \neq \alpha \in E$ such that $\alpha I = 0$ for some essential tight ideal I of E. Since IA is A-essential by (a) and $\alpha(IA) = 0$, we obtain $R_A(A/IA) \neq A/IA$, a contradiction.

From this point on, we assume that $R = \mathbb{Z}$, although some results will carry over to modules over integral domains. We first investigate how non-singularity, H-torsion-freeness and $\mathbb{Z}$-torsion-freeness are related for right R-modules of the form $H_A(G)$.

Theorem 4 *Let A be an Abelian group which faithfully flat as an E-module.*

(a) $Q \otimes_r E$ is semi-simple Artinian if and only if, for all A-generated groups G, $H_A(G)$ is non-singular if and only if G is torsion-free.

(b) E is a p.p.-ring and $\mathbb{Q}E$ is semi-simple Artinian if and only if,

(i) E has no infinite family of orthogonal idempotents and,

(ii) for all A-generated groups G, $H_A(G)$ is H-torsion-free as a right E-module if and only if G is torsion-free.

Proof (a) Suppose that, for all A-generated groups G, $H_A(G)$ is non-singular if and only if G is torsion-free. Assume that there is a proper essential right ideal I of E such that E/I is torsion-free. Since A is faithfully flat, $A/IA \cong T_A(E/I)$ is torsion-free. Moreover, the faithful flatness of A guarantees that the first two vertical maps in the next diagram are isomorphisms:

$$\begin{array}{ccccccccc} 0 & \longrightarrow & I & \longrightarrow & E & \longrightarrow & E/I & \longrightarrow & 0 \\ & & \cong \downarrow \Phi_I & & \cong \downarrow \Phi_E & & \downarrow \Phi_{E/I} & & \\ 0 & \longrightarrow & H_AT_A(I) & \longrightarrow & H_AT_A(E) & \longrightarrow & H_AT_A(E/I) & & \end{array}$$

A simple diagram chase shows that $\Phi_{E/I}$ is a monomorphism. Since

$$H_A T_A(E/I) \cong H_A(A/IA),$$

we have that $H_A T_A(E/I)$ is non-singular, and the same holds for E/I. In view of the fact that I is essential, this is only possible if $E = I$.
The converse is obvious.

(b) Suppose that E has no infinite family of orthogonal idempotents and, for all A-generated groups G, $H_A(G)$ is H-torsion-free if and only if G is torsion-free. We first show that every right ideal I of E is H-torsion-free. Since A is faithfully flat, $I = H_A(IA)$, and the latter is H-torsion-free. By [14], submodules of H-torsion-free E-modules are H-torsion-free.
Observe that multiplication by a nonzero integer n is a monomorphism $E \to E$. If X is any H-torsion-free right E-module, then we obtain an exact sequence

$$0 = \mathrm{Tor}_1^E(X, E/En) \to X \otimes_E E \xrightarrow{n\cdot} X \otimes_E E$$

from which it follows that X has a torsion-free additive group. Thus, if M_i is a collection of H-torsion-free E-modules, then each M_i has a torsion-free additive group. The same holds for $M = \Pi_I M_i$ and $T_A(M)$. Thus, $H_A T_A(M)$ is an H-torsion-free module. However, M is isomorphic to a submodule of $H_A T_A(M)$ since A is faithfully flat. Thus, M is H-torsion-free. By [14, Theorem 4.6], E is a left p.p.-ring. Since E has no infinite families of orthogonal idempotents, E is a right p.p.-ring too.

Now suppose that I is any right essential right ideal of E with E/I torsion-free as an Abelian group. Since $T_A(E/I)$ is torsion-free, and E/I is isomorphic to a submodule of $H_A T_A(E/I)$, we obtain that E/I is H-torsion-free. By [6], cyclic H-torsion-free modules are projective. The resulting contradiction shows that $\mathbb{Q}E$ is semi-simple.

Conversely, E clearly has no infinite family of orthogonal idempotents, and non-singularity and H-torsion-freeness are equivalent by [6]. But non-singularity is equivalent to $\mathbb{Z}$-torsion-freeness in this case.

Corollary 4 *Let A be an Abelian group which is faithfully flat as an E-module such that E is a p.p.-ring and $\mathbb{Q}E$ is semi-simple Artinian. The following are equivalent for an A-generated group G:*

(a) G is torsion-free.
(b) $H_A(G)$ is non-singular.
(c) $H_A(G)$ is H-torsion-free.

However, similar to the case of A-solvability, $\phi(A)$ need not be A-projective for all $\phi \in H_A(G)$ whenever G is an A-generated torsion-free group. We want to remind the reader that a torsion-free Abelian group A of finite rank is a *Murley group* if $r_p(A) \leq 1$ for all primes p. Dubois called an Abelian group A *cohesive* if A/U is divisible for all nonzero pure subgroups U of A [15]. Although all cohesive groups are Murley groups, there exist (strongly) indecomposable Murley groups which are not cohesive [7].

Theorem 5 *The following are equivalent for a strongly indecomposable group A of finite rank:*

(a) If G is a torsion-free reduced group, then $\phi(A)$ is A-projective for all $\phi \in H_A(G)$.
(b) A is a cohesive group.

Moreover, if A has the additional property $r_0(E) < r_0(A)$, then J_p is A-generated, but not A-solvable for all primes p with $A \neq pA$.

Proof $(a) \Rightarrow (b)$: Consider a pure subgroup U of A such that $A/U \neq 0$ is A-projective. Let $\pi : A \to A/U$ be the canonical projection. Since A has finite rank, there is $n < \omega$ such that A/U is isomorphic to a direct summand of A^n. In particular, $H_A(A/U)$ is a finitely generated right E-module. For $M = im\ H_A(\pi)$, we consider the induced diagram

$$\begin{array}{ccccccc} T_A(M) & \longrightarrow & T_AH_A(A/U) & \longrightarrow & T_A(H_A(A/U)/M) & \longrightarrow & 0 \\ \downarrow \theta & & \wr \downarrow \theta_{A/U} & & & & \\ A/U & \xrightarrow{1} & A/U & & & & \end{array}$$

in which the evaluation map θ is onto. Thus, $N = H_A(A/U)/M$ is a finitely generated right E-module with $T_A(N) = 0$.

We now show that N^+ is torsion. For this, we may assume that N is cyclic, so that we can find a right ideal I of E with $N \cong E/I$ which yields $A = IA$. Since A is strongly indecomposable, every endomorphism of A is either a monomorphism or nilpotent. If all elements of I are nilpotent, then I is nilpotent since it is contained in $J(\mathbb{Q}E)$, say $I^n = 0$. But then $A = I^nA = 0$, a contradiction. Thus, I contains a monomorphism, and E/I is torsion since A has finite rank.

Hence, $H_A(A/U)/M$ is torsion, and $M \doteq H_A(A/U)$. Since $H_A(A/U)$ is projective, $H_A(\pi)$ quasi-splits, and the same holds for $T_AH_A(\pi)$. We consider the commutative diagram

$$\begin{array}{ccc} T_AH_A(A) & \xrightarrow{T_AH_A(\pi)} & T_AH_A(A/U) \\ \wr \downarrow \theta_A & & \wr \downarrow \theta_{A/U} \\ A & \xrightarrow{\pi} & A/U \end{array}$$

which yields that π quasi-splits. Since A is strongly indecomposable, $U = 0$. In particular, $r_p(A) \leq 1$ for all primes p. Thus, A is cohesive. In particular, it is a strongly indecomposable Murley group, and E is a principal ideal domain.

$(b) \Rightarrow (a)$ is obvious since $A/\ker\ \alpha$ would be divisible for all $0 \neq \alpha \in H_A(G)$.

Finally, assume $r_0(E) < r_0(A)$. Let p be a prime with $A \neq pA$, and choose a p-basic subgroup $F \neq 0$ of A. For every $0 \neq x \in J_p$, there is a nonzero $\alpha : F \to J_p$ with $x \in im\,\alpha$. Since F is p-pure in A, α extends to $\beta : A \to J_p$ from which we obtain that $S_A(J_p) = J_p$. If $r_0(E) < r_0(A)$, then ${}_EA$ cannot contain an essential submodule isomorphic to E. Thus, there is a left E-module morphism $E \oplus E \to A$ since E is a principal ideal domain. In particular, $H_A(J_p)$ is flat as an E-module.

If J_p were A-solvable, then we would get a monomorphism

$$H_A(J_p) \oplus H_A(J_p) \to T_A H_A(J_p) \cong J_p.$$

Since $r_p(A) = 1$, $H_A(J_p) \cong J_p$. Thus, J_p contains a subgroup isomorphic to $J_p \oplus J_p$, which is not possible.

Corollary 5 *A Murley group A whose type-set only contains idempotent types is cohesive if and only if A is homogeneous.*

Proof Suppose that A has the stated property. If U is a nonzero pure subgroup of A, then $r_p(U) = 1$ if and only if $r_p(A) = 1$ for all primes p. Thus, $type(U)_p$ is ∞ if and only if $A = pA$. Since $type(U)$ is idempotent, all rank 1 subgroups of A are isomorphic.

For the converse, let U be any pure rank 1 subgroup of A. Since A is homogeneous, $r_p(U) = 1$ if and only if $r_p(A/U) = 0$. Hence, A/U is divisible by all primes for which $r_p(A) = 1$. A simple induction argument shows that A/V is divisible for all nonzero pure subgroups V of A.

For instance, any rank 2 p-local group A with $E = \mathbb{Z}_p$ is a homogeneous Murley group. If A is a strongly homogeneous Murley group, then all reduced A-generated groups are A-solvable, and the converse holds if A is strongly indecomposable [9]. Theorem 5 shows that a group need not be strongly homogeneous to ensure that $H_A(G)$ is H-torsion-free for all reduced torsion-free groups G.

References

1. Albrecht, U.: Faithful abelian groups of infinite rank. Proc. Amer. Math. Soc. **103**, 21–26 (1988)
2. Albrecht, U.: Endomorphism rings and a generalization of torsion-freeness and purity. Commun. Algebra **17**(5), 1101–1135 (1989)
3. Albrecht, U.: Abelian groups, A, such that the category of A-solvable groups is preabelian. Contemp. Math. **87**, 117–131 (1989)
4. Albrecht, U.: Two-sided essential submodules of $Q^r(R)$. Houston J. Math. **33**(1), 103–123 (2007)
5. Albrecht, U.: Non-singular rings of injective dimension 1, pp. 421–430. Models, Modules and Abelian Groups, De Gruyter (2008)
6. Albrecht, U., Dauns, J., Fuchs, L.; Torsion-freeness and non-singularity over right p.p.-rings. J. Algebra 285, 98–119 (2005)
7. Albrecht, U., Friedenberg, S.: Murley Groups and the torsion-freeness of Ext. J. Algebra **331**, 378–387 (2011)

8. Albrecht, U., Goeters, H.P.: Flatness and the ring of quasi-endomorphisms. Quest. Math. **19**, 379–396 (1996)
9. Albrecht, U., Goeters, H.P.: Strong S-groups. Colloq. Math. **80**, 97–105 (1999)
10. Arnold, D.M., Lady, L.: Endomorphism rings and direct sums of torsion-free Abelian groups. Trans. Amer. Math. Soc. **211**, 225–237 (1975)
11. Arnold, D.M., Murley, C.E.: Abelian groups, A, such that Hom(A)-preserves direct sums of copies of A. Pac. J. Math. **56**, 7–20 (1975)
12. Baer, R.: Abelian groups without elements of finite order. Duke J. Math. **3**, 68–122 (1937)
13. Chatters, A.W., Hajarnavis, C.R.: Rings with Chain Conditions, vol. 44. Pitman Advanced Publishing, Boston, London, Melbourne (1980)
14. Dauns, J., Fuchs, L.: Torsion-freeness for rings with zero-divisors. J. Algebra Appl. **3**, 221–238 (2004)
15. Dubois, D.W.: Cohesive groups and p-adic integers. Publ. Math. Debrecen **12**, 51–58 (1965)
16. Faticoni, T.: Each countable reduced torsion-free commutative ring is a pure subring of an E-ring. Comm. Algebra **15**(12), 2545–2564 (1987)
17. Fuchs, L.: Abelian Groups. Springer (2015)
18. Fuchs, L., Saclce, L.: Modules over Non-Noetherian domains. Amer. Math. Soc. Monogr. **84**, (2000)
19. Goodearl, K.: Ring Theory. Marcel Dekker, New York, Basel (1976)
20. Hattori, A.: A foundation of torsion theory for modules over general rings. Nagoya Math. J. **17**, 147–158 (1960)
21. McQuaig, B.; Morita-equivalence between strongly non-singular rings and the structure of the maximal ring of quotients. Dissertation, Auburn University (2017)
22. Stenström, B.; Rings of Quotients. Lecture Notes in Mathematics, vol. 217. Springer Verlag, Berlin, Heidelberg, New York (1975)
23. Ulmer, F.: A flatness criterrion in Groethendick categories. Invent. Math. **19**, 331–336 (1973)

Regularity of Edge Ideals and Their Powers

Arindam Banerjee, Selvi Kara Beyarslan and Hà Huy Tài

Abstract We survey recent studies on the Castelnuovo–Mumford regularity of edge ideals of graphs and their powers. Our focus is on bounds and exact values of $\text{reg}\, I(G)$ and the asymptotic linear function $\text{reg}\, I(G)^q$, for $q \geq 1$, in terms of combinatorial data of the given graph G.

Keywords Monomial ideals · Edge ideals · Regularity · Asymptotic linearity

1 Introduction

Monomial ideals are classical objects that live at the crossroad of three areas in mathematics: algebra, combinatorics, and topology. Investigating monomial ideals has led to many important results in these areas. The new construction of edge ideals of (hyper)graphs has again resurrected much interest in and regenerated a large amount of work on this class of ideals (cf. [22, 44, 51] and references therein). In this paper, we survey recent works on the Castelnuovo–Mumford regularity of edge ideals of graphs and their powers.

Castelnuovo–Mumford regularity is an important algebraic invariant which, roughly speaking, measures the complexity of ideals and modules. Restricting to the class of edge ideals of graphs, our focus is on studies that relate this algebraic invariant to the combinatorial data of given graphs. Our interest in powers of edge

A. Banerjee
Ramakrishna Mission Vivekananda University, Belur, West Bengal, India
e-mail: 123.arindam@gmail.com

S. K. Beyarslan (✉)
Department of Mathematics and Statistics, University of South Alabama,
Mobile, AL 36688, USA
e-mail: selvi@southalabama.edu

H. Huy Tài
Department of Mathematics, Tulane University, 6823 St. Charles Ave.,
New Orleans, LA 70118, USA
e-mail: tha@tulane.edu

J. Feldvoss et al. (eds.), *Advances in Algebra*, Springer Proceedings
in Mathematics & Statistics 277, https://doi.org/10.1007/978-3-030-11521-0_2

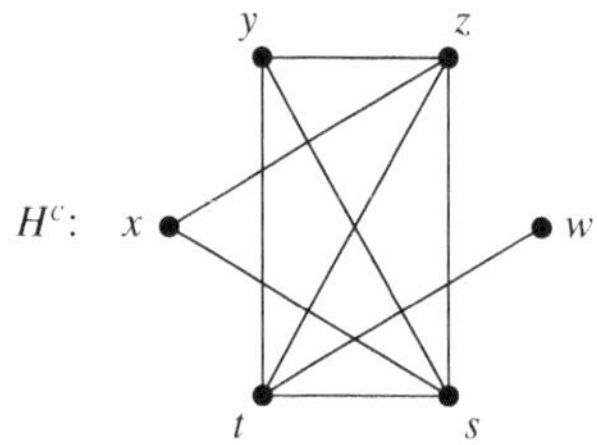

Fig. 2 Complement graph

For any vertex x in a graph G the *degree* of x, denoted by $d(x)$ is the number of vertices connected to x. For any vertex x in a graph G the *neighborhood of* x, denoted by $N_G[x]$ is the set consisting of x and all its neighbors. For the graph H'' above, $d(x) = 2$ and $N_G[x] = \{x, y, w\}$. For any graph G and any vertex x, by $G \setminus x$ we denote the induced subgraph on $V(G) \setminus \{x\}$.

For any graph G, the *complement graph*, denoted by G^c, is the graph whose vertices are the vertices of G and whose edges are the non-edges of G, i.e., for $a, b \in V(G)$, ab is an edge in G^c if and only if ab is *not* an edge in G. For the graph H in Fig. 1, the complement graph H^c is given in Fig. 2.

A *cycle* of length n in a graph G is a closed walk along its edges, $x_1x_2, x_2x_3, \ldots, x_{n-1}x_n, x_nx_1$, such that $x_i \neq x_j$ for $i \neq j$. We denote the cycle on n vertices by C_n. A *chord* in the cycle C_n is an edge x_ix_j where $x_j \neq x_{i-1}, x_{i+1}$. A graph is said to be *chordal* if for any cycle of length greater than or equal to 4 there is a chord. A graph is said to be *co-chordal* if the complement of G is chordal. The graph H in Fig. 2 is chordal and H^c is co-chordal.

A *forest* is a graph without any cycles. A *tree* is a connected forest.

A *complete* graph (or *clique*) on n vertices is a graph where for any two vertices there is an edge joining them. It is denoted by K_n.

A *bipartite graph* is a graph whose vertices can be split into two groups such that there is no edge between vertices of same group; edges are between vertices coming from different groups. It is easy to see that a graph is bipartite if and only if it is without any cycle of odd length. We use $K_{m,n}$ to denote the complete bipartite graph with m vertices on one side, and n on the other. We observe that H'' in Fig. 1 has no cycle, so it is a tree. As a consequence it is also a bipartite graph with bipartition $\{x, z, s\}$ and $\{y, w\}$.

Let G be a graph. We say two disjoint edges uv and xy form a *gap* in G if G does not have an edge with one endpoint in $\{u, v\}$ and the other in $\{x, y\}$. A graph without an induced gap is called *gap-free*. Equivalently, G is gap-free if and only if G^c contains no induced C_4. The graph H in Fig. 1 is gap-free. In Fig. 3 $\{x, y\}$ and $\{z, w\}$ form a gap.

A *matching* in a graph is a set of pairwise disjoint edges. A matching is called an *induced matching* if the induced subgraph on the vertices of the edges forming the matching has no other edge. We observe that $\{x, y\}$ and $\{z, w\}$ forms an induced matching for graph G in Fig. 3.

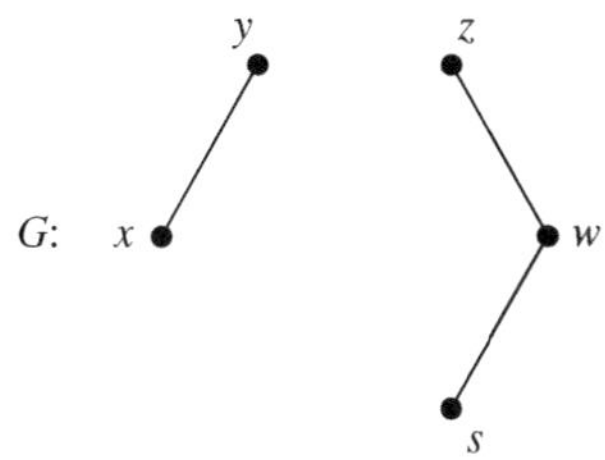

Fig. 3 A graph which is not gap-free

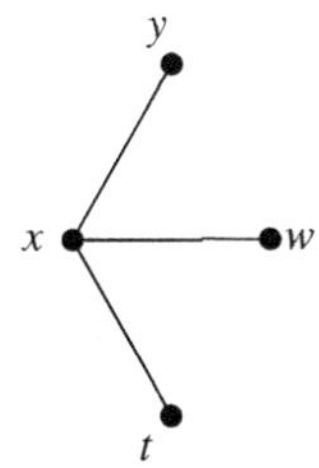

Fig. 4 A 3-claw (or simply, a claw)

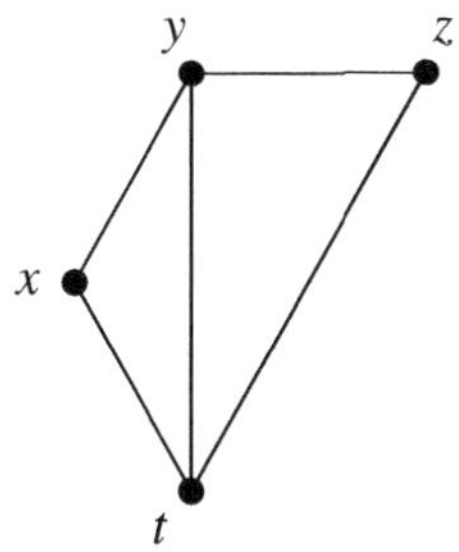

Fig. 5 A diamond

Any graph isomorphic to $K_{1,3}$ is called a *claw*. Any graph isomorphic to $K_{1,n}$ is called an *n-claw*. If $n > 1$, the vertex with degree n is called the root in $K_{1,n}$. A graph without an induced claw is called *claw-free*. A graph without an induced *n-claw* is called *n-claw-free* (Fig. 4).

A graph is called a *diamond* if it is isomorphic to the graph with vertex set $\{a, b, c, d\}$ and edge set $\{ab, bc, ac, ad, cd\}$. A graph without an induced diamond is called *diamond-free*.

A graph is said to be *planar* if as a 1-dimensional topological space it can be embedded in the complex plane, i.e., if it can be drawn in the plane in such a way that no pair of edges cross. The following graph in Fig. 5 is a diamond. We observe that it is also planar.

Any graph isomorphic to the graph with set of vertices $\{w_1, w_2, w_3, w_4, w_5\}$ and set of edges $\{w_1w_3, w_2w_3, w_3w_4, w_3w_5, w_4w_5\}$ is called a *cricket*. A graph without an induced cricket is called *cricket-free*. It is easy to see that a claw-free graph is also cricket-free (Fig. 6).

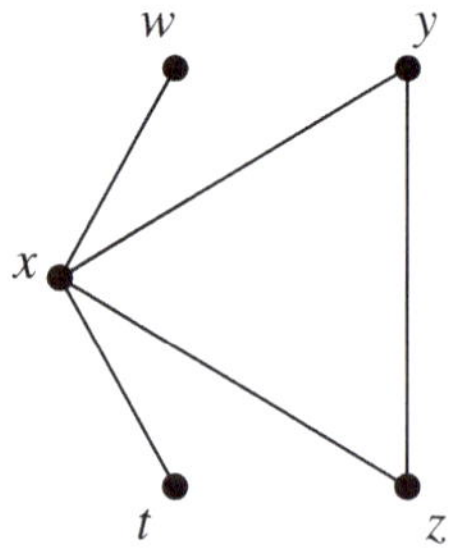

Fig. 6 A cricket

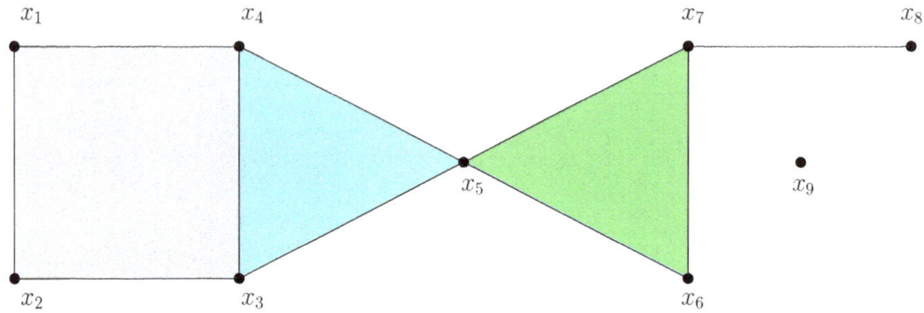

Fig. 7 A hypergraph

An edge in a graph is called a *whisker* if one of its vertices has degree one. A graph is called an *anticycle* if its complement is a cycle.

A *hypergraph* is the natural higher degree analogue of graphs, in the sense that we allow an edge to be a collection of any number of vertices. Formally speaking, a hypergraph $H = (V(H), E(H))$ consists of the *vertices* $V(H) = \{x_1, \ldots, x_n\}$ and the *edges* $E(H)$, where $E(H)$ is a collection of nonempty subsets of the vertices. With this notation, a graph G is a hypergraph whose edges are subsets of cardinality 2. By abusing notation, we often identify an edge $\{x_{i_1}, \ldots, x_{i_r}\} \in E(H)$ with the monomial $x_{i_1}, \ldots, x_{i_r}$. Figure 7 depicts a hypergraph with edges $\{x_1x_2x_3x_4, x_3x_4x_5, x_5x_6x_7, x_7x_8, x_9\}$. In this example, the vertex x_9 can also be viewed as an *isolated vertex*.

A *clutter* is a hypergraph none of whose edges contains any other edge as a subset. The hypergraph in Fig. 7 is a clutter.

A *simplicial complex* with vertices $\{x_1, \ldots, x_n\}$ is subset of the power set of $\{x_1, \ldots, x_n\}$ which is closed under the subset operation. The sets that constitute a simplicial complex are called its *faces*. Maximal faces under inclusion are called *facets*. Figure 8 gives a simplicial complex with facets $\{a, b, f\}$, $\{c, d, e\}$, $\{f, e\}$, $\{b, c\}$.

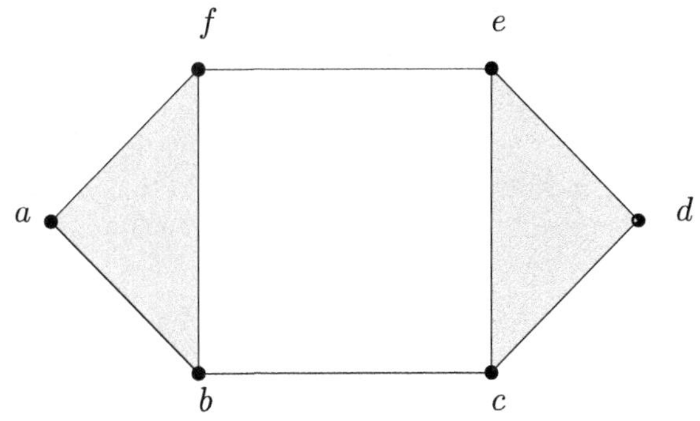

Fig. 8 A simplicial complex

An *independent set* in a graph G is a set of vertices no two of which forms an edge. The independence complex of a graph G, denoted by $\Delta(G)$, is the simplicial complex whose faces are independent sets in G.

Let Δ be a simplicial complex, and let $\sigma \in \Delta$. The *deletion* of σ in Δ, denoted by $\operatorname{del}_\Delta(\sigma)$, is the simplicial complex obtained by removing σ and all faces containing σ from Δ. The *link* of σ in Δ, denoted by $\operatorname{link}_\Delta(\sigma)$, is the simplicial complex whose faces are

$$\{F \in \Delta \mid F \cap \sigma = \emptyset, \sigma \cup F \in \Delta\}.$$

A simplicial complex Δ is recursively defined to be *vertex decomposable* if either

(i) Δ is a simplex; or
(ii) there is a vertex v in Δ such that both $\operatorname{link}_\Delta(v)$ and $\operatorname{del}_\Delta(v)$ are vertex decomposable, and all facets of $\operatorname{del}_\Delta(v)$ are facets of Δ.

A vertex satisfying condition (2) is called a *shedding vertex*, and the recursive choice of shedding vertices is called a *shedding order* of Δ.

Recall that a simplicial complex Δ is said to be *shellable* if there exists a linear order of its facets $F_1, F_2, \ldots, F_t$ such that for all $k = 2, \ldots, t$, the subcomplex $\left(\bigcup_{i=1}^{k-1} \overline{F_i}\right) \bigcap \overline{F_k}$ is pure and of dimension ($\dim F_k - 1$). Here $\overline{F}$ represents the simplex over the vertices of F. It is a celebrated fact that *pure* shellable complexes give rise to *Cohen–Macaulay Stanley–Reisner rings*. Recall also that a ring or module is *sequentially Cohen–Macaulay* if it has a filtration in which the factors are Cohen–Macaulay and their dimensions are increasing. This property corresponds to (*nonpure*) shellability in general. Vertex decomposability can be thought of as a combinatorial criterion for shellability and sequentially Cohen–Macaulayness. In particular, for a simplicial complex Δ,

$$\Delta \text{ vertex decomposable } \Rightarrow \Delta \text{ shellable } \Rightarrow \Delta \text{ sequentially Cohen-Macaulay.}$$

2.2 Algebraic Preliminaries

Let $S = K[x_1, \dots, x_n]$ be a polynomial ring over a field K. Let M be a finitely generated $\mathbb{Z}^n$-graded S-module. It is known that M can be successively approximated by free modules. Formally speaking, there exists an exact sequence of minimal possible length, called *a minimal free resolution* of M

$$0 \longrightarrow \mathbb{F}_p \xrightarrow{d_p} \mathbb{F}_{p-1} \cdots \xrightarrow{d_2} \mathbb{F}_1 \xrightarrow{d_1} \mathbb{F}_0 \xrightarrow{d_0} M \longrightarrow 0 \qquad (*)$$

Here, $\mathbb{F}_i = \bigoplus_{\sigma \in \mathbb{Z}^n} S(-\sigma)^{\beta_{i,\sigma}}$, where $S(-\sigma)$ denotes the free module obtained by shifting the degrees in S by σ. The numbers $\beta_{i,\sigma}$'s are positive integers and are called the multigraded Betti numbers of M. We often identify σ with the monomial whose exponent vector is σ. For example, over $K[x, y]$, we may write $\beta_{i,x^2y}(M)$ instead of $\beta_{i,(2,1)}(M)$.

For every $j \in \mathbb{Z}$, $\beta_{i,j} = \sum_{\{\sigma \mid |\sigma|=j\}} \beta_{i,\sigma}$ is called the (i, j)-th standard graded Betti number of M. Three very important homological invariants that are related to these numbers are the Castelnuovo–Mumford regularity, or simply regularity, the depth and the projective dimension, denoted by $\mathrm{reg}(M)$, $\mathrm{depth}(M)$ and $\mathrm{pd}(M)$ respectively:

$$\begin{aligned} \mathrm{reg} M &= \max\{|\sigma| - i \mid \beta_{i,\sigma} \neq 0\} \\ \mathrm{depth} M &= \inf\{i \mid \mathrm{Ext}^i(K, M) \neq 0\} \\ \mathrm{pd} M &= \max\{i \mid \text{there is a } \sigma, \beta_{i,\sigma} \neq 0\}. \end{aligned}$$

If S is viewed as a standard graded K-algebra and M is a graded S-module, then the graded Betti numbers of M are also given by $\beta_{i,j}(M) = \dim_k \mathrm{Tor}_i(M, K)_j$, and so we have

$$\begin{aligned} \mathrm{reg} M &= \max\{j - i \mid \mathrm{Tor}_i(M, K)_j \neq 0\} \\ \mathrm{pd} M &= \max\{i \mid \mathrm{Tor}_i(M, K) \neq 0\} \\ \mathrm{depth} M &= n - \mathrm{pd} M. \end{aligned}$$

In practice, we often work with short exact sequences, so it is worthwhile to mention that the regularity can also be defined via the vanishings of *local cohomology* modules with respect to the "irrelevant maximal ideal" $m = (x_1 \dots, x_n)$. Particularly, for $i \geq 0$, define

$$a_i(M) := \begin{cases} \max\{j \mid H^i_m(M)_j \neq 0\} & \text{if } H^i_m(M) \neq 0 \\ -\infty & \text{otherwise.} \end{cases}$$

Then, the regularity of M is also given by

$$\operatorname{reg} M = \max_i \{a_i(M) + i\}.$$

Example 1 Let $M = \mathbb{Q}[x_1, \ldots, x_5]/(x_1x_2, x_2x_3, x_3x_4, x_4x_5, x_5x_1)$. Then the minimal free resolution of M is

$$0 \longrightarrow \mathbb{F}_3 \xrightarrow{d_3} \mathbb{F}_2 \xrightarrow{d_2} \mathbb{F}_1 \xrightarrow{d_1} \mathbb{F}_0 \xrightarrow{d_0} M \longrightarrow 0$$

Here:

$$\beta_{0,\sigma} = 1 \text{ if } \sigma = 1, \text{ and } \beta_{0,\sigma} = 0 \text{ otherwise}$$

$$\beta_{1,\sigma} = 1 \text{ if } \sigma = x_1x_2, x_2x_3, x_3x_4, x_4x_5, x_5x_1, \text{ and } \beta_{1,\sigma} = 0 \text{ otherwise}$$

$$\beta_{2,\sigma} = 1 \text{ if } \sigma = x_1x_2x_3, x_2x_3x_4, x_1x_2x_5, x_1x_4x_5, x_3x_4x_5, \text{ and } \beta_{2,\sigma} = 0 \text{ otherwise}$$

$$\beta_{3,\sigma} = 1 \text{ if } \sigma = x_1x_2x_3x_4x_5, \text{ and } \beta_{3,\sigma} = 0 \text{ otherwise.}$$

An ideal I in S is said to be *unmixed* if all its associated primes are minimal of same height. We say I (or equivalently S/I) is *Cohen–Macaulay* if the Krull dimension and depth are equal. Cohen–Macaulay ideals are always unmixed and known to have many nice geometric properties. It can be checked that the module M in Example 1 is Cohen–Macaulay.

2.3 *Algebraic Objects with Underlying Combinatorial Structures*

For any (hyper)graph H over the vertex set $V(H) = \{x_1, \ldots, x_n\}$, its *edge ideal* is defined as follows:

$$I(H) = \left(\prod_{x \in e} x \;\middle|\; e \in E(H) \right) \subseteq S = K[x_1, \ldots, x_n].$$

Example 2 The edge ideal of the hypergraph, say H, in Fig. 7, is

$$I(H) = (x_1x_2x_3x_4, x_3x_4x_5, x_5x_6x_7, x_7x_8, x_9).$$

For any simplicial complex Δ over the vertex set $V = \{x_1, \ldots, x_n\}$, its *Stanley–Reisner ideal* is defined as follows:

$$I_\Delta = \left(\prod_{x \in e} x \;\middle|\; e \text{ is a minimal non-face of } \Delta \right) \subseteq S.$$

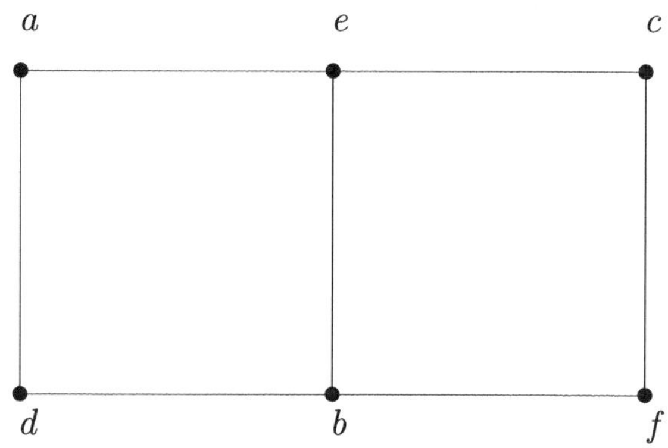

Fig. 9 A simple graph whose independence complex is in Fig. 8

Example 3 The Stanley–Reisner ideal of the simplicial complex in Fig. 8 is

$$I_\Delta = (ac, ad, ae, bd, be, cf, df).$$

As we have seen before, (hyper)graphs are related to simplicial complexes via the notion of independence complex. The algebraic view of this relation exposes the following equality:

Lemma 1 *Let G be a simple (hyper)graph and let $\Delta(G)$ be its independence complex. Then $I(G) = I_{\Delta(G)}$.*

Example 4 The edge ideal of the graph G in Fig. 9 is the same as the Stanley–Reisner ideal of the simplicial complex Δ in Fig. 8. Note that $\Delta = \Delta(G)$.

$$I(G) = (ae, ad, bd, be, bf, ec, ef).$$

For any simple graph G with $V(G) = \{x_1, \ldots, x_n\}$, we define the *t-path ideal* as:

$$I_t(G) = (x_{i_1}, \ldots, x_{i_t} \mid i_k \neq i_l \text{ for } k \neq l, x_{i_k}x_{i_{k+1}} \text{ an edge of } G).$$

Example 5 The 3-path ideal of the five cycle $C_5 : x_1x_2x_3x_4x_5$ is

$$I_3(C_5) = (x_1x_2x_3, x_2x_3x_4, x_3x_4x_5, x_4x_5x_1, x_5x_1x_2).$$

Finally, as a matter of convention, for any (hyper)graph G, by reg$G$ we mean reg$I(G)$.

3 Formulas and Inductive Approaches

Computing non-vanishings of local cohomology modules or Betti numbers of an ideal can be quite complicated. As a result, inductive techniques that relate regularity of edge ideals with simplicial complexes and smaller ideals are employed as a common tool in the literature. In this context, induced structures such as induced subcomplexes and induced subgraphs have proven to be significant objects to investigate regularity of an edge ideal and its powers.

In this section, we focus on the methods that enable us to bound and compute regularity of an edge ideal and its powers. We start the section by recalling two important formulas. Then we address a few inductive bounds that are widely used in the literature.

3.1 *Hochster's and Takayama's Formulas*

Hochster's formula has been a significant tool in the study of squarefree monomial ideals due to its power to relate the multigraded Betti numbers of a simplicial complex Δ to the non-vanishings of the reduced simplicial homology groups of Δ and its induced subcomplexes.

Theorem 1 (Hochster's Formula [29]) *Let Δ be a simplicial complex on the vertex set V and let I_Δ be its Stanley–Reisner ideal. Then*

$$\beta_{i,j} I_\Delta = \sum_{W \subseteq V,\ |W|=j} \dim_K \left(\widetilde{H}_{j-i-2}(\Delta_W; K)\right)$$

where Δ_W is the restriction of Δ to the vertex set W.

For an arbitrary monomial ideal, Hochster's formula can not be employed. Takayama's formula given in [48] can perform a similar task as Hochster's formula for this class of ideals.

Let $I \subseteq S = K[x_1, \ldots, x_n]$ be a monomial ideal. Takayama's formula provides a combinatorial description for the non-vanishings of $\mathbb{Z}^n$-graded component $H^i_{\mathfrak{m}}(S/I)_{\mathbf{a}}$ for $\mathbf{a} \in \mathbb{Z}^{\mathbf{n}}$, and this description is given in terms of certain simplicial complexes $\Delta_{\mathbf{a}}(I)$ related to I. Note that S/I is an $\mathbb{N}^n$-graded algebra, and so $H^i_{\mathfrak{m}}(S/I)$ is a $\mathbb{Z}^n$-graded module over S/I.

The simplicial complex $\Delta_{\mathbf{a}}(I)$ is called the *degree complex* of I. The construction of $\Delta_{\mathbf{a}}(I)$ was first given in [48] and then simplified in [41]. We recall the construction from [41].

For $\mathbf{a} = (a_1, \ldots, a_n) \in \mathbb{Z}^n$, set $x^a = x_1^{a_1}, \ldots, x_n^{a_n}$ and $G_{\mathbf{a}} := \{j \in \{1, \ldots, n\} \mid a_j < 0\}$. For every subset $F \subseteq \{1, \ldots, n\}$, let $S_F = S[x_j^{-1} \mid j \in F]$. Define

$$\Delta_{\mathbf{a}}(I) = \{F \setminus G_{\mathbf{a}} \mid G_{\mathbf{a}} \subseteq F, x^{\mathbf{a}} \notin IS_F\}.$$

Theorem 2 (Takayama's Formula [48]) *Let $I \subseteq S$ be a monomial ideal, and let $\Delta(I)$ denote the simplicial complex corresponding to $\sqrt{I}$. Then*

$$\dim_k H^i_{\mathfrak{m}}(S/I)_{\mathbf{a}} = \begin{cases} \dim_k \left(\widetilde{H}_{i-|G_{\mathbf{a}}|-1}(\Delta_{\mathbf{a}}(I), K\right) & \text{if } G_{\mathbf{a}} \in \Delta(I), \\ 0 & \text{otherwise.} \end{cases}$$

The formula stated in [41] and here is different than the original formula introduced in [48]. The original formula has additional conditions on **a** for $H^i_{\mathfrak{m}}(S/I)_{\mathbf{a}} = 0$. It follows from the proof in [48] that those conditions can be omitted.

Due to the equality between edge ideals of graphs and Stanley–Reisner ideals of their independence complexes, we can employ Hochster's and Takayama's formulas in the study of regularity of edge ideals.

In order to deal with arbitrary monomial ideals, polarization is proved to be a powerful process to obtain a squarefree monomial ideal from a given monomial ideal. For details of polarization we refer the reader to [26].

Definition 1 Let $M = x_1^{a_1} \dots x_n^{a_n}$ be a monomial in $S = K[x_1, \dots, x_n]$. Then we define the squarefree monomial $P(M)$ (*polarization* of M) as

$$P(M) = x_{11} \dots x_{1a_1} x_{21} \dots x_{2a_2} \dots x_{n1} \dots x_{na_n}$$

in the polynomial ring $R = K[x_{ij} \mid 1 \le i \le n, 1 \le j \le a_i]$. If $I = (M_1, \dots, M_q)$ is an ideal in S, then the polarization of I, denoted by I^{pol}, is defined as $I^{\text{pol}} = (P(M_1), \dots, P(M_q))$.

The regularity is preserved under polarization.

Corollary 1 ([26, Corollary 1.6.3.d]) *Let $I \subset S$ be a monomial ideal and $I^{\text{pol}} \subset R$ be its polarization. Then $\text{reg}(S/I) = \text{reg}(R/I^{\text{pol}})$.*

3.2 *Inductive Techniques*

We start the section with an easy yet essential consequence of Hochster's formula that links the regularity of a graph with regularity of its induced subgraphs.

Lemma 2 *Let G be a simple graph. Then $\text{reg} H \le \text{reg} G$ for any induced subgraph H of G.*

For any homogeneous ideal $I \subseteq S$ and any homogeneous element m of degree d, the following short exact sequences are used as standard tools in commutative algebra and also proved to be very useful in computing regularity of edge ideals and their powers.

$$0 \longrightarrow \frac{R}{I : m}(-d) \xrightarrow{\cdot m} \frac{R}{I} \longrightarrow \frac{R}{I + (m)} \longrightarrow 0 \tag{1}$$

Let I and J be ideals in S. Another useful exact sequence is

$$0 \longrightarrow \frac{R}{I \cap J} \to \frac{R}{I} \oplus \frac{R}{J} \longrightarrow \frac{R}{I+J} \longrightarrow 0 \tag{2}$$

We can see how regularity changes in a short exact sequence by taking the associated long exact sequence of local cohomology modules. Particularly, we have the following useful inductive bound (note that the second statement is the content of [14, Lemma 2.10]).

Lemma 3 *Let $I \subseteq S$ be a monomial ideal, and let m be a monomial of degree d. Then*

$$reg I \leq \max\{reg(I : m) + d, reg(I, m)\}.$$

Furthermore, if x is a variable appearing in I, then

$$reg I \in \{reg(I : x) + 1, reg(I, x)\}.$$

In the statement of Lemma 3, the expression x is a variable appearing in I means some of the minimal generators of I is divisible by x. Note that if x is a variable not appearing in I, then $\operatorname{reg}(I, x) = \operatorname{reg} I$. In case of edge ideals, if x is an isolated vertex in G, we can drop the vertex x when computing regularity. Thus we have $\operatorname{reg}(I(G) : x) = \operatorname{reg} I(G \setminus N_G[x])$ and $\operatorname{reg}(I(G), x) = \operatorname{reg}(I(G \setminus x))$ for any vertex x in G. Then Lemma 3 can be restated in terms of edge ideals.

Lemma 4 *Let x be a vertex in G. Then*

$$reg G \in \{reg(G \setminus N_G[x]) + 1, reg(G \setminus x)\}.$$

Kalai and Meshulam [33] proved the following result for squarefree monomial ideals and Herzog [25] generalized it to any monomial ideal.

Theorem 3 *Let $I_1, \ldots, I_s$ be squarefree monomial ideals in S. Then*

$$reg \left(S/\sum_{i=1}^{s} I_i \right) \leq \sum_{i=1}^{s} reg S/I_i.$$

In the case of edge ideals, we have the following bound.

Corollary 2 *Let G be a simple graph. If $G_1, \ldots, G_s$ are subgraphs of G such that $E(G) = \bigcup_{i=1}^{s} E(G_i)$ then*

$$reg\Big(S/I(G)\Big) \leq \sum_{i=1}^{s} reg S/I(G_i).$$

If G is the disjoint union of graphs $G_1, \ldots, G_s$ then the above equality is achieved by using Künneth formula in algebraic topology.

Corollary 3 *Let G be a simple graph. If G can be written as a disjoint union of graphs $G_1, \ldots, G_s$ then*

$$reg\Big(S/I(G)\Big) = \sum_{i=1}^{s} regS/I(G_i).$$

In the study of powers of edge ideals, Banerjee developed the notion of even-connection and gave an important inductive inequality in [4].

Theorem 4 *For any simple graph G and any $s \geq 1$, let the set of minimal monomial generators of $I(G)^s$ be $\{m_1, \ldots, m_k\}$. Then*

$$regI(G)^{s+1} \leq \max\{reg(I(G)^{s+1} : m_l) + 2s, 1 \leq l \leq k, regI(G)^s\}.$$

In particular, if for all $s \geq 1$ and for all minimal monomial generators m of $I(G)^s$, $reg(I(G)^{s+1} : m) \leq 2$ and $regI(G) \leq 4$, then $regI(G)^{s+1} = 2s + 2$ for all $s \geq 1$; as a consequence $I(G)^{s+1}$ has a linear minimal free resolution.

Working with the above inequality requires a good understanding of the ideal $(I(G)^{s+1} : m)$ and its generators when m is a minimal monomial generator of $I(G)^s$. The even-connection definition is key to attain this goal. We recall the definition of even-connectedness and its important properties from [4].

Definition 2 Let $G = (V, E)$ be a graph. Two vertices u and v (u may be the same as v) are said to be even-connected with respect to an s-fold product $e_1, \ldots, e_s$ where e_i's are edges of G, not necessarily distinct, if there is a path $p_0, p_1, \ldots, p_{2k+1}, k \geq 1$ in G such that

1. $p_0 = u$, $p_{2k+1} = v$.
2. For all $0 \leq l \leq k-1$, $p_{2l+1}p_{2l+2} = e_i$ for some i.
3. For all i, $|\,\{l \geq 0 \mid p_{2l+1}p_{2l+2} = e_i\}\,| \;\leq\; |\,\{j \mid e_j = e_i\}\,|$.
4. For all $0 \leq r \leq 2k$, $p_r p_{r+1}$ is an edge in G.

Fortunately, the colon ideal $(I(G)^{s+1} : m)$ is generated by monomials in degree 2.

Theorem 5 ([4, Theorem 6.1 and Theorem 6.7]) *Let G be a graph with edge ideal $I = I(G)$, and let $s \geq 1$ be an integer. Let m be a minimal generator of I^s. Then $(I^{s+1} : m)$ is minimally generated by monomials of degree 2, and uv (u and v may be the same) is a minimal generator of $(I^{s+1} : m)$ if and only if either $\{u, v\} \in E(G)$ or u and v are even-connected with respect to m.*

After polarization, the ideal $(I(G)^{s+1} : m)$ can be viewed as the edge ideal of a graph that is obtained from G by adding even-connected edges with respect to m. Some of the combinatorial properties of this ideal in relation to G are studied in [32].

4 Regularity of Edge Ideals

Computing and/or bounding the regularity of edge ideals is the foundation in the study of regularity of powers of edge ideals. In this section, we identify the combinatorial structures that are related to regularity of edge ideals. First, we collect the general upper and lower bounds given for the regularity of edge ideals, then we present the list of known classes of graphs where the regularity is computed explicitly.

Note that regularity of an edge ideal is bounded below by 2, which is the generating degree of an edge ideal. Thus, identifying combinatorial structures of a graph with regularity 2 can be considered as the base case of results in this section. The following combinatorial characterization of such graphs is nowadays often referred to as *Fröberg's characterization*. It was, in fact, given first in topological language by Wegner [53] and later, independently, by Lyubeznik [39] and Fröberg [20] in monomial ideals language.

Theorem 6 ([20, Theorem 1]) *Let G be a simple graph. Then $regI(G) = 2$ if and only if G is a co-chordal graph.*

4.1 Lower and Upper Bounds

One of the graph-theoretical invariants that can be related to the regularity is the induced matching number. The first result revealing this relation is due to Katzman and it provides a general lower bound on the regularity of edge ideals.

Theorem 7 ([34, Lemma 2.2]) *Let G be a simple graph and $\nu(G)$ be the maximum size of an induced matching in G. Then*

$$regI(G) \geq \nu(G) + 1.$$

Proof Let $\{e_1, \ldots, e_r\}$ be an induced matching of maximal size in G. Suppose H is the induced subgraph of G with $E(H) = \{e_1, \ldots, e_r\}$. Note that all the edges in H are disjoint. Thus $\text{reg}(H) = r + 1$. By Lemma 2, $\text{reg}(G) \geq \nu(G) + 1$.

Another graph-theoretical invariant of interest is the matching number and this invariant actually emerges as a general upper bound for any graph.

Theorem 8 ([23, Theorem 6.7], [55, Theorem 11]) *Let G be a simple graph. Let $\beta(G)$ be the minimum size of a maximal matching in G. Then*

$$regI(G) \leq \beta(G) + 1.$$

Example 6 Let G be the graph given in Fig. 10. It is clear that $\nu(G) = \beta(G) = 3$. Then $\text{reg}G = 4$ by Theorem 7 and Theorem 8. In general, if G is a simple graph with disjoint edges, then the above bounds coincide and become an equality.

Fig. 10 A graph with disjoint edges

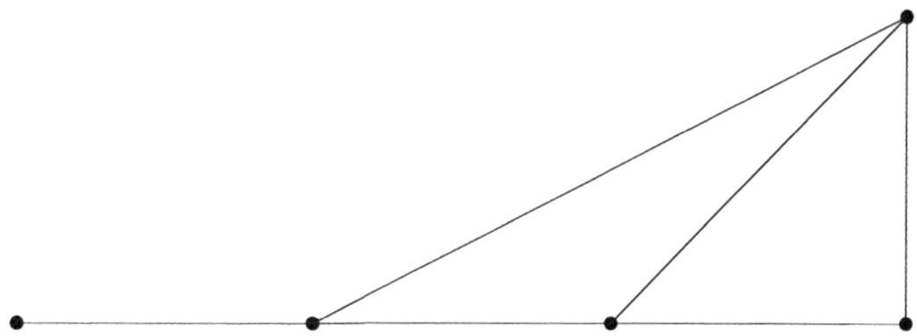

Fig. 11 A co-chordal graph

Comparing the existing lower and upper bounds yields interesting classifications, particularly when these bounds coincide. For example, Cameron and Walker [10, Theorem 1] gave the first classification of graphs G with $\nu(G) = \beta(G)$. Then Hibi et al. [28] modified their result slightly and gave a full generalization with some corrections.

Example 7 Let G be the graph given in Fig. 11. One can easily verify that $\beta(G) = 2$. However $\text{reg}G = 2$ by Fröberg's characterization. In general, we can find examples of graphs where $\beta(G) + 1$ can be arbitrarily large compared to $\text{reg}G$.

The above upper bound is strengthened by making use of co-chordal subgraph covers of a graph and this bound is proved by Woodroofe. Recall that a graph is co-chordal if its complement is chordal and co-chordal(G), the *co-chordal number* of G, denotes the least number of co-chordal subgraphs of G whose union is G. The graph in Fig. 11 is an example of a co-chordal graph.

Let $\{e_1, \ldots, e_r\}$ be a maximal matching of minimal size in G. For each i, let G_i be the subgraph of G with edges $e_i \cup \{\text{edges in } G \text{ adjacent to } e_i\}$. Note that $G_1, \ldots, G_r$ forms a co-chordal subgraph cover of G, thus co-chordal$(G) \leq \beta(G)$ and the bound in Theorem 9 improves the bound in Theorem 8.

Theorem 9 ([55, Lemma 1]) *Let G be a simple graph. Then*

$$\text{reg}I(G) \leq \text{co-chordal}(G) + 1.$$

Proof Let co-chordal$(G) = r$ and let $G_1, \ldots, G_r$ be a co-chordal cover of G. It follows from Fröberg's characterization of regularity 2 graphs, Theorem 6, that $\text{reg}(G_i) = 2$ for each $i \in \{1, \ldots, r\}$. Then the result follows immediately from Corollary 2.

Gap-free graphs have also been of interest in the investigation of regularity. These graphs arise naturally since their induced matching number is 1 which is the smallest possible number it could be. However, computing or bounding the regularity of this class of graphs is not so easy. Furthermore, there are very few examples on how large the regularity of such edge ideals can be, see [46] by Nevo and Peeva for an example of a gap-free graph G in 12 variables with $\text{reg}G = 4$. Putting an additional condition on the gap-freeness of G may result with an upper bound and it is indeed achieved in [45].

Theorem 10 ([45, Theorem 1.2], [14, Proposition 19]) *If G is gap-free and claw-free, then $\text{reg}I(G) \leq 3$.*

Proof Let x be a vertex in G with the maximum possible degree. By Lemma 4, we have $\text{reg}(G) \leq \max\{\text{reg}(G \setminus N_G[x]) + 1, \text{reg}(G \setminus x)\}$. Note that induced subgraphs $G \setminus N_G[x]$ and $G \setminus x$ of G are both gap-free and claw-free. It follows from the induction on the number of vertices that $\text{reg}(G \setminus x) \leq 3$. Thus it suffices to show that $\text{reg}(G \setminus N_G[x]) \leq 2$. By Fröberg's characterization, it is enough to prove that $(G \setminus N_G[x])^c$ is chordal and it is proved by contradiction:

1. Suppose $(G \setminus N_G[x])^c$ has an induced cycle on $w_1, w_2, \ldots, w_n$ (in order) of length at least 4.
2. Any vertex of G is distance 2 from x in G by [14, Proposition 3.3], then $\{x, y\}$ and $\{y, w_1\}$ are edges in G for some vertex y.
3. Note that either $\{y, w_2\}$ or $\{y, w_n\}$ must be an edge in G. Otherwise edges $\{w_2, w_n\}$ and $\{x, y\}$ form a gap in G.
4. Without loss of generality, suppose $\{y, w_2\}$ is an edge. Then the induced subgraph on $\{x, y, w_1, w_2\}$ is a claw in G, a contradiction.

Example 8 There are examples of gap-free- and claw-free graphs where both values in Theorem 10 can be attained for regularity. For example, if G^c is a tree then $\text{reg}G = 2$ and if G^c is C_n for $n \geq 5$ then $\text{reg}G = 3$.

This result is generalized to n-claw-free- and gap-free graphs by Banerjee in [4]. The proof of the general case follows similarly and uses induction on n.

Theorem 11 ([4, Theorem 3.5]) *If G is gap-free- and n-claw-free graph, then $\text{reg}I(G) \leq n$.*

Another special class of graphs are planar graphs. This class of graphs emerges frequently in applications since they can be drawn in the plane without edges crossing. It is given in [55] that even though regularity of a planar graph may be arbitrarily large, regularity of its complement can be bounded above by 4.

Theorem 12 ([55, Theorem 3.4]) *If G is a planar graph, then $\text{reg}I(G^c) \leq 4$.*

4.2 Exact Values

Computing the regularity for special classes of graphs has been an attractive research topic in the recent years. To that extend characterizing edge ideals with certain regularity has been of interest as well. However, very little is known in the latter case.

A combinatorial characterization of edge ideals with regularity 3 is still not known. However, a partial result for bipartite graphs is achieved by Fernández-Ramos and Gimenez in [17]. Recall that a graph G is bipartite if the vertices V can be partitioned into disjoint subsets $V = X \cup Y$ such that $\{x, y\}$ is an edge in G only if $x \in X$ and $y \in Y$ or vice versa. The *bipartite complement* of a bipartite graph G, denoted by G^{bc}, is the bipartite graph over the same partition of vertices and $\{x, y\} \in G^{bc}$ if and only if $\{x, y\} \notin G$ for $x \in X$, $y \in Y$.

Theorem 13 ([17, Theorem 3.1]) *Let G be a connected bipartite graph. Then $\operatorname{reg} I(G) = 3$ if and only if G^c has no induced cycles of length ≥ 4 and G^{bc} has no induced cycles of length ≥ 6.*

In most of the known cases, it turns out that regularity can be expressed in terms of the induced matching number. We first recall the results when the regularity is one more than the induced matching number.

Theorem 14 *Let G be a simple graph and $\nu(G)$ be the induced matching number of G. Then*

$$\operatorname{reg} I(G) = \nu(G) + 1$$

in the following cases:

1. *G is a chordal graph (see [23]);*
2. *G is a weakly chordal graph (see [55]) (Fig. 12);*
3. *G is sequentially Cohen–Macaulay bipartite graph (see [52]);*
4. *G is unmixed bipartite graph (see [37]);*
5. *G is very well-covered graph (see [40]);*
6. *G is vertex decomposable graph and has no closed circuit of length 5 (see [35]);*
7. *G is a (C_4, C_5)-free vertex decomposable graph (see [7]);*
8. *G is a unicyclic graph with cycle C_n when*
    a. *$n \equiv 0, 1 \ (mod\ 3)$ (see [3, 6, 30]) or*
    b. *$\nu(G \setminus \Gamma(G)) < \nu(G)$ where $\Gamma(G)$ is the collection of all neighbors of the roots in the rooted trees attached to C_n (see [3]).*

Remark 1 Chordal graphs are vertex decomposable and vertex decomposable graphs are sequentially Cohen–Macaulay (see [18, 54]). Sequentially Cohen–Macaulay bipartite graphs are vertex decomposable (see [52]). Also if G is a very well-covered graph then G is unmixed and if G is chordal, then it is weakly chordal. Thus, (2) implies (1), and (5) implies (4). Note that bipartite graphs have no odd cycles. Hence, (6) implies (3).

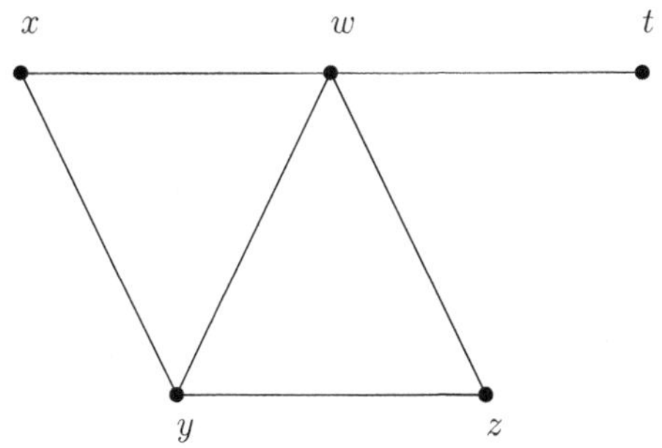

Fig. 12 A vertex decomposable and a (weakly) chordal graph

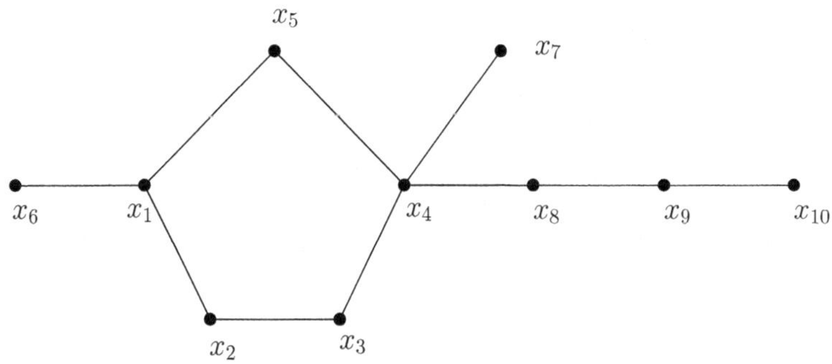

Fig. 13 A graph satisfying (8) (b) in Theorem 14

Example 9 Let G be the graph given in Fig. 13. Then $\Gamma(G) = \{x_6, x_7, x_8\}$ and $\nu(G \setminus \Gamma(G)) = 2 < \nu(G) = 3$. Furthermore, this graph does not belong to any of the classes described in (1)–(7).

It is of interest to find different expressions for the regularity of edge ideals. The next result collects all known cases in which the regularity has a different expression than the above classes and it is still in terms of induced matching.

Theorem 15 *Let G be a simple graph and $\nu(G)$ be the induced matching number of G. Then*

$$reg I(G) = \nu(G) + 2$$

in the following cases:

1. *G is an n-cycle C_n when $n \equiv 2 \ (mod\ 3)$ (see [6, 30]);*
2. *G is a unicyclic graph with cycle C_n when $n \equiv 2 \ (mod\ 3)$ and $\nu(G \setminus \Gamma(G)) = \nu(G)$ where $\Gamma(G)$ is the collection of all neighbors of the roots in the rooted trees attached to C_n (see [3]).*

5 Regularity of Powers of Edge Ideals

The regularity of powers of an edge ideal is considerably harder to compute than that of the edge ideal itself. However, in many cases, known bounds and exact formulas for the regularity of $I(G)$ inspire new bounds and exact formulas for the regularity of $I(G)^q$, for $q \geq 1$. This section is divided into two parts, where in the first subsection we list a number of lower and upper bounds, and in the second subsection we give the exact values of $\text{reg} I(G)^q$, for $q \geq 1$, for special classes of graphs.

5.1 Lower and Upper Bounds

Just like with studies on the regularity of edge ideals, the induced matching number of a graph is ultimately connected to the regularity of powers of its edge ideal. The following general lower bound generalizes that of Theorem 7 for an edge ideal to its powers:

Theorem 16 ([6, Theorem 4.5]) *Let G be any graph and let $I = I(G)$ be its edge ideal. Then for all $q \geq 1$, we have*

$$reg I^q \geq 2q + \nu(G) - 1.$$

Proof The proof is based on the following observations:

1. If H is an induced subgraph of G then for any $i, j \in \mathbb{Z}$, we have
$$\beta_{i,j}(I(H)^q) \leq \beta_{i,j}(I(G)^q). \tag{3}$$
In particular, this gives $\text{reg} I(H)^q \leq \text{reg} I(G)^q$.
2. If H is the induced subgraph of G consisting of a maximal induced matching of $\nu(G)$ edges then for any $q \geq 1$, we have
$$\text{reg} I(H)^q = 2q + \nu(G) - 1. \tag{4}$$

To establish the first observation, recall that the *upper Koszul simplicial complex* $K^\alpha(I)$ associated to a monomial ideal $I \subseteq S = K[x_1, \ldots, x_n]$ at degree $\alpha \in \mathbb{Z}^n$ consists of faces

$$\left\{ W \subseteq \{1, \ldots, n\} \,\middle|\, \frac{x^\alpha}{\prod_{w \in W} w} \in I \right\},$$

and a variation of Hochster's formula (Theorem 1) gives

$$\beta_{i,\alpha}(I) = \dim_k \widetilde{H}_{i-1}(K^\alpha(I); K) \text{ for all } i \geq 0.$$

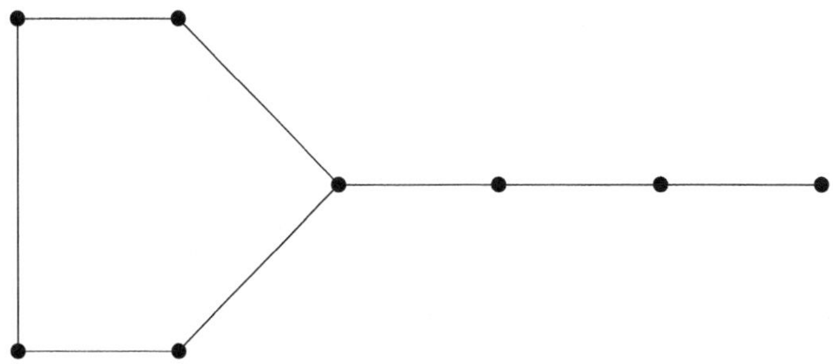

Fig. 14 A unicyclic graph

The inequality (3) then follows by noting that for $\alpha \in \mathbb{Z}^n$ with $\text{supp}(\alpha) \subseteq V_H$, $K^\alpha(I(H)^q) = K^\alpha(I(G)^q)$.

The second observation is proved by induction, noting that in this case $I(H)$ is a complete intersection.

Example 10 Though there are many classes of graphs for which the lower bound given in Theorem 16 is attained, there are also classes of graphs where the asymptotic linear function $\text{reg} I(G)^q$ is strictly bigger than $2q + \nu(G) - 1$ for all $q \gg 0$. Let G be the graph depicted in Fig. 14. Then it is easy to see that $\nu(G) = 2$, whereas $\text{reg} I(G)^q = 2q + 2$ for $q \geq 1$.

A similar general upper bound generalizing that of Theorem 9 is, unfortunately, not available. It is established, by Jayanthan, Narayanan, and Selvaraja [32], only for a special class of graphs—bipartite graphs.

Theorem 17 ([32, Theorem 1.1]) *Let G be a bipartite graph and let $I = I(G)$ be its edge ideal. Then for all $q \geq 1$, we have*

$$\textit{reg} I^q \leq 2q + \textit{co-chordal}(G) - 1.$$

Proof The statement is proved by induction utilizing Theorem 4. The crucial step in the proof is to show that for any $q \geq 1$ and any collection of edges $e_1, \ldots, e_q$ of G (not necessarily distinct), we have

$$\text{reg}(I^{q+1} : e_1 \ldots e_q) \leq \text{co-chordal}(G) + 1. \tag{5}$$

Observe that when G is a bipartite graph, by [2, Lemma 3.7],

$$I^{q+1} : e_1 \ldots e_q = (((I^2 : e_1)^2 : \ldots)^2 : e_q),$$

and so (5) itself can be obtained by induction. To this end, let G' be the graph associated to the polarization of $I^2 : e$ for an edge e in G (which is generated in degree 2 by Theorem 5). The proof is completed by establishing the following facts:

1. $\operatorname{reg} I(G') \leq \text{co-chord}(G') + 1$. This inequality was proved to hold for any graph in [55].
2. $\text{co-chord}(G') \leq \text{co-chord}(G)$. This is combinatorial statement, which can be shown by analyzing how G' is constructed from G.

Remark 2 Let G be a 4-cycle. We can easily verify that $\nu(G) = 1$ and $\text{co-chord}(G) = 1$. Since the lower bound in Theorem 16 coincides with the upper bound in Theorem 17, we have $\operatorname{reg} I(G)^q = 2q = 2q + \nu(G) - 1 = 2q + \text{co-chord}(G) - 1$ for all $q \geq 1$. Classes of graphs for which the two upper and lower bounds agree were discussed in [32, Corollary 5.1].

On the other hand, the upper bound given in Theorem 17 can be strict. For example, if G is C_8, then $\nu(G) = 2$ and $\text{co-chord}(G) = 3$. By [6, Theorem 5.2], it is known that $\operatorname{reg} I(G)^q = 2q + 1 < 2q + \text{co-chord}(G) - 1$ for all $q \geq 1$.

For a special class of graphs—gap-free graphs—there is another upper bound that was proved by Banerjee [4].

Theorem 18 ([4, Theorem 6.19]) *Let G be a gap-free graph and let $I = I(G)$ be its edge ideal. Then for all $q \geq 2$, we have*

$$reg I^q \leq 2q + reg I - 1.$$

Remark 3 The bound in Theorem 18 is slightly weaker than the conjectural bound of Conjecture 1.

5.2 Exact Values

In most of the known cases where the asymptotic linear function $\operatorname{reg} I(G)^q$ can be computed explicitly, the lower bound in Theorem 16 turns out to give the exact formula. In this subsection, we describe those instances.

In [27, Theorem 3.2], the authors showed that $I(G)$ has a linear resolution if and only if $I(G)^q$ has a linear resolution for all $q \geq 1$. Combination of their result with Fröberg's characterization yields to the exact value of regularity of powers of co-chordal graphs.

Theorem 19 *Let G be a co-chordal graph and $I = I(G)$ be its edge ideal. Then for all $q \geq 1$, we have*

$$reg I^q = 2q.$$

As often the case when dealing with graphs, one of the first classes to consider is that of trees and forests.

Theorem 20 ([6, Theorem 4.7]) *Let G be a forest and let $I = I(G)$ be its edge ideal. Then for all $q \geq 1$, we have*

$$\mathrm{reg} I^q = 2q + \nu(G) - 1.$$

Proof Thanks to the general lower bound in Theorem 16, it remains to establish the upper bound

$$\mathrm{reg} I^q \leq 2q + \nu(G) - 1.$$

In fact, a more general inequality can be proved for induced subgraphs of G. Let H and K be induced subgraphs of K such that

$$E_H \cup E_K = E_G \text{ and } E_H \cap E_K = \emptyset.$$

Then the required upper bound follows from the following inequality (by setting H to be the empty graph):

$$\mathrm{reg}(I(H) + I(K)^q) \leq 2q + \nu(G) - 1. \tag{6}$$

The inequality (6) is proved by induction on $q + |V_K|$, making use of the short exact sequence arising by taking quotient and colon with respect to a leaf xy in K. Note that, by [43, Lemma 2.10], we have

$$(I(H) + I(K)^q) : xy = (I(H) : xy) + (I(K)^q : xy) = (I(H) : xy) + I(K)^{q-1}.$$

The next natural class of graphs to consider is that of cycles and graphs containing exactly one cycle (i.e., *unicyclic* graphs).

Theorem 21 ([6, Theorem 5.2]) *Let C_n denote the n-cycle and let $I = I(C_n)$ be its edge ideal. Let $\nu = \lfloor \frac{n}{3} \rfloor$ be the induced matching number of C_n. Then*

$$\mathrm{reg} I = \begin{cases} \nu + 1 \text{ if } n \equiv 0, 1 \pmod 3 \\ \nu + 2 \text{ if } n \equiv 2 \pmod 3 \end{cases}$$

and for any $q \geq 2$, we have

$$\mathrm{reg} I^q = 2q + \nu - 1.$$

Proof The first statement was already proved in [30]. To prove the second statement, again thanks to the general lower bound of Theorem 16, it remains to establish the upper bound

$$\mathrm{reg} I^q \leq 2q + \nu - 1. \tag{7}$$

The inductive method of Theorem 4 once again is invoked, and it reduces the problem to showing that for any collection of edges $e_1, \ldots, e_q$ of G (not necessarily distinct), we have

$$\mathrm{reg}(I^{q+1} : e_1 \ldots e_q) \leq \nu + 1.$$

To this end, let J be the polarization of $I^{q+1} : e_1 \ldots e_q$. It can be seen that

$$J = I(H) + (x_{i_1} y_{i_1}, \ldots, x_{i_t} y_{i_t}),$$

where H is a graph over the vertices $\{x_1, \ldots, x_n\}$ and $x_{i_1}^2, \ldots, x_{i_t}^2$ are the non-squarefree generators of $I^{q+1} : e_1 \ldots e_q$. By using standard short exact sequences, it can be shown that

$$\operatorname{reg}(J) = \operatorname{reg} I(H).$$

Observe further that H contains C_n as a subgraph, so H has a Hamiltonian cycle. Thus, the required upper bound now follows from a more general upper bound for the regularity of a graph admitting either a Hamiltonian path or a Hamiltonian cycle (which is the content of [6, Theorems 3.1 and 3.2]).

The following theorem was proved in the special case of *whiskered* cycles by Moghimian, Seyed Fakhari, and Yassemi [42], and then in more generality for unicyclic graphs by Alilooee, Beyarslan, and Selvaraja [3].

Theorem 22 ([3, Theorem 1.2], [42, Proposition 1.1]) *Let G be a unicyclic graph and let $I = I(G)$ be its edge ideal. Then for all $q \geq 1$, we have*

$$reg I^q = 2q + reg I - 2.$$

Proof The proof is based on establishing the conjectural bound of Conjecture 1. (2)

$$\operatorname{reg}(I^q) \leq 2q + \operatorname{reg} I - 2. \tag{8}$$

The inequality (8) is proved by using induction on q. It also requires a good understanding of $\operatorname{reg}(I(C_n)^q, f_1, \ldots, f_k)$ where C_n is the cycle in G and $f_1, \ldots, f_k$ are the edges of G that are not in C_n. By making use of the general lower bound of Theorem 16 and the first main result of [3], namely,

$$\operatorname{reg} I = \begin{cases} \nu(G) + 2 \text{ if } n \equiv 2 (\text{mod } 3) \text{ and } \nu(G \setminus \Gamma(G)) = \nu(G) \\ \nu(G) + 1 \text{ otherwise,} \end{cases}$$

equality is achieved for the latter case. (Here, $\Gamma(G)$ is a well described subset of the vertices in G.) The proof is completed by showing $\operatorname{reg}(I(G \setminus \Gamma(G))^q) = 2q + \nu(G)$ and using inequality (3).

Example 11 Let G be the graph depicted in Fig. 15. Macaulay 2 computations show that

$$\operatorname{reg} I(G) = 5,$$

$$\operatorname{reg} I(G)^2 = 6, \quad \operatorname{reg} I(G)^3 = 8,$$

$$\operatorname{reg} I(G)^4 = 10, \quad \operatorname{reg} I(G)^5 = 12.$$

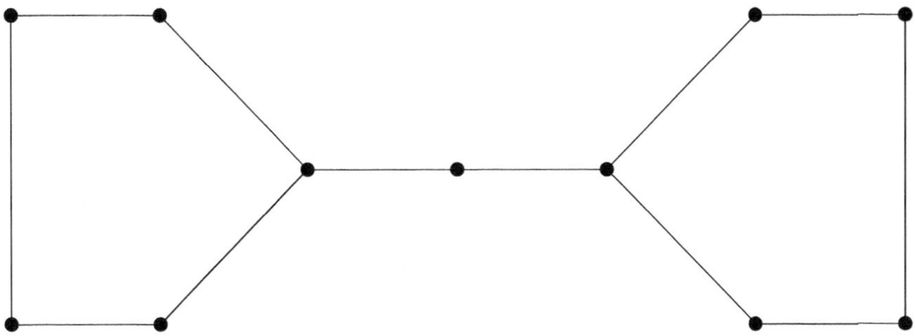

Fig. 15 A bicyclic graph

Thus, the formula given in Theorem 22 does not necessarily hold for a graph containing more than one cycle.

A particular interesting class of graph is those for which $\nu(G) = 1$. It is expected that for such a graph G, powers of its edge ideal should asymptotically have linear resolutions. This is examined in the next two theorems under various additional conditions.

Theorem 23 ([4, Theorem 1.2]) *Let G be a gap-free- and cricket-free graph, and let $I = I(G)$ be its edge ideal. Then for all $q \geq 2$, we have*

$$reg I^q = 2q;$$

i.e., I^q has a linear resolution.

Proof Since I^q is generated in degree $2q$, it remains to show that for $q \geq 2$, $\text{reg} I^q \leq 2q$. By induction, making use of the inductive techniques of Theorem 4 (and [4, Theorem 3.4] which proves that $\text{reg} I \leq 3$), it suffices to show that for any collection of edges $e_1, \dots, e_q$ in G, $\text{reg}(I^{q+1} : e_1 \dots e_q) \leq 2$. Let J be the polarization of $I^{q+1} : e_1 \dots e_q$ and let H be the simple graph associated to J. The statement is reduced to showing that $\text{reg} J = 2$, or equivalently (by Theorem 6), that H^c is a chordal graph. That is, H does not have any induced anticycle of length at least 4.

By contradiction, suppose that H has an anticycle $w_1, \dots, w_s$ of length at least 4. Since an induced anticycle of length 4 gives a gap, we may assume further that $s \geq 5$. Suppose also that $e_1 = xy$. The proof follows from the following observations:

1. There must be an edge between $\{x, y\}$ and $\{w_1, w_3\}$; otherwise xy and w_1w_3 form a gap in G.
2. Suppose that $xw_1 \in E_G$. Then neither of w_2 nor w_n can coincide with x.
3. Neither of w_2 nor w_n can coincide with y; otherwise the other vertex (among $\{w_2, w_n\}$) and w_1 would be even-connected implying that w_1w_2 or w_1w_n is an edge in H.
4. Either w_2 or w_n must be connected to x; otherwise, xy and w_2w_n form a gap in G.

5. Suppose that $xw_2 \in E_G$. Now, by the same line of arguments applying to $\{w_3, w_n\}$, we deduce that either w_3 or w_n must be a neighbor of x.
6. Suppose that $xw_3 \in E_G$. We arrive at a contradiction that $\{w_1, w_3, x, y, w_2\}$ forms a cricket in G.

Theorem 24 ([16, Theorem 4.9]) *Let G be a gap-free- and diamond-free graph, and let $I = I(G)$ be its edge ideal. Then for all $q \geq 2$, we have*

$$reg I^q = 2q;$$

i.e., I^q has a linear resolution.

Proof The proof of this theorem is based on a good understanding of the combinatorial structures of gap-free- and diamond-free graphs. Let $\omega(G)$ denote the largest size of a complete subgraph in G (the *clique number* of G). If $\omega(G) < 3$ then G is cricket-free, and the assertion follows from Theorem 23. If $\omega(G) \geq 4$, then it is shown that the complement of G is chordal, and the conclusion follows from Theorem 6.

Consider the case where $\omega(G) = 3$. It is shown that if, in addition, G is also C_5-free then the complement of G is either chordal or C_6, and the assertion again follows from known results. The essential part of the proof is then to examine the structures of G when G is gap-free, diamond-free and contains a C_5. This is where novel and interesting combinatorics happen. It is shown that, in this case, G can be obtained from a list of ten specific graphs via a so-called process of *multiplying vertices*. The proof is completed with a careful analysis of each of these ten graphs to show that $\text{reg}(I(G)^{q+1} : e_1 \dots e_q) \leq 2$ and to employ Banerjee's result, Theorem 4.

Another large class of graphs for which the regularity of powers of their edge ideals can be computed explicitly is that of very well-covered graphs. The following theorem was proved by Norouzi, Seyed Fakhari, and Yassemi [47] for very well-covered graphs with an additional condition, and then by Jayanthan and Selvaraja [31] in full generality for any very well-covered graph.

Theorem 25 ([31, Theorem 5.3], [47, Theorem 3.6]) *Let G be a very well-covered graph and let $I = I(G)$ be its edge ideal. Then for all $q \geq 1$, we have*

$$reg I^q = 2q + \nu(G) - 1.$$

Proof The proof also uses inductive techniques given by Theorem 4 and goes along the same line as that of previous theorems in this section. The heart of the arguments is to verify that for any collection of edges $e_1, \dots, e_q$ in G, we have

$$\text{reg}(I^{q+1} : e_1 \dots e_q) \leq \nu(G) + 1.$$

If $J = (I^{q+1} : e_1 \dots e_q)$ is squarefree then this is achieved by letting H be the graph associated to J and establishing the following facts:

1. H is also very well-covered;
2. $\nu(H) \leq \nu(G)$;
3. $\operatorname{reg} I(H) = \nu(H) + 1$ and $\operatorname{reg} I(G) = \nu(G) + 1$ (this is the content of [40, Theorem 4.12]).

The arguments are much more involved for the case where J is not squarefree. Let H be the graph corresponding to the polarization of J. The proof is completed by an ingenious induction on the number of vertices added to G in order to obtain H.

6 Higher Dimension

In this section, we discuss the regularity of squarefree monomial ideals generated in degree more than two. The goal, as in the case of the edge ideals, is to find bounds and/or formulas for regularity of the ideal in terms of "its combinatorics". As mentioned in the preliminaries, one can view these ideals both as Stanley–Reisner ideals or edge ideals of hypergraphs. If interpreted as Stanley–Reisner ideals, via Hochster's formula one can potentially get all Betti numbers of a given ideal in terms of the combinatorics of the underlying complex. The hypergraph case, however, is far less understood compared to either the edge ideal case or the Stanley–Reisner ideal interpretation. However, in the last decade, some results have been proven in that direction. In some cases, general squarefree monomial ideals can be viewed as path ideals of simple graphs, and there has been some progress for those cases in the last few years.

Since the main topic of this survey is the edge ideals, we are not trying to be comprehensive in this section. Our aim is to give the reader an idea of possible approaches to generalize the results on edge ideals to higher dimensions. We split the section into three subsections; devoted to Stanley–Reisner ideals, edge ideals of hypergraphs and path ideals respectively.

6.1 Stanley–Reisner Ideals

As mentioned earlier, any squarefree monomial ideal can be viewed as the Stanley–Reisner ideal of some simplicial complex. This combinatorial interpretation is by far the most studied one among monomial ideals generated in higher degrees.

The first result we mention is a famous theorem by Eagon and Reiner which establishes the relation between minimal regularity and maximal depth as well as relation between regularity of a Stanley–Reisner ideal and the structure of a special kind of dual complex, namely, the Alexander dual. In a sense it is similar to Fröberg's result; this gives a classification of linear resolution. The *Alexander dual* of a simplicial complex Δ is the simplicial complex whose faces are the complements of the non-faces of Δ. If I is a Stanley–Reisner ideal then by $I^\vee$ we denote the Stanley–Reisner ideal of the Alexander dual of the simplicial complex of I.

Theorem 26 ([15, Theorem 3]) *Let I be a Stanley–Reisner ideal in $S = K[x_1, \ldots, x_n]$. Then I has q-linear resolution if and only if depth of $S/I^\vee$ has depth $n - q$. In particular, I has linear resolution if and only if the Alexander dual of $\Delta(I)$ is Cohen–Macaulay.*

The next result bounds regularity with another important algebraic invariant, the arithmetic degree. For a squarefree monomial ideal I, its *arithmetic degree*, denoted by $\mathrm{adeg}(I)$, is given by number of facets in the corresponding simplicial complex.

Theorem 27 ([19, Theorem 3.8]) *Suppose I is a Stanley–Reisner ideal with codimension ≥ 2. Then, $\mathrm{reg} I \leq \mathrm{adeg}(I)$.*

6.2 Hypergraphs

Any squarefree monomial ideal is the edge ideal of a hypergraph. The corresponding simplicial complex is the independence complex of this hypergraph and computing that is an NP-hard problem in general.

The problem of finding bounds for the regularity of graphs can be extended to hypergraphs. In [11] the authors provided a sufficient condition for a hypergraph to have regularity ≤ 3. For every vertex x of a hypergraph $\mathcal{H}$, let $\mathcal{H} : x$ denote the simple hypergraph of all minimal subsets $A \subset V \setminus \{x\}$ such that A or $A \cup \{x\}$ is an edge of $\mathcal{H}$.

Theorem 28 ([11, Theorem 6.4]) *Let $\mathcal{H}$ be a simple hypergraph such that $\mathcal{H} : x$ is a graph whose complement is chordal for all vertices of $\mathcal{H}$. Then, $\mathrm{reg}\mathcal{H} \leq 3$.*

Any hypergraph whose all edges have same cardinality is called *uniform*. A d-uniform hypergraph (i.e., a uniform hypergraph whose all edges are represented by monomials of degree d) is called *properly connected* if for any two edges E, E' sharing at least one vertex, the length of the shortest path between E and E' is $d - |E \cap E'|$. If the length of the shortest path between two edges is $\geq t$ then they are called t-disjoint (for relevant definitions see [23]).

Example 12 Consider the 4-uniform hypergraph H with edge set:

$$\{x_1x_2x_3x_4, x_1x_2x_3x_7, x_1x_2x_6x_7, x_1x_5x_6x_7, x_1x_5x_6x_8\}.$$

There is a proper irredundant chain of length 4 from the edge $E = x_1x_2x_3x_4$ to $E_0 = x_1x_5x_6x_8$. Furthermore, there is no shorter such chain. But these edges have a nonempty intersection. So H is not properly connected since the distance between them is 4 which is not same as $4 - |E \cap E_0| = 3$. This hypergraph is not properly connected. On the other hand every finite simple graph is properly connected.

A d-uniform properly connected hypergraph H is called *triangulated* if for any subset A of vertices the induced subgraph H' on A has a vertex x such that the induced subgraph on its neighborhood $N(x) \cup \{x\}$ is complete. For a more formal definition and relevant notions see [23].

Example 13 Simple graphs that are triangulated are precisely the chordal graphs due to Theorem 5.2 of [23].

The following result by Hà and Van-Tuyl gives a lower bound for regularity of properly connected hypergraphs and the bound becomes an equality if the hypergraph is triangulated.

Theorem 29 ([23, Theorem 6.5 and 6.8]) *Let H be a properly connected hypergraph. Suppose d is the common cardinality of the edges in H. Let c be the maximal number of pairwise $d+1$ disjoint edges in H. Then, reg$H \geq (d-1)c+1$, and the equality occurs if H is triangulated.*

A 2-*collage* for a hypergraph is a subset C of the edges with the property that for every edge E of the hypergraph, we can delete a vertex v of E so that $E \setminus \{v\}$ is contained in some edge of C. For uniform hypergraphs, the condition for a collection C of the edges to be a 2-collage is equivalent to requiring that for any edge E not in C, there exists $F \in C$ such that the symmetric difference of E and F consists of exactly two vertices. When H is a graph, it is straightforward to see that for any minimal 2-collage, there is a maximal matching of the same or lesser cardinality.

The following theorem gives a formula for regularity of uniform hypergraphs in terms of its collages.

Theorem 30 ([24, Theorem 1.1]) *Let H be a simple d-uniform hypergraph and let c be the minimum size of a 2 collage in H. Then, reg$I(H) \leq (d-1)c+1$.*

The next theorem is a generalization of the previous that work for all simple hypergraph.

Theorem 31 ([24, Theorem 1.2]) *Let H be a simple hypergraph and let $\{m_1, \ldots, m_c\}$ be a 2-collage. Then, reg$I(H) \leq \sum_{i=1}^{c} |m_i| - c + 1$.*

The next theorem gives a formula for regularity of edge ideals of *clutters*, which are hypergraphs where no edge contains any other edge as a subset. A monomial ideal I has *linear quotients* if the monomials that generate I can be ordered $g_1, \ldots, g_q$ such that for all $1 \leq i \leq q-1$, $((g_1, \ldots, g_i) : g_{i+1})$ is generated by linear forms. For further details see [44].

Theorem 32 ([44, Corollary 3.35]) *Let H be a clutter such that $I(H)$ has linear quotients. Then, reg$I(H) = \max\{|E| : E \in E(H)\}$.*

Taylor resolutions, introduced in [49], are free resolutions of monomial ideals constructed in a natural way that makes computations easy to deal with. These are in general not minimal. The following theorem gives a formula for regularity when Taylor resolutions are minimal. A hypergraph H is called *saturated* when the Taylor resolution of $I(H)$ is minimal. For further details see [38].

Theorem 33 ([38, Proposition 4.1]) *For a saturated hypergraph H, we have* $\operatorname{reg} I(H) = |X| - |L| + 1$. *Here L is the minimal monomial generating set of $I(H)$ and X is the number of variables that divide some minimal monomial generator.*

6.3 Path Ideals

Path ideals of finite simple graphs are interesting generalizations of edge ideals, see Sect. 2.2 to recall the definition of path ideals. For gap-free graphs their regularity too tend to behave nicely.

Theorem 34 ([5, Theorem 1.1]) *Let G be a gap-free- and claw-free graph. Then for all $t \leq 6$, I_t has linear resolution. If further G is induced whiskered-K_4 free then for all t, I_t has linear resolution.*

Let L_n denote a line of length n. The regularity of path ideals of lines have been computed by Alilooee and Faridi in the next theorem.

Theorem 35 ([1, Theorem 3.2]) *Let n, t, p and d be integers such that $n \geq 2$, $2 \leq t \leq n$, $n = (t+1)p + d$, where $p \geq 0$, $0 \leq d \leq t$. Then, $\operatorname{reg} I_t(L_n)$ is $p(t-1)+1$ for $d < t$ and $(p+1)(t-1)+1$ for $d = t$.*

We end this section by a result regarding a somewhat different kind of path ideals. A tree is a graph in which there exists a unique path between every pair of distinct vertices; a rooted tree is a tree together with a fixed vertex called the root. In particular, in a rooted tree there exists a unique path from the root to any given vertices. We can also view a rooted tree as a directed graph by assigning to each edge the direction that goes "away" from the root. Let Γ be a rooted tree and $I_t(\Gamma)$ be the squarefree monomial ideal generated by all"directed paths" of length $t-1$ in the above sense. The following theorem gives regularity of such ideals.

In this theorem, we define $l_t(\Gamma)$ to be the number of leaves in Γ whose level is at least $t-1$ and $p_t(\Gamma)$ to be the maximal number of pairwise disjoint paths of length t in Γ (i.e., $p_t(\Gamma) = \max\{|D| \mid D$ is a set of disjoint paths of length t in $\Gamma\}$).

Theorem 36 ([9, Theorem 3.4]) *Let Γ be a rooted tree on n vertices. Then,* $\operatorname{reg} I_t(\Gamma) \leq (t-1)[l_t(\Gamma) + p_t(\Gamma)] + 1$.

7 Open Problems and Questions

We end the paper by listing a number of open problems and questions in the research area which we hope to be solved.

Our first problem is inspired by Theorems 6 and 13. Since graphs with regularity 2 are classified. The next class of graph to examine is that of regularity 3.

Problem 1 Characterize graphs G for which $\text{reg}I(G) = 3$.

In various results, for example Theorems 10 and 28, "local" condition on the regularity of $G : x$ for all vertices x lead to a"global" statement on the regularity of G. We ask if similar local conditions on $G : x$, for all vertices x, would also lead to a statement on the asymptotic linear function $\text{reg}I(G)^q$.

Question 1 Let G be a simple graph and let $I = I(G)$ be its edge ideal. Suppose that for any vertex x in G, we have $\text{reg}(I : x) \leq r$. Does this imply that for any $q \geq 1$,

$$\text{reg}I^q \leq 2q + r - 1?$$

As noted in Lemma 4, for any vertex x, the regularity of $I(G)$ is always equal to either the regularity of $I(G) : x$ or the regularity of $(I(G), x)$. It would be interesting to know for which vertex x the equality happens one way or another.

Problem 2 Let G be a graph and let $I = I(G)$ be its edge ideal. Find conditions on a vertex x of G such that

1. $\text{reg}I = \text{reg}(I, x)$.
2. $\text{reg}I = \text{reg}(I : x)$.

The regularity of $I(G)$ has been computed for several special classes of graphs (see Theorem 14). A particular class of graphs which is of interest is that of vertex decomposable graphs. For a vertex decomposable graph G, the statement of Lemma 4 can be made slightly more precise (see [24]), namely, there exists a vertex x such that

$$\text{reg}G = \max\{\text{reg}(G \setminus N_G[x]) + 1, \text{reg}(G \setminus x)\}.$$

For such a graph G, it is also known that the independent complex $\Delta(G)$ is shellable and the quotient ring $S/I(G)$ is sequentially Cohen–Macaulay.

Problem 3 Let G be a vertex decomposable graph. Compute $\text{reg}I(G)$ via combinatorial invariant of G.

As noted throughout Sects. 4 and 5, the induced matching number of a graph G is closely related to the regularity of $I(G)$. In fact, $\nu(G) + 1$ gives a lower bound for $\text{reg}I(G)$ and, more generally, $2q + \nu(G) - 1$ gives a lower bound for $\text{reg}I(G)^q$ for any $q \geq 1$. Moreover, for many special classes of graphs the equality has been shown to hold. Thus, it is desirable to characterize all graphs for which the equality is attained.

Problem 4 Characterize graphs G for which the edge ideals $I = I(G)$ satisfy

1. $\text{reg}I = \nu(G) + 1$.
2. $\text{reg}I^q = 2q + \nu(G) - 1 \ \forall\, q \gg 0$.

When $\nu(G) = 1$, the answer to Problem 4. (2) is predicted in the following open problem.

Problem 5 (*Francisco-Hà-Van Tuyl and Nevo-Peeva*) Suppose that $\nu(G) = 1$, i.e., G^c has no induced 4-cycle and let $I = I(G)$.

1. Prove (or disprove) that $\text{reg}I^q = 2q$ for all $q \gg 0$.
2. Prove (or disprove) that $\text{reg}I^{q+1} = \text{reg}I^q + 2$ for all $q \geq \text{reg}(I) - 1$.

Note that examples exist in which $\nu(G) = 1$ and $\text{reg}I^q \neq 2q$ for small values of q (cf. [46]), so in Problem 5 it is necessary to consider $q \gg 0$. A satisfactory solution to Problem 1 would be a good starting point to tackle Problem 5, since regularity 3 is the first open case of the problem. In fact, in this case, $\text{reg}I(G)^q$ is expected to be linear starting at $q = 2$.

Problem 6 (*Nevo-Peeva*) Suppose that $\nu(G) = 1$ and $\text{reg}I(G) = 3$. Then is it true that for all $q \geq 2$,

$$\text{reg}I(G)^q = 2q?$$

In many known cases where the asymptotic linear function $\text{reg}I(G)^q$ can be computed, it happens to be $\text{reg}I(G)^q = 2q + \nu(G) - 1$. We would like to see if this is the case when the equality is already known to hold for small values of q. If this is indeed the case then how far one must go before concluding that the equality holds for all $q \geq 1$?

Problem 7 Let G be a graph. Find a number N such that if $\text{reg}I(G)^q = 2q + \nu(G) - 1$ for all $1 \leq q \leq N$ then, for all $q \geq 1$, we have

$$\text{reg}I(G)^q = 2q + \nu(G) - 1.$$

A particularly related question is whether for special classes of graphs satisfying $\text{reg}I(G) = \nu(G) + 1$ one would have $\text{reg}I(G)^q = 2q + \nu(G) - 1$ for all $q \geq 1$ (or for all $q \gg 0$). Inspired by Theorem 14, we raise the following question.

Question 2 Let G be a graph and let $I = I(G)$ be its edge ideal. Suppose that G is of one of the following types:

1. G is chordal;
2. G is weakly chordal;
3. G is sequentially Cohen–Macaulay bipartite;
4. G is vertex decomposable and contains no closed circuit of length 5;
5. G is (C_4, C_5)-free vertex decomposable.

Is it true that for all $q \gg 0$,

$$\text{reg}I^q = 2q + \nu(G) - 1?$$

Note that chordal graphs and sequentially Cohen–Macaulay bipartite graphs are vertex decomposable. Also bipartite graphs contain no closed circuit of length 5. Thus, in Question 2, an affirmative answer to part (4) would imply that for part (3).

It is well known that the resolution of a monomial ideal is dependent on the characteristic of the ground field. And yet, in all known cases where the regularity of powers of an edge ideal can be computed, it is characteristic-independent. We would like to see examples where this is not the case, or a confirmation that this is always the case if the regularity of the edge ideal itself is characteristic-independent.

Problem 8

1. Find examples of graphs G for which the asymptotic linear function $\operatorname{reg} I(G)^q$, for $q \gg 0$, is characteristic-dependent.
2. Suppose that $\operatorname{reg} I(G)$ is independent of the characteristic of the ground field. Is the asymptotic linear function $\operatorname{reg} I(G)^q$, for $q \gg 0$, necessarily also characteristic-independent?

When an exact formula may not be available, it is of interest to find a lower and an upper bound. Since the lower bound in Theorem 16 holds for any graph, it is desirable to find an upper bound to couple with this general lower bound. In this direction, one may either try to prove the bound in Theorem 17 for any graph or to relate the regularity of powers of $I(G)$ to the regularity of $I(G)$ itself.

Conjecture 1 *(Alilooee, Banerjee, Beyarslan and Hà)* Let G be any graph and let $I = I(G)$ be its edge ideal. Then for all $q \geq 1$, we have

1. $\operatorname{reg} I^q \leq 2q + \text{co-chordal}(G) - 1.$
2. $\operatorname{reg} I^q \leq 2q + \operatorname{reg} I - 2.$
3. $\operatorname{reg} I^{q+1} \leq \operatorname{reg} I^q + 2.$

Note that by Theorem 9, Conjecture 1.(3) $\Rightarrow$ Conjecture 1.(2) $\Rightarrow$ Conjecture 1.(1).

Even though, in general, $\operatorname{reg} I^q$ is asymptotically a linear function, for small values of q, there are examples for which $\operatorname{reg} I^q > \operatorname{reg} I^{q+1}$. We ask if this would *not* be the case for edge ideals of graphs.

Question 3 Let G be a graph and let $I = I(G)$ be its edge ideal. Is the function $\operatorname{reg} I^q$ increasing for all $q \geq 1$?

In investigating the asymptotic linear function $\operatorname{reg} I^q$, it is also of interest to know the smallest value q_0 starting from which $\operatorname{reg} I^q$ attains its linear form. In known cases where the regularity of $I(G)^q$ can be computed explicitly for all $q \geq 1$, we have $q_0 \leq 2$.

Problem 9 Find a number N (which may depend on n and m) such that for any graph G over n vertices and m edges, we have $q_0 \leq N$.

Acknowledgements The authors would like to thank the organizers of SRAC (Southern Regional Algebra Conference) 2017 for their encouragement, which led us to writing this survey. The last named author is partially supported by the Simons Foundation (grant #279786) and Louisiana Board of Regents (grant #LEQSF(2017-19)-ENH-TR-25).

References

1. Alilooee, A., Faridi, S.: On the resolutions of path ideals of cycles. Commun. Algebra **43**(12), 5413–5433 (2015)
2. Alilooee, A., Banerjee, A.: Powers of edge ideals of regularity three bipartite graphs. J. Commun. Algebra **9**(4), 441–454 (2017)
3. Alilooee, A., Beyarslan, S., Selvaraja, S.: Regularity of powers of unicyclic graphs. Rocky Mt. J. Math. (2017). arXiv:1702.00916
4. Banerjee, A.: The regularity of powers of edge ideals. J. Algebraic Combin. **41**(2), 303–321 (2015)
5. Banerjee, A.: Regularity of path ideals of gap free graphs. J. Pure Appl. Alg. **221**(10), 2409–2419 (2017)
6. Beyarslan, S., Hà, H.T., Trung, T.N.: Regularity of powers of forests and cycles. J. Algebraic Combin. **42**(4), 1077–1095 (2015)
7. Biyikoglu, T., Civan, Y.: Vertex decomposable graphs, codismantlability, Cohen-Macaulayness and Castelnuovo-Mumford regularity. Electron. J. Combin. **21**(1), 1–1 (2014)
8. Björner, A.: Topological methods. In: Handbook of Combinatorics, vol. 1, 2, pp. 1819–1872. Elsevier, Amsterdam (1995)
9. Bouchat, R., Hà, H.T., O'Keefe, A.: Path ideals of rooted trees and their graded Betti numbers. J. Combin. Theory Ser. A **118**, 2411–2425 (2011)
10. Cameron, K., Walker, T.: The graphs with maximum induced matching and maximum matching the same size. Discrete Math. **299**, 49–55 (2005)
11. Caviglia, G., Hà, H.T., Herzog, J., Kummini, M., Terai, N., Trung, N.V.: Depth and regularity modulo a principal ideal. J. Algebraic. Combin. (2017). arXiv:1706.09675
12. Chardin, M.: Some results and questions on Castelnuovo-Mumford regularity. In: Syzygiez and Hilbert functions. Lecture Notes in Pure and Applied Mathematics, vol. 254, pp. 1–40. Chapman & Hall/CRC (2007)
13. Cutkosky, S.D., Herzog, J., Trung, N.V.: Asymptotic behaviour of the Castelnuovo-Mumford regularity. Compos. Math. **118**, 243–261 (1999)
14. Dao, H., Huneke, C., Schweig, J.: Bounds on the regularity and projective dimension of ideals associated to graphs. J. Algebraic Combin. **38**(1), 37–55 (2013)
15. Eagon, J.A., Reiner, V.: Resolutions of Stanley-Reisner rings and Alexander duality. J. Pure Appl. Algebra **130**(3), 265–267 (1998)
16. Erey, N.: Powers of edge ideals with linear resolutions. Commun. Algebra **46**(9), 4007–4020 (2018)
17. Fernàndez-Ramos, O., Gimenez, P.: Regularity 3 in edge ideals associated to bipartite graphs. J. Algebraic Combin. **39**(4), 919–937 (2014)
18. Francisco, C.A., Tuyl, A.V.: Sequentially Cohen-Macaulay edge ideals. Proc. Am. Math. Soc. **135**(8), 23–37 (2007) (electronic)
19. Frühbis-Krüger, A., Terai, N.: Bounds for the regularity of monomial ideals. Le Matematiche Supplemento 83–97 (2007)
20. Fröberg, R.: On Stanley-Reisner rings. In: Topics in Algebra, Part 2 (Warsaw, 1988), vol. 26, pp. 57–70, Banach Center Publ., Part 2, PWN, Warsaw (1990)
21. Hà, H.T.: Regularity of squarefree monomial ideals. In: Connections Between Algebra, Combinatorics, and Geometry. Springer Proceedings in Mathematics and Statistics, vol. 76, pp. 251–276. Springer, New York (2014)
22. Hà, H.T., Van Tuyl, A.: Resolutions of square-free monomial ideals via facet ideals: a survey. In: Algebra, Geometry and Their Interactions. Contemporary Mathematics, vol. 448, pp. 91–117. American Mathematical Society, Providence, RI (2007)
23. Hà, H.T., Van Tuyl, A.: Monomial ideals, edge ideals of hypergraphs, and their minimal graded free resolutions. J. Algebraic Combin. **27**(2), 215–245 (2008)
24. Hà, H.T., Woodroofe, R.: Results on the regularity of square-free monomial ideals. Adv. Appl. Math. **58**, 21–36 (2014)

25. Herzog, J.: A generalization of the Taylor complex construction. Commun. Algebra **35**(5), 1747–1756 (2007)
26. Herzog, J., Hibi, T.: Monomial Ideals. GTM, vol. 260. Springer-Verlag (2011)
27. Herzog, J., Hibi, T., Zheng, X.: Monomial ideals whose powers have a linear resolution. Math. Scand. **95**(1), 23–32 (2004)
28. Hibi, T., Higashitani, A., Kimura, K., O'Keefe, A.B.: Algebraic study on Cameron-Walker graphs. J. Algebra **422**, 257–269 (2015)
29. Hochster, M.: Cohen-Macaulay rings, combinatorics, and simplicial complexes. In: Proceedings of the 2nd Oklahoma Conference. Lecture Notes in Pure and Applied Mathematics, vol. 26, pp. 171–224. Marcel Dekker (1977)
30. Jacques, S.: Betti numbers of graph ideals. Ph.D. Thesis, University of Sheffield (2004). arXiv:math.AC/0410107
31. Jayanthan, A.V., Selvaraja, S.: Linear Polynomials for the Regularity of Powers of Edge Ideals of Very Well-covered Graphs. To appear in J. Commut. Algebra
32. Jayanthan, A.V., Narayanan, N., Selvaraja, S.: Regularity of powers of bipartite graphs. J. Algebraic Combin. **47**(1), 17–38 (2018)
33. Kalai, G., Meshulam, R.: Intersections of Leray complexes and regularity of monomial ideals. J. Combin. Theory Ser. A **113**(7), 1586–1592 (2006)
34. Katzman, M.: Characteristic-independence of Betti numbers of graph ideals. J. Combin. Theory Ser. A **113**(3), 435–454 (2006)
35. Khosh-Ahang, F., Moradi, S.: Regularity and projective dimension of edge ideal of C_5-free vertex decomposable graphs. Proc. Am. Math. Soc. **142**(5), 1567–1576 (2014)
36. Kodiyalam, V.: Asymptotic behaviour of Castelnuovo-Mumford regularity. Proc. Am. Math. Soc. **128**(2), 407–411 (1999)
37. Kummini, M.: Regularity, depth and arithmetic rank of bipartite edge ideals. J. Algebraic Combin. **30**(4), 429–445 (2009)
38. Lin, K.-N., McCullough, J.: Hypergraphs and regularity of square-free monomial ideals. Int. J. Algebra Comput. **23**, 1573–1590 (2013)
39. Lyubeznik, G.: The minimal non-Cohen-Macaulay monomial ideals. J. Pure Appl. Algebra **51**, 261–266 (1988)
40. Mahmoudi, M., Mousivand, A., Crupi, M., Rinaldo, G., Terai, N., Yassemi, S.: Vertex decomposability and regularity of very well-covered graphs. J. Pure Appl. Algebra **215**(10), 2473–2480 (2011)
41. Minh, N.C., Trung, N.V.: Cohen-Macaulayness of monomial ideals and symbolic powers of Stanley-Reisner ideals. Adv. Math. **226**, 1285–1306 (2011)
42. Moghimian, M., Seyed Fakhari, S.A., Yassemi, S.: Regularity of powers of edge ideal of whiskered cycles. Comm. Algebra **45**(3), 1246–1259 (2017)
43. Morey, S.: Depths of powers of the edge ideal of a tree. Comm. Algebra **38**(11), 4042–4055 (2010)
44. Morey, S., Villarreal, R.H.: Edge ideals: algebraic and combinatorial properties. In: Progress in Commutative Algebra, vol. 1, pp. 85–126. de Gruyter, Berlin (2012)
45. Nevo, E.: Regularity of edge ideals of C_4-free graphs via the topology of the lcm-lattice. J. Combin. Theory Ser. A **118**, 491–501 (2011)
46. Nevo, E., Peeva, I.: C_4-free edge ideals. J. Algebraic Combin. **37**(2), 243–248 (2013)
47. Norouzi, P., Seyed Fakhari, S.A., Yassemi, S.: Regularity of powers of edge ideals of very well-covered graphs (2017). arXiv:1707.04874
48. Takayama, Y.: Combinatorial characterizations of generalized Cohen-Macaulay monomial ideals. Bull. Math. Soc. Sci. Math. Roumanie (N.S.) **48**, 327–344 (2005)
49. Taylor, D.K.: Ideals generated by monomials in an R-sequence. Ph.D. Thesis, The University of Chicago, ProQuest LLC, Ann Arbor, MI (1996)
50. Trung, N.V., Wang, H.: On the asymptotic behavior of Castelnuovo-Mumford regularity. J. Pure Appl. Algebra **201**(1–3), 42–48 (2005)
51. Villarreal, R.H.: Monomial Algebras. Monographs and Textbooks in Pure and Applied Mathematics, vol. **238**. Marcel Dekker, Inc., New York (2001)

52. Van Tuyl, A.: Sequentially Cohen-Macaulay bipartite graphs: vertex decomposability and regularity. Arch. Math. (Basel) **93**(5), 451–459 (2009)
53. Wegner, G.: d-collapsing and nerves of families of convex sets. Arch. Math. (Basel) **26**, 317–321 (1975)
54. Woodroofe, R.: Vertex decomposable graphs and obstructions to shellability. Proc. Amer. Math. Soc. **137**(10), 3235–3246 (2009)
55. Woodroofe, R.: Matchings, coverings, and Castelnuovo-Mumford regularity. J. Commut. Algebra **6**(2), 287–304 (2014)

A Survey of Rings Satisfying Annihilator or Extending Conditions on Projection Invariant Ideals

Gary F. Birkenmeier, Yeliz Kara and Adnan Tercan

Abstract In this paper, we survey results involving the projection invariant condition on one-sided ideals of rings. We focus on rings satisfying the right projection invariant extending condition (denoted right π-extending) or the projection invariant Baer condition (denoted π-Baer). Examples are provided to illustrate and delimit the results.

Keywords Baer ring · Quasi-Baer ring · Extending module · Extending ring · FI-extending · Projection invariant · Ring extension

2010 AMS Subject Classification: 16D10 · 16S36 · 16S50 · 16N60 · 16S70 · 16S75 · 16W60

1 Introduction

All rings are associative with unity, subrings contain the same unity as the overring, and modules are unital right modules unless indicated otherwise. We use R to denote such a ring and M to denote such a module.

In this paper, we survey results which generalize the concepts of a Baer ring and of a right extending ring through the use of projection invariant one-sided ideals. Both of these concepts evolved in the 1930s from the papers of Murray and von Neumann [22] and von Neumann [23]. Building on the ideas of Rickart in [25], Kaplansky

G. F. Birkenmeier (✉)
Department of Mathematics, University of Louisiana at Lafayette,
Lafayette, LA 70504-1010, USA
e-mail: gfb1127@louisiana.edu

Y. Kara
Department of Mathematics, Bursa Uludağ University, 16059 Bursa, Turkey
e-mail: yelizkara@uludag.edu.tr

A. Tercan
Department of Mathematics, Hacettepe University, 06532 Ankara, Turkey
e-mail: tercan@hacettepe.edu.tr

J. Feldvoss et al. (eds.), *Advances in Algebra*, Springer Proceedings
in Mathematics & Statistics 277, https://doi.org/10.1007/978-3-030-11521-0_3

defined a Baer ring in [18]. The following definition is equivalent to Kaplansky's: R is a *Baer* ring if the right annihilator of a left ideal is generated by an idempotent. Note that the Baer condition is left–right symmetric [18, Theorem 3]. In [28], Utumi said R satisfies the *right* C_1 *condition* (equivalently, *right extending condition*) if every right ideal is essential in an idempotent generated right ideal of R. Both of these concepts have been extended to modules and generalized and studied by many authors. For further details see [3, 6, 14, 18, 21, 27]. In Chatters and Khuri [12, Theorem 2.1], the concepts of a Baer ring and a right extending ring are linked.

Unfortunately, the Baer property and right extending property do not transfer from a base ring to various ring extensions without adding further conditions to the base ring and/or the ring extension. For example, if R is Baer or right extending, then, in general, the n-by-n ($n > 1$) full or upper triangular matrix rings are neither Baer nor right extending (e.g., take R to be a commutative non-Prüfer domain). In general, if R is Baer or right extending, then $R[x]$ is neither Baer [2] nor right extending. However, generalizations of these conditions have been found which transfer easily from a base ring to a ring extension. Clark [13] called a ring *quasi-Baer* if the right annihilator of an ideal is generated as a right ideal by an idempotent. From [4], R is called *right* FI*-extending* if every ideal is right essential in an idempotent generated right ideal. From [24], it can be seen that the quasi-Baer condition is a Morita invariant, thus it transfers to full matrix extensions. Also, in [24], the quasi-Baer condition is closed with respect to forming triangular matrix rings. From [6], one can see that the quasi-Baer condition transfers from R to $R[x]$ and to many other ring extensions. In [4], the right FI-extending property is shown to be closed for full and upper triangular matrix extensions. Since the Baer and right extending conditions are defined in terms of one-sided ideals, whereas the quasi-Baer and right FI-extending conditions are defined in terms of ideals, then the Baer and right extending conditions give more precise structural information about a ring than do the quasi-Baer and right FI-extending conditions. For example, a ring R is a domain if and only if it is Baer and $\{0, 1\}$ are its only idempotents, whereas R is prime if and only if R is quasi-Baer and $\{0, 1\}$ are its only semicentral idempotents. Therefore, it is natural to ask, is there a condition intermediate between Baer and quasi-Baer (resp., right extending and right FI-extending) that has some of the salient features of each of the Baer and quasi-Baer (resp., right extending and right FI-extending) conditions?

Fuchs [16, p. 50] introduced the projection invariance concept for a subgroup of an Abelian group. Megibben in [20] showed that projection invariant subgroups are fully invariant under some conditions. Abdioğlu et al. [1] extended the projection invariant notion to modules and investigated the behavior of the projection invariant condition with respect to direct sums. Moreover, they defined the projection invariant lifting property and studied finite direct sums of projection invariant lifting modules. Sikander et al. discuss projection invariant submodules of QTAG modules in [26]. In [9, 10], Birkenmeier et al. defined projection invariant extending modules and rings; and in [11], Birkenmeier et al. defined and investigated projection invariant Baer rings.

In Sect. 1, we discuss the projection invariance of one-sided ideals in a ring. Sections 2 and 3 provide results and examples of right projection invariant extending rings and projection invariant Baer rings. Moreover, these rings provide an affirmative answer to the question posed above.

For notation, we use $\mathbf{I}(R)$, $\mathbf{T}_n(R)$, $\mathbf{M}_n(R)$, $Z(R_R)$, $Q(R)$, $S_l(R)$, $S_r(R)$, CFM_Γ, $CRFM_\Gamma$, and $S(R, M)$ for the subring of R generated by idempotents, the n-by-n upper triangular matrix ring over R, the n-by-n full matrix ring over R, the right singular ideal of R, the maximal right ring of quotients of R, the set of left semicentral idempotents of R, the set of right semicentral idempotents of R, the column finite $\Gamma \times \Gamma$ matrices over R, the column and row finite $\Gamma \times \Gamma$ matrices over R, and the split-null extension (or trivial extension) of R by M, where M is an (R, R)-bimodule, respectively. For a nonempty subset X of R, $X \trianglelefteq R$, $X_R \trianglelefteq_p R_R$, $X_R \leq R_R$, $X_R \leq^{ess} R_R$, $r_R(X)$, and $l_R(X)$ denote X is an ideal of R, X is a projection invariant right ideal of R, X is a right ideal of R, X is an essential right ideal of R, the right and the left annihilators of X, respectively. Recall that a ring R is called *Abelian* if every idempotent of R is central. Let e be an idempotent of R. Then, $e \in R$ is called *left* (resp., *right*) *semicentral* if $xe = exe$ (resp., $ex = exe$) for all $x \in R$. Also, a ring R is called *semicentral reduced* if $S_l(R) = \{0, 1\}$ (equivalently $S_r(R) = \{0, 1\}$). Other terminology and notation can be found in [6, 14, 19, 27].

2 Projection Invariance

In this section, we provide basic results and examples of projection invariance.

Definition 1 A left (resp., right) ideal Y of R is called *projection invariant* if $Ye \subseteq Y$ (resp., $eY \subseteq Y$) for all $e = e^2 \in R$ denoted by ${}_RY \trianglelefteq_p {}_RR$ (resp., $Y_R \trianglelefteq_p R_R$).

Note that every ideal of R is projection invariant but not conversely. Observe that the set of projection invariant left (right) ideals forms a complete sublattice of the lattice of left (right) ideals of R. Recall that a module M is π-injective (also called quasi-continuous) if and only if it is a projection invariant submodule of its injective hull [14, p. 14] and [21, p. 20].

Lemma 1 ([11, Lemma 2.1]) *Let Y and X be left and right ideals of R, respectively, which are projection invariant. Then,*

(i) $r_R(Y)$ is a projection invariant right ideal.
(ii) $l_R(X)$ is a projection invariant left ideal.
(iii) If $c = c^2$ and cR is a projection invariant right ideal, then $c \in S_l(R)$.
(iv) If $e = e^2$ and Re is a projection invariant left ideal, then $e \in S_r(R)$.

Lemma 2 ([11, Lemma 2.3]) *The following conditions are equivalent:*

(i) R is Abelian.
(ii) Every one-sided ideal is projection invariant.
(iii) Every right ideal generated by an idempotent is projection invariant.
(iv) Every right ideal generated by an idempotent is an ideal.

Proposition 1

(i) *Let $e = e^2 \in R$. Then, $eR_R \trianglelefteq_p R_R$ if and only if $eR \trianglelefteq R$.*
(ii) *If $R = \mathbf{I}(R)$ (e.g., $R = \mathbf{M}_n(A)$ where A is a ring), then every projection invariant one-sided ideal is an ideal.*
(iii) *If R is a regular ring, then every finitely generated projection invariant one-sided ideal is an ideal.*
(iv) *If R is a semiprime Artinian ring, then every projection invariant one-sided ideal is an ideal.*

Proof This result is an immediate consequence of Lemmas 1 and 2.

Example 1 Let A be an Abelian ring (e.g., a reduced ring) with a right (left) ideal X (Y) which is not an ideal (e.g., a simple domain that is not a division ring). Let $R = \mathbf{T}_2(A)$. Then, $\begin{bmatrix} X & X \\ 0 & 0 \end{bmatrix}$, $\begin{bmatrix} 0 & AX \\ 0 & X \end{bmatrix}$, and $\begin{bmatrix} 0 & X \\ 0 & 0 \end{bmatrix}$ are projection invariant right ideals of R that are not ideals of R. Similarly, $\begin{bmatrix} 0 & Y \\ 0 & Y \end{bmatrix}$, $\begin{bmatrix} Y & YA \\ 0 & 0 \end{bmatrix}$, and $\begin{bmatrix} 0 & Y \\ 0 & 0 \end{bmatrix}$ are projection invariant left ideals of R that are not ideals of R. Note that $\begin{bmatrix} 0 & X \\ 0 & X \end{bmatrix}$ and $\begin{bmatrix} Y & Y \\ 0 & 0 \end{bmatrix}$ are right and left ideals of R, respectively, that are not projection invariant.

Proposition 2

(i) *If Y is a left (right) ideal of $\mathbf{I}(R)$, then YR (RY) is a projection invariant right (left) ideal of R.*
(ii) *If X is a projection invariant one-sided ideal of R, then $X \cap \mathbf{I}(R)$ is an ideal in $\mathbf{I}(R)$.*

Proof The proof is routine.

3 Projection Invariant Extending

Definition 2 ([9, 10]) A module M is *projection invariant extending* (denoted by π-extending) if each projection invariant submodule X (i.e., X is a submodule of M such that $eX \subseteq X$ for all $e = e^2 \in \text{End}(M_R)$) is essential in a direct summand of M. A ring R is *right projection invariant extending* if R_R is π-extending. Analogously, R is *left projection invariant extending* if ${}_RR$ is π-extending.

Originally in [9, 27] "(PI)" was used in place of "π". Since this may cause confusion when considering polynomial identity rings, the "(PI)" was replaced by "π" in subsequent papers (e.g., [10]). Our first result shows that the π-extending condition lies strictly between the extending and FI-extending conditions.

Theorem 1 *Consider the following conditions on R:*

(i) right extending,
(ii) right π-extending, and
(iii) right FI-extending.

Then $(i) \Rightarrow (ii) \Rightarrow (iii)$, but the implications are not reversible.

Proof $(i) \Rightarrow (ii) \Rightarrow (iii)$ These implications follow from [9, Proposition 3.7].

$(ii) \not\Rightarrow (i)$ Let $R = \mathbf{T}_2(\mathbb{Z})$. It is well known that R is not right extending (e.g., see [6, Theorem 9.1.11]), but R is right π-extending by [10, Corollary 13] or see Corollary 3.

$(iii) \not\Rightarrow (ii)$ Let $R = \mathbb{Z}\langle x, y\rangle$, the free ring on two noncommuting indeterminates. Since R is a domain, then 0,1 are its only idempotents. By Lemma 2, every one-sided ideal is projection invariant. But xR is not essential in R_R. Hence, R is not π-extending. However, since every ideal of R is essential in R, then R is right FI-extending.

Proposition 3 *Let R be a domain. Then, the following conditions are equivalent:*

(i) R_R is uniform (i.e., R is a right Ore domain).
(ii) R is right extending.
(iii) R is right π-extending.

Proof This result is a consequence of [9, Proposition 3.8].

Proposition 4 ([10, Proposition 2])

(i) Let R be a prime right nonsingular ring. Then, R_R is π-extending if and only if $0 \neq X \trianglelefteq_p R$ implies $X_R \leq^{ess} R_R$.
(ii) If R_R is π-extending, right nonsingular, and semicentral reduced, then R is a prime ring.

Proposition 5 ([10, Proposition 3]) *Let R be a semiprime right nonsingular right π-extending ring. Assume $S_R \leq R_R$. Then, there exists $X \trianglelefteq_p R$ such that $S = S_1 \oplus S_2$, where $X \leq^{ess} (S_1)_R$ and S_2 contain no nonzero projection invariant right ideals of R.*

Recall from [15], a ring extension T of R is said to be *right (left) intrinsic* over R if $X \cap R \neq 0$ for each nonzero right (left) ideal X of T, denoted $R \leq_r^{int} T$ ($R \leq_l^{int} T$), see [7, 8] for more results and examples of intrinsic extensions.

Proposition 6 ([10, Proposition 4]) *If $\mathbf{I}(R) \leq_r^{int} R$ and $\mathbf{I}(R)$ is right FI-extending, then $\mathbf{I}(R)$ and R are right π-extending.*

Some examples of rings for which $\mathbf{I}(R) \leq_r^{int} R$ are rings such that every nonzero right ideal contains a nonzero idempotent such as regular, π-regular rings with zero nil radical, and semiprime rings with essential socle [6, Example 3.1.27].

Corollary 1 ([10, Corollary 5]) *Let R be a semiprime Baer ring such that $\mathbf{I}(R) \leq_r^{int}$ R. Then, R is right π-extending. In particular, every regular Baer ring is right and left π-extending.*

Observe that if R is both left and right extending, then R is directly finite [19, Theorem 6.48]. Thus, if R is a regular Baer ring which is not directly finite, then R is right and left π-extending by Corollary 1 but not extending on at least one side.

Example 2 ([10, Example 6])

(i) Let $R = R_1 \oplus R_2$ where $R_1 = End(V_F)$ and $R_2 = End({}_K W)$ where V_F is an infinite-dimensional right vector space over a field F and W is an infinite-dimensional left vector space over a field K. From [17, Theorem 9.12], R_1 is a prime regular right self-injective ring and R_2 is a prime regular left self-injective ring and both have a nonzero proper socle. Thus, neither ring is simple, so neither ring is directly finite [17, Corollary 9.26]. Moreover, R_1 is not left self-injective and R_2 is not right self-injective [17, Corollary 9.30]. From [17, Corollaries 13.4 and 13.20], R_1 is not left extending and R_2 is not right extending. Thus, R is neither right extending nor left extending, but R is a regular Baer ring [19, Corollary 7.53]. However, by Corollary 1, R is right and left π-extending.

(ii) The ring $\mathbf{M}_2(R)$ in [17, Example 13.8] is regular but not Baer. Hence, it satisfies neither the left nor the right extending condition by [12, Theorem 2.1]. However, since R is continuous, it is left and right π-extending. From [5, Example 2.7] and [9, Proposition 3.7], $\mathbf{M}_2(R)$ is right and left π-extending.

Proposition 7 ([10, Proposition 7]) *Let $R = \prod_{j\in J} R_j$ be a direct product of rings. Then, R is π-extending if and only if each R_j is π-extending.*

Proposition 8 ([10, Proposition 8]) *Let S be a right essential overring of R (i.e., $R_R \leq^{ess} S_R$).*

(i) If R_R is π-extending, then S_R and S_S are π-extending.

(ii) If S is a subring of $Q(R)$, then S_R is π-extending if and only if S_S is π-extending.

Theorem 2 ([10, Theorem 9 (i), (iii)]) *Let R be a ring, M an ideal of R, and $S = S(R, M)$.*

(i) If S_S is π-extending and $MX \subseteq X$ for all $X \trianglelefteq_p R$, then R_R is π-extending.

(ii) If ${}_R M$ is faithful and R_R is π-extending, then S_S is π-extending

Theorem 3 ([10, Theorem 11]) *Let S and R be rings, M an (S, R)-bimodule and $T = \begin{bmatrix} S & M \\ 0 & R \end{bmatrix}$ the 2-by-2 generalized triangular matrix ring. Then, T_T is π-extending if and only if all of the following conditions hold:*

(i) $l_S(M) = eS$, where $e = e^2 \in S$ and eS_S is π-extending.

(ii) For $N_R \leq M_R$ with $aN \leq N$ for all $a = a^2 \in S$, there exists $f = f^2 \in S$ such that $N_R \leq^{ess} fM_R$.

(iii) R_R is π-extending.

Corollary 2 ([10, Corollary 12]) *Let $T = \begin{bmatrix} End(M_R) & M \\ 0 & R \end{bmatrix}$, where M_R is a right R-module. Then, T_T is right π-extending if and only if M_R and R_R are π-extending.*

Corollary 3 ([10, Corollary 13]) *Let A be a ring. The following statements are equivalent:*

(i) A_A *is* π*-extending.*
(ii) $\boldsymbol{T}_n(A)$ *is right* π*-extending for every positive integer* n*.*
(iii) $\boldsymbol{T}_n(A)$ *is right* π*-extending for some integer* $n \geq 1$*.*

Theorem 4 ([10, Theorem 16]) *Let* R_R *be a* π*-extending ring.*

(i) $End(F_R)$ *is right* π*-extending, where* F_R *is a free right* R*-module.*
(ii) $CFM_\Gamma(R)$ *and* $CRFM_\Gamma(R)$ *are right* π*-extending.*

Corollary 4 ([10, Corollary 18])

(i) Assume $R = \boldsymbol{I}(R)$*. Then,* R_R *is* π*-extending if and only if* R_R *is FI-extending.*
(ii) Assume n *is a positive integer with* $n \geq 1$*. Then,* $\boldsymbol{M}_n(R)$ *is right* π*-extending if and only if* $\boldsymbol{M}_n(R)$ *is right FI-extending.*

The next example provides a class of rings that are right π-extending but not extending.

Example 3 ([10, Example 19]) Let R be a right Ore domain. Then, R_R is uniform, and hence extending.

(i) From Corollary 3, $\mathbf{T}_n(R)$ is right π-extending. However, if $n > 1$ and R is not a division ring, then $\mathbf{T}_n(R)$ is not extending by [12, Theorem 2.1] and [18, p. 16, Exercise 2].
(ii) From Theorem 4, $\mathbf{M}_n(R)$ is right π-extending. If R is commutative but not Prüfer, then $\mathbf{M}_n(R)$ is not right extending by [14, pp. 108–109].

4 Projection Invariant Baer Rings

Definition 3 ([11, Definition 2.2]) A ring R (with unity) is projection invariant Baer (denoted π-Baer) if for each projection invariant left ideal Y there exists $c = c^2 \in R$ such that $r_R(Y) = cR$.

Our first result of this section shows that the π-Baer condition is left–right symmetric.

Proposition 9 ([11, Proposition 2.1]) *Let R be a ring not necessarily with unity. Any two of the following conditions imply the remaining condition.*

(i) R has a unity.
(ii) For each projection invariant left ideal Y of R there exists $c = c^2 \in R$ *such that* $r_R(Y) = cR$*.*
(iii) For each projection invariant right ideal X of R there exists $e = e^2 \in R$ *such that* $l_R(X) = Re$*.*

Proposition 10 ([11, Proposition 2.2]) *For a ring R, the following conditions are equivalent:*

(i) *R is π-Baer.*
(ii) *For each nonempty set S such that $Se \subseteq S$ for all $e = e^2 \in R$, there exists $c = c^2 \in R$ such that $r_R(S) = cR$.*
(iii) *Every projection invariant right annihilator is generated, as a right ideal, by an idempotent element of R.*

Proposition 11 ([11, Proposition 2.3]) *A ring R is π-Baer if and only if whenever I is a projection invariant right ideal of R, there exists $e \in S_l(R)$ such that $I \subseteq eR$ and $l_R(I) \cap eR = eR(1-e)$.*

Theorem 5 ([11, Theorem 2.1]) *Consider the following conditions:*

(i) *R is Baer.*
(ii) *R is π-Baer.*
(iii) *R is quasi-Baer.*

Then, $(i) \Rightarrow (ii) \Rightarrow (iii)$, but the implications are not reversible.

Corollary 5 ([11, Corollary 2.1]) *The center of a π-Baer ring is a Baer ring.*

Corollary 6 [11, Corollary 2.2]) *Let S be a subring of R.*

(i) *If Y is a projection invariant left (right) ideal of R, then $Y \cap S$ is a projection invariant left (right) ideal of S.*
(ii) *Assume that $\boldsymbol{I}(R) \subseteq S$. If X is a projection invariant left (right) ideal of S, then RX (XR) is a projection invariant left (right) ideal of R. Moreover, if R is π-Baer, then S is π-Baer.*
(iii) *Assume that $R = \boldsymbol{I}(R)$. Then, every projection invariant one-sided ideal of R is an ideal of R, and R is π-Baer if and only if R is quasi-Baer.*

Since Baer rings are right and left nonsingular, one might expect that π-Baer rings have the same property in terms of nonsingularity. However, the following example eliminates this situation.

Example 4 ([11, Example 2.2]) Let A be a prime ring with $Z(A_A) \neq 0$ (see the proof of [11, Theorem 2.1]). Let $R = \mathbf{M}_n(A)$ for $n > 1$. Since R is prime then it is quasi-Baer. Also, $R = \mathbf{I}(R)$. By Corollary 6(iii), R is π-Baer but $Z(R_R) \neq 0$. Note that from the proof of [11, Theorem 2.1], there are right nonsingular π-Baer rings which are not Baer. Thus, it is of interest to consider right (or left) nonsingular π-Baer rings.

Proposition 12 ([11, Proposition 2.6]) *Let R be a π-Baer ring. If every essential right ideal of R is an essential extension of a projection invariant right ideal of R, then R is right nonsingular.*

Proposition 13 ([11, Proposition 2.4]) *Let M be a nonsingular π-extending right R-module. Then, $End(M_R)$ is a π-Baer ring.*

Theorem 6 ([11, Theorem 2.2]) *Let R be a right nonsingular ring. Then, R_R is π-extending if and only if R is π-Baer and $A_R \leq^{ess} r_R(l_R(A))$ for every projection invariant right ideal A of R.*

Recall a ring R is *right cononsingular* if $l_R(X) = 0$ with $X_R \leq R_R$ implies $X_R \leq^{ess} R_R$.

Lemma 3 ([11, Lemma 2.4]) *Assume R_R is nonsingular. Then, the following conditions are equivalent:*

(i) Every closed right ideal of R is a right annihilator.
(ii) R is right cononsingular.
(iii) For each $A_R \leq R_R$, $A_R \leq^{ess} r_R(l_R(A))$.

Our next result is an analogue to the Chatters–Khuri Theorem [12, Theorem 2.1].

Corollary 7 ([11, Corollary 2.3]) *Assume R is right nonsingular and right cononsingular. Then, R is π-Baer if and only if R_R is π-extending.*

Proposition 14 ([11, Proposition 2.7]) *Let S be a subring of R with the same unity which is left and right intrinsic in R. If S is π-Baer, then so is R.*

Proposition 15 ([11, Proposition 2.8])

(i) Let R be a semiprime π-Baer ring such that $\mathbf{I}(R) \leqslant_{\mathbf{r}}^{\mathbf{int}} R$. Then, R is right π-extending.
(ii) Let R be a regular ring. Then, R is π-Baer if and only if R is right π-extending if and only if R is left π-extending.

Proposition 16 ([11, Proposition 2.9]) *Let R be a prime ring.*

(i) R is π-Baer if and only if $r(Y) = 0$ for each $0 \neq {}_RY \trianglelefteq_p {}_RR$ if and only if $l(X) = 0$ for each $0 \neq X_R \trianglelefteq_p R_R$.
(ii) Additionally, assume $soc(R_R) \neq 0$. Then, the following conditions are equivalent:
    (a) R is π-Baer.
    (b) R is right π-extending.
    (c) R is left π-extending.
    (d) ${}_RY \leq^{ess} {}_RR$ for each ${}_RY \trianglelefteq_p {}_RR$.

Theorem 7 ([11, Theorem 2.3]) *Assume that R is π-Baer and $e = e^2 \in R$. If $(1-e)R$ is a projection invariant right ideal of R, then $eR = eRe$ and $R(1-e) = (1-e)R(1-e)$ are both π-Baer rings and Re is a projection invariant left ideal of R.*

Corollary 8 ([11, Corollary 2.4]) *Assume that R is π-Baer and $e \in S_l(R) \cup S_r(R)$. Then eRe is π-Baer.*

Proposition 17 ([11, Proposition 2.10]) *Let $R = \prod_{j \in J} R_j$ be a direct product of rings. Then, R is π-Baer if and only if each R_j is π-Baer.*

The following two results give us information about projection invariant left ideals of generalized triangular matrix rings and annihilator conditions. In the remainder of this paper, T denotes a generalized triangular matrix ring of the form $\begin{bmatrix} S & M \\ 0 & R \end{bmatrix}$, where S and R are rings and M is an (S, R)-bimodule.

Lemma 4 ([11, Lemma 3.1]) *Let I, N, and J be additive subgroups of S, M, and R, respectively. Then, $K = \begin{bmatrix} I & N \\ 0 & J \end{bmatrix}$ is a projection invariant left ideal of T if and only if the following conditions hold:*

(i) I and J are projection invariant left ideals of S and R, respectively.
(ii) $SN + MJ \subseteq N$.
(iii) N is a $(S, \mathbf{I}(R))$-bisubmodule of M.
(iv) $IM \subseteq N$.

Proposition 18 ([11, Proposition 3.1]) *Let $\begin{bmatrix} I & N \\ 0 & J \end{bmatrix}$ be a projection invariant left ideal of T. Then $r_T(\begin{bmatrix} I & N \\ 0 & J \end{bmatrix}) = \begin{bmatrix} r_S(I) & r_M(I) \\ 0 & r_R(J) \cap r_R(N) \end{bmatrix}$.*

Now, we characterize π-Baer 2-by-2 generalized triangular matrix rings and m-by-m upper triangular matrix rings. Also, we show that an m-by-m matrix ring over a π-Baer ring is π-Baer.

Theorem 8 ([11, Theorem 3.1]) *T is π-Baer if and only if all of the following conditions are satisfied:*

(i) S and R are π-Baer.
(ii) $r_M(I) = (r_S(I))M$ for all projection invariant left ideals I of S.
(iii) If ${}_SN_{\mathbf{I}(R)} \leq {}_SM_{\mathbf{I}(R)}$, then $r_R(N) = aR$ for some $a^2 = a \in R$.

Corollary 9 ([11, Corollary 3.1]) *Let $S = \mathbb{Z}$. Then, T is π-Baer if and only if all of the following conditions are satisfied:*

(i) R is π-Baer.
(ii) ${}_{\mathbb{Z}}M$ is torsion free.
(iii) If $N_{\mathbf{I}(R)} \leq M_{\mathbf{I}(R)}$, then $r_R(N) = aR$ for some $a^2 = a \in R$.

Corollary 10 ([11, Corollary 3.2]) *Let $S = End(M_R)$. Then, T is π-Baer if and only if all of the following conditions are satisfied:*

(i) R is π-Baer.
(ii) $r_M(I)$ is a direct summand of M, where I is a projection invariant left ideal of S.
(iii) If ${}_SN_{\mathbf{I}(R)} \leq {}_SM_{\mathbf{I}(R)}$, then $r_R(N) = aR$ for some $a^2 = a \in R$.

Corollary 11 ([11, Corollary 3.3]) *Let M_R be a nonsingular π-extending module and $S = End(M_R)$. Then, T is π-Baer if and only if all of the following conditions are satisfied:*

(i) R is π-Baer.
(ii) If ${}_S N_{I(R)} \leq {}_S M_{I(R)}$, then $r_R(N) = aR$ for some $a = a^2 \in R$.

Corollary 12 ([11, Corollary 3.4]) *Assume R is a ring. Then, the following statements are equivalent:*

(1) R is π-Baer.
(2) $\mathbf{T}_n(R)$ is π-Baer for every positive integer n.
(3) $\mathbf{T}_n(R)$ is π-Baer for some positive integer $n > 1$.

Observe that a right Ore domain R which is not a division ring is both Baer and right extending. However, $\mathbf{T}_n(R)$ $(n > 1)$ is neither Baer nor right extending (see [18, p. 16] and [6, Theorem 9.1.1]). By Proposition 3, Corollaries 3 and 12, $\mathbf{T}_n(R)$ is right π-extending and π-Baer.

Corollary 13 ([11, Corollary 3.5]) *Let R be a π-Baer ring, $M = \bigoplus_{i \in \Gamma} R_i$ (external direct sum) where each $R_i = R$ and Γ is an infinite ordered set, and $S = CRFM_\Gamma(R)$ (i.e., the ring of column and row finite Γ-by-Γ matrices over R). Then, T is π-Baer. If R is Baer but not semisimple Artinian, then T is π-Baer but not Baer.*

Proposition 19 ([11, Proposition 3.2]) *Let R be a π-Baer ring. Then, $\mathbf{M}_n(R)$ is a π-Baer ring for every positive integer n.*

Note that the π-Baer property is not a Morita invariant. To see this, let R be a prime ring with R_R uniform and $Z(R_R) \neq 0$ (see the proof of [11, Theorem 2.1]). Then, R is quasi-Baer so $\mathbf{M}_n(R)$ is quasi-Baer [6, Theorem 3.2.12]. By Corollary 6 (iii), $\mathbf{M}_n(R)$ is π-Baer. But R is not π-Baer.

In general, polynomial extensions of Baer rings are not Baer. In the following results, we investigate the behavior of the π-Baer condition with respect to polynomial extensions. In contrast to the Baer condition, the π-Baer condition does transfer from a base ring to many of its polynomial extensions.

Example 5 ([11, Example 4.1]) Let $R = \mathbf{M}_2(\mathbb{Z})$. From [18, p. 17], R is a Baer ring. Note that $\begin{bmatrix} 2 & 0 \\ 0 & 0 \end{bmatrix} + \begin{bmatrix} 0 & 0 \\ 1 & 0 \end{bmatrix} x = \begin{bmatrix} 2 & 0 \\ x & 0 \end{bmatrix} \in R[x]$. But for $S = R[x]$, $l_S\left(\begin{bmatrix} 2 & 0 \\ x & 0 \end{bmatrix}\right)$ is not generated by an idempotent. Hence, neither $R[x]$ nor $R[[x]]$ are Baer. Therefore, by [12, Theorem 2.1], neither $R[x]$ nor $R[[x]]$ are right extending.

In the following polynomial extensions, the coefficients are on the left side of the indeterminates.

Theorem 9 ([11, Theorem 4.1]) *Let R be a π-Baer ring. Then, the following extension rings are π-Baer rings, where X is an arbitrary nonempty set of not necessarily commuting indeterminates, α is a ring automorphism of R and σ is an α derivation of R:*

(i) $R[X]$,
(ii) $R[[X]]$,
(iii) $R[x;\alpha,\sigma]$*(i.e., the Ore extension of R),*
(iv) $R[[x;\alpha]]$,
(v) $R[x,x^{-1};\alpha]$,
(vi) $R[[x,x^{-1};\alpha]]$.

Note that in Example 5, R is Baer but neither $R[x]$ nor $R[[x]]$ are Baer. However, by Theorems 5 and 9, $R[x]$ and $R[[x]]$ are π-Baer. The next lemma seems to be a hitherto unknown fact about the coefficients of an idempotent polynomial or of an idempotent formal power series.

Lemma 5 ([11, Lemma 4.2]) *Let* $e(x) \in R[x]$ *(resp.* $R[[x]]$*), where* $e(x) = e_0 + e_1x + \cdots + e_nx^n$ *(resp.,* $e(x) = e_0 + e_1x + \cdots$*).*

(i) $e(x) = (e(x))^2$ *if and only if* $e_k = \sum_{i+j=k} e_ie_j$*, for* $0 \le k$ *and* $0 \le i, j \le k$*. Hence,* $e(x) = 0$ *if and only if* $e_0 = 0$.
(ii) *If* $e(x) = e_0 + e_1x + \cdots + e_nx^n = (e(x))^2$*, then* $\sum_{i=0}^{n} e_i = \left(\sum_{i=0}^{n} e_i\right)^2 \in R$.
(iii) *Assume* $e(x) = (e(x))^2$*. Then,*

  (a) $e_i \in \boldsymbol{I}(R)$*, for all* $i \ge 0$*. Hence,* $\boldsymbol{I}(R[x]) \subseteq \boldsymbol{I}(R)[x]$ *and* $\boldsymbol{I}(R[[x]]) \subseteq \boldsymbol{I}(R)[[x]]$.
  (b) $e_i, e_0e_i, e_ie_0 \in \boldsymbol{N}(R)$*, for all* $i \ge 1$*, where* $\boldsymbol{N}(R)$ *denotes the subring of R (not necessarily with unity) generated by the nilpotent elements of R.*

(iv) *Let I be a right ideal of R. Then, I is projection invariant if and only if* $I[x]$ *(resp.,* $I[[x]]$*) is a projection invariant right ideal in* $R[x]$ *(resp.,* $R[[x]]$*).*

Theorem 10 ([11, Theorem 4.2]) *The following conditions are equivalent:*

(i) *R is* π*-Baer.*
(ii) $R[x]$ *is* π*-Baer.*
(iii) $R[[x]]$ *is* π*-Baer.*

Acknowledgements We wish to thank the referee and the editor for their efforts.

References

1. Abdioğlu, C., Koşan, M.T., Şahinkaya, S.: On modules for which all submodules are projection invariant and the lifting condition. Southeast Asian Bull. Math. **5**, 807–818 (2010)
2. Armendariz, E.P.: A note on extensions of Baer and pp-rings. J. Austral. Math. Soc. **18**, 470–473 (1974)
3. Berberian, S.K.: Baer *-Rings. Grundlehren der mathematischen Wissenschaften, vol. 195. Springer, Berlin (1972)
4. Birkenmeier, G.F., Müller, B.J., Rizvi, S.T.: Modules in which every fully invariant submodule is essential in a direct summand. Commun. Algebra **30**, 1395–1415 (2002)
5. Birkenmeier, G.F., Tercan, A.: When some complement of a submodule is a direct summand. Commun. Algebra **35**, 597–611 (2007)

6. Birkenmeier, G.F., Park, J.K., Rizvi, S.T.: Extensions of Rings and Modules. Birkhäuser, New York (2013)
7. Birkenmeier, G.F., Lennon, M.J.: Extending sets of idempotents to ring extensions. Commun. Algebra **42**, 5134–5151 (2014)
8. Birkenmeier, G.F., Lennon, M.J.: Dense intrinsic extensions. Houston J. Math. **40**, 21–42 (2014)
9. Birkenmeier, G.F., Tercan, A., Yücel, C.C.: The extending condition relative to sets of submodules. Commun. Algebra **42**, 764–778 (2014)
10. Birkenmeier, G.F., Tercan, A., Yücel, C.C.: Projection invariant extending rings. J. Algebra Appl. **115**, 1650121 (2016)
11. Birkenmeier, G.F., Kara, Y., Tercan, A.: π-Baer rings. J. Algebra Appl. **17**(2), 1850029 (2018)
12. Chatters, A.W., Khuri, S.M.: Endomorphism rings of modules over nonsingular CS-rings. J. London Math. Soc. **21**, 434–444 (1980)
13. Clark, W.E.: Twisted matrix units semigroup algebras. Duke Math. J. **34**, 417–423 (1967)
14. Dung, N.V., Van Huynh, D., Smith, P.F., Wisbauer, R.: Extending Modules. Longman, England (1994)
15. Faith, C., Utumi, Y.: Intrinsic extensions of rings. Pacific J. Math. **14**, 505–512 (1964)
16. Fuchs, L.: Infinite Abelian Groups I. Academic Press, New York (1970)
17. Goodearl, K.R.: Von Neumann Regular Rings. Krieger, Malabar (1991)
18. Kaplansky, I.: Rings of Operators. Benjamin, New York (1968)
19. Lam, T.Y.: Lectures on Modules and Rings. Springer, Berlin (1999)
20. Megibben, C.K.: Projection invariant subgroups of Abelian groups. Tamkang J. Math. **2**, 177–182 (1977)
21. Mohamed, S.H., Muller, B.J.: Continuous and Discrete Modules. Cambridge Univ. Press, Cambridge (1990)
22. Murray, F.J., von Neumann, J.: On rings of operators. Ann. Math. **37**, 116–229 (1936)
23. von Neumann, J.: Continuous geometry. Proc. Math. Acad. Sci. **22**, 92–100 (1936)
24. Pollingher, A., Zaks, A.: On Baer and quasi-Baer rings. Duke Math. J. **37**, 127–138 (1970)
25. Rickart, C.E.: Banach Algebras with an adjoint operation. Ann. Math. **47**, 528–550 (1946)
26. Sikander, F., Mehdi, A., Naji, S.A.R.K.: On projection invariant submodules of QTAG-modules. J. Egyptian Math. Soc. **24**, 156–159 (2016)
27. Tercan, A., Yücel, C.C.: Module Theory, Extending Modules and Generalizations. Birkhäuser, Basel (2016)
28. Utumi, Y.: On continuous rings and self injective rings. Trans. Amer. Math. Soc. **118**, 158–173 (1965)

Classification of Reductive Monoid Spaces over an Arbitrary Field

Mahir Bilen Can

Abstract In this semi-expository paper, we review the notion of a spherical space. In particular, we present some recent results of Wedhorn on the classification of spherical spaces over arbitrary fields. As an application, we introduce and classify reductive monoid spaces over an arbitrary field.

Keywords Spherical spaces · Colored fans · Reductive monoid spaces

1 Introduction

The classification of spherical homogenous varieties, more generally of spherical actions, is an important, lively, and very interesting chapter in modern algebraic geometry. It naturally encompasses the classification theories of toric varieties, horospherical varieties, symmetric varieties, wonderful compactifications, as well as that of reductive monoids. Our goal in this paper is to give an expository account of some recent work in this field. As far as we are aware, the broadest context in which a classification of such objects is achieved, is in the theory of algebraic spaces. This accomplishment is due to Wedhorn [34]. Here, we will follow Wedhorn's footsteps closely to derive some conclusions. At the same time, our intention is to provide enough detail to make basic definitions accessible to beginners. We will explain a straightforward application of Wedhorn's progress to monoid schemes.

It is not completely wrong to claim that the origins of our story go back to Legendre's work [18], where he analyzed the gravitational potential of a point surrounded by a spherical surface in 3-space. To describe the representative functions of his enterprise, he found a clever change-of-coordinates argument and introduced the set of orthogonal polynomials $\{P_n(x)\}_{n\geq 0}$ via $\frac{1}{\sqrt{1-2hx+h^2}} = \sum_{n\geq 0} P_n(x)h^n$, which are now known as Lagrange polynomials. It has eventually been understood that the $P_n(x)$'s are the eigenfunctions of the operator $\Delta = \frac{\partial^2}{\partial x^2} + \frac{\partial^2}{\partial y^2} + \frac{\partial^2}{\partial z^2}$ restricted to the space of $\mathscr{C}^\infty$ functions on the unit 2-sphere. Nowadays, bits and pieces of these elementary

M. B. Can (✉)
Mathematics Department, Tulane University, New Orleans, LA 70118, USA
e-mail: mahirbilencan@gmail.com

J. Feldvoss et al. (eds.), *Advances in Algebra*, Springer Proceedings
in Mathematics & Statistics 277, https://doi.org/10.1007/978-3-030-11521-0_4

facts can be found in every standard calculus textbook but their generalization to higher dimensions can be explained in a conceptual way by using transformation groups.

Let n be a positive integer, and let Q_n denote the standard quadratic form on $\mathbb{R}^n$,

$$Q_n(x) := x_1^2 + \cdots + x_n^2, \quad x = (x_1, \ldots, x_n) \in \mathbb{R}^n.$$

The orthogonal group, denoted by $\mathbf{O}(Q_n)$, consists of linear transformations $L : \mathbb{R}^n \to \mathbb{R}^n$ such that $Q_n(L(x)) = Q_n(x)$ for all $x \in \mathbb{R}^n$. It acts transitively on the $n-1$ sphere $S^{n-1} = \{x \in \mathbb{R}^n : Q_n(x) = 1\}$, and the isotropy subgroup in $\mathbf{O}(Q_n)$ of a point $x \in S^{n-1}$ is isomorphic to $\mathbf{O}(Q_{n-1})$. It is not difficult to write down a Lie group automorphism $\sigma : \mathbf{O}(Q_n) \to \mathbf{O}(Q_n)$ of order two such that the fixed point subgroup $\mathbf{O}(Q_n)^\sigma$ is isomorphic to $\mathbf{O}(Q_{n-1})$. In other words, S^{n-1} has the structure of a *symmetric space*, that is, a quotient manifold of the form G/K, where G is a Lie group and $K = \{g \in G : \sigma(g) = g\}$ is the fixed subgroup of an automorphism $\sigma : G \to G$ with $\sigma^2 = id$. It is known that the Laplace–Beltrami operator $\Delta_n := \sum_{i=1}^n \frac{\partial^2}{\partial x_i^2}$ generates the algebra of $\mathbf{O}(Q_n)$-invariant differential operators on S^{n-1}. Moreover, for each $k \in \mathbb{N}$, there is one eigenspace E_k of Δ_n with eigenvalue $-k(k+n-2)$; each E_k defines a finite dimensional and irreducible representation of $\mathbf{O}(Q_n)$. In addition, the Hilbert space of square integrable functions on S^{n-1} has an orthogonal space decomposition, $L^2(S^{n-1}) = \sum_{k=0}^\infty E_k$. The last point of this example is the most important for our purposes; the representation of $\mathbf{O}(Q_n)$ on the polynomial functions on S^{n-1} is multiplicity-free! All these facts are well known and can be found in classical textbooks such as [14] or [7].

We will give another example to indicate how often we run into such multiplicity-free phenomena in the theory of Lie groups. This time we start with an arbitrary compact Lie group, denoted by K, and consider $C^0(K, \mathbb{C})$, the algebra of continuous functions on K with complex values. The doubled group $K \times K$ acts on K by translations:

$$(g, h) \cdot x = gxh^{-1} \text{ for all } g, h, x \in K.$$

In particular, we have a representation of $K \times K$ on $C^0(K, \mathbb{C})$, which is infinite dimensional unless K is a finite group. Let $\mathscr{F}(K, \mathbb{C})$ denote the subalgebra of *representative functions*, which, by definition, are the functions $f \in C^0(K, \mathbb{C})$ such that $K \times K \cdot f$ is contained in a finite-dimensional submodule of $C^0(K, \mathbb{C})$. The theorem of Peter and Weyl states that $\mathscr{F}(K, \mathbb{C})$ is dense in $C^0(K, \mathbb{C})$. Moreover as a representation of $K \times K$, $\mathscr{F}(K, \mathbb{C})$ has an orthogonal space decomposition into finite-dimensional irreducible $K \times K$-representations, each irreducible occurring with multiplicity at most one.

We have a quite related, analogous statement on the multiplicity-freeness of the $G \times G$-module structure of the coordinate ring $\mathbb{C}[G]$ of a reductive complex algebraic group G. Indeed, it is a well-known fact that on every compact Lie group K there exists a unique real algebraic group structure, and furthermore, its complexification $K(\mathbb{C})$ is a complex algebraic group which is reductive. Conversely, any reductive complex algebraic group has an algebraic compact real form and this real form has the structure of a compact Lie group. Two compact Lie groups are isomorphic as Lie groups if and only if the corresponding reductive complex algebraic groups are isomorphic (see [8, 22]).

The unifying theme of these examples, as we alluded to before, is the multiplicity-freeness of the action on the function space of the underlying variety or manifold. It turns out that the multiplicity-freeness is closely related to the size of orbits of certain subgroups. To explain this more clearly, for the time being, we confine ourselves to the setting of affine algebraic varieties that are defined over $\mathbb{C}$. But we have a disclaimer: irrespective of the underlying field of definition, our tacit assumption throughout this paper is that all reductive groups are connected unless otherwise noted.

Now, let G be a reductive complex algebraic group, and let X be an affine variety on which G acts algebraically. We fix a Borel subgroup B, that is, a maximal connected solvable subgroup of G. It is well known that every other Borel subgroup of G is conjugate to B. The action of G on X gives rise to an action of G, hence of B, on the coordinate ring $\mathbb{C}[X]$. Let us assume that B has finitely many orbits in X, so, in the Zariski topology, one of the B-orbits is open. Let χ be a character of B and x_0 be a general point from the dense B-orbit. Let f be a regular function that is only defined on the open orbit. Hence, we view f as an element of $\mathbb{C}(X)$, the field of rational functions on X. If f is an χ-eigenfunction for the B-action, that is,

$$b \cdot f = \chi(b) f \text{ for all } b \in B,$$

then the value of f on the whole orbit is uniquely determined by χ and the base point x_0. Indeed, any point x from the open orbit has the form $x = b \cdot x_0$ for some $b \in B$, therefore,

$$f(x) = b^{-1} \cdot f(x_0) = \chi(b^{-1}) f(x_0).$$

This simple argument shows that there exists at most one χ-eigenvector in $\mathbb{C}[X]$ whose restriction to the open B-orbit equals f. As irreducible representations of reductive groups are parametrized by the "highest" B-eigenvectors, now we understand that the number of occurrence of the irreducible representation corresponding to the character χ in $\mathbb{C}[X]$ cannot exceed 1. In other words, $\mathbb{C}[X]$ is a multiplicity-free G-module. The converse of this statement is true as well; if a linear and algebraic action $G \times \mathbb{C}[X] \to \mathbb{C}[X]$ is multiplicity-free, then a Borel subgroup of G has finitely many orbits in X, and one of these orbits is open and dense in X (see [3], as well as [24, 32]). This brings us to the (special case of a) fundamental definition that will occupy us in the rest of our paper.

Definition 1 Let k be an algebraically closed field, and let G be a reductive algebraic group defined over k. Let X be a normal variety that is defined over k, and finally, let $\psi : G \times X \to X$ be an algebraic action of G on X. If the restriction of the action to a Borel subgroup has finitely many orbits, then the action is called *spherical*, and X is called a *spherical G-variety*.

Let $H \subset G$ be a closed subgroup in a reductive group G. The homogenous space G/H is called spherical if BH is an open dense subvariety in G for some Borel subgroup $B \subset G$. A G-variety X is called an equivariant embedding of G/H if X has an open orbit that is isomorphic to G/H. In particular, spherical varieties are normal G-equivariant embeddings of spherical homogenous G-varieties. To see this, let x_0 be a general point from the open B-orbit in a spherical variety X, and let H denote the isotropy subgroup $H = \{g \in G : g \cdot x_0 = x_0\}$ in G. Clearly, G/H is isomorphic to the open G-orbit, and it is a spherical subvariety of X. It follows that X is a G-equivariant embedding of G/H.

It goes without saying that this theory, as we know it, owes its existence to the work of Luna and Vust [20] who classified the equivariant embeddings of spherical homogenous varieties over algebraically closed fields of characteristic 0. Their results are extended to all characteristics by Knop in [16]. What is left is the classification of spherical subgroups (over arbitrary fields) and this program is well on its way; see the recent paper [2] and the references therein. For a good and broad introduction to the field of equivariant embeddings, up to 2011, we recommend the encyclopedic treatment [31] of Timashev.

We mentioned in the first paragraph of this introduction that the examples of spherical embeddings include algebraic monoids. By definition, an algebraic monoid over an algebraically closed field is an algebraic variety M endowed with an associative multiplication morphism $m : M \times M \to M$ and there is a neutral element for the multiplication. The foundations of these monoid varieties are secured mainly by the efforts of Renner and Putcha, who chiefly developed the theory for linear (affine) algebraic monoids (see [25, 28]). From another angle, Brion and Rittatore looked at the general structure of an algebraic monoid. Amplifying the importance of linear algebraic monoids, Brion and Rittatore showed that any irreducible normal algebraic monoid is a homogenous fiber bundle over an abelian variety, where the fiber over the identity element is a normal irreducible linear algebraic monoid (see Brion's lecture notes [5]). In this regard, let us recite a result of Mumford about the possible monoid structure on a complete irreducible variety ([21, Chap. II]): if a complete irreducible variety X has a (possibly nonassociative) composition law $m : X \times X \to X$ with a neutral element, then X is an abelian variety with group law m. In other words, an irreducible and complete algebraic monoid is an abelian variety. This interesting result of Mumford is extended to families by Brion in [5].

A G-equivariant embedding is said to be simple if it has a unique closed G-orbit. A *reductive monoid* is an irreducible algebraic monoid whose unit-group is a reductive group. The role of such monoids for the theory of equivariant embeddings was understood very early; Renner recognized in [26, 27] that the normal reductive monoids are simple $G \times G$-equivariant embeddings of reductive groups. Explicating this observation, Rittatore showed in [29, Theorem 1] that every irreducible algebraic monoid M is a simple $G(M) \times G(M)$-equivariant embedding of its unit-group $G(M)$. In the same paper, by using Knop's work on colored fans, which we will describe in the sequel, Rittatore described a classification of reductive monoids in terms of colored cones. This classification is a generalization of the earlier classification of the reductive normal monoids by Renner [27] and Vinberg [33].

Let k be an algebraically closed field. We define a *toric variety over* k as a normal algebraic variety on which the torus $(k^*)^n$ acts (faithfully) with an open orbit. Toric varieties are prevalent in the category of spherical varieties in the following sense: if X is a spherical G-variety over k, then the closure in X of the T-orbit of a general point from the open B-orbit will be a toric variety. Here, T stands for a maximal torus contained in B. Also, let us not forget the fact that the affine toric varieties are precisely the commutative reductive monoids, [29, Theorem 3].

Now let k be an arbitrary field, and let T be a torus that is isomorphic to $(k^*)^n$, where both the isomorphism and the torus T are defined over k. The technical term for such a torus is *k-split* or *split torus over k*. Toric varieties defined by split tori are parametrized by combinatorial objects, the so-called fans. This fact is due to Demazure in the smooth case [12] and Danilov for all toric varieties [10]. A *fan* $\mathscr{F}$ in $\mathbb{Q}^n$ is a finite collection of strictly convex cones such that (1) every face of an element $\mathscr{C}$ from $\mathscr{F}$ lies in $\mathscr{F}$; (2) the intersection of two elements of $\mathscr{F}$ is a face of both of the cones. Here, by a *cone* we mean a subset of $\mathbb{Q}^n$ that is closed under addition and the scaling action of $\mathbb{Q}_{\geq 0}$. A *face* of a cone $\mathscr{C}$ is a subset of the form $\{v \in \mathscr{C} : \alpha(v) = 0\}$, where α is a linear functional on $\mathbb{Q}^n$ that takes nonnegative values on $\mathscr{C}$. Let us define two more notions that will be used in the sequel. A cone is called *strictly convex*, if it does not contain a line. The *relative interior* of $\mathscr{C}$, denoted by $\mathscr{C}^0$, is what is left after removing all of its proper faces.

The toric varieties defined by nonsplit tori are quite interesting and their classification is significantly more intricate. Parametrizing combinatorial objects in this case, as shown by Huruguen in [15] are fans that are stable under the Galois group of a splitting field. To explain this, we extend our earlier definition of toric varieties as follows. Let k be a field, and let $\bar{k}$ denote an algebraic closure of k. Let T be a torus defined over k. A normal T-variety Y is called a *toric variety over k* if $Y(\bar{k})$ is a toric variety with respect to $T(\bar{k})$. (This notation will be made precise in Sect. 2.) Let us continue with the assumption that Y is a toric variety with respect to a k-split torus T, and let $k' \subseteq k$ be a field extension with T not necessarily split over k'. Let Γ denote the Galois group of the extension. Since all tori become split over a finite separable field extension, we will assume also that $k' \subset k$ is a finite extension. Of course, it may happen that Y is not defined over k'. If it is defined, then $Y(k')$ is called the(!) k'-form of Y. In [15, Theorem 1.22], Huruguen gives two necessary and sufficient conditions for the existence of a k'-form. The first of these two conditions is rather natural in the sense that the fan of $Y(k)$ is stable under the action of Γ. The second condition is also concrete, however, it is more difficult to check. Actually, it is a criterion about the quasi-projectiveness of an equivariant embedding in terms of the fan of the embedding. Its colored version, namely a quasi-projective colored fan, which is also used by Huruguen, is introduced by Brion in [4]. We postpone the precise definition of a quasi-projective colored fan to Sect. 4.1, but let us mention that, in [15], Huruguen shows by an example that quasi-projectiveness is an essential requirement for Y to have a k'-form.

Using the underlying idea that worked for toric varieties, in the same paper, Huruguen proves more. Let Y be a spherical homogenous variety for a reductive group G defined over a perfect field k. (The perfectness assumption is a minor glitch; it will not be needed once we start to work with the algebraic spaces as Wedhorn does. Indeed, it is required here so that the homogenous varieties have rational points.) Let

$\overline{k}$ be an algebraic closure of k, and let Γ denote the Galois group of $k \subset \overline{k}$. The introduction of the absolute Galois group, which is often too big, is not a serious problem since in the situations that we are interested in, the absolute Galois group factors through a finite quotient. Let $\overline{Y}$ be a G-equivariant embedding of a spherical homogenous G-variety Y that is defined over $\overline{k}$. In [15, Theorem 2.26], Huruguen shows that a G-equivariant embedding $\overline{Y}$ of Y has a k-form if and only if the colored fan of $\overline{Y}(\overline{k})$ is Γ-stable and it is "quasiprojective with respect to Γ." Once again, Huruguen shows by examples that the failure of the second condition implies the nonexistence of k-forms.

As Wedhorn shows in [34] via algebraic spaces, as soon as we get over the contrived emotional barrier set in front of us by algebraic varieties, we happily see that the existential questions (about k-forms) disappear. Roughly speaking, in some sense, algebraic spaces are to schemes, what schemes are to algebraic varieties. Such is the transition from spherical varieties to spherical spaces.

Definition 2 For an arbitrary base scheme S and a reductive group scheme G over S, a *spherical G-space over S* is a flat separated algebraic space of finite presentation over S with a G-action such that the geometric fibers are spherical varieties.

Notice that we have not defined reductive S-groups yet. But timeliness may not be the only problematic aspect of Definition 2. Understandably, it may look overly general at first sight. Nevertheless, the definition has many remarkable consequences. For example, according to this definition, the property for a flat finitely presented G-space with normal geometric fibers to be spherical is open and constructible on the base scheme. Moreover, for a flat finitely presented subgroup scheme H of G the property to be spherical is open and closed on the base scheme. Most importantly for our purposes, the specialization of the base scheme to the spectrum of a field in Definition 2 yields a general classification of spherical G-spaces over arbitrary fields in terms of colored fans that are stable under the Galois group. In fact, for spaces over fields, Wedhorn's definition is much easier to use.

Since explaining the concepts associated with spherical algebraic spaces and colored fans will take a bulky portion of our paper, we postpone giving the ballistics of Wedhorn's theory to Sect. 4.2. Nearing the end of our lengthy introduction, let us mention that as a rather straightforward consequence of Wedhorn's deus ex machina we obtain a classification of "reductive monoid spaces," the only new definition that we offer in this paper.

The organization of our paper is as follows. In Sect. 2, we introduce what is required to explain spherical algebraic spaces in the following order: Sect. 2.1 is on the fundamentals; we introduce the notion of scheme as a functor. Section 2.2 sets up the notation for algebraic spaces. In Sect. 2.3, we talk about group schemes, which is followed by Sect. 2.4 on reductive group schemes. Section 2.5 is devoted to parabolic subgroups. In Subsection 2.6, we briefly discuss the actions of group schemes. In Sect. 3, we discuss affine monoid schemes and prove a souped-up version of a result of Rittatore on the unit-dense algebraic monoids. The beginning of Sect. 4 is devoted to the review of colored fans. In Sect. 4.1, we review the quasi-projectiveness criterion of Brion and mention Huruguen's theorem on the k-forms of spherical varieties over perfect fields. In Sect. 4.2, we present Wedhorn's generalization of Huruguen's results. In particular, we talk about his colored fans for spherical algebraic spaces. In the subsequent Sect. 5, we introduce reductive monoid spaces and apply Wedhorn's and Huruguen's theorems. Finally, we close our paper in Sect. 5.1 by presenting an application of our observations to the lined closures of representations of reductive groups, whose geometry is investigated by De Concini in [11].

2 Notation and Preliminaries

The purpose of this section is to set up our notation and provide some background on algebraic spaces, reductive k-groups, and reductive k-monoids. We tried to give most of the necessary definitions for explaining the logical dependencies. For standard algebraic geometry facts, we recommend the books [13, 17]. (It seems to us that the Stacks Project [30] will eventually replace all standard references.) In addition, we find that Brion's lecture notes in [6] are exceptionally valuable as a resource for the background on algebraic groups.

Notation: Throughout our paper k will stand for a field, as G does for a group. k is called a perfect field if every algebraic extension of k is separable.

2.1 Schemes

A point that we want to make in this section is that the only way to know a group scheme is to know all (affine) group schemes related to it. In fact, this statement is a theorem.

Terminology: Let $(X, \mathscr{F})$ be a pair of a topological space X and a sheaf of rings $\mathscr{F}$ on X. If for each point $x \in X$, the corresponding stalk $\mathscr{F}_x$ is a local ring, then the pair $(X, \mathscr{F})$ is called a *locally ringed space*. An *affine scheme* is a locally ringed space which is of the form $(\mathrm{Spec}(R), \mathscr{O}_{\mathrm{Spec}(R)}$, where R is a commutative ring, and $\mathrm{Spec}(R)$ is its spectrum endowed with the Zariski topology. The sheaf $\mathscr{O}_{\mathrm{Spec}(R)}$ is the structure sheaf of $\mathrm{Spec}(R)$. A *scheme* is a locally ringed space $(X, \mathscr{O}_X)$ which has a covering by open subsets $X = \bigcup_{i \in I} U_i$ such that each pair $(U_i, \mathscr{O}_X|_{U_i})$ is an affine scheme.

There is a tremendous advantage of using categorical language while studying a scheme in relation with others. Therefore, we proceed with the identification of schemes with their functors of points. In this regard, we will use the following standard notation throughout:

Obj($\mathscr{C}$)	the class of objects of a category $\mathscr{C}$
Mor($\mathscr{C}$)	the class of morphisms between the objects of $\mathscr{C}$

Fig. 1 A morphism in (***R*-schemes**)

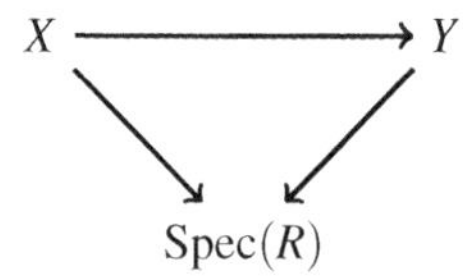

$\mathrm{Mor}(X, Y)$	the set of morphisms from X to Y, where $X, Y \in \mathbf{Obj}(\mathscr{C})$
$\mathscr{C}^o$	the category opposite to $\mathscr{C}$
(**schemes**)	the category of schemes
(**sets**)	the category of sets

The *functor of points* of a scheme X is the functor $h_X : (\mathbf{schemes})^o \to (\mathbf{sets})$ that is defined by the following assignments:

1. if $Y \in \mathbf{Obj}((\mathbf{schemes})^o)$, then $h_X(Y) = \mathrm{Mor}(Y, X)$;
2. if $f \in \mathrm{Mor}(Y, Z)$, then $h_X(f)$ is the set map

$$\begin{aligned} h_X(f) : h_X(Z) &\longrightarrow h_X(Y) \\ g &\longmapsto f \circ g \end{aligned}$$

Let h denote the following natural transformation:

$$\begin{aligned} h : (\mathbf{schemes}) &\to \mathbf{functors}((\mathbf{schemes})^o, (\mathbf{sets})) \\ X &\mapsto h_X \end{aligned} \tag{1}$$

Here, **functors(-, -)** stands for the category whose objects are functors, and its morphisms are the natural transformations between functors. It follows from Yoneda's Lemma that (1) is an equivalence onto a full subcategory of the target category. In other words, a scheme X is uniquely represented by its functor of points h_X.

Terminology: In the presence of a morphism $Y \to X$ between two schemes X and Y, we will occasionally say that Y *is a scheme over* X or that Y *is an* X*-scheme*. If there is no danger for confusion, we will write $X(Y)$ for $h_X(Y)$, which is the set of all morphisms from Y to X. If, in addition, Y is an affine scheme of the form $Y = \mathrm{Spec}(R)$, then we often write $X(R)$ instead of $X(Y)$, and we say that X *is a scheme over* R.

We want to show that the functor h in (1) behaves well upon restriction to the category of schemes over an affine scheme. To this end, let R be a commutative ring, and denote the category of R-schemes by (***R*-schemes**). A *morphism* in this category is a commutative diagram of morphisms as in Fig. 1.

It is well known that the category (***R*-schemes**) is equivalent to the opposite category of commutative R-algebras, denoted by (***R*-algebras**). The highlight of this subsection is the following result whose proof easily follows from the definitions.

Proposition 1 *The functor*

$$h : (\textbf{\textit{R-schemes}}) \longrightarrow \textbf{\textit{functors}}((\textbf{\textit{R-algebras}}), (\textbf{\textit{sets}})),$$

*which is obtained from (1) by restriction to the subcategory (**R-schemes**), is an equivalence onto a full subcategory of the target category. In particular, a scheme over R is determined by the restriction of its functor of points to the category of affine schemes over R.*

For our purposes, the most important consequence of Proposition 1 is that an R-scheme can be thought of as a sheaf of sets on the category of commutative R-algebras with respect to the Zariski topology. Closely related to this sheaf realization of schemes is the notion of an "algebraic space." Concisely, and very roughly speaking, an algebraic space is a sheaf of sets on the category of commutative R-algebras with respect to the "étale topology." We will give a more precise definition of algebraic spaces in the next subsection. Next, we remind ourselves of some basic notions regarding the morphisms between schemes.

1. A ring homomorphism $f : R \to S$ is called *flat* if the associated induced functor

$$f_* : (\textbf{\textit{R}}\textbf{-modules}) \to (\textbf{\textit{S}}\textbf{-modules})$$

 is exact, that is to say, it maps short-exact sequences to short-exact sequences. In this case, the morphism $f^* : \mathrm{Spec}(S) \to \mathrm{Spec}(R)$ is called flat. More generally, a map $f : X \to Y$ of schemes is called *flat* if the induced functor

$$f^* : (\textbf{Quasicoherent sheaves on } Y) \to (\textbf{Quasicoherent sheaves on } X)$$

 is exact.
2. A ring homomorphism $f : R \to S$ is called *of finite presentation* if S is isomorphic (via f) to a finitely generated polynomial algebra over R and the ideal of relations among the generators is finitely generated. A map $f : X \to Y$ of schemes is called *a map of finite presentation at* $x \in X$ if there exist an affine open neighborhood $U = \mathrm{Spec}(S)$ of x in X and an affine open neighborhood $V = \mathrm{Spec}(R)$ of Y with $f(U) \subseteq V$ such that the induced ring map $R \to S$ is of finite presentation. More generally, a map $f : X \to Y$ of schemes is called *of locally finite presentation* if it is of finite presentation at every point $x \in X$.
3. A map $f : X \to Y$ of schemes is called *separated* if the induced diagonal morphism $X \to X \times_Y X$ is a closed immersion.
4. A map $f : X \to Y$ of schemes is called *unramified* if the induced diagonal morphism $X \to X \times_Y X$ is an open immersion.
5. A map $f : X \to Y$ of schemes is called *étale* if it is unramified, flat, and of locally finite presentation.

Associated with these types of maps, we have two important topologies. These are

1. The topology on the category of schemes associated with the étale morphisms; this topology is called the *étale topology.*
2. The topology on the category of schemes associated with the set of maps, flat and of locally finite presentation; this topology is called the *fppf topology.*

2.2 Algebraic Spaces

Let S be a scheme. An *algebraic space X over S* is a functor

$$X : (\mathbf{S\text{-schemes}})^o \longrightarrow (\mathbf{sets})$$

satisfying the following properties:

(i) X is a sheaf in the fppf topology.
(ii) The diagonal morphism $X \to X \times_S X$ is representable by a morphism of schemes.
(iii) There exists a surjective étale morphism $\widetilde{X} \to X$, where $\widetilde{X}$ is an S-scheme.

It turns out that, in this definition, specifically in part (i), replacing the fppf topology by the étale topology on the category of schemes does not cause any harm. In other words, the resulting functor describes the same algebraic space (see [30, Tag 076L]). Let us mention in passing that the category of schemes is a full subcategory of the category of algebraic spaces.

We will follow the standard assumption, as in [34] (and as in [17]), that all algebraic spaces are quasi-separated over some scheme, hence they are Zariski locally quasi-separated. This implies that our algebraic spaces are reasonable and decent in the sense of [30, Tag 03I8] and [30, Tag 03JX].

Our final note in this subsection is that if k is a separably closed field, then an algebraic space over k is a k-scheme.

2.3 Group Schemes

Let R be a commutative ring. A *group scheme* over R is an R-scheme whose functor of points factors through the forgetful functor from the category of groups to the category of sets. In a nutshell, an R-scheme G is called a group scheme over R if for every R-scheme S, there is a natural group structure on $G(S)$ which is functorial with respect to the morphisms $R \to S$.

Terminology: If G is a (group) scheme over R, then the morphism $G \to \text{Spec}(R)$ is called the *structure morphism.*

Definition 3 Let k be a field and let G be a group scheme over k. We will call G a *k-group* (or, an *algebraic group over k*) if G is of finite type as a scheme over k. If k' is a subfield of k, then G is said to be *defined over k'* if G, as a scheme, and all of its group operations as morphisms are defined over k'.

Let S be a scheme, and let G be a group scheme over S. G is called *affine, smooth, flat,* or *separated*, respectively, if the structure morphism $G \to S$ is affine, smooth, flat, or separated, respectively. We have some remarks regarding these properties:

1. Any k-group G is separated as an algebraic scheme. Indeed, the diagonal in $G \times G$ is closed as being the inverse image of the "identity" point of $G(k)$ under the morphism $t : G \times G \to G$ defined by $t(g, h) = gh^{-1}$.
2. Let k be a perfect field, and let G be a k-group scheme. If the underlying scheme of G is reduced, that is to say, the structure ring has no nilpotents, then G is smooth (see [13, p. 287]).
3. Cartier's theorem states that if the characteristic of the underlying field k is 0, then a k-group scheme is reduced (see [21, p. 101]). It follows that, in characteristic 0, all k-groups are smooth.

Next, we will define a particular k-group that is of fundamental importance for the whole development of algebraic group theory.

Example 1 Let V be a finite-dimensional vector space over k. The *general linear group*, denoted by $GL(V)$, is the group functor such that

$$GL(V)(S) = \text{ the automorphism group of the sheaf of } \mathscr{O}_S - \text{modules } \mathscr{O}_S \otimes_k V,$$

where S is a k-scheme. By choosing a basis for V, we see that $GL(V)(S)$ is isomorphic to the group of invertible $n \times n$ matrices with coefficients in the algebra $\mathscr{O}_S(S)$, hence, $GL(V)$ is represented by the open affine scheme

$$\mathrm{GL}_n = \{P \in \mathbb{A}^{n^2} : \det P \neq 0\}.$$

It follows that $GL(V)$ is smooth and connected.

Definition 4 A group scheme G is called *linear* if it is isomorphic to a closed subgroup scheme of GL_n for some positive integer n. If V is a vector space, then a homomorphism $\rho : G \to GL(V)$, which is a morphism of schemes, is called a *linear representation* of G on V. In this case, V is called a *G-module*.

Note that if a k-group scheme is linear, then it is an affine scheme. Note conversely that every k-group scheme which is affine and of finite type is a linear group scheme (see [6, Proposition 3.1.1]).

In some sense, atomic pieces of k-groups are given by the following two very special k-groups:

Example 2 Let $+$ and $\cdot$ denote the addition and the multiplication operations on the field k.

- The additive one-dimensional k-group, denoted by $\mathbf{G}_a$, is the affine line $\mathbb{A}_k^1$ considered with the group structure $(k, +)$.
- The multiplicative one-dimensional k-group, denoted by $\mathbf{G}_m$, is $\mathbb{A}_k^1 - \{0\}$ considered with the group structure $(k^*, \cdot)$.

Definition 5 A k-group G is called a *torus* if there exists an isomorphism $\zeta : G \to \mathbf{G}_m \times \cdots \times \mathbf{G}_m$ (n copies, for some $n \geq 1$) over some field containing k. Assuming that G is a torus defined over k', and that k' is a subfield of k, G is called k-split if both of G and ζ are defined over k.

A modern proof of the following basic result can be found in [9, Appendix A].

Theorem 1 (Grothendieck) *Let k' and k be two fields such that $k' \subset k$. Let G be a smooth connected affine k-group. If G is defined over k', then G contains a maximal k'-split torus T such that $T(\overline{k})$ is a maximal torus of $G(\overline{k})$.*

2.4 Reductive Group Schemes

Almost any fact about reductive group schemes can be found in Conrad's SGA3 replacement [9].

Let G be a k-group, and let $\rho : G \hookrightarrow GL(V)$ be a finite-dimensional faithful linear representation. An element g from G is called *semisimple* (respectively, *unipotent*) if the linear operator $\rho(g)$ on V is diagonalizable (respectively, unipotent). It is not difficult to check that these notions (semisimplicity and unipotency) are independent of the faithful representation, and they are preserved under k-homomorphisms. Therefore, the following definition is unambiguous:

Definition 6 Let G be a linear algebraic k-group. A k-subgroup U of G is called *unipotent* if every element of U is unipotent.

Next, we give the definition of a reductive group. However, we will do this in the opposite of the chronological development of the subject to emphasize the differences. So, let us start with the definition of the relative reductive group schemes. This is most useful for studying properties that are preserved in families over a commutative ring.

Let S be a scheme, and let G be a smooth S-affine group scheme over S. Let s be a point from S, and denote by $G_{\overline{s}}$ the geometric fiber $G \times_S \text{Spec}(\overline{k(s)})$ of $G \to S$. Here, $k(s)$ is the residue field of s. If for each $s \in S$ the fiber $G_{\overline{s}}$ is a connected reductive group, then G is called a *reductive S-group*.

So, what is a reductive k-group over an algebraically closed field? We take this opportunity to define the "unipotent radical" and the "radical" of an algebraic group. Let G be a k-group (affine or not). There is a maximal connected solvable normal linear algebraic subgroup, denoted by $\mathscr{R}(G(\overline{k}))$, and it is called the *radical* of G (see [6, Lemma 3.1.4]). If $\mathscr{R}(G(\overline{k}))$ is trivial, then $G(\overline{k})$ is called *semisimple*. The *unipotent radical* of $G(\overline{k})$, denoted by $\mathscr{R}_u(G(\overline{k}))$, is the maximal connected normal unipotent subgroup of $G(\overline{k})$. If $\mathscr{R}_u(G(\overline{k}))$ is trivial, then $G(\overline{k})$ is called *reductive*. Clearly, semisimplicity implies reductivity. Notice also that we have no connectedness assumption here. If the characteristic of k is 0, then the property of reductiveness of the identity component of G is equivalent to the semisimplicity of all linear representations of G. This equivalence fails in positive characteristics (see [9, Remark 1.1.13.]).

In the rest of our paper, we will focus mainly on the "relative" reductive group schemes over fields.

Definition 7 Let k be a field. A k-group G is called *reductive* if the geometric fiber $G_{\overline{k}}$ (which we take to be equal to $G(\overline{k})$) is a connected reductive group in the sense of the previous paragraph.

2.5 *Parabolic Subgroups*

Let G be a reductive k-group. A subgroup $P(\overline{k})$ of $G(\overline{k})$ is called *parabolic* if $G(\overline{k})/P(\overline{k})$ has the structure of a projective variety. More generally, a smooth affine k-subgroup P of G is called a *parabolic subgroup* if $P(\overline{k})$ is a parabolic subgroup of $G(\overline{k})$.

A *Borel subgroup* in $G(\overline{k})$ is a maximal connected solvable subgroup. More generally, a parabolic k-subgroup B of G is called a *Borel subgroup* if $B(\overline{k})$ is a Borel subgroup in $G(\overline{k})$.

A fundamentally important result that is due to Borel (see [1, Theorem 11.1]) states that any two Borel subgroups are conjugate in $G(k)$, and furthermore, for any Borel k-subgroup $B(k)$ the quotient $G(\overline{k})/B(\overline{k})$ is projective. Of course, according to the above definition of parabolic k-subgroups, a Borel subgroup in $G(k)$ is a (minimal) parabolic k-subgroup.

Example 3 The upper triangular subgroup T_n of GL_n is a Borel subgroup. As a consequence of the Lie–Kolchin Theorem (see [1, Corollary 10.2]) any connected solvable group admits a faithful representation with image in T_n.

It may happen that $G(k)$ is defined over a field but has no nontrivial Borel subgroup. This holds true, even for some classical groups, as we will demonstrate in the next classical example (from [1]).

Example 4 Let k be a field whose characteristic is not 2, and let V be k-vector space. Let Q be a nondegenerate quadratic form on V, and let F denote its symmetric bilinear form. We assume that Q is *isotropic*, that is to say, there exists a nonzero vector $v \in V$ such that $Q(v) = 0$. A subspace is called *isotropic* if it contains an isotropic vector; a subspace is called *anisotropic* if it contains no nonzero isotropic vector; a subspace is called *totally isotropic* if it consists of isotropic vectors only. A *hyperbolic plane* is a two-dimensional subspace E of V with a basis $\{e, f\}$ with respect to which the restriction of F has the form $F(x_1e + x_2f, y_1e + y_2f) = x_1y_2 + x_2y_1$.

By Witt's Decomposition Theorem, we know that the dimension q of a maximal totally isotropic subspace is an invariant of Q. More precisely, it states that V contains q linearly independent hyperbolic planes $H_1, \ldots, H_q$, such that V is an orthogonal direct sum of the form

$$V \cong V_o \oplus \bigoplus_{i=1}^{q} H_i,$$

where V_o is an anisotropic subspace. Let Q_o denote the restriction of Q to V_o.

For $i = 1, \ldots, q$ we choose a basis $\{e_i, e_{n-q+i}\}$ for H_i in such a way that the following identities are satisfied:

$$F(e_i, e_i) = F(e_{n-q+i}, e_{n-q+i}) = 0 \text{ and } F(e_i, e_{n-q+i}) = 1.$$

Here n is the dimension of V. Let $\{e_j : j = q+1, \ldots, n-q\}$ denote a basis for V_o. For each pair e_i, e_{n-q+i} $(i = 1, \ldots, q)$ of basis elements and $x \in \overline{k}$, we have a linear map $s_i(x) : V(\overline{k}) \to V(\overline{k})$ defined by

$$s_i(x)e_i = xe_i, \quad s_i(x)e_{n-q+i} = x^{-1}e_{n-q+i}, \quad \text{and } s_i(x)f = f \text{ for } f \in H_i^{\perp}.$$

Clearly, the $s_i(x)$'s are semisimple and generate a diagonal torus, denoted by T, whose elements expressed in the basis $\{e_1, \ldots, e_n\}$ of V are of the form

$$\text{diag}(a_1, \ldots, a_q, 1, \ldots, 1, a_1^{-1}, \ldots, a_q^{-1}), \quad \text{where } a_i \in \overline{k}.$$

If G denotes $\mathbf{SO}(Q)$, the k-group consisting of linear automorphisms of V that preserves the quadratic form Q and of determinant 1, then $S := T \cap G$ is in fact a maximal torus of G. Note that the centralizer of S in G, denoted by $\mathscr{Z}_G(S)$, is isomorphic to the product $S \times \mathbf{SO}(Q_o)$.

The group G is a reductive k-group, and since S is a maximal torus, by definition, $\mathscr{Z}_G(S)$ is a Cartan subgroup of G. Therefore $\mathscr{Z}_G(S)$ is contained in a minimal parabolic k-subgroup P of G. In fact, $\mathscr{Z}_G(S)$ is equal to the Levi component of P. Observe that, if $n > 2q + 2$, then $\mathbf{SO}(Q_o)$ is not commutative, hence $\mathscr{Z}_G(S)$ is not contained by a Borel subgroup. We conclude that if $n > 2q + 2$, then G does not have any Borel subgroups.

For connected semisimple k-groups, where k is a perfect field, the question of the existence of Borel subgroups has a nice answer.

Theorem 2 (Ono) *Let k' and k be two fields such that $k' \subset k$ and k' is perfect. If G is a connected semisimple k-group, then G possesses a Borel subgroup B defined over k' if and only if a maximal torus T of G is k'-split. In this case, all Borel subgroups containing $T(k')$ are conjugate by the elements of the group $N_G(T)(k')$, where $N_G(T)$ denotes the normalizer subgroup of T in G.*

Proof See [23].

Finally, we finish this section by mentioning another important related result.

Theorem 3 (Chevalley) *The parabolic subgroups of any connected linear algebraic k-group are connected, and moreover, the normalizer of a parabolic subgroup in $G(k)$ is equal to $P(k)$.*

Proof See [1, Sect. 11].

2.6 Actions of Group Schemes

Let G be a group scheme over S. An action of G on an S-scheme X is an S-morphism

$$a : G \times_S X \to X$$

such that for any S-scheme T the morphism $a(T) : G(T) \times X(T) \to X(T)$ is an action of $G(T)$ on $X(T)$. In this case, X is called a *G-scheme*. If the action map a is clear from the context we will denote $a(g, x)$ by $g \cdot x$, where $g \in G,\ x \in G(S)$.

Let X be a G-scheme with respect to an action $a : G \times X \to X$. If X is an affine scheme, then for any scheme S and $g \in G(S)$, we have an $\mathcal{O}_S$-algebra automorphism defined as follows:

$$\begin{aligned} \rho(g) :\ \mathcal{O}_S \otimes_k \Gamma(X, k) &\longrightarrow \mathcal{O}_S \otimes_k \Gamma(X, k) \\ 1 \otimes f(-)) &\longmapsto 1 \otimes (f \circ a(g^{-1}, -)), \end{aligned}$$

where $\Gamma(X, k)$ is the global section functor applied to X. These automorphisms patch up to give a linear representation

$$\rho : G \to GL(\Gamma(X, k)). \tag{2}$$

Remark 1 If X is a G-scheme of finite type, then the representation (2) decomposes $\Gamma(X, k)$ into a union of finite- dimensional G-submodules ([6, Proposition 2.3.4]). Furthermore, if X is affine, then there exists a closed G-equivariant immersion of X into a finite-dimensional representation of G ([6, Proposition 2.3.5]).

We close this section by mentioning an important theorem of Brion on projective equivariant embeddings of algebraic groups. Let G be a k-group, and let $H \subseteq G$ be a subgroup scheme. An *equivariant compactification* of the homogenous space G/H is a proper G-scheme X equipped with an open equivariant immersion $G/H \to X$ with schematically dense image.

Theorem 4 (Brion) *Let G be a k-group, and let $H \subseteq G$ be a subgroup scheme. Then G/H has an equivariant compactification by a projective scheme.*

Proof See [6, Sect. 5.2].

Some remarks are in order:

1. The homogenous space G/H of Theorem 4 is quasi-projective; this is a well-known theorem for algebraic groups in the classical sense. But notice here that Brion proves the result for not necessarily reduced algebraic groups; of course, if the ground field is of characteristic zero, then Cartier's theorem implies that G is reduced.
2. If in addition G is smooth, then G/H has an equivariant compactification by a normal projective scheme.
3. If the characteristic of k is 0, then every homogenous space has a smooth projective equivariant compactification.
4. Over any imperfect field, there exists smooth connected algebraic groups having no smooth compactification (see [6, Remark 5.2.3] for an example).

3 Algebraic Monoids

For an algebro-geometric introduction to the theory of (not necessarily affine) monoid and semigroup schemes we recommend Brion's lecture notes [5] (see also [13, Chapter II]).[1]

Let us define a monoid scheme by relaxing the condition of invertibility in the definition of group schemes. More precisely, let R be a commutative ring. A monoid scheme over R is an R-scheme whose functor of points factors through the forgetful functor from the category of monoids to the category of sets.

Definition 8 Let k be an algebraically closed field. An *algebraic monoid over k*, also called a *k-monoid*, is a monoid scheme over k whose underlying scheme is separated and of finite type.

Remark 2 Our definition of k-monoids is somewhat more general than the one that is used by Brion in [5] since we do not assume reducedness of the underlying scheme structure.

[1]This book is one of the few if not the only book in algebraic geometry that acknowledges monoid schemes as part of the theory of group schemes.

It is not difficult to see that the category of k-groups forms a full subcategory of the category of k-monoids. We are mainly interested in the affine k-monoids but let us first mention a general result.

Let M be a k-monoid, and let $G = G(M)$ denote its unit-group. Then M is called *unit-dense* if G is dense in M. A (weaker) form of the following theorem was first proven by Rittatore. Here, we will make use of Brion's proof from [5, Sect. 3].

Theorem 5 *Let M be a unit-dense irreducible k-monoid, and let G denote the unit-group of M. If G is affine, then so is M.*

Proof If M is reduced, then the result follows from [5, Theorem 2]. If M is not reduced, then we pass to a normalization $\widetilde{M}$ which factors through the reduction M_{red}. This follows from the fact that M is irreducible (hence Noetherian). Therefore, we can apply [30, Lemma 28.52.2]. But a normalization is a finite morphism, therefore $M_{red} \to M$ is a finite morphism. Since M_{red} is affine (once again by [5, Theorem 2]), and since the image of an affine scheme under a finite morphism is affine, M is an affine scheme as well.

4 From Spherical Varieties to Spherical Spaces

We start with reviewing the classification schematics of the spherical embeddings over an algebraically closed field. Throughout this section k will denote an algebraically closed field and we assume that all varieties are defined over k. As usual, let G be a reductive group, let B be a Borel subgroup, and let T be a maximal torus contained in B. If K is an algebraic group, we will denote by $X^*(K)$ the group of characters of K. Note that $X^*(B) = X^*(T)$. This is because the commutator subgroup of B coincides with the unipotent radical $\mathscr{R}_u(B)$, and $B = \mathscr{R}_u(B) \rtimes T$.

Let Y denote G/H, where H is a spherical subgroup of G. Quotients of affine groups, in particular Y, have the structure of a quasi-projective variety. Recall that Y is a spherical G-variety if and only if B has only finitely many orbits with respect to the left multiplication action on Y. By a theorem of Brion, this is equivalent to the statement that B has an open orbit in Y. Thus, it should come as no surprise that the B-invariant rational functions on Y are among the main players in this game.

The space of B-semiinvariant rational functions on Y is denoted by $k(Y)^{(B)}$. In other words,

$$k(Y)^{(B)} = \{f \in k(Y) : \ b \cdot f = \chi(b) f \ \text{ for all } b \in B \text{ and for some character } \chi \text{ of } B\}. \quad (3)$$

The gist of the classification schematics for spherical varieties will take place inside the vector space $\mathbf{Hom}_{\mathbb{Z}}(X^*(B), \mathbb{Q})$. Indeed, it is easy to check that $k(Y)^{(B)}$ is a subgroup of $k(Y)$ and that the assignment

$$\begin{aligned} k(Y)^{(B)} &\longrightarrow X^*(B) \\ f &\rightsquigarrow \chi_f, \end{aligned} \quad (4)$$

where χ_f is a character as in (3) is an injective group homomorphism. We will denote the image of (4) by Ω_Y, and we will denote the $\mathbb{Q}$-vector space associated with the dual of Ω_Y by $\mathcal{Q}_Y$. In other words,

- $\Omega_Y := \{\chi \in X^*(B) : b \cdot f = \chi(b) f \text{ for all } b \in B \text{ and for some } f \in k(Y)^{(B)}\}$;
- $\mathcal{Q}_Y := \mathbf{Hom}_{\mathbb{Z}}(\Omega_Y, \mathbb{Q})$.

We occasionally refer to Ω_Y as the character group of the homogenous variety Y since in the special case, where $Y = G \times G/\text{diag}(G) \cong G$ viewed as a spherical $G \times G$-variety, the character group of Y is isomorphic to the "ordinary" character group, $\Omega_Y \cong X^*(B)$.

We look closely at the divisors and their invariants on Y. A function $\nu : k(Y) \to \mathbb{Q}$ is called a $\mathbb{Q}$-valued discrete valuation on $k(Y)$ if for every $a, b \in k(Y)^*$, we have:

1. $\nu(k(Y)^*) \cong \mathbb{Z}$ and $\nu(k) = \{0\}$;
2. $\nu(ab) = \nu(a) + \nu(b)$;
3. $\nu(a + b) \geq \min\{\nu(a), \nu(b)\}$ provided $a + b \neq 0$.

We notice here that every $\mathbb{Q}$-valued discrete valuation ν on $k(Y)$ defines a function on Ω_Y. More precisely, there is a map

$$\begin{aligned} \rho : \{\mathbb{Q}\text{-valued discrete valuations on } k(Y)\} &\longrightarrow \mathcal{Q}_Y \\ \nu &\longmapsto \rho_\nu \end{aligned}$$

such that ρ_ν is the map that sends an element $\chi = \chi_f$ of Ω_Y to $\nu(f)$, where f is the B-semiinvariant that specifies χ (so we wrote $\chi = \chi_f$). Indeed, since χ is defined uniquely by f (up to a scalar), ρ_ν is well defined. Moreover, it is easy to check that $\rho_\nu(\chi_f \chi_g) = \rho_\nu(\chi_f) + \rho_\nu(\chi_g)$, hence $\rho_\nu \in \mathcal{Q}_Y$.

- A valuation ν is called G-invariant if $\nu(g \cdot a) = \nu(a)$ for every $g \in G$ and $a \in k(Y)$. We will denote the set of all $\mathbb{Q}$-valued G-invariant discrete valuations on Y by $\mathcal{V}_Y$.

It is not completely obvious, but nevertheless true, that the restriction

$$\rho|_{\mathcal{V}_Y} : \mathcal{V}_Y \longrightarrow \mathcal{Q}_Y$$

is an injective map.

The reason for which we started looking at discrete valuations defined on Y in the first place is because many important discrete valuations come from irreducible hypersurfaces. From now on, we will refer to irreducible hypersurfaces as prime divisors. For any spherical action of G on Y, we will consider the set of all B-stable prime divisors in Y. More generally, if Y' is a normal spherical G-variety, then *a color of* Y' is defined as a B-stable, but not G-stable, prime divisor. In our case, since G acts transitively on $Y = G/H$, any B-stable prime divisor in Y is non-G-stable.

- The set of all B-stable, but not G-stable, prime divisors in Y is denoted by $\mathscr{D}_Y$. The elements of $\mathscr{D}_Y$ are called the colors of Y.

Let us point out that on a noetherian integral separated scheme, which is regular in codimension one, the local ring associated with each prime divisor is a discrete valuation ring (DVR). In particular, since our Y is a smooth variety, the local ring of a color of Y at its generic point is a DVR, and we have the map

$$\widetilde{\rho} : \mathscr{D}_Y \to \mathscr{Q}_Y, \tag{5}$$

which is defined as the composition of ρ with the map that assigns a color to the corresponding discrete valuation.

Definition 9 The *colored cone* of $Y = G/H$ is the pair $(\mathscr{C}_Y, \mathscr{D}_Y)$, where $\mathscr{C}_Y$ is the cone in $\mathscr{Q}_Y$ that is generated by $\rho(\mathscr{V}_Y)$ and $\widetilde{\rho}(\mathscr{D}_Y)$.

So far what we have are some "birational invariants" that are defined solely for $Y = G/H$, and we have not given any indication of how they are related to its embeddings. To see how all these basic ingredients come together to play a role, next, we introduce the notion of a colored fan. This will give us a generalization of the combinatorial classification of toric varieties.

Let $\overline{Y}$ be a G-equivariant embedding of Y. Let $\mathscr{D}_{\overline{Y}}$ denote the set of B-stable, but not G-stable, prime divisors of $\overline{Y}$. Clearly, $\mathscr{D}_Y$ is a subset of $\mathscr{D}_{\overline{Y}}$. Since Y is the open G-orbit in $\overline{Y}$, we have $k(Y) = k(\overline{Y})$. In particular, there is an extension of (5) to a map $\overline{\rho} : \mathscr{D}_{\overline{Y}} \to \mathscr{Q}_Y$. Let

$$\pi : \mathscr{D}_{\overline{Y}} \to \mathscr{D}_Y$$

denote the partially defined map $\pi(S) = S \cap Y$, whenever $S \cap Y$ is an element of $\mathscr{D}_Y$. Let us mention in passing that both of the sets $\mathscr{D}_Y$ and $\mathscr{D}_{\overline{Y}}$ contain a finite number of elements since $\overline{Y}$ is spherical. Note also that the set of G-invariant discrete valuations on $\overline{Y}$ is equal to $\mathscr{V}_Y$.

Y'	a G-orbit in $\overline{Y}$
ν_S	the discrete valuation in $k(Y)$ associated with a prime divisor S of $\overline{Y}$
$\mathscr{V}_{Y' \hookrightarrow \overline{Y}}$	the set of G-invariant valuations in $k(Y)$ of the form ν_S with $Y' \subset S \subset \overline{Y}$
$\mathscr{D}_{Y' \hookrightarrow \overline{Y}}$	the set of colors $D \in \mathscr{D}_Y$ such that $\exists S \in \mathscr{D}_{\overline{Y}}$ with $Y' \subset S$ and $D = \pi(S)$
$\mathscr{C}_{Y' \hookrightarrow \overline{Y}}$	the cone in $\mathscr{Q}_Y$ that is generated by the images $\rho(\mathscr{V}_{Y' \hookrightarrow \overline{Y}})$ and $\widetilde{\rho}(\mathscr{D}_{Y' \hookrightarrow \overline{Y}})$

- The pair $(\mathscr{C}_{Y' \hookrightarrow \overline{Y}}, \mathscr{D}_{Y' \hookrightarrow \overline{Y}})$ is called the *colored cone* of the G-orbit Y'. A *face* of $(\mathscr{C}_{Y' \hookrightarrow \overline{Y}}, \mathscr{D}_{Y' \hookrightarrow \overline{Y}})$ is a pair of the form $(\mathscr{C}, \mathscr{D})$, where $\mathscr{C}$ is a face of $\mathscr{C}_{Y' \hookrightarrow \overline{Y}}$ such that

  (i) $\mathscr{C} \cap (\mathscr{C}_{Y' \hookrightarrow \overline{Y}})^0 \neq \emptyset$;
  (ii) $\widetilde{\rho}^{-1}(\mathscr{C}) \cap \mathscr{D}_{Y' \hookrightarrow \overline{Y}} \neq \emptyset$.

Any colored cone satisfies the following defining properties:

C1. The cone $\mathscr{C}_{Y' \hookrightarrow \overline{Y}}$ is generated by $\widetilde{\rho}(\mathscr{D}_{Y' \hookrightarrow \overline{Y}})$ and a finite number of vectors of the form $\rho(v_S)$, where $S \in \mathscr{V}_{Y' \hookrightarrow \overline{Y}}$.
C2. The relative interior of $\mathscr{C}_{Y' \hookrightarrow \overline{Y}}$ has a nonempty intersection with the set $\rho(\mathscr{V}_{Y' \hookrightarrow \overline{Y}})$.
C3. $\mathscr{C}_{Y' \hookrightarrow \overline{Y}}$ is strictly convex, that is to say, $\mathscr{C}_{Y' \hookrightarrow \overline{Y}} \cap (-\mathscr{C}_{Y' \hookrightarrow \overline{Y}}) = \{0\}$.
C4. 0 is not an element of $\widetilde{\rho}(\mathscr{D}_{Y' \hookrightarrow \overline{Y}})$.

Now, we are ready to introduce the combinatorial objects which parametrize the G-equivariant embeddings of Y.

Definition 10 The following (finite) set is called the *colored fan* of $\overline{Y}$:

$$\mathscr{F}_{\overline{Y}} := \{(\mathscr{C}_{Y' \hookrightarrow \overline{Y}}, \mathscr{D}_{Y' \hookrightarrow \overline{Y}}) : \ Y' \text{ is a } G - \text{orbit in } \overline{Y}\}.$$

The colored fans satisfy the following defining properties:

F1. Every face of a colored cone in $\mathscr{F}_{\overline{Y}}$ is an element of $\mathscr{F}_{\overline{Y}}$.
F2. For every G-invariant valuation ν in $\mathscr{V}_Y$, there exists at most one colored cone $(\mathscr{C}, \mathscr{D})$ in $\mathscr{F}_{\overline{Y}}$ such that $\upsilon \in \mathscr{C}^0$.

It is easy to make abstract versions of colored fans. Let V be a finite-dimensional vector space over $\mathbb{Q}$. Starting with a subset $\mathscr{V}$ of V and a finite set $\mathscr{D}$ together with a set map $\widetilde{\rho} : \mathscr{D} \to V$, we define a *colored fan associated with* $(V, \mathscr{V}, \mathscr{D}, \widetilde{\rho})$ as a finite collection of pairs $(\mathscr{C}, \mathscr{E})$, where $\mathscr{C}$ is a cone in V and $\mathscr{E}$ is a subset of $\mathscr{D}$ satisfying the properties **F1**, **F2**, **C1**–**C4**. Of course, $\mathscr{V}$ plays the role of $\rho(\mathscr{V}_Y)$ in $\mathscr{Q}_Y$ and $\widetilde{\rho} : \mathscr{D} \to V$ plays the role of $\widetilde{\rho} : \mathscr{D}_Y \to \mathscr{Q}_Y$.

Let H and H' be two closed subgroups in G such that the homogenous varieties

$$Y := G/H \text{ and } Z := G/H'$$

are spherical. Let

$$\phi : Y \to Z$$

be a morphism of varieties. If ϕ is G-equivariant, then the resulting map on the character groups $\phi^* : X^*(Z) \to X^*(Y)$ is injective, hence, the "dual" linear map $\phi_* : \mathscr{Q}_Y \to \mathscr{Q}_Z$ is surjective. Furthermore, we have $\phi_*(\mathscr{V}_Y) = \mathscr{V}_Z$. Let $\mathscr{D}_\phi^c$ denote the set of colors of Y that are mapped into Z dominantly,

$$\mathscr{D}_\phi^c := \{D \in \mathscr{D}_Y : \ \phi(D) \text{ is dense in } Z\}.$$

In other words, $\mathscr{D}_\phi^c$ is the set of colors of Y which are too big, so, we may ignore(!) them in the combinatorial setup. We set

$$\mathscr{D}_\phi := \mathscr{D}_Y - \mathscr{D}_\phi^c.$$

Definition 11 Let $\phi : Y \to Z$ be a G-equivariant morphism between two spherical homogenous G-varieties. Let $\overline{Y}$ and $\overline{Z}$ denote two equivariant embeddings of Y

and Z, respectively. Let Y' and Z' be two G-orbits in $\overline{Y}$ and $\overline{Z}$, respectively. The map ϕ is said to be a *morphism* between the colored cones $(\mathscr{C}_{Y' \hookrightarrow \overline{Y}}, \mathscr{D}_{Y' \hookrightarrow \overline{Y}})$ and $(\mathscr{C}_{Z' \hookrightarrow \overline{Z}}, \mathscr{D}_{Z' \hookrightarrow \overline{Z}})$ if we have

1. $\phi_*(\mathscr{C}_{Y' \hookrightarrow \overline{Y}}) \subseteq \mathscr{C}_{Z' \hookrightarrow \overline{Z}}$, and
2. $\phi(\mathscr{D}_\phi \cap \mathscr{D}_{Y' \hookrightarrow \overline{Y}}) \subseteq \mathscr{D}_{Z' \hookrightarrow \overline{Z}}$.

The map ϕ is said to be a *morphism* between the colored fans $\mathscr{F}_{\overline{Y}}$ and $\mathscr{F}_{\overline{Z}}$ if for every cone $(\mathscr{C}, \mathscr{D})$ in $\mathscr{F}_{\overline{Y}}$ there exists a cone $(\mathscr{C}', \mathscr{D}')$ in $\mathscr{F}_{\overline{Z}}$ such that $\phi : (\mathscr{C}, \mathscr{D}) \to (\mathscr{C}', \mathscr{D}')$ is a morphism of cones.

The following result, which is proven by Knop in [16], is a generalization of the classification result of Luna and Vust for simple embeddings.

Theorem 6 (Knop) *Let Y be a spherical homogenous G-variety, and let B be a Borel subgroup in G. The assignment $\overline{Y} \rightsquigarrow \mathscr{F}_{\overline{Y}}$ is a bijective correspondence between the isomorphism classes of G-equivariant embeddings of Y and the isomorphism classes of colored fans associated with $(\mathscr{Q}_Y, \mathscr{V}_Y, \mathscr{D}_Y, \widetilde{\rho})$. In fact, this assignment is an equivalence between the category of equivariant embeddings of Y and the category of colored fans associated with $(\mathscr{Q}_Y, \mathscr{V}_Y, \mathscr{D}_Y, \widetilde{\rho})$.*

Remark 3 As we mentioned before the theorem of Knop, the role of colored fans for simple embeddings was already known. In fact, Luna and Vust had shown in [20] that the colored cone $(\mathscr{C}_{Z \hookrightarrow \overline{Y}}, \mathscr{D}_{Z \hookrightarrow \overline{Y}})$, where $Z \hookrightarrow \overline{Y}$ is the closed orbit of $\overline{Y}$, uniquely determines $\overline{Y}$.

Remark 4 It is not difficult to check that all definitions pertaining to the colored cones make sense (definable) if we use a separably closed field instead of an algebraically closed field.

4.1 Quasi-projective Colored Fans

It is useful to know when an equivariant embedding of a spherical homogenous variety is affine, projective, or more generally quasi-projective. Such criteria are found by Brion in [4]. Here we only give Brion's criterion for quasi-projectiveness.

Theorem 7 (Brion) *Let $\mathscr{F}_{\overline{Y}}$ be the colored fan of an equivariant embedding $\overline{Y}$ of a spherical homogenous G-variety Y. In this case, $\overline{Y}$ is quasi-projective if and only if for each colored cone $C_Z := (\mathscr{C}_{Z \hookrightarrow \overline{Y}}, \mathscr{D}_{Z \hookrightarrow \overline{Y}})$ in $\mathscr{F}_{\overline{Y}}$ there exists a linear form, denoted by ℓ_Z, on $\mathscr{Q}_{\overline{Y}}$ such that the following two conditions are satisfied:*

1. *If $C_Z = (\mathscr{C}_{Z \hookrightarrow \overline{Y}}, \mathscr{D}_{Z \hookrightarrow \overline{Y}})$ and $C_{Z'} = (\mathscr{C}_{Z' \hookrightarrow \overline{Y}}, \mathscr{D}_{Z' \hookrightarrow \overline{Y}})$ are two elements from $\mathscr{F}_{\overline{Y}}$, then the restrictions of the corresponding linear forms onto $\mathscr{C}_{Z \hookrightarrow \overline{Y}} \cap \mathscr{C}_{Z' \hookrightarrow \overline{Y}}$ are the same.*
2. *If $C_Z = (\mathscr{C}_{Z \hookrightarrow \overline{Y}}, \mathscr{D}_{Z \hookrightarrow \overline{Y}})$ and $C_{Z'} = (\mathscr{C}_{Z' \hookrightarrow \overline{Y}}, \mathscr{D}_{Z' \hookrightarrow \overline{Y}})$ are two distinct elements from $\mathscr{F}_{\overline{Y}}$, and if a vector $\chi \in \mathscr{Q}_{\overline{Y}}$ lies in the intersection of the interior of $\mathscr{C}_{Z \hookrightarrow \overline{Y}}$ with the image $\rho(\mathscr{V}_{\overline{Y}})$, then $\ell_Z(\chi) > \ell_{Z'}(\chi)$.*

Following Huruguen, we call a colored fan whose cones satisfy the requirements of Theorem 7 a *quasiprojective colored fan*. We know from (the remarks following) Theorem 4 that there are plenty of quasi-projective colored fans, especially over perfect fields.

With this precise definition of quasi-projectiveness at hand, we are now able to state Huruguen's result.

Theorem 8 (Huruguen) *Let k be a perfect field, let G be a connected reductive group that is defined over k, and let $\overline{Y}(\overline{k})$ be an embedding of a spherical homogenous spherical G-variety Y defined over $\overline{k}$. We assume that the fan of $\overline{Y}$ is Γ-stable. In this case, $\overline{Y}$ admits a k-form if and only if for every cone $C_Z := (\mathscr{C}_{Z\hookrightarrow\overline{Y}}, \mathscr{D}_{Z\hookrightarrow\overline{Y}})$ in $\mathscr{F}_{\overline{Y}}$, the colored fan consisting of the cones $(\sigma(\mathscr{C}_{Z\hookrightarrow\overline{Y}}), \sigma(\mathscr{D}_{Z\hookrightarrow\overline{Y}}))$, $\sigma \in \Gamma$ as well as all of its faces are quasi-projective.*

Proof See [15, Theorem 2.26].

4.2 Spherical Spaces over Arbitrary Fields

We will start with giving a brief summary of Wedhorn's work on the classification of spherical spaces. For all unjustified claims (and for some definitions) we refer the reader to [34] and to the references therein.

Definition 12 Let k be a field, and let G be a reductive k-group. Recall that this amounts to the requirement that $G_{\overline{k}}$ is a connected reductive group. According to [34, Remark 2.2], an algebraic space X over k with an action of G is G-spherical if $X_{\overline{k}}$ is a spherical $G_{\overline{k}}$-variety.

Let $\overline{k}$ denote a fixed algebraic closure of k, let k_s denote a separable closure of k, and let us denote by Γ the Galois group of the extension k_s/k. (Here, we are intentionally vague about our choices because it does not matter which separable closure we choose.) In the sequel, we will look at continuous and linear actions of Γ on some structures. When we speak of a continuous action of Γ on a set X, we will treat X with the discrete topology. The important point here is that if X is a finite set, or, if the action of Γ is linear on some finite-dimensional vector space X, then the action is continuous if and only if it factors through some finite discrete quotient of Γ. This fact should alleviate a possible pain of confronting a large absolute Galois group such as Γ of $\mathbb{Q}_s/\mathbb{Q}$.

If $\overline{Y}$ is a spherical G-space, then there exists a homogenous spherical G-space Y such that $\overline{Y}$ is a spherical embedding of Y. This actually amounts to the statement that Y is the unique open minimal G-invariant subspace of $\overline{Y}$. By definition, a G-invariant subspace in an algebraic space $\overline{Y}$ is *minimal* if there exists no proper non-empty G-invariant subspace of $\overline{Y}$.

Theorem 9 (Wedhorn) *Let G be a reductive k-group, and let $\overline{Y}$ be a spherical G-scheme viewed as an equivariant embedding of the spherical homogenous scheme*

G/H. If k is separably closed, then the assignment $\overline{Y} \rightsquigarrow \overline{Y}_{\overline{k}}$ induces a bijection between the isomorphism classes of spherical embeddings of G/H over k and the isomorphism classes of spherical embeddings of G/H over $\overline{k}$.

Notice that the bijection between isomorphism classes that is mentioned in Theorem 9 is essentially the application of the base change functor from k to $\overline{k}$. In general, this does not give an equivalence of categories. A straightforward example is produced by the left translation action of $G = \mathbf{G}_m$ on $Y = \mathbf{G}_m$. Luckily, since the definition of colored fans work over separably closed fields, and since we have faithfully flat descent upon restriction, the classification reduces to the classification over algebraically closed fields. The caveat is that one needs to consider all G-invariant minimal subschemes of the spherical space.

Corollary 1 *Let G be a reductive k-group, and let $\overline{Y}$ be a spherical G-space viewed as an equivariant embedding of the spherical homogenous space $Y := G/H$. If k is separably closed, then the assignment $\overline{Y} \rightsquigarrow (\mathscr{C}_{Y' \hookrightarrow \overline{Y}}, \mathscr{D}_{Y' \hookrightarrow \overline{Y}})_{Y'}$, where Y' runs over all minimal G-invariant subschemes of $\overline{Y}$, is an equivalence between the category of equivariant embeddings of Y (over k) and the category of colored fans associated with $(\mathscr{Q}_Y, \mathscr{V}_Y, \mathscr{D}_Y, \widetilde{\rho})$.*

Of course, the theorem and its corollary that we just presented here give us something new (compared to Knop's theorem) only when the characteristic of k is nonzero. Now, we proceed with the general case and assume that G is a reductive k-group. Let Y be a spherical homogenous G-variety, and let $\overline{Y}$ be a spherical embedding of Y. Both of Y and $\overline{Y}$ is assumed to be defined over k. Note that Borel subgroups always exist over separably closed fields, whence we fix a Borel subgroup B in G despite the fact that B may not have any k-rational points. This is where we start to notice a departure from Huruguen's work.

There is a natural action of the Galois group Γ on the space of B-semi invariants $k_s(Y)^{(B)}$. In particular, Γ acts on the k_s-vector space $\Omega_{Y_{k_s}}$ continuously and linearly. Moreover, it acts continuously on the valuation cone $\mathscr{V}_{Y_{k_s}}$ as well as on the set of colors $\mathscr{D}_{Y_{k_s}}$, and the maps $\rho : \mathscr{V}_{Y_{k_s}} \to \mathscr{Q}_{Y_{k_s}}$ and $\widetilde{\rho} : \mathscr{D}_{Y_{k_s}} \to \mathscr{Q}_{Y_{k_s}}$ are Γ-equivariant.

- A colored fan $\mathscr{F}_{\overline{Y}_{k_s}}$ is said to be *Γ-invariant* if its colored cones are permuted by the action of Γ.

Theorem 10 (Wedhorn) *Let G be a reductive k-group, and let $\overline{Y}$ be a spherical G-space viewed as an equivariant embedding of the spherical homogenous G-space $Y := G/H$, which is defined over k. Then, the assignment $\overline{Y}_{k_s} \rightsquigarrow (\mathscr{C}_{Y' \hookrightarrow \overline{Y}_{k_s}}, \mathscr{D}_{Y' \hookrightarrow \overline{Y}_{k_s}})_{Y'}$, where Y' runs over all minimal G-invariant subschemes of $\overline{Y}_{k_s}$, induces an equivalence between the category of equivariant embeddings of Y over k and the category of Γ-invariant colored fans associated with $(\mathscr{Q}_{Y_{k_s}}, \mathscr{V}_{Y_{k_s}}, \mathscr{D}_{Y_{k_s}}, \widetilde{\rho})$.*

5 Reductive Monoids over Arbitrary Fields

In this section, we will consider the reductive monoids that are defined over arbitrary fields. We will show how Wedhorn's theorems are applicable to the algebraic monoid setting.

Definition 13 A *reductive* k-monoid is a k-monoid whose unit-group is a reductive k-group in the sense of Definition 7.

In particular, according to our Definition 13, the unit-group of a reductive $\overline{k}$-monoid is a connected reductive monoid, conforming with our tacit assumption from the introduction as well as with that of [29].

Remark 5 Let M be a reductive monoid defined over an algebraically closed field, and let G denote its unit-group. The following results are recorded in [29]:

1. G is dense in M;
2. M is affine;
3. the reductive monoids are exactly the affine $G \times G$-embeddings of reductive groups;
4. the commutative reductive monoids are exactly the affine embeddings of tori;
5. the isomorphism classes of reductive monoids with unit-group G are in bijection with the strictly convex polyhedral cones of $\mathscr{Q}_G$ generated by all of the colors and a finite set of elements from $\mathscr{V}_G$.

Now we propose a definition for "monoid algebraic spaces." Probably this definition exists in the literature, however, we could not locate it. For our monoid space definition, once again, we will relax the definition of a group algebraic space (as given in Stacks Project Tag 043G).

Definition 14 Let S be a scheme, and let B be an algebraic space that is separated over S.

- A *monoid algebraic space over* B is a pair (M, m), where M is a separated algebraic space over B and $m : M \times_B M \to M$ is a morphism of algebraic spaces over B with the property that, for every scheme T over B, the pair $(M(T), m)$ is a monoid.
- A *morphism* $\psi : (M, m) \to (M', m')$ of monoid algebraic spaces over B is a morphism $\psi : M \to M'$ of algebraic spaces such that, for every T/B, the induced map $\psi : M(T) \to M'(T)$ is a homomorphism of monoids.

Definition 15 A *reductive monoid space* over a scheme S is a monoid algebraic space M over S such that $M \to S$ is flat, of finite presentation over S, and for all $s \in S$ the geometric fiber $M_{\overline{s}}$ is a reductive $\overline{k}$-monoid.

Clearly, if a monoid algebraic space M over a field k is a scheme, then M is a k-monoid in the sense of Definition 8 but the converse is not true. Indeed, in [15], Huruguen has found an example of a smooth toric variety of dimension 3 that is split over a quadratic extension of k, having no k-forms. This pathological example shows that even for the purposes of classifying reductive monoids over an arbitrary field one needs to venture into the category of algebraic spaces.

Extending Rittatore's classification to reductive monoid spaces, we record the following observations which are simple corollaries of Wedhorn's theorems combined with Rittatore's results.

Recall that the reductive monoids with unit-group G are $G \times G$-equivariant embeddings of G. When we speak of "colors" in this context, we always mean the colors of G as a $G \times G$-spherical k-group.

Theorem 11 *Let k be a field, and let G denote a reductive k-group. Let M be a reductive monoid space with G as the group of invertible elements.*

1. *If k is separably closed, then the assignment $M \rightsquigarrow (\mathscr{C}_{Y' \hookrightarrow M}, \mathscr{D}_G)_{Y'}$, where Y' runs over all minimal $G \times G$-invariant subschemes of M, is an equivalence between the category of reductive monoid spaces over k and the category of strictly convex colored polyhedral cones of $\mathscr{Q}_G$ generated by all of the colors of G and a finite set of elements from $\mathscr{V}_G$.*
2. *Let k_s be a separable closure of k. If k is properly contained in k_s, then the following categories are equivalent:*
   a. *the category of reductive monoid spaces over k with unit-group G;*
   b. *the category of Γ-invariant strictly convex colored polyhedral cones of $\mathscr{Q}_{G_{k_s}}$ generated by all of the colors of G_{k_s} and a finite set of elements of $\mathscr{V}_{G_{k_s}}$. Here, Γ is the Galois group of the extension k_s/k.*

It is now desirable to know exactly which reductive monoid schemes over a field have a k-form.

Theorem 12 *Let k be a perfect field, let M be a reductive monoid defined over $\overline{k}$ with unit-group G, and assume that G is defined over k. In this case, M has a k-form if and only if its colored fan, which is a strictly convex polyhedral cone, is stable under the action of the absolute Galois group of $k \subset \overline{k}$.*

Proof By Theorem 5, we know that M is affine, therefore, its colored fan is automatically quasi-projective. Now our result follows from Theorem 8.

5.1 k-Forms of Lined Closures

For the next application, we restrict our attention to the field of complex numbers, and we assume that the reader is familiar with the highest weight theory.

It is well known that any complex irreducible affine monoid M admits a faithful finite-dimensional rational monoid representation. In other words, there exists a finite-dimensional vector space V and an injective monoid homomorphism

$$\rho : M \to End(V),$$

which is a morphism of varieties (see [25]). We notice, in the light of Remark 1, that this fact holds true more generally for all irreducible affine k-monoids, where k is an arbitrary field. In particular, $(V, \rho|_G)$ is a faithful rational representation of the unit-group G, and $M \cong \overline{\rho(G)}$ in $End(V)$. In this case, we will write

$$M = M_V. \tag{6}$$

Now, let V_λ denote the irreducible representation of G with highest weight λ. The *saturation of* λ, denoted by Σ_λ, is the set of all dominant weights that are less than or equal to λ,

$$\Sigma_\lambda := \{\mu : \mu \text{ is dominant and } \mu \leq \lambda\}.$$

Let V_{Σ_λ} denote the representation $\oplus_{\mu \in \Sigma} V_\mu$, and let $\mathscr{M}_\lambda$ denote the reductive monoid defined by V_{Σ_λ} as in in (6). In a similar manner, we will denote M_{V_λ} by M_λ. (These are special cases of the "multi-lined closure" construction of Li and Putcha in [19].) Clearly, both of the monoids M_λ and $\mathscr{M}_\lambda$ are reductive.

In [11], De Concini analyzed the geometric properties of $\mathscr{M}_\lambda$ in relation with that of M_λ, and he proved the following theorem.

Theorem 13 (DeConcini)

1. *$\mathscr{M}_\lambda$ is a normal variety with rational singularities.*
2. *$\mathscr{M}_\lambda$ is the normalization of M_λ.*
3. *$\mathscr{M}_\lambda$ and M_λ are equal if and only if λ is minuscule, that is to say, $\Sigma_\lambda = \{\lambda\}$.*

We finish our paper with a theorem whose proof will be given somewhere else.

Theorem 14 *The reductive monoid M_λ, hence its normalization $\mathscr{M}_\lambda$ have an $\mathbb{R}$-form if and only if there exists an involutory automorphism θ of the reductive unit-group G of M_λ such that $\theta^*\lambda = -\lambda$.*

Acknowledgements I thank the organizers of the 2017 Southern Regional Algebra Conference: Laxmi Chataut, Jörg Feldvoss, Lauren Grimley, Drew Lewis, Andrei Pavelescu, and Cornelius Pillen. I am grateful to Jörg Feldvoss, Lex Renner, Soumya Dipta Banerjee, and to the anonymous referee for their very careful reading of the paper and for their suggestions, which improved the quality of the article. This work was partially supported by a grant from the Louisiana Board of Regents.

References

1. Borel, A.: Linear Algebraic Groups, 2nd edn. Graduate Texts in Mathematics, vol. 126. Springer-Verlag, New York (1991)
2. Bravi, P., Pezzini, G.: Primitive wonderful varieties. Math. Z. **282**(3–4), 1067–1096 (2016)
3. Brion, M.: Quelques propriétés des espaces homogènes sphériques. Manuscr. Math. **55**(2), 191–198 (1986)
4. Brion, M.: Groupe de Picard et nombres caractéristiques des variétés sphériques. Duke Math. J. **58**(2), 397–424 (1989)

5. Brion, M.: On algebraic semigroups and monoids. In: Algebraic Monoids, Group Embeddings, and Algebraic Combinatorics. Fields Institute Communications, vol. 71, pp. 1–54. Springer, New York (2014)
6. Brion, M.: Some structure theorems for algebraic groups. In: Algebraic Groups: Structure and Actions. Proceedings of Symposia in Pure Mathematics, vol. 94, pp. 53–126. American Mathematical Society, Providence, RI (2017)
7. Bröcker, T., tom Dieck, T.: Representations of Compact Lie Groups. Graduate Texts in Mathematics, vol. 98. Springer-Verlag, New York (1995). Translated from the German manuscript, Corrected reprint of the 1985 translation
8. Chevalley, C.: Theory of Lie Groups I. Princeton Mathematical Series, vol. 8. Fifteenth printing, Princeton Landmarks in Mathematics. Princeton University Press, Princeton, NJ (1999)
9. Conrad, B.: Reductive group schemes. In: Autour des Schémas en Groupes, vol. I. Panor. Synthèses, vol. 42/43, pp. 93–444. Société mathématique de France, Paris (2014)
10. Danilov, V.I.: The geometry of toric varieties. Uspekhi Mat. Nauk **33**(2), 85–134, 247 (1978)
11. De Concini, C.: Normality and non normality of certain semigroups and orbit closures. In: Algebraic Transformation Groups and Algebraic Varieties. Encylopaedia of Mathematical Sciences, vol. 132, pp. 15–35. Springer, Berlin (2004)
12. Demazure, M.: Sous-groupes algébriques de rang maximum du groupe de Cremona. Ann. Sci. École Norm. Sup. **4**(3), 507–588 (1970)
13. Demazure, M., Gabriel, P.: Introduction to Algebraic Geometry and Algebraic Groups. North-Holland Mathematics Studies, vol. 39. North-Holland Publishing Co., Amsterdam-New York (1980). Translated from the French by J. Bell
14. Helgason, S.: Groups and Geometric Analysis. Mathematical Surveys and Monographs, vol. 83. American Mathematical Society, Providence, RI (2000). Integral geometry, invariant differential operators, and spherical functions, Corrected reprint of the 1984 original
15. Huruguen, M.: Toric varieties and spherical embeddings over an arbitrary field. J. Algebra **342**, 212–234 (2011)
16. Knop, F.: The Luna-Vust theory of spherical embeddings. In: Proceedings of the Hyderabad Conference on Algebraic Groups, Hyderabad, pp. 225–249 (1989). Manoj Prakashan, Madras (1991)
17. Knutson, D.: Algebraic Spaces. Lecture Notes in Mathematics, vol. 203. Springer-Verlag, Berlin-New York (1971)
18. Legendre, A.M.: Recherches sur l'attraction des spheroides homogenes. In: Mémoires de mathématique et de physique : prés. à l'Académie Royale des Sciences, par divers savans, et lûs dans ses assemblées, vol. 1785, pp. 411–434 (2007)
19. Li, Z., Putcha, M.: Types of reductive monoids. J. Algebra **221**(1), 102–116 (1999)
20. Luna, D., Vust, T.: Plongements d'espaces homogènes. Comment. Math. Helv. **58**(2), 186–245 (1983)
21. Mumford, D.: Abelian Varieties. Tata Institute of Fundamental Research Studies in Mathematics, vol. 5. Published for the Tata Institute of Fundamental Research, Bombay; by Hindustan Book Agency, New Delhi (2008). With appendices by C.P. Ramanujam and Y. Manin, Corrected reprint of the second edition (1974)
22. Onishchik, A.L., Vinberg, È.B.: Lie Groups and Algebraic Groups. Springer Series in Soviet Mathematics. Springer-Verlag, Berlin (1990). Translated from the Russian and with a preface by D.A. Leites
23. Ono, T.: On the field of definition of Borel subgroups of semi-simple algebraic groups. J. Math. Soc. Jpn. **15**, 392–395 (1963)
24. Popov, V.L.: Contractions of actions of reductive algebraic groups. Mat. Sb. (N.S.) **130(172)**(3(7)), 310–334, 431 (1986)
25. Putcha, M.S.: Linear Algebraic Monoids. London Mathematical Society Lecture Note Series, vol. 133. Cambridge University Press, Cambridge (1988)
26. Renner, L.E.: Reductive monoids are von Neumann regular. J. Algebra **93**(2), 237–245 (1985)
27. Renner, L.E.: Classification of semisimple varieties. J. Algebra **122**(2), 275–287 (1989)

28. Renner, L.E.: Linear Algebraic Monoids. Encyclopaedia of Mathematical Sciences, vol. 134. Springer-Verlag, Berlin (2005). Invariant Theory and Algebraic Transformation Groups, V
29. Rittatore, A.: Algebraic monoids and group embeddings. Trans. Groups **3**(4), 375–396 (1998)
30. The Stacks Project Authors. Stacks project (2017). http://stacks.math.columbia.edu
31. Timashev, D.A.: Homogeneous Spaces and Equivariant Embeddings. Encyclopaedia of Mathematical Sciences, vol. 138. Springer, Heidelberg (2011). Invariant Theory and Algebraic Transformation Groups, 8
32. Vinberg, È.B.: Complexity of actions of reductive groups. Funktsional. Anal. i Prilozhen. **20**(1), 1–13, 96 (1986)
33. Vinberg, È.B.: On reductive algebraic semigroups. In: Lie Groups and Lie Algebras: E. B. Dynkin's Seminar. American Mathematical Society Translations: Series 2, vol. 169, pp. 145–182. American Mathematical Society, Providence, RI (1995)
34. Wedhorn, T.: Spherical spaces. Ann. l'Institut Fourier **68**(1), 229–256 (2018)

Computing with Matrix and Basic Algebras

Jon F. Carlson

Abstract This is a survey of my lecture delivered at the Southern Regional Algebra Conference in March 2017. It is meant to demonstrate some of the methods and technology that can be used to investigate examples in the theory of representations of groups and algebras. The concentration is on using computational methods to determine the structure of a matrix algebra as in the theorems of Wedderburn and investigating homological properties of the algebra by way of its basic algebra. For every split finite-dimensional associative algebra, there is a basic algebra, one whose simple modules have dimension one, that has the same representation theory. In the computational setting, it is usually much more practical to first find the basic algebra when studying representation theory.

Keywords Matrix algebra · Basic algebra · Computational methods · Idempotent decomposition

1 Introduction

Computational methods for experimentation in mathematics have become commonplace in many areas. The computer has been used to compute examples that were unapproachable even a few years ago. Certainly, the impact of computational methods on modern number theory has been huge. Another of the mathematical subjects that have been greatly affected by the development of computer applications has been group representation theory. Ordinary and modular characters have been computed for a vast number of finite groups. In the area of modular representation theory,

Research partially funded by Simonds Foundation Grant No. 054813-01

J. F. Carlson (✉)
Department of Mathematics, University of Georgia, Athens, GA 30602, USA
e-mail: jfc@math.uga.edu

J. Feldvoss et al. (eds.), *Advances in Algebra*, Springer Proceedings
in Mathematics & Statistics 277, https://doi.org/10.1007/978-3-030-11521-0_5

where the group algebras are not semi-simple, computational techniques have been employed to reveal the more complicated structure of the algebras and their blocks.

Personally, I became involved in computational projects almost 30 years ago when it became evident that the computer could be used to obtain some amazing results in the area of group cohomology. I was interested in the ring structure of group cohomology with coefficients in finite fields. It was an area in which there were very few known examples and those that were known seemed to be too simple to reveal many of the complexities of the subject. I succeeded in computing the mod-2 cohomology rings of all of the groups of order 64 (see [2, 3, 5]). Others, particularly David Green, have carried cohomology projects much further (see [9, 10] for a small sampling of results). My early experiences led naturally to an interest in using computational techniques to approach representations of groups and algebras.

The group algebra of a p-group with coefficients in a field of characteristic p is an example of a split basic algebra, an algebra whose simple modules all have dimension one. If one wishes to study the representation theory of an algebra by computational means, it is most efficient to condense the algebra to an algebra of small dimension if possible. That is, every algebra is Morita equivalent to a basic algebra having, possibly, much smaller dimension. This means that the original algebra and its basic algebra have exactly the same representation theory, and their module categories are equivalent. Thus, the structure of modules can be more easily studied over the algebra of smaller dimension. In this paper, we present a sketch of how this is done.

This paper is a report of an invited lecture presented at the Southeast Algebra Conference at the University of South Alabama in March 2017. The original invitation was to speak on computational methods of algebra. It being impossible in a single lecture, to adequately survey the vast array of uses of the computer in algebraic research, I have chosen the usual approach of staying close to the areas that I know best. My purpose in the lecture was to present a brief sketch of what is available, with the hope of showing how much fun, inspiration, and edification can be found in such projects.

The examples of computer code that appear in the paper all use the language Magma [1]. There are several high-level computer algebra systems available. Many of the features that are highlighted here are available in packages such as GAP [21] or Sage [20]. The reason for choosing Magma for this presentation is mostly that this is what I know, and this should not be taken as a value judgement.

I want to thank the organizers of the Southeast Regional Algebra Conference for the honor of the invitation to give the talk on which this paper is based.

2 Basics and Algebras

Throughout this paper, we let k denote a field of prime characteristic $p > 0$. For computational purposes, it can also be assumed that k is finite. All algebras that we consider are finite-dimensional k-algebras that are associative and have unit elements. Modules over an algebra A are finite-dimensional left modules. If G is a finite group, then kG denotes its group algebra with coefficients in k.

The basic definition that we want is the following.

Definition 1 A basic algebra A over k is a finite-dimensional associative algebra for which every simple module M has dimension one as a module over its endomorphism ring $\mathrm{End}_A(M)$. A split basic algebra is a basic algebra for which every simple module has dimension one over k.

Schur's Lemma tells us the endomorphism ring $\mathrm{End}_A(M)$ of a simple A-module M is a division ring. Under the assumption that M is finitely generated, $\mathrm{End}_A(M)$ is a finite-dimensional algebra over k. If k is finite, then $\mathrm{End}_A(M)$ is a finite division algebra and by Wedderburn's theorem, it is a field, a finite extension of k.

The following describes some of the structure of a split basic algebra.

Lemma 1 (See [13]) *Suppose that A is a split basic algebra. Let $S_1, \ldots, S_r$ be a complete set of representatives of the isomorphism classes of simple A-modules. For each i, let P_i be the projective cover of S_i. Then as a module over itself, $A \cong \sum_{i=1}^{r} P_i$.*

The proof is based on the analysis that $A/\operatorname{Rad}(A)$ is semi-simple, and since each simple module has dimension one, $A/\operatorname{Rad}(A)$ is a direct sum of simple algebras of dimension one. Each simple module corresponds to a simple module S_i and contains an idempotent e_i. Here, $1 = \sum e_i$ and hence $A = \sum Ae_i$ and $P_i \cong Ae_i$.

For more general algebras, consider the following examples. Suppose that $\mathbf{Mat}_n$ is the full matrix algebra of all $n \times n$ matrices over the field k. It is easily seen that the representation theory of $\mathbf{Mat}_n$ is the same as the representation theory of k. That is, the category of $\mathbf{Mat}_n$-modules and k-vector spaces are basically the same. The point is that $\mathbf{Mat}_n$ acts on its natural module, which is a vector space V of dimension n, and every $\mathbf{Mat}_n$-module is a direct sum of t copies of V for some natural number t. The isomorphism type of the module is determined entirely by the number t. Every k-module, i.e. every k-vector space is a direct sum of t copies of k where t is the dimension. So, the functor giving the equivalence of module categories takes V to k and preserves direct sums.

The functor can be made more precise. Take a primitive idempotent e. For example, in $\mathbf{Mat}_n$ let $e = e_{1,1}$, the matrix with with 1 as the entry in the upper left corner and zero elsewhere. Note that $(\mathbf{Mat}_n)e \simeq V$, and that $e(\mathbf{Mat}_n)e \simeq k$. Then, the *functor* from $\mathbf{Mat}_n$-modules to k-modules is given by $X \mapsto eX$. The inverse functor from k-modules to $\mathbf{Mat}_n$-modules is given by the tensor product $Y \mapsto (\mathbf{Mat}_n)e \otimes_k Y$.

The above example can be generalized to other algebras. Suppose that A is the subalgebra of $\mathbf{Mat}_5$ consisting of all 5×5 matrices having the form

$$\begin{bmatrix} * & * & * & * & * \\ * & * & * & * & * \\ * & * & * & * & * \\ 0 & 0 & 0 & * & * \\ 0 & 0 & 0 & * & * \end{bmatrix}$$

over k. Here, each asterisk can be filled in with any element of the field k. That is, every element of A has a 2×3 block in the lower left hand corner that is identically zero.

Let B be the subalgebra of $\mathbf{Mat}_2$ consisting of all upper triangular 2×2 matrices:

$$\begin{bmatrix} * & * \\ 0 & * \end{bmatrix}$$

Note that B is a split basic algebra.

We claim that A has the same representation theory as B, The functor from the category of A-modules to B-modules can be defined by $X \mapsto eX$ where e is the idempotent

$$e = \begin{bmatrix} 1 & 0 & 0 & 0 & 0 \\ 0 & 0 & 0 & 0 & 0 \\ 0 & 0 & 0 & 0 & 0 \\ 0 & 0 & 0 & 1 & 0 \\ 0 & 0 & 0 & 0 & 0 \end{bmatrix}.$$

The two examples above are the examples of Morita equivalences. The technical definition of a Morita equivalence involves something called a Morita context that for our algebras is essentially the information in the idempotents as in the example (see [13] for a formal definition). The theorem that we want is the following.

Theorem 1 (See [13]), p. 439) *Every algebra A is Morita equivalent to a basic algebra. If the algebra is split, meaning that the endomorphism rings of the simple A-modules have dimension one over the base field, then the basic algebra is likewise split.*

For any algebra A, its basic algebra (which is unique up to isomorphism) is

$$B \cong \mathrm{Hom}_A(P, P)$$

where P is a direct sum of one copy of each of the representatives of the isomorphism classes of projective indecomposable modules. Any projective indecomposable module has the form Ae for some primitive idempotent e. Now note that, any homomorphism φ from Ae_1 to Ae_2 is determined completely by the element $x_{12} = \varphi(e_1) = e_1\varphi(e_1)$ in $e_1 Ae_2$. Consequently, a map $\phi \in \mathrm{Hom}_A(P, P)$ is determined by an element in eAe where $e = \sum e_i$ for $P = \sum Ae_i$. So

$$B \cong eAe.$$

Let $M = Ae$, be an A-eAe-bimodule. Then

$$eA \otimes_A Ae \cong eAe \quad \text{and} \quad Ae \otimes_{eAe} eA \cong A$$

More generally, given any idempotent e in an algebra A, there are functors

$$\mathbf{mod}(A) \to \mathbf{mod}(eAe) \quad \text{given by} \quad M \mapsto eA \otimes_A M, \quad \text{and}$$

$$\mathbf{mod}(eAe) \to \mathbf{mod}(A) \quad \text{given by} \quad N \mapsto Ae \otimes_{eAe} N.$$

This process is known in the world of computational algebra as *condensation*. In the case that $e = \sum e_i$ where $P_i \cong Ae_i$ runs through a complete set of representatives of the isomorphism classes of projective indecomposable modules, then the condensation functor is an equivalence. Actually, it is an equivalence provided $eM \neq 0$ for all simple A-modules M.

Here is an example of a routine in the computer algebra system Magma. The actual data structures and how they are computed are discussed later in this paper.

```
> A := BasicAlgebraFromGroup("2A7", 2, 1);
> B := BasicAlgebraFromGroup("2A9", 2, 2);
> A;
Basic algebra of dimension 38 over GF(2)
Numberofprojectivemodules : 3
Numberofgenerators : 8
> B;
Basic algebra of dimension 38 over GF(2)
Numberofprojectivemodules : 3
Numberofgenerators : 8
> time a, b := IsIsomorphic(A, B);
Time : 0.360
> a;
true
```

In the first two lines, the routine is calling two basic algebras out of a library of basic algebras that is in Magma. The inputs identify a specific data set in the library. The first input, "$2A7$", is the Atlas notation [6] for the double cover of the alternating group A_7 on seven letters. This group has a normal subgroup of order 2 and the quotient by that normal subgroup is A_7. The notation "$2A9$" means the same except for A_9 instead of A_7. The second input in the first two lines is "2", indication that the field is $GF(2)$ with two elements. The third input ("1" in the first line, "2" in the second) indicates the number of the block. The group algebra kG of a finite group G is a direct sum of algebras that are two-sided ideals and are called the blocks. The first block is always the principal block, meaning the block that contains the trivial module k. So for $2A_7$ we have chosen the principal block, while for $2A_9$, it is a nonprincipal block.

The third line asks Magma to print the object A, and the three lines follow are the output of that request. It says that the object is the type of a basic algebra and gives the dimension and the number of generators. We see from the output that the two algebras have the same dimension and number of generators. So we ask if they are isomorphic. The answer, called "a", is yes. The output "b", which we have not printed, is a matrix that gives the isomorphism.

Here is another, somewhat similar example. This time we examine the dimensions of the projective modules and note that they are the same. Here, "S45" is the symplectic group Sp(4, 5) of dimension 4 with coefficients in the field of five elements, and "M22" is the Mathieu group on 22 letters.

```
> A := BasicAlgebraFromGroup("S45", 3, 1);
> B := BasicAlgebraFromGroup("M22", 3, 1);
> DimensionsOfProjectiveModules(A);
[10, 10, 11, 11, 9]
> DimensionsOfProjectiveModules(B);
[10, 11, 11, 10, 9]
> time a, b := IsIsomorphic(A, B);
Time : 0.640
> a;
false
```

The natural question is why the two blocks are not isomorphic.

```
> CartanMatrix(A);
[4 1 2 2 1]
[1 4 2 2 1]
[2 2 4 1 2]
[2 2 1 4 2]
[1 1 2 2 3]
> CartanMatrix(B);
[3 2 2 2 1]
[2 3 2 2 2]
[2 2 3 2 2]
[2 2 2 3 1]
[1 2 2 1 3]
```

The (i, j) entry in the Cartan matrix is the number of composition factors of the projective cover of the ith simple module that is isomorphic to the jth simple module. So we see that while the projective modules could be renumbered so that their dimensions match up, their composition factors cannot.

We present one more slightly different example. Here, $J2$ is the second Janko group, also called the Hall–Janko group. The symbol "3A7" denotes the triple cover of A_7.

```
> A := BasicAlgebraFromGroup("J2", 2, 2);
> B := BasicAlgebraFromGroup("3A7", 2, 2);
> A;
Basic algebra of dimension 18 over GF(2^2)
Number of projective modules : 3
```

```
Number of generators : 7
> B;
Basic algebra of dimension 18 over GF(2)
Number of projective modules : 3
Number of generators : 7
> IsIsomorphic(A, B);
....
Runtime error : Arguments 1 and 2 have
incompatible coefficient rings
```

Well, of course, they are not isomorphic. Algebra A is defined over the field with 4 elements, while B is over $GF(2)$. But that does not mean that they are not essentially the same. We extend the field on B so it is an algebra over the same field as A, and try again.

```
> C := ExtendField(B, GF(4));
> a, b := IsIsomorphic(A, C);
> a;
true
```

The reason for A being defined over $GF(4)$ is that the simple kJ_2-modules were not absolutely irreducible in the case that $k = GF(2)$. It was necessary to extend the field to $GF(4)$ in order to make the block algebra Morita equivalent to a split basic algebra. It is currently an open problem to find the smallest field over which an algebra can be written.

3 How Do We Do Condensation?

First, it should be pointed out that in the computational setting, an algebra is usually given only as a collection of generators. That is, a matrix algebra is defined by giving only a few, perhaps only two or three, matrices that generate the algebra. Then, an element of the algebra is any matrix that can be obtained by a sequence of additions and multiplications of the generators. We say that an element is a "word" in the generators. A representation of a group is usually defined by giving a matrix for each of the generators of the group. As a consequence, the algorithms that are used to assume this sort of structure.

One of the most fundamental tools in computational representation theory is the Irreducibility Criterion discovered by Simon Norton (see [18] for a description of the criterion and the algorithm). If we have a module M over an algebra A, it tells us if the module is simple, and if it is not simple, then it finds a proper nontrivial submodule. In the following, we equate A with its image in $\mathrm{End_k(M)}$. Taking the transpose is is essentially the same as taking the dual, so that a submodule of $A^{tr}m$ translates to a quotient module of the original module M.

Theorem 2 *Given an A-module M, and a singular element x in A, at least one of the following holds:*

- *There is m in the kernel of x such that* $Am \neq M$.
- *There is m in the kernel of* x^{tr} *such that* $A^{tr}m \neq M$.
- *M is a simple A-module.*

To make use of the criterion, we need the notion of a *peakword*, an idea due to Richard Parker. The point is to find a singular element in A that will make the meataxe work efficiently. A peakword is an element x of A such that $\varphi(x)$ has corank 1, where $\varphi : A \to \mathbf{Mat}(M)$. That is, the image of x as an operation on M has rank one less than the dimension of M. Then Parker wrote the fundamental algorithm. The trick was figuring out that it could actually be made to work.

Meataxe Algorithm: Find a peakword by "random" search, and apply the irreducibility criterion.

The point is that if x is a peakword, then the kernels of x and x^{tr} have dimension one. So any nonzero element that we choose in the appropriate kernel serves the purpose.

The idea of "random" in this context, is not very close to what a statistician would call random. Indeed, finding a statistically random element might be extremely difficult. Instead, the idea is to simply take a small degree polynomial in the generators of the algebra. For example, if x and y are generators of the algebra, then a "random" element might be $xy + x$. It is surprising that this method works very well in many cases. It often turns out that such an element is a peakword. Newer versions of the algorithm remove some of the ambiguities. The algorithm assumes that the ranks of the generators are large compared to the dimension of the module and it requires that the field be finite. Indeed, the smaller the field is, the better the algorithm works.

There are numerous efficient implementations of the Meataxe Algorithm that are available in the major computer algebra systems and in independent systems. Given a module M over an algebra A, we can find a full composition series for the module by repeated application of the algorithm.

The next thing that we need to be able to do is to find idempotent elements in the algebra A. Remember that given any idempotent e in an algebra A, there is a functor

$$\mathbf{mod}(A) \to \mathbf{mod}(eAe) \quad \text{given by} \quad M \mapsto eA \otimes_A M$$

This is condensation! So how do we find idempotents?

If $A = kG$, H is a subgroup having order prime to p, and χ is a character of a simple kH-module, then

$$e = \frac{1}{|H|} \sum_{h \in H} \chi(h^{-1})h$$

is an idempotent. This method for finding idempotents is often used in group representation computations. There are many variations on this basic technique that have been extremely effective on problems in modular representation theory. But keep in mind that for such an idempotent e, the functor $M \mapsto eM$ may not be faithful. That is, there may be modules M such that $eM = 0$. This can sometimes be a problem. Numerous variations on the method exploit the special architecture of group algebras. See [14, 15, 17] for a small collection of examples.

For a process that works with more general algebras, there is the following. Suppose that $x \in A$ has the property that its minimal polynomial factors $f(t) = g(t)h(t)$ where g and h are relatively prime. Write $1 = a(t)g(t) + b(t)h(t)$ for polynomials a and b. Then, $a(x)g(x)$ is an idempotent. The reason is that

$$(a(x)g(x))^2 = a(x)g(x)(1 - b(x)h(x)) = a(x)g(x) \quad \text{since} \quad g(x)h(x) = 0.$$

Suppose that A is a matrix algebra and $x \in A$ has corank one (a peakword). Then its minimal polynomial has the form $t \cdot h(t)$. Then $1 = a(t)t + b(t)h(t)$ and $b(t)h(t)$ is an idempotent of rank one! Hence a primitive idempotent.

For the sake of computations, a matrix algebra A is an algebra generated by a collection of $n \times n$ matrices $\alpha_1, \alpha_2, \ldots, \alpha_t$, for some n. The algebra is the collection of all polynomials polynomials in these elements.

The Meataxe gives us (by decomposing the natural module) a set of irreducible modules $M_1, \ldots, M_r$. Each simple module M_i is presented as the matrix algebra A_i of the action of A on M, that is, as a set of matrices $\alpha_{i,1}, \ldots, \alpha_{i,t}$, which are the images of the original generators in A_i.

We have $\psi_i : A \longrightarrow A_i$ for $i = 1, \ldots, r$. We want: $E_1, \ldots, E_r$ such that $\psi_j(E_i) = \delta_{ij} \operatorname{Id}_{A_i}$ and $E_1 + E_2 + \cdots + E_r = \operatorname{Id}_A$. And we want: $\beta_1, \ldots, \beta_r$ and $\tau_1, \ldots, \tau_r$ such that

$$\psi_i(\beta_i) = \begin{pmatrix} t_i & 0 & \ldots & 0 \\ 0 & 0 & \ldots & 0 \\ & & & \\ 0 & 0 & \ldots & 0 \end{pmatrix}$$

Here t_i is a generator for K_i over k, $\psi_j(\beta_i) = 0$, and

$$\psi_i(\tau_i) = \begin{pmatrix} 0 & 1 & 0 & \ldots & 0 \\ 0 & 0 & 1 & \ldots & 0 \\ & & & & \\ 0 & 0 & 0 & \ldots & 1 \\ 1 & 0 & 0 & \ldots & 0 \end{pmatrix}$$

with $\psi_j(\tau_i) = 0$.

Then we proceed through several steps.

1. For $j \neq i$, find minimal polynomial f_j of $\psi_j(E_i)$ and replace E_i by $f_j(E_i)E_i$.
2. Resulting E_i has $\psi_j(E_i) = 0$. Hope that $\psi_i(E_i) \neq 0$.
3. Get $\psi_i(E_i)$ to be nonsingular.
4. Get $\psi_i(E_i) = \operatorname{Id}_{A_i}$.

5. Replace E_i by $E_i^{p^n}$ so that $E_i^2 = E_i$. Then
6. Find idempotents in each A_i as word in the generators.
7. Lift them to elements of $E_i A E_i$ by evaluating the words.
8. Take pth powers until getting primitive idempotents.

See [4] for more details. In the end, we have a complete set of mutually orthogonal primitive idempotents. The next problem is to obtain generators and relations for the radical. But we have the following results that rely on being able to obtain a complete set of primitive idempotents.

Lemma 2 (see [4]) *The generators for the radical can be taken to be in eAe where $e = \sum_{i=1}^{r} e_i$ and e_i is a primitive idempotent corresponding to A_i.*

Along the way, we have also computed a large amount of basic information including:

1. the condensed algebra eAe (Morita equivalent to A),
2. the dimension of A,
3. the Cartan matrix for A, and
4. the dimensions of the projective modules for A.

In addition, we can define modules over A, and we can test for membership in the algebra.

Here is another example of Magma output. We construct an algebra A that is the algebra of the action of group algebra on a sum of projective indecomposable modules. The group is the alternating group on eight letters and the coefficient field k is the field with three elements.

```
> prj := ProjectiveIndecomposableModules(AlternatingGroup
  (8), GF(3));
> [Dimension(x) : x in prj];
[135, 153, 90, 63, 162, 225, 90]
> M := DirectSum([prj[i] : i in [1..4]]);
> A := Action(M);
```

We have taken the direct sum of the first four projective indecomposable modules. The direct sum has dimension 441. So, the algebra A is the action of the group algebra on this direct sum and is a subalgebra of the algebra of 441×441 matrices over k.

So, now we check the dimension of A by the usual method in Magma and in other computer algebra systems. It is called "spinning" vectors or generators. It works by starting with the generators of A. Then create another element by multiplying two together. If it is a new element, then append it to the list. Multiply two more elements in the list and append the product if it is linearly independent of the subspace generated by the list. Continue until we have a complete basis. The reported time is in seconds of CPU time.

```
> time X := Basis(A);
Time : 461.910
```

Next, we compute a basis by constructing a complete collection of idempotents, getting a basis for the radical, etc., as outlined above.

```
> time Y := BasisOfAlgebra(A);
Time : 17.140
```

So, we see that the seemingly more complicated method takes only a fraction of the time of the default method for computing bases. In addition, by this method, we have computed a great deal more information about the algebra.

Just to be sure, we check that we have the same number of elements in the basis.

```
> #X, #Y;
11357 11357
```

Here is another example of the use of this technology. We consider the Schur algebra A of type (4, 7) over a finite field with five elements [16]. Magma has a program that will compute it directly. Then we take its Ext algebra, $\mathrm{Ext}_A^*(S, S)$, where S is a direct sum of one copy of each of the simple A-modules. Because the Schur algebra has finite global dimension, this is a finite dimensional algebra. The point of this exercise is that if the algebra A is isomorphic to its double Ext algebra, then it is a Koszul algebra [19]. We show this to be the case for the particular algebra.

```
> A := BasicAlgebraFromSchur(4, 7, 5);
> A;
Basic algebra of dimension 26 over GF(5)
Number of projective modules : 11
Number of generators : 21
> B := BasicAlgebraOfExtAlgebra(A, 8);
> B;
Basic algebra of dimension 48 over GF(5)
Number of projective modules : 11
Number of generators : 21
> C := BasicAlgebraOfExtAlgebra(B, 8);
> C;
Basic algebra of dimension 26 over GF(5)
Number of projective modules : 11
Number of generators : 21
> IsIsomorphic(A, C);
true Mapping from : AlgBas : A to AlgBas : C
```

4 What Is a Basic Algebra in the Computational Setting?

Remember that a module over an algebra is a sequence of matrices, each giving the action of a generator of the algebra. The computational data (see [2]) defining a basic algebra A is a sequence of projective modules together with a sequence of "path trees", one for each projective module. The projective modules form a complete set of representatives of the isomorphism classes of indecomposable projective modules for the algebra. The direct sum of these modules is isomorphic to the right regular module for the algebra. The path trees tell how the projective modules are "built".

The generators of the algebra are first a sequence of idempotents, followed by a sequence of nonidempotent generators. For each nonidempotent generator, there is a unique idempotent that acts as the identity by multiplication on the right side and another unique idempotent that is the identity on the left. We should think of this as a quiver with relations.

Indeed, from a basic algebra, it is possible to write out its quiver with relations as in the following example.

```
> G := AlternatingGroup(9);
> P := PermutationModule(G, Normalizer(G, Sylow(G, 3)), GF(3));
> A := BasicAlgebraOfEndomorphismAlgebra(P);
> a, b, c := QuiverAndRelations(A);
> a;
[< 3, 3 >, < 3, 4 >, < 3, 4 >, < 4, 3 >, < 4, 3 >, < 4, 4 >, < 4, 4 >,
   < 4, 4 >, < 4, 4 >, < 5, 5 >]
> c;
[a.1^2 + a.3 * a.5, a.1 * a.2, a.1 * a.3, a.2 * a.4,
a.2 * a.5, a.2 * a.6, a.2 * a.7, a.2 * a.8, a.2 * a.9,
a.3 * a.4 + a.3 * a.5, a.3 * a.6, a.3 * a.7, a.3 * a.8,
a.3 * a.9, a.4 * a.1, a.4 * a.2, a.4 * a.3 + a.9^2, ...
```

Here, A is the basic algebra of the endomorphism ring of a module over the alternating group on nine letters with coefficients in the field with three elements. The output called "a" is the quiver, which consists 5 nodes and 10 arrows. The list is only the beginning and end point of each arrow. That is, the pair $< 4, 3 >$ indicates an arrow from node 4 to node 3. The output "b" which is not printed, is the set of all relations among the arrows. The output "c" is the same as b except that it omits the obvious relations. An example of an obvious relation is $a.9 * a.10$ since $a.9$ ends in node 4 while $a.10$ begins at node 5. Note that we have not printed all of output "c", because the complete list of relations is quite long.

Given a basic algebra A with primitive idempotent generators $e_1, \ldots, e_t$, and nonidempotent generators $\alpha_{t+1}, \ldots, \alpha_n$, the path tree for the projective module $P_i = Ae_i$ is a sequence of pairs: $[\langle 1, i\rangle, \ldots, \langle u, v\rangle, \ldots]$. It tells you that P_i has a basis $b_1, \ldots, b_s$ where $b_1 = e_i$, and if the jth element of the path tree is $\langle u, v\rangle$, then $b_j = \alpha_v b_u$.

The data structure is optimized to solve the homomorphism lifting problem: Given a map μ and a surjective map θ find ψ as in the following diagram:

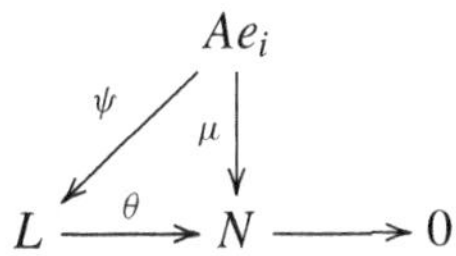

Hence, the data structure is optimized for computing cohomology and homological algebra. An example of how this might help is the following. The example comes from Burkhard Külshammer and Radha Kessar (see [11] and Proposition 7 of [12]).

```
> F < x, y, z >:= FreeAlgebra(GF(3), 3);
> R := [y * x + x * y, y^2 - x^2, z^2 - x^4, x * z, z * x, z * y, y * z, x^3 * y - x^5];
> A := BasicAlgebra(F, R);
> A;
Basic algebra of dimension 9 over GF(3)
Number of projective modules : 1
Number of generators : 4
> kk := SimpleModule(A, 1);
> time cpx := CompactProjectiveResolution(kk, 6);
Time : 0.060
> time cpx := CompactProjectiveResolution(kk, 10);
Time : 98.450
> cpx`BettiNumbers;
[17711], [6765], [2584], [987], [377],
[144], [55], [21], [8], [3], [1]
```

In this example, we form the local algebra with three generators x, y and z and relations $xy = -yx$, $y^2 = x^2$, $z^2 = x^4$, $xz = zx = zy = yz = 0$ and $x^3y = x^5$ over the field with three elements. We see that it has dimension 9. The question is whether this could be the block algebra of a finite group. If it were a block algebra, then any projective resolution of any module would necessarily have polynomial growth rate. So we compute the projective resolution in a compact form. The first six steps take less than a tenth of a second to compute. However, then the next four steps, going out to degree 10 take almost 100 s. This is an indication that the projective resolution is getting very large. Checking the Betti numbers, we see that the tenth term in the projective resolution is a direct sum of 17711 copies of the free module A. This is a clear indication that the answer to the question of whether this is a block algebra is negative. That is, the growth rate of the projective resolution appears to be exponential as the dimensions of the terms more than double at each step.

5 A Couple of Problems

There remain many areas where the computational algorithms are sometimes inadequate or perhaps only disappointing. Many people are working on these problems. We mention only a couple of issues relevant to the subjects discussed above.

Generation is the hobgoblin of condensation. The problem is that given an algebra A generated by $\alpha_1, \ldots, \alpha_t$, and an idempotent e, the algebra eAe is not, in general, generated by $e\alpha_1 e, \ldots, e\alpha_t e$. In fact, it seldom happens, even when the condensation functor defined by e is faithful. There has been a huge amount of work on this problem (see [17] for one of the best results), and in practice the problem is solved by a variety of means, some of which are *ad hoc*.

The best situation is the one in which we can compute a complete set of orthogonal primitive idempotents [4]. In such a case, there is a finite list of explicitly computed generators, and the only problem is the easily solved one of paring down the list to a minimal or near minimal collection of elements.

We mention one last problem which is that of invariants. Given a kG-module M and a subspace $V \subseteq M$, find the stabilizer of V in the group G. This problem is extremely important in many computations, including the programs that test isomorphisms (see for example [7, 8]). Such programs rely on being able to find a minimal element in the orbit of such a subspace V under the action of the group. An improvement in the methods for finding stabilizers could greatly extend the range of applicability of the isomorphism testing routines.

References

1. Bosma, W., Cannon, J., Fieker, C., Steel, A. (eds.): Handbook of Magma Functions, V2.23 (2017). http://magma.maths.usyd.edu.au/magma/
2. Carlson, J.F.: Problems in the calculation of group cohomology. In: Computational Methods for Representations of Groups and Algebras. Essen, (1997); Progr. Math. **173**, 107–120 (1999)
3. Carlson, J.F.: Calculating group cohomology: tests for completion. In: Computational Algebra and Number Theory. Milwaukee, WI (1996); J. Symbolic Comput. **31**(1–2), 229–242 (2001)
4. Carlson, J.F., Matthews, G.: Generators and relations for matrix algebras. J. Algebra **300**, 134–159 (2006)
5. Carlson, J.F., Townsley, L., Valeri-Elizondo, L., Zhang, M.: Cohomology rings of finite groups. With an appendix: calculations of cohomology rings of groups of order dividing 64. Algebras and Applications, vol. 3. Kluwer, Dordrecht (2003)
6. Conway, J.H., Curtis, R.T., Norton, S.P., Parker, R.A., Wilson, R.A.: ATLAS of Finite Groups. Oxford University Press, Oxford (1985)
7. Eick, B.: Computing automorphism groups and testing isomorphisms for modular group algebras. J. Algebra **320**(11), 3895–3910 (2008)
8. Eick, B., King, S.: The isomorphism problem for graded algebras and its application to mod-p cohomology rings of small p-groups. J. Algebra **452**, 487–501 (2016)
9. Green, D.J., King, S.A.: The computation of the cohomology rings of all groups of order 128. J. Algebra **325**, 352–363 (2011)
10. King, S.A., Green, D.J., Ellis, G.: The mod-2 cohomology ring of the third Conway group is Cohen-Macaulay. Algebr. Geom. Topol. **11**(2), 719–734 (2011)

11. Kessar, R.: On blocks stably equivalent to a quantum complete intersection of dimension 9 in characteristic 3 and a case of the abelian defect group conjecture. J. London Math. Soc. **2**(85), 491–510 (2012)
12. Külshammer, B., Landrock, P., Reinhardt, S.: Algebras related to a class of 3-blocks of defect 2 (2016)
13. Lam, T.Y.: Lectures on Modules and Rings. Graduate Texts in Mathematics, vol. 189. Springer-Verlag, New York (1999)
14. Lux, K., Wiegelmann, M.: Condensing tensor product modules. In: The Atlas of Finite Groups: Ten Years On, pp. 174–190. Birmingham (1995). London Mathematical Society Lecture Note series, vol. 249. Cambridge University Press, Cambridge (1998)
15. Lux, K., Müller, J., Ringe, M.: Peakword condensation and submodule lattices: an application of the MEAT-AXE. J. Symb. Comput. **17**(6), 529–544 (1996)
16. Martin, S.: Schur algebras and representation theory. In: Cambridge Tracts in Mathematics, vol. 112. Cambridge University Press, Cambridge (1993)
17. Noeske, F.: Tackling the generation problem in condensation. J. Algebra **309**(2), 711–722 (2007)
18. Parker, R.: The computer calculation of modular characters (The Meat-Axe). In: Atkinson, M. (ed.) Computational Group Theory, pp. 267–274. Academic Press, London (1984)
19. Priddy, S.B.: Koszul resolutions. Trans. Amer. Math. Soc. **152**, 39–60 (1970)
20. SageMath: The Sage Mathematics Software System. The Sage Developers (2017). http://www.sagemath.org
21. The GAP Group: GAP—Groups, Algorithms, and Programming, Version 4.8.7 (2017). http://www.gap-system.org

A Product Formula for the Normalized Volume of Free Sums of Lattice Polytopes

Tianran Chen and Robert Davis

Abstract The free sum is a basic geometric operation among convex polytopes. This note focuses on the relationship between the normalized volume of the free sum and that of the summands. In particular, we show that the normalized volume of the free sum of full-dimensional polytopes is precisely the product of the normalized volumes of the summands.

Keywords Lattice polytopes · Free sum · BKK bound

1 Introduction

A *(convex) polytope* $P \subset \mathbb{R}^n$ is the convex hull of finitely many points in $\mathbb{R}^n$. Equivalently, P is the bounded intersection of finitely many closed half-spaces in $\mathbb{R}^n$. We call a polytope *lattice* if its vertices are elements of $\mathbb{Z}^n$. Polytopes are fascinating combinatorial objects, and lattice polytopes are especially interesting due to their appearance in many contexts such as combinatorics, algebraic statistics, and physics; see, for example, [2, 26, 27].

Given polytopes $P \subset \mathbb{R}^m$ and $Q \subset \mathbb{R}^n$, set

$$P \oplus Q := \operatorname{conv}\{(P, \mathbf{0}) \cup (\mathbf{0}, Q)\} \subset \mathbb{R}^{m+n}.$$

If P and Q each contain the origin, then we call $P \oplus Q$ the *free sum* of P and Q. It will be convenient to use the notation P', Q' for $(P, \mathbf{0})$, $(\mathbf{0}, Q) \in \mathbb{R}^{m+n}$, respectively, so that we simply write $P \oplus Q = \operatorname{conv}\{P' \cup Q'\}$. In the present contribution, we show

T. Chen (✉)
Department of Mathematics and Computer Science, Auburn University Montgomery, Montgomery, AL 36124, USA
e-mail: ti@nranchen.org

R. Davis
Department of Mathematics, Michigan State University, East Lansing, MI 48824, USA
e-mail: davisr@math.msu.edu

J. Feldvoss et al. (eds.), *Advances in Algebra*, Springer Proceedings in Mathematics & Statistics 277, https://doi.org/10.1007/978-3-030-11521-0_6

that the normalized volume of the free sum $P \oplus Q$ is precisely the product of the normalized volumes of P and Q.

Given a lattice polytope $P \subset \mathbb{R}^n$, its *lattice point enumerator*, defined on the positive integers, is

$$L_P(m) := |mP \cap \mathbb{Z}^n|.$$

Ehrhart [12] showed that $L_P(m)$ agrees with a polynomial of degree dim P with rational coefficients called the *Ehrhart polynomial*. Thus, the *Ehrhart series*

$$E_P(t) := 1 + \sum_{m \geq 1} L_P(m) t^m$$

is a rational function of the form

$$E_P(t) = \frac{h_P^*(t)}{(1-t)^{\dim P + 1}},$$

where $h_P^*(t) = \sum_{i=0}^{d} h_i^* t^i$ has degree $d \leq \dim P$. The vector $h^*(P) = (h_0^*, \ldots, h_d^*)$ is called the *h^*-vector* of P, and Stanley [24] proved that it consists of nonnegative integers. We refer the reader to [4] for background on Ehrhart theory.

The choice of notation h^* evokes parallels with combinatorial h-vectors of simplicial complexes. For example, under certain conditions, the h^*-vector of a lattice polytope is the h-vector of a simplicial polytope [24]. This is not to be expected, in general, since such a correspondence can fail even in $\mathbb{R}^3$.

Example 1 Let T_r be the tetrahedron

$$T_r = \text{conv}\{(0, 0, 0), (1, 0, 0), (0, 1, 0), (1, 1, r)\},$$

where r is a positive integer. It is not difficult to verify that that $h^*(T_r) = (1, 0, r-1)$, but this is not the h-vector for a $(d-1)$-dimensional simplicial complex whenever $r > 1$.

Another basic result of Ehrhart theory is that if P is a lattice polytope, then $h_P^*(1)$ is the *normalized volume*

$$\text{Vol } P := (\dim P)! \text{ vol } P,$$

where vol P is the relative volume of P. Additionally, for all lattice polytopes, $h_0^* = 1$, and there are simple combinatorial interpretations for h_1^* as well as h_d^*. However, the remaining coefficients do not have such elementary descriptions in general. This has led to a large amount of research being conducted to better understand the behavior of h^*-vectors for lattice polytopes, even for quite restricted classes of polytopes. See [8] for a recent survey regarding h^*-vectors of lattice polytopes.

Define the *(polar) dual* of $P \subset \mathbb{R}^m$ to be

$$P^\vee := \{x \in \mathbb{R}^m \mid x^T y \le 1 \text{ for all } y \in P\}.$$

The dual $P^\vee$ is again a polytope if and only if $\mathbf{0}$ is an interior point of P; otherwise, $P^\vee$ is an unbounded polyhedron. If P and $P^\vee$ are both lattice polytopes, then P is called *reflexive*. A lattice translate of a reflexive polytope is also called reflexive. Reflexive polytopes were originally introduced by Batyrev [2] to describe mirror symmetry phenomena in supersymmetric string theory. Since their introduction, many fascinating combinatorial and arithmetic properties of theirs have been discovered [9, 11, 13, 15, 16, 18, 23]. An increasingly well-known result in Ehrhart theory related to reflexive polytopes is the following, also known as Braun's formula.

Theorem 1 (Braun's formula [7]) *If $P \subset \mathbb{R}^m$ is a reflexive polytope and $Q \subset \mathbb{R}^n$ is a lattice polytope containing the origin in its interior, then*

$$E_{P\oplus Q}(t) = (1-t)E_P(t)E_Q(t).$$

Consequently,

$$h^*_{P\oplus Q}(t) = h^*_P(t)h^*_Q(t).$$

The conditions needed for Braun's formula to hold were significantly relaxed in [3], including a multivariate generalization. However, we will not need such power here.

Because $h^*_P(1)$ gives the normalized volume of P, it follows that if $P \oplus Q$ is a free sum satisfying the conditions of Braun's formula, then

$$\mathrm{Vol}_{m+n}(P \oplus Q) = \mathrm{Vol}_m(P)\,\mathrm{Vol}_n(Q).$$

If $\mathbf{0}$ is merely on the boundary of P or Q, then $P \oplus Q$ is still called a free sum but its h^*-polynomial may not factor. In this note, we shall examine instances in which the h^*-polynomial may fail to factor yet the normalized volume of the free sum does factor as the product of normalized volumes of the summands.

Example 2 Let $P = T_2$ as in Example 1 and let $Q = P \oplus P \subset \mathbb{R}^6$. It is easy to verify that P contains no interior lattice point, and so is not reflexive. Moreover, a routine computation shows that the h^*-vectors of P and $P \oplus P$ are $(1, 0, 1)$ and $(1, 0, 2, 1)$, respectively. Therefore, the h^*-polynomial of $P \oplus P$ is distinct from $(h^*_P(t))^2$. However, the normalized volume factoring

$$\mathrm{Vol}_6(P \oplus P) = 4 = \mathrm{Vol}_3(P)\,\mathrm{Vol}_3(P)$$

remains valid.

The main goal of this article is to establish the general fact that the (top dimensional) normalized volume of a free sum is always the product of the normalized volumes of the summands.

Theorem 2 *Given full-dimensional convex polytopes $P \subset \mathbb{R}^m$ and $Q \subset \mathbb{R}^n$, if both P and Q contain the origin (of $\mathbb{R}^m$ and $\mathbb{R}^n$, respectively), then*

$$\mathrm{Vol}_{m+n}(P \oplus Q) = \mathrm{Vol}_m(P)\,\mathrm{Vol}_n(Q).$$

This theorem demonstrates that even if the distribution of lattice points in the various dilations of $P \oplus Q$ does not carry over from that of P and Q, we can still easily recover geometric information about the polytope.

It is easy to see that if P' and Q' do not intersect, then there is no hope of a similar product formula for $\mathrm{Vol}_{m+n}(P \oplus Q)$ to occur in general. As a simple example, note that if $P = [k, k+1] \subset \mathbb{R}$ for some positive integer k and $Q = [0, 1] \subset \mathbb{R}$, then the product of their normalized volumes is 1 but the normalized volume of $P \oplus Q$ increases as $k \to \infty$.

Section 2 recalls important results related to lattice polytopes and roots of Laurent polynomial systems. We then present the proof of Theorem 2 in Sect. 3.

We conclude the introduction by remarking that Theorem 2 is implicit in the work of [25], which considers a *weighted h^*-polynomial.* While Stapledon's results are more general (by considering rational dilations of polytopes), our alternate proof for integer dilations approaches the question via bounds on isolated points of complex varieties. Through personal communication, we also learned that Tyrrell McAllister produced (but did not publish) yet another proof of this result via elementary convex-analytic techniques in 2015 [21]. In comparison, our approach highlights the close connection between algebra and convex geometry. Finally, we point out that a similar statement can be made about co-convex bodies [1].

2 Preliminaries

This section briefly reviews notations and concepts to be used. Given a polytope $P \subset \mathbb{R}^n$ and a linear form $l(\mathbf{x})$, we call

$$l_b(\mathbf{x}) := \{\mathbf{x} \in \mathbb{R}^n \mid l(\mathbf{x}) = b\}$$

a *supporting hyperplane* of P if either

$$l(\mathbf{x}) \leq b \text{ for all } \mathbf{x} \in P$$

or

$$l(\mathbf{x}) \geq b \text{ for all } \mathbf{x} \in P.$$

A *face* of P is a set of the form $l_b(\mathbf{x}) \cap P$ for some supporting hyperplane $l_b(\mathbf{x})$ of P. The empty set and P itself are faces of P, corresponding to a hyperplane that does not intersect P in the former case, and the "empty" hyperplane in the latter. The *dimension* of a face F is defined as $\dim \operatorname{aff} F$, where $\operatorname{aff} F$ is the affine span of

F. By convention, $\dim \emptyset := -1$. A face of dimension 0 is called a *vertex*, a face of dimension 1 is called an *edge*, and a face of dimension $\dim P - 1$ is called a *facet*.

Next, recall that the *Minkowski sum* of two sets $A, B \subset \mathbb{R}^n$, is

$$A + B = \{\mathbf{a} + \mathbf{b} \in \mathbb{R}^n \mid \mathbf{a} \in A, \mathbf{b} \in B\}.$$

Given polytopes $Q_1, \ldots, Q_n \subset \mathbb{R}^n$, and positive scalars $\lambda_1, \ldots, \lambda_n$, the Minkowski sum $\lambda_1 Q_1 + \cdots + \lambda_n Q_n$ is also a polytope, and its volume is a homogeneous polynomial in $\lambda_1, \ldots, \lambda_n$. In it, the coefficient of $\lambda_1, \ldots, \lambda_n$ is the *mixed volume*[1] [22] of $Q_1, \ldots, Q_n$, denoted $\mathrm{MV}(Q_1, \ldots, Q_n)$. It agrees with normalized volume "on the diagonal," that is, $\mathrm{MV}(Q, \ldots, Q) = \mathrm{Vol}_n(Q)$.

Though the main results to be established in this note concern a geometric property of polytopes, our proofs take a decidedly algebraic approach via the theory for counting roots of Laurent polynomial systems. A *Laurent polynomial* in $x_1, \ldots, x_n$ is an expression of the form

$$p = \sum_{k=1}^{m} c_k x_1^{a_{k,1}}, \ldots, x_n^{a_{k,n}},$$

where $c_k \in \mathbb{C}^* = \mathbb{C} \setminus \{0\}$ and $a_{k,j} \in \mathbb{Z}$. Thus, the Laurent polynomials are exactly the expressions obtained when localizing the usual polynomial ring $\mathbb{C}[x_1, \ldots, x_n]$ at each indeterminate. We adopt the compact notation $p = \sum_{\mathbf{a} \in S} c_{\mathbf{a}} \mathbf{x}^{\mathbf{a}}$ where each $\mathbf{a} = (a_1, \ldots, a_n) \in \mathbb{Z}^n$ encodes the exponents of a term, $\mathbf{x}^{\mathbf{a}} := x_1^{a_1}, \ldots, x_n^{a_n}$, and the set $S \subset \mathbb{Z}^n$ collecting all such exponents is known as the *support* of p, denoted $\mathrm{supp}(p)$.

For a Laurent polynomial system $P = (p_1, \ldots, p_m)$ in $\mathbf{x} = (x_1, \ldots, x_n)$, we define $\mathbb{V}^*(P) := \{\mathbf{x} \in (\mathbb{C}^*)^n \mid P(\mathbf{x}) = \mathbf{0}\}$, which is the common zero set of P within $(\mathbb{C}^*)^n$. Here, the dimension n (of the ambient space) is generally clear from context, and hence does not directly appear in the notation $\mathbb{V}^*(P)$. Such a zero set has rich internal structures (e.g., the structure of a quasi-projective algebraic set [14]). In particular, $\mathbb{V}^*(P)$ consists of finitely many irreducible components each with a well-defined dimension. Of special interest to us are the 0-dimensional components of $\mathbb{V}^*(P)$, which are the *isolated points*: elements $\mathbf{x}$ for which there is an open set that contains $\mathbf{x}$ but does not contain any other points in $\mathbb{V}^*(P)$. This subset will be denoted by $\mathbb{V}_0^*(P)$. Our proof makes frequent use of the following two theorems:

Theorem 3 (Kushnirenko's Theorem [20]) *Given a Laurent polynomial system $P = (p_1, \ldots, p_n)$ in $\mathbf{x} = (x_1, \ldots, x_n)$, if the supports of $p_1, \ldots, p_n$ are identical, that is, if $S := \mathrm{supp}(p_1) = \cdots = \mathrm{supp}(p_n)$, then*

$$|\mathbb{V}_0^*(P)| \le n! \,\mathrm{vol}_n(\mathrm{conv}(S)).$$

[1] An alternative definition for mixed volume is the coefficient of $\lambda_1, \ldots, \lambda_n$ in the above polynomial divided by $n!$.

Theorem 4 (Bernshtein's First Theorem [5]) *For a Laurent polynomial system* $P(x_1, \ldots, x_n) = (p_1, \ldots, p_n)$,

$$|\mathbb{V}_0^*(P)| \leq \mathrm{MV}(\mathrm{conv}(\mathrm{supp}(p_1)), \ldots, \mathrm{conv}(\mathrm{supp}(p_n))).$$

Remark 1 Implicit in the above theorems is the fact that the upper bounds for $|\mathbb{V}_0^*(P)|$ given in both statements are "generically exact" in the sense that if the coefficients are chosen at random, the probability of picking coefficients for which the bounds are not exact is zero. Stated more precisely, the set of coefficients for which the bounds are exact form a nonempty Zariski open set among the set of all possible choices of coefficients. The generically exact solution bound given by Theorem 4 has since been known as the BKK bound after the circle of works by Bernshtein [5], Kushnirenko [19, 20], and Khovanskii [17].

Our main proof additionally relies on the following result on the connection between mixed volume and normalized volume established independently in [6, 10].

Theorem 5 ([10, Theorem 2]) *Given nonempty finite sets* $S_1, \ldots, S_n \subset \mathbb{Q}^n$, *let* $\tilde{S} = S_1 \cup \cdots \cup S_n$. *If every positive-dimensional face* F *of* $\mathrm{conv}(\tilde{S})$ *satisfies one of the following conditions:*

(A) $F \cap S_i \neq \varnothing$ *for all* $i \in \{1, \ldots, n\}$;
(B) $F \cap S_i$ *is a singleton for some* $i \in \{1, \ldots, n\}$;
(C) For each $i \in I := \{i \mid F \cap S_i \neq \varnothing\}$, $F \cap S_i$ *is contained in a common coordinate subspace of dimension* $|I|$, *and the projection of* F *in this subspace is of dimension less than* $|I|$;

then $\mathrm{MV}(\mathrm{conv}(S_1), \ldots, \mathrm{conv}(S_n)) = \mathrm{Vol}_n(\mathrm{conv}(\tilde{S}))$.

Note that if $\mathrm{conv}(\tilde{S})$ is not full dimensional, then both sides of the above equations will be zero.

3 Proofs of the Main Results

We now turn to the main result, Theorem 2, which we shall recall here:

Theorem 2 *Given full-dimensional convex polytopes* $P \subset \mathbb{R}^m$ *and* $Q \subset \mathbb{R}^n$, *if both* P *and* Q *contain the origin (of* $\mathbb{R}^m$ *and* $\mathbb{R}^n$, *respectively), then*

$$\mathrm{Vol}_{m+n}(P \oplus Q) = \mathrm{Vol}_m(P)\,\mathrm{Vol}_n(Q).$$

Proof First, we consider the cases where P and Q are both lattice polytopes. Let $S \subset \mathbb{Z}^m$ and $T \subset \mathbb{Z}^n$ be the set of vertices of P and Q, respectively. Also, let $P' = \{(\mathbf{p}, \mathbf{0}) \in \mathbb{R}^{m+n} \mid \mathbf{p} \in P\}$ and $Q' = \{(\mathbf{0}, \mathbf{q}) \in \mathbb{R}^{m+n} \mid \mathbf{q} \in Q\}$. The finite sets $S', T' \subset$

$\mathbb{Z}^{m+n}$ are the sets of vertices of P' and Q', respectively. Consider the two sets of Laurent monomials $\mathbf{x}^S = \{\mathbf{x}^{\mathbf{a}} \mid \mathbf{a} \in S\}$ and $\mathbf{y}^T = \{\mathbf{y}^{\mathbf{b}} \mid \mathbf{b} \in T\}$ in the variables $\mathbf{x} = (x_1, \ldots, x_m)$ and $\mathbf{y} = (y_1, \ldots, y_n)$. We take the linear combinations

$$f_i(\mathbf{x}) := \sum_{\mathbf{a} \in S} c_{i,\mathbf{a}} \mathbf{x}^{\mathbf{a}} \quad \text{for } i = 1, \ldots, m \qquad g_j(\mathbf{y}) := \sum_{\mathbf{b} \in T} c'_{j,\mathbf{b}} \mathbf{y}^{\mathbf{b}} \quad \text{for } j = 1, \ldots, n$$

where the coefficients $c_{i,\mathbf{a}}$ and $c'_{j,\mathbf{b}}$ can be taken generically, which is possible by the discussion in Remark 1. We can now form a system of $m + n$ Laurent polynomial equations

$$H(\mathbf{x}, \mathbf{y}) := (f_1(\mathbf{x}), \ \ldots, \ f_m(\mathbf{x}), \ g_1(\mathbf{y}), \ \ldots, \ g_n(\mathbf{x}))$$

in the $m + n$ variables $(\mathbf{x}, \mathbf{y}) = (x_1, \ldots, x_n, y_1, \ldots, y_n)$. It is easy to verify that

$$\operatorname{supp}(H) = (\underbrace{S', \ldots, S'}_{m}, \underbrace{T', \ldots, T'}_{n}).$$

Since the $c_{i,\mathbf{a}}$ and $c'_{j,\mathbf{b}}$ are generic, the BKK bound for H given by Theorem 4 is exact. In that case,

$$|\mathbb{V}_0^*(H)| = \mathrm{MV}(\mathrm{conv}(S'), \ldots, \mathrm{conv}(S'), \mathrm{conv}(T'), \ldots, \mathrm{conv}(T')).$$

We shall now show, using Theorem 5, that this quantity is also the normalized volume of

$$\mathrm{conv}(S' \cup T') = \mathrm{conv}(P' \cup Q') = P \oplus Q.$$

Let F be a positive-dimensional proper face of $P \oplus Q$. We shall show F always satisfy one of the conditions listed in Theorem 5.

(Case I) First, suppose F contains P' or Q'. Then, by assumption, F intersects both P' and Q' since the two must intersect at the origin according to the definition of a free sum. Therefore, F satisfies condition (A) in Theorem 5.

(Case II) If F is a proper face of P' and $F \cap Q' = \varnothing$, then

$$F \subset P' \subset \{(\mathbf{p}, \mathbf{0}) \in \mathbb{R}^{m+n} \mid \mathbf{p} \in \mathbb{R}^m\}.$$

In other words, F is contained in a coordinate subspace of dimension m and its projection in this subspace is of dimension less than m. Therefore, F satisfies condition (C) in Theorem 5. By the same argument, F would satisfy the same condition if F is a proper face of Q' instead and $F \cap P' = \varnothing$.

(Case III) Finally, suppose F does not contain P' or Q' nor is it contained in P' or Q'. Since a face of $P \oplus Q$ must contain vertices of $P \oplus Q$ which are necessarily of the form $(\mathbf{p}, \mathbf{0})$ or $(\mathbf{0}, \mathbf{q})$ for $\mathbf{p} \in P$ or $\mathbf{q} \in Q$, F must intersect both P' and Q'. That is, it satisfies condition (A) in Theorem 5.

The above cases exhausts all the possibilities, and therefore by Theorem 5,

$$\begin{aligned}\mathrm{MV}(\mathrm{conv}(S'), \ldots, \mathrm{conv}(S'), \mathrm{conv}(T'), \ldots, \mathrm{conv}(T')) &= \mathrm{Vol}(\mathrm{conv}(S' \cup T')) \\ &= \mathrm{Vol}(\mathrm{conv}(P' \cup Q')) \\ &= \mathrm{Vol}(P \oplus Q).\end{aligned}$$

Now focusing on $|\mathbb{V}_0^*(H)|$, we can see that the first m equations in $H = \mathbf{0}$ only involve variables $\mathbf{x} = (x_1, \ldots, x_m)$ and the remaining equations only involve variables $\mathbf{y} = (y_1, \ldots, y_n)$. So the solutions to $H(\mathbf{x}, \mathbf{y}) = \mathbf{0}$ are precisely the points of the form $(\mathbf{x}, \mathbf{y})$ such that $F(\mathbf{x}) = \mathbf{0}$ and $G(\mathbf{y}) = \mathbf{0}$, where

$$F(\mathbf{x}) = (f_1(\mathbf{x}), \ldots, f_m(\mathbf{x})) \text{ and } G(\mathbf{y}) = (g_1(\mathbf{y}), \ldots, g_n(\mathbf{y})).$$

Consequently,

$$|\mathbb{V}_0^*(H)| = |\mathbb{V}_0^*(F)| \cdot |\mathbb{V}_0^*(G)| = \mathrm{Vol}_m(P) \cdot \mathrm{Vol}_n(Q)$$

by Kushnirenko's Theorem (Theorem 3). Therefore, we have a chain of equalities

$$\mathrm{Vol}(P \oplus Q) = \mathrm{MV}(P', \ldots, P', Q', \ldots, Q') = |\mathbb{V}_0^*(H)| = \mathrm{Vol}_m(P) \cdot \mathrm{Vol}_n(Q).$$

The above shows the statement holds for two lattice polytopes P and Q. Since the volume forms vol_{m+n}, vol_m, and vol_n are homogeneous of degree $m+n$, m, and n, respectively, under uniform scaling, this result directly extends to polytopes with vertices in $\mathbb{Q}^m$ and $\mathbb{Q}^n$, respectively. Then, by the continuity of volume forms, the statement must also hold for polytopes with vertices of real coordinates.

References

1. Alilooee, A., Soprunov, I., Validashti, J.: Generalized multiplicities of edge ideals. J. Algebraic Comb. 1–32 (2016)
2. Batyrev, V.V.: Dual polyhedra and mirror symmetry for Calabi-Yau hypersurfaces in toric varieties. J. Algebraic Geom. **3**, 493–535 (1994)
3. Beck, M., Jayawant, P., McAllister, T.B.: Lattice-point generating functions for free sums of convex sets. J. Comb. Theory Ser. A **120**, 1246–1262 (2013). https://doi.org/10.1016/j.jcta.2013.03.007, http://www.sciencedirect.com/science/article/pii/S0097316513000599
4. Beck, M., Robins, S.: Computing the Continuous Discretely. Undergraduate Texts in Mathematics. Springer, New York (2007)
5. Bernshtein, D.N.: The number of roots of a system of equations. Funct. Anal. Appl. **9**, 183–185 (1975)
6. Bihan, F., Soprunov, I.: Criteria for strict monotonicity of the mixed volume of convex polytopes (2017). arXiv:1702.07676 [math], http://arxiv.org/abs/1702.07676.pdf
7. Braun, B.: An Ehrhart series formula for reflexive polytopes. Electron. J. Combin. **13**(15), 5 pp. (electronic) (2006)

8. Braun, B.: Unimodality Problems in Ehrhart Theory, pp. 687–711. Springer International Publishing, Cham (2016). https://doi.org/10.1007/978-3-319-24298-9_27
9. Braun, B., Davis, R., Solus, L.: Detecting the integer decomposition property and Ehrhart unimodality in reflexive simplices (2016). http://arxiv.org/abs/1608.01614
10. Chen, T.: Unmixing the mixed volume computation (2017). arXiv:1703.01684 [math], http://arxiv.org/abs/1703.01684
11. Conrads, H.: Weighted projective spaces and reflexive simplices. Manuscripta Math. **107**, 215–227 (2002). https://doi.org/10.1007/s002290100235
12. Ehrhart, E.: Sur les polyèdres rationnels homothétiques à n dimensions. C. R. Acad. Sci. Paris **254**, 616–618 (1962)
13. Haase, C., Melnikov, I.V.: The reflexive dimension of a lattice polytope. Ann. Comb. **10**, 211–217 (2006)
14. Hartshorne, R.: Algebraic Geometry, no. 52. Springer (1977)
15. Hibi, T.: Ehrhart polynomials of convex polytopes, h-vectors of simplicial complexes, and nonsingular projective toric varieties. In: Discrete and Computational Geometry: Papers from the DIMACS Special Year, vol. 6, pp. 165–177 (1991)
16. Hibi, T.: Algebraic Combinatorics on Convex Polytopes. Carslaw Publications, Australia (1992)
17. Khovanskii, A.G.: Newton polyhedra and the genus of complete intersections. Funct. Anal. Appl. **12**, 38–46 (1978). https://doi.org/10.1007/BF01077562
18. Kreuzer, M., Skarke, H.: Classification of reflexive polyhedra in three dimensions. Adv. Theor. Math. Phys. **2**, 853–871 (1998)
19. Kushnirenko, A.G.: A Newton polyhedron and the number of solutions of a system of k equations in k unknowns. Usp. Math. Nauk **30**, 266–267 (1975)
20. Kushnirenko, A.G.: Newton polytopes and the Bezout theorem. Funct. Anal. Appl. **10**, 233–235 (1976). https://doi.org/10.1007/BF01075534, https://link.springer.com/article/10.1007/BF01075534
21. McAllister, T.: Private Communication (2018)
22. Minkowski, H.: Theorie der konvexen Korper, insbesondere Begrundung ihres Oberflachenbegriffs. Gesammelte Abhandlungen von Hermann Minkowski **2**, 131–229 (1911)
23. Payne, S.: Ehrhart series and lattice triangulations. Discrete Comput. Geom. **40**, 365–376 (2008)
24. Stanley, R.P.: Decompositions of rational convex polytopes. Ann. Discrete Math. **6**, 333–342 (1980); Combinatorial mathematics, optimal designs and their applications. In: Proceedings of the Symposium of Combinatorial Mathematics, Optimal Designs, Colorado State University, Fort Collins, Colorado (1978)
25. Stapledon, A.: Counting lattice points in free sums of polytopes. J. Comb. Theory Ser. A (2017). To appear
26. Sturmfels, B.: Gröbner Bases and Convex Polytopes. University Lecture Series, vol. 8. American Mathematical Society, Providence, RI (1996)
27. Ziegler, G.M.: Lectures on Polytopes. Graduate Texts in Mathematics, vol. 152. Springer-Verlag, New York (1995)

On the Cohomological Spectrum and Support Varieties for Infinitesimal Unipotent Supergroup Schemes

Christopher M. Drupieski and Jonathan R. Kujawa

Abstract We show that if G is an infinitesimal elementary supergroup scheme of height $\leq r$, then the cohomological spectrum $|G|$ of G is naturally homeomorphic to the variety $\mathcal{N}_r(G)$ of supergroup homomorphisms $\rho : \mathbb{M}_r \to G$ from a certain (non-algebraic) affine supergroup scheme $\mathbb{M}_r$ into G. In the case $r = 1$, we further identify the cohomological support variety of a finite-dimensional G-supermodule M as a subset of $\mathcal{N}_1(G)$. We then discuss how our methods, when combined with recently announced results by Benson, Iyengar, Krause, and Pevtsova, can be applied to extend the homeomorphism $\mathcal{N}_r(G) \simeq |G|$ to arbitrary infinitesimal unipotent supergroup schemes.

Keywords Support varieties · Infinitesimal group schemes · One-parameter subgroups · Supergroups

1 Introduction

1.1 Overview

Let k be a field of positive characteristic p. For more than thirty years, support varieties have played a prominent role in relating the p-modular representation theory of

---

Drupieski was supported in part by a Simons Collaboration Grant for Mathematicians, and by NSF Grant DMS-1440140 while he was in residence at the Mathematical Sciences Research Institute in Berkeley, CA, during the Spring 2018 semester.
Kujawa was supported in part by NSA grant H98230-16-0055 and in part by a Simons Collaboration Grant for Mathematicians.

C. M. Drupieski (✉)
Department of Mathematical Sciences, DePaul University, Chicago, IL 60614, USA
e-mail: c.drupieski@depaul.edu

J. R. Kujawa
Department of Mathematics, University of Oklahoma, Norman, OK 73019, USA
e-mail: kujawa@math.ou.edu

J. Feldvoss et al. (eds.), *Advances in Algebra*, Springer Proceedings in Mathematics & Statistics 277, https://doi.org/10.1007/978-3-030-11521-0_7

a finite group G (and related algebraic structures) to the ambient geometry encoded by the spectrum $|G|$ of the cohomology ring $\mathrm{H}^\bullet(G,k)$. Unipotent subgroups have frequently been important intermediate actors in this relationship. For example, Quillen [18] showed that the cohomological spectrum of a finite group is stratified by pieces coming from its elementary abelian p-subgroups. Later, building on work by Friedlander and Parshall [10, 11] for finite-dimensional restricted Lie algebras, Suslin, Friedlander, and Bendel (SFB) [19, 20] showed that if G is an infinitesimal k-group scheme of height $\leq r$, then $|G|$ is homeomorphic to the variety of infinitesimal one-parameter subgroups $\nu : \mathbb{G}_{a(r)} \to G$, i.e., the variety of homomorphisms into G from the r-th Frobenius kernel of the additive group scheme $\mathbb{G}_a$. More generally, Friedlander and Pevtsova [12] related the cohomological spectrum of an arbitrary finite k-group scheme G to the space $\Pi(G)$ of π-points of G, which consists of equivalence classes of flat K-algebra maps $\alpha_K : K[t]/(t^p) \to KG$ (for K a field extension of k) that factor through the group algebra of an abelian unipotent subgroup scheme of G_K.

Now suppose $p \geq 3$. This paper is a continuation of our work in [9] investigating the cohomology of finite k-supergroup schemes. Recall that an affine k-supergroup scheme is a representable functor from the category $\mathfrak{csalg}_k$ of commutative k-superalgebras to groups. An affine k-supergroup scheme G is finite if its representing Hopf superalgebra $k[G]$ is finite-dimensional, and is infinitesimal if it is finite and if the augmentation ideal of $k[G]$ is nilpotent. In [9] we introduced a family $\mathbb{M}_{r;f,\eta}$ of infinitesimal supergroup schemes, which we called multiparameter supergroups, that are parametrized by an integer $r \geq 1$, a p-polynomial $0 \neq f \in k[T]$ without constant term, and a scalar $\eta \in k$. The definitions of these multiparameter supergroups are recalled in Sect. 1.3. If $f = T^{p^s}$ is a single monomial (and if $r \geq 2$ when $\eta \neq 0$), then $\mathbb{M}_{r;s,\eta} := \mathbb{M}_{r;T^{p^s},\eta}$ is unipotent, but in general the group algebra $k\mathbb{M}_{r;f,\eta}$ has a nontrivial semisimple subalgebra. Following Suslin, Friedlander, and Bendel [19], in [9] we defined characteristic extension classes for the general linear supergroup $GL_{m|n}$, and calculated how the classes restricted along homomorphisms $\rho : \mathbb{M}_{r;f,\eta} \to GL_{m|n}$. We then applied our calculations to describe (for k algebraically closed) the maximal ideal spectrum $|GL_{m|n(r)}|$ of the cohomology ring for the r-th Frobenius kernel of $GL_{m|n}$. Roughly speaking (and modulo a finite morphism), we showed that $|GL_{m|n(r)}|$ is equal to $\bigcup_{f,\eta} \mathbf{Hom}(\mathbb{M}_{r;f,\eta}, GL_{m|n(r)})(k)$, where $\mathbf{Hom}(\mathbb{M}_{r;f,\eta}, GL_{m|n(r)})(k)$ is the variety of homomorphisms $\phi : \mathbb{M}_{r;f,\eta} \to GL_{m|n(r)}$.

In the present work we focus our attention on the *unipotent* multiparameter supergroups, or what is more-or-less the same, the infinitesimal objects in the family of *elementary* finite supergroup schemes, as recently classified (for k perfect) by Benson, Iyengar, Krause, and Pevtsova (BIKP) [3]. The only elementary supergroup schemes that are not enumerated among our multiparameter supergroups are $\mathbb{G}_{a(r)}$ for $r \geq 1$, and the odd additive group scheme $\mathbb{G}_a^-$, though one has $\mathbb{M}_{r;1} := \mathbb{M}_{r;T^p,0} = \mathbb{G}_{a(r)} \times \mathbb{G}_a^-$. As we point out in Lemma 2.1.4, the (height-r) infinitesimal elementary supergroup schemes are all quotients of a common supergroup scheme, $\mathbb{M}_r$, which we initially defined in [9, Sect. 3.1]. The supergroup $\mathbb{M}_r$ then plays a central role in the rest of the paper.

One of our main results, presented in Corollary 3.1.2, is that if k is algebraically closed, and if G is a height-r infinitesimal elementary supergroup scheme, then the cohomological spectrum $|G|$ of G is naturally homeomorphic to the variety

$$\mathcal{N}_r(G) := \mathbf{Hom}(\mathbb{M}_r, G)(k) := \mathrm{Hom}_{Grp/k}(\mathbb{M}_r, G)$$

of all k-supergroup scheme homomorphisms $\rho : \mathbb{M}_r \to G$. For $r = 1$ (and G still infinitesimal elementary) we further identify, in Theorem 3.4.1, the support variety $|G|_M$ of a finite-dimensional G-supermodule M with the set

$$\begin{aligned} \mathcal{N}_1(G)_M : &= \left\{\phi \in \mathcal{N}_1(G) : \mathrm{id}_{\mathbb{M}_1}(\phi^* M) = \infty\right\} \\ &= \left\{\phi \in \mathcal{N}_1(G) : \mathrm{Ext}^i_{\mathbb{M}_1}(M, M) \neq 0 \text{ for infinitely many } i \geq 1\right\}. \end{aligned}$$

Here $\phi^* M$ denotes the pullback of M along the homomorphism $\phi : \mathbb{M}_1 \to G$, and $\mathrm{id}_{\mathbb{M}_1}(\phi^* M)$ denotes the injective dimension of $\phi^* M$ in the category of rational $\mathbb{M}_1$-supermodules. Our definition of the support set $\mathcal{N}_1(G)_M$ is inspired by similar definitions appearing in the literature in the context of commutative local rings (cf. [1, 2, 15]), and which were brought to our attention by way of a talk by Srikanth Iyengar at the Conference on Groups, Representations, and Cohomology, held at Sabal Mòr Ostaig, Isle of Skye, Scotland, in June 2015. At present we do not understand how the existing support theory for local commutative rings relates to the rational cohomology of $\mathbb{M}_1$, but an obvious intermediate actor is the (infinite-dimensional) group algebra $k\mathbb{M}_1 := k[\mathbb{M}_1]^\#$ and its "polynomial subalgebra" $\mathbb{P}_1$. We expect that a better understanding of the relationship will help answer, for example, the question of whether or not the support sets of Sect. 3.3 are Zariski closed subsets of $\mathcal{N}_r(G)$.

At the end of the paper we discuss how our methods can be extended to arbitrary infinitesimal unipotent supergroup schemes when they are combined with recently announced results by Benson, Iyengar, Krause, and Pevtsova concerning the detection of nilpotent elements in cohomology. The first anticipated application, which we state as Conjecture 3.5.2, is that for an arbitrary infinitesimal unipotent supergroup scheme G of height $\leq r$, there exists a natural homeomorphism

$$|G| \simeq \mathcal{N}_r(G) = \mathrm{Hom}_{Grp/k}(\mathbb{M}_r, G).$$

This suggests that, at least for infinitesimal unipotent supergroups, $\mathcal{N}_r(G)$ is the correct analogue of SFB's variety of infinitesimal one-parameter subgroups. We state this result as a conjecture rather than as a theorem only because, at the time of writing this article, the BIKP detection theorem has been publicly announced but has not yet appeared in preprint form; otherwise we provide all of the details to justify the conjecture. The second anticipated application, stated as Conjecture 3.5.6, is an extension of Theorem 3.4.1 to arbitrary height-1 infinitesimal unipotent supergroup schemes. To justify Conjecture 3.5.6 we must assume that the support set $\mathcal{N}_1(G)_M$ is a Zariski closed subset of $\mathcal{N}_1(G)$, and we must also assume a technical condition concerning the ideal of functions defining $\mathcal{N}_1(G)_M$ as a subset of $\mathcal{N}_1(G)$.

We do not make any particular speculations concerning support varieties for infinitesimal unipotent supergroups of height greater than one, though we expect that it should be possible to use methods like those employed by Suslin, Friedlander, and Bendel [20] to bootstrap up from height-1 unipotent supergroups to arbitrary infinitesimal unipotent supergroups. For example, for $r \geq 2$ and $0 \neq \eta \in k$, there are superalgebra (though not Hopf superalgebra) isomorphisms

$$k\mathbb{M}_{r;s,\eta} \cong k\mathbb{M}_{r-1;s+1} \quad \text{and} \quad k\mathbb{M}_{r;s} \cong (k\mathbb{G}_{a(1)})^{\otimes(r-1)} \otimes k\mathbb{M}_{1;s},$$

which should enable one, as in [20, Proposition 6.5], to reduce calculations for arbitrary infinitesimal elementary supergroup schemes to calculations for $\mathbb{G}_{a(1)}$ and $\mathbb{M}_{1;s}$. More generally, if $\mathbb{M}_{r;f,\eta}$ is an arbitrary multiparameter supergroup and if $A = k\mathbb{M}_{r;s,\eta}$, then there exists a semisimple Hopf subalgebra $B \subset A$ such that the Hopf quotient $A/\!/B$ is isomorphic as a superalgebra to $k\mathbb{M}_{r';s'}$ for some integers $r', s' \geq 0$. Thus, one could expect to reduce support variety calculations for arbitrary multiparameter supergroups to the unipotent case.

1.2 Organization of the Paper

Section 2 is devoted to foundational results concerning the set $\mathcal{N}_r(G)$ of supergroup scheme homomorphisms $\rho : \mathbb{M}_r \to G$. In Sect. 2.2 we significantly extend the calculations of [9, Lemma 3.3.2] by computing for each purely even reduced commutative k-algebra A and for each infinitesimal elementary supergroup scheme G the set

$$\mathbf{Hom}(\mathbb{M}_r, G)(A) = \mathrm{Hom}_{Grp/A}(\mathbb{M}_r \otimes_k A, G \otimes_k A)$$

of all A-supergroup scheme homomorphisms $\rho : \mathbb{M}_r \otimes_k A \to G \otimes_k A$. In Sect. 2.3 we show that if the field k is algebraically closed, and if G is an algebraic k-supergroup scheme, then the set $\mathcal{N}_r(G)$ admits the structure of an affine algebraic variety. In particular, $\mathcal{N}_r(GL_{m|n})$ may be identified with a certain variety of commuting nilpotent supermatrices. Then following SFB, in Sect. 2.4 we define for each algebraic k-supergroup scheme G the universal supergroup homomorphism from $\mathbb{M}_r$ to G, and use it to define a homomorphism of graded k-algebras $\psi_r : H(G,k) \to k[\mathcal{N}_r(G)]$.

In Sect. 3 we investigate the map of varieties $\Psi : \mathcal{N}_r(G) \to |G|$ induced by the homomorphism ψ_r. In Sect. 3.1 we show that Ψ is a homeomorphism if G is a height-r infinitesimal elementary supergroup scheme. Then after making some observations in Sect. 3.2 arising from various Lyndon–Hochschild–Serre spectral sequences, in Sect. 3.3 we define the support sets $\mathcal{N}_r(G)_M$, and in Sect. 3.4 we show that $\mathcal{N}_1(G)_M$ identifies with the cohomological support variety $|G|_M$ when G is a height-one infinitesimal elementary supergroup scheme. Finally, in Sect. 3.5 we discuss applications of the detection theorem recently announced by BIKP.

1.3 Conventions

We generally follow the conventions of our previous work [8, 9], to which we refer the reader for any unexplained terminology or notation. For additional standard terminology, notation, and conventions, the reader may consult Jantzen's book [14].

Except when indicated otherwise, k will denote a field of characteristic $p \geq 3$, r will denote a positive integer, and $\mathbb{N} = \{0, 1, 2, 3, \dots\}$ will denote the set of nonnegative integers. All vector spaces will be k-vector spaces, and all unadorned tensor products will denote tensor products over k. Given a k-vector space V, let $V^\#$ be its k-linear dual $\mathrm{Hom}_k(V, k)$. Set $\mathbb{Z}_2 = \mathbb{Z}/2\mathbb{Z} = \{\overline{0}, \overline{1}\}$. Following the literature, we use the prefix 'super' to indicate that an object is $\mathbb{Z}_2$-graded. We denote the decomposition of a vector superspace (superalgebra, etc.) V into its $\mathbb{Z}_2$-graded homogeneous components by $V = V_{\overline{0}} \oplus V_{\overline{1}}$, calling $V_{\overline{0}}$ the even subspace of V and $V_{\overline{1}}$ the odd subspace of V, and write $\overline{v} \in \mathbb{Z}_2$ to denote the $\mathbb{Z}_2$-degree of a homogeneous element $v \in V$. We use the symbol $\cong$ to denote even (i.e., degree-preserving) isomorphisms of superspaces, and reserve the symbol $\simeq$ for odd (i.e., degree-reversing) isomorphisms of superspaces.

For the reader's convenience we recall here the definitions of the multiparameter supergroups introduced in [9]. First, $\mathbb{M}_r$ is the affine k-supergroup scheme whose coordinate superalgebra $k[\mathbb{M}_r]$ is the commutative k-superalgebra generated by the odd element τ and the even elements θ and σ_i for $i \in \mathbb{N}$, such that $\tau^2 = 0$, $\sigma_0 = 1$, $\theta^{p^{r-1}} = \sigma_1$, and $\sigma_i \sigma_j = \binom{i+j}{i}\sigma_{i+j}$ for $i, j \in \mathbb{N}$:

$$k[\mathbb{M}_r] = k[\tau, \theta, \sigma_1, \sigma_2, \dots]/\langle \tau^2 = 0, \theta^{p^{r-1}} = \sigma_1, \text{ and } \sigma_i\sigma_j = \tbinom{i+j}{i}\sigma_{i+j} \text{ for } i, j \in \mathbb{N}\rangle.$$

If $0 \leq i < p^s$ and $i = \sum_{\ell=0}^{s-1} i_\ell p^\ell$ with $0 \leq i_\ell < p$ for each ℓ, then

$$\sigma_i = \frac{(\sigma_1)^{i_0}(\sigma_p)^{i_1} \cdots (\sigma_{p^{s-1}})^{i_{s-1}}}{(i_0!)(i_1!) \cdots (i_{s-1}!)}. \tag{1}$$

The coproduct Δ and the antipode S on $k[\mathbb{M}_r]$ are given by the formulas

$$\begin{aligned}
\Delta(\tau) &= \tau \otimes 1 + 1 \otimes \tau, & S(\tau) &= -\tau, \\
\Delta(\theta) &= \theta \otimes 1 + 1 \otimes \theta, & S(\theta) &= -\theta, \\
\Delta(\sigma_i) &= \textstyle\sum_{u+v=i} \sigma_u \otimes \sigma_v + \sum_{u+v+p=i} \sigma_u\tau \otimes \sigma_v\tau, & S(\sigma_i) &= (-1)^i\sigma_i.
\end{aligned}$$

The coproduct on $k[\mathbb{M}_r]$ induces a superalgebra structure on its k-linear dual $k\mathbb{M}_r := k[\mathbb{M}_r]^\#$. With this structure, we call $k\mathbb{M}_r$ the *group algebra* of $\mathbb{M}_r$. By [9, Proposition 3.1.4], the group algebra $k\mathbb{M}_r$ is given by

$$k\mathbb{M}_r = k[[u_0, \dots, u_{r-1}, v]]/\langle u_0^p, \dots, u_{r-2}^p, u_{r-1}^p + v^2\rangle, \tag{2}$$

where the superdegrees of elements are given by $\overline{u_i} = \overline{0}$ for each i and $\overline{v} = \overline{1}$.

Let $\mathbb{P}_r$ be the "polynomial subalgebra" of $k\mathbb{M}_r$,

$$\mathbb{P}_r = k[u_0, \dots, u_{r-1}, v]/\langle u_0^p, \dots, u_{r-2}^p, u_{r-1}^p + v^2 \rangle. \tag{3}$$

By [9, Remark 3.1.3(3)], the $\mathbb{Z}_2$-grading on $k[\mathbb{M}_r]$ lifts to a $\mathbb{Z}$-grading, which makes $k[\mathbb{M}_r]$ into a graded Hopf algebra of finite type in the sense of Milnor and Moore [16]. Then $\mathbb{P}_r$ is the graded dual of $k[\mathbb{M}_r]$, and hence $\mathbb{P}_r$ inherits by duality a Hopf superalgebra structure [16, Proposition 4.8]. The elements u_0, u_{r-1}^p, and v are each primitive with respect to the coproduct on $\mathbb{P}_r$. Then given a nonzero p-polynomial $f \in k[T]$ without constant term, and given a scalar $\eta \in k$, the element $f(u_{r-1}) + \eta \cdot u_0$ is primitive in $\mathbb{P}_r$, and hence the quotient

$$k\mathbb{M}_{r;f,\eta} := \mathbb{P}_r / \langle f(u_{r-1}) + \eta \cdot u_0 \rangle$$

is a finite-dimensional (super)cocommutative Hopf superalgebra. Now the multiparameter supergroup $\mathbb{M}_{r;f,\eta}$ is the affine k-supergroup scheme such that $k[\mathbb{M}_{r;f,\eta}]^\# = k\mathbb{M}_{r;f,\eta}$. Set $\mathbb{M}_{r;f} = \mathbb{M}_{r;f,0}$, and given an integer $s \geq 1$ set $\mathbb{M}_{r;s,\eta} = \mathbb{M}_{r;T^{p^s},\eta}$ and $\mathbb{M}_{r;s} = \mathbb{M}_{r;T^{p^s},0}$. Then $k[\mathbb{M}_{r;s}]$ is the subalgebra of $k[\mathbb{M}_r]$ generated by τ, θ, and σ_i for $1 \leq i < p^s$ (equivalently, by τ, θ, and $\sigma_1, \sigma_p, \dots, \sigma_{p^{s-1}}$).

2 Varieties of Homomorphisms

2.1 *Elementary Supergroup Schemes*

The following definition (stated using slightly different though equivalent terminology) is due to Benson, Iyengar, Krause, and Pevtsova (BIKP) [3].

Definition 2.1.1 (*Elementary supergroup schemes*) A finite supergroup scheme is *elementary* if it is isomorphic to a quotient of $\mathbb{M}_{r;s} \times (\mathbb{Z}/p\mathbb{Z})^{\times t}$ for some integers $r, s, t \geq 1$.

In the preceding definition, $\mathbb{Z}/p\mathbb{Z}$ denotes the finite supergroup scheme whose group algebra is $k(\mathbb{Z}/p\mathbb{Z})$, the ordinary (purely even) group algebra over k of the finite cyclic group $\mathbb{Z}/p\mathbb{Z}$. The next theorem was announced (using slightly different terminology) by Julia Pevtsova at the Mathematical Congress of the Americas in Montreal in July 2017, in the talk she gave in the Special Session on Cohomology of Groups. By convention, if $r = 0$ then $\mathbb{G}_{a(r)}$ is the trivial group.

Theorem 2.1.2 (BIKP) *Suppose k is perfect. Then each elementary finite k-supergroup scheme is isomorphic to one of the following (unipotent) k-supergroup schemes:*

- $\mathbb{G}_{a(r)} \times (\mathbb{Z}/p\mathbb{Z})^{\times t}$ *for some integers $r, t \geq 0$;*
- $\mathbb{G}_{a(r)} \times \mathbb{G}_a^- \times (\mathbb{Z}/p\mathbb{Z})^{\times t}$ *for some integers $r, t \geq 0$;*

- $\mathbb{M}_{r;s} \times (\mathbb{Z}/p\mathbb{Z})^{\times t}$ *for some integers* $r, s \geq 1$ *and* $t \geq 0$; *or*
- $\mathbb{M}_{r;s,\eta} \times (\mathbb{Z}/p\mathbb{Z})^{\times t}$ *for some integers* $r \geq 2$, $s \geq 1$, $t \geq 0$, *and some* $0 \neq \eta \in k$.

Corollary 2.1.3 *Suppose k is perfect. Then each infinitesimal elementary k-supergroup scheme is isomorphic to one of the following infinitesimal (unipotent) k-supergroup schemes:*

- $\mathbb{G}_{a(r)}$ *for some integer* $r \geq 0$;
- $\mathbb{G}_{a(r)} \times \mathbb{G}_a^-$ *for some integer* $r \geq 0$;
- $\mathbb{M}_{r;s}$ *for some integers* $r, s \geq 1$; *or*
- $\mathbb{M}_{r;s,\eta}$ *for some integers* $r \geq 2$, $s \geq 1$, *and some* $0 \neq \eta \in k$.

Recall that a k-supergroup scheme G is *algebraic* if $k[G]$ is finitely generated as a k-algebra.

Lemma 2.1.4 *Suppose k is perfect, and let G be an algebraic k-supergroup scheme. Then G is a quotient of $\mathbb{M}_r$ for some $r \geq 1$ if and only if G is an infinitesimal elementary supergroup scheme.*

Proof First suppose $\phi : \mathbb{M}_r \to G$ is a quotient homomorphism. By [9, Remark 3.1.3(4)], ϕ factors through the canonical quotient map $\pi_{r;s} : \mathbb{M}_r \twoheadrightarrow \mathbb{M}_{r;s}$ for some $s \geq 1$. Then G is a quotient of $\mathbb{M}_{r;s}$ for some $s \geq 1$, and hence G is an elementary supergroup scheme. Conversely, each supergroup listed in Corollary 2.1.3 is a quotient of $\mathbb{M}_r$ for some $r \geq 1$; cf. Remark 2.2.3(1) below. □

2.2 Homomorphisms Between Infinitesimal Elementary Supergroups

Recall that if G is an affine k-supergroup scheme and if $A \in \mathfrak{csalg}_k$ is a commutative k-superalgebra, then $G \otimes_k A$ denotes the affine A-supergroup scheme with coordinate superalgebra $A[G] := k[G] \otimes_k A$. Given affine supergroup schemes G and G' over k, the k-superfunctor $\mathbf{Hom}(G, G') : \mathfrak{csalg}_k \to \mathfrak{sets}$ is defined by

$$\mathbf{Hom}(G, G')(A) = \mathrm{Hom}_{Grp/A}(G \otimes_k A, G' \otimes_k A),$$

the set of A-supergroup scheme homomorphisms $\rho : G \otimes_k A \to G' \otimes_k A$. In [9, Lemma 3.3.2] we calculated $\mathbf{Hom}(G, G')(A)$ in the special case $G = G' = \mathbb{M}_{r;s}$ under the assumption that A is purely even and reduced. In this section we extend those calculations to the case when G' is an arbitrary height-r infinitesimal elementary supergroup scheme and $G = \mathbb{M}_r$.

Let $s \geq 1$ and let $\eta \in k$. As observed in [9, Lemma 3.1.7], the group algebras $k\mathbb{M}_{r;s,\eta}$ and $k\mathbb{M}_{r;s}$ are isomorphic as k-supercoalgebras, and hence $k[\mathbb{M}_{r;s,\eta}]$ and $k[\mathbb{M}_{r;s}]$ are isomorphic as k-superalgebras. Under this identification, the coproduct on $k[\mathbb{M}_{1;s,\eta}]$ is identified in [9, Lemma 3.1.9]. More generally, for $r \geq 2$ one has:

Lemma 2.2.1 *Let $r \geq 2$, let $s \geq 1$, and let $\eta \in k$. Then identifying $k[\mathbb{M}_{r;s,\eta}]$ and $k[\mathbb{M}_{r;s}]$ as k-superalgebras, the coproduct on $k[\mathbb{M}_{r;s,\eta}]$ satisfies the formulas*

$$\begin{aligned}
&\Delta(\tau) = \tau \otimes 1 + 1 \otimes \tau, \\
&\Delta(\sigma_\ell) = \textstyle\sum_{i+j=\ell} \sigma_i \otimes \sigma_j + \sum_{i+j+p=\ell} \sigma_i \tau \otimes \sigma_j \tau, \quad \text{and} \\
&\Delta(\theta) = \theta \otimes 1 + 1 \otimes \theta - \eta \cdot \left(\textstyle\sum_{i=1}^{p^s-1} \sigma_i \otimes \sigma_{p^s-i} + \sum_{i+j+p=p^s} \sigma_i \tau \otimes \sigma_j \tau \right).
\end{aligned}$$

Proof The stated formulas can be deduced via duality from the algebra structure of the group algebra $k\mathbb{M}_{r;s,\eta}$ in the same manner as the proof of [9, Lemma 3.1.9]. □

Portions of the next proposition are contained already in [9, Lemma 3.3.2]. Recall from [9, Sect. 3.1] that $k[\mathbb{G}_{a(r)}] = k[\theta]/\langle \theta^{p^r} \rangle$ and $k[\mathbb{G}_a^-] = \Lambda(\tau)$. Also recall for $t \geq 1$ that the canonical quotient maps $\mathbb{M}_r \twoheadrightarrow \mathbb{M}_{r;t} \twoheadrightarrow \mathbb{G}_{a(r)}$ are defined via the subalgebra inclusions $k[\mathbb{G}_{a(r)}] \hookrightarrow k[\mathbb{M}_{r;t}] \hookrightarrow k[\mathbb{M}_r]$, and similarly for the other quotient maps referenced in the next proposition.

Proposition 2.2.2 *Let $A = A_{\overline{0}} \in \mathfrak{calg}_k$ be a purely even commutative k-algebra.*

1. *Let $t \geq 1$. The canonical quotient maps $\mathbb{M}_r \twoheadrightarrow \mathbb{M}_{r;t} \twoheadrightarrow \mathbb{M}_{r;1}$ induce identifications*

$$\begin{aligned}
\mathbf{Hom}(\mathbb{M}_{r;1}, \mathbb{M}_{r;1})(A) &= \mathbf{Hom}(\mathbb{M}_{r;t}, \mathbb{M}_{r;1})(A) = \mathbf{Hom}(\mathbb{M}_r, \mathbb{M}_{r;1})(A) \\
&= \left\{ (\mu, a_0, \ldots, a_{r-1}) \in A^{r+1} \right\}.
\end{aligned}$$

   Given $(\mu, a_0, \ldots, a_{r-1}) \in A^{r+1}$, the corresponding comorphism $\phi : A[\mathbb{M}_{r;1}] \to A[\mathbb{M}_r]$ is determined by the formulas $\phi(\tau) = \mu \cdot \tau$ and $\phi(\theta) = \sum_{i=0}^{r-1} a_i \cdot \theta^{p^i}$.
2. *Let $2 \leq s \leq t$, and suppose A is reduced. The canonical quotient maps $\mathbb{M}_r \twoheadrightarrow \mathbb{M}_{r;t} \twoheadrightarrow \mathbb{M}_{r;s}$ induce identifications*

$$\begin{aligned}
\mathbf{Hom}(\mathbb{M}_{r;s}, \mathbb{M}_{r;s})(A) &= \mathbf{Hom}(\mathbb{M}_{r;t}, \mathbb{M}_{r;s})(A) = \mathbf{Hom}(\mathbb{M}_r, \mathbb{M}_{r;s})(A) \\
&= \left\{ (\mu, a_0, \ldots, a_{r-1}, b_s) \in A^{r+2} : \mu^2 = a_0^{p^r} \right\}.
\end{aligned}$$

   Given $(\mu, a_0, \ldots, a_{r-1}, b_s) \in A^{r+2}$ such that $\mu^2 = a_0^{p^r}$, the corresponding comorphism $\phi : A[\mathbb{M}_{r;s}] \to A[\mathbb{M}_r]$ is determined[1] *by $\phi(\tau) = \mu \cdot \tau$, $\phi(\theta) = \sum_{i=0}^{r-1} a_i \cdot \theta^{p^i}$, $\phi(\sigma_i) = a_0^{ip^{r-1}} \cdot \sigma_i$ for $1 \leq i < p^{s-1}$, and $\phi(\sigma_{p^{s-1}}) = a_0^{p^{r+s-2}} \cdot \sigma_{p^{s-1}} + b_s \cdot \sigma_1$.*
3. *Suppose $r \geq 2$, let $1 \leq s < t$, and let $0 \neq \eta \in k$. Suppose A is reduced. Then the canonical quotient maps $\mathbb{M}_r \twoheadrightarrow \mathbb{M}_{r;t} \twoheadrightarrow \mathbb{M}_{r;s+1}$ induce identifications*

$$\begin{aligned}
\mathbf{Hom}(\mathbb{M}_{r;s+1}, \mathbb{M}_{r;s,\eta})(A) &= \mathbf{Hom}(\mathbb{M}_{r;t}, \mathbb{M}_{r;s,\eta})(A) = \mathbf{Hom}(\mathbb{M}_r, \mathbb{M}_{r;s,\eta})(A) \\
&= \left\{ (\mu, a_0, \ldots, a_{r-1}) \in A^{r+1} : \mu^2 = a_0^{p^r} \right\}
\end{aligned}$$

[1] cf. eqn. (1).

Given $(\mu, a_0, \ldots, a_{r-1}) \in A^{r+1}$ such that $\mu^2 = a_0^{p^r}$, the corresponding comorphism $\phi : A[\mathbb{M}_{r;s,\eta}] \to A[\mathbb{M}_r]$ is determined by the formulas $\phi(\tau) = \mu \cdot \tau$,

$$\phi(\theta) = (\sum_{i=0}^{r-1} a_i \cdot \theta^{p^i}) - \eta \cdot (a_0^{p^{r+s-1}} \sigma_{p^s}),$$

and $\phi(\sigma_i) = a_0^{ip^{r-1}} \cdot \sigma_i$ for $0 \leq i < p^s$.

4. *Let $t \geq 1$. The canonical quotient maps $\mathbb{M}_r \twoheadrightarrow \mathbb{M}_{r;t} \twoheadrightarrow \mathbb{G}_{a(r)}$ induce identifications*

$$\mathbf{Hom}(\mathbb{G}_{a(r)}, \mathbb{G}_{a(r)})(A) = \mathbf{Hom}(\mathbb{M}_{r;t}, \mathbb{G}_{a(r)})(A) = \mathbf{Hom}(\mathbb{M}_r, \mathbb{G}_{a(r)})(A) = A^r$$

Given $(a_0, \ldots, a_{r-1}) \in A^r$, the corresponding comorphism $\phi : A[\mathbb{G}_{a(r)}] \to A[\mathbb{M}_r]$ is determined by the formula $\phi(\theta) = \sum_{i=0}^{r-1} a_i \cdot \theta^{p^i}$.

5. *Let $t \geq 1$. The canonical quotient maps $\mathbb{M}_r \twoheadrightarrow \mathbb{M}_{r;t} \twoheadrightarrow \mathbb{G}_a^-$ induce identifications*

$$\mathbf{Hom}(\mathbb{G}_a^-, \mathbb{G}_a^-)(A) = \mathbf{Hom}(\mathbb{M}_{r;t}, \mathbb{G}_a^-)(A) = \mathbf{Hom}(\mathbb{M}_r, \mathbb{G}_a^-)(A) = A$$

Given $\mu \in A$, the corresponding comorphism $\phi : A[\mathbb{G}_a^-] \to A[\mathbb{M}_r]$ is determined by $\phi(\tau) = \mu \cdot \tau$.

Proof Recall that specifying a homomorphism of A-supergroup schemes $\rho : G \to G'$ is equivalent to specifying a homomorphism of A-Hopf superalgebras $\rho^* : A[G'] \to A[G]$. The proof strategy for each of the statements is to first identify the set of all Hopf superalgebra homomorphisms from $A[\mathbb{M}_{r;1}]$ (resp. from $A[\mathbb{M}_{r;s}]$, $A[\mathbb{M}_{r;s,\eta}]$, $A[\mathbb{G}_{a(r)}]$, $A[\mathbb{G}_a^-]$) into $A[\mathbb{M}_r]$. Using the explicit identifications, one can then observe that the image of each Hopf superalgebra homomorphism is contained in the Hopf subalgebra $A[\mathbb{M}_{r;1}]$ (resp. $A[\mathbb{M}_{r;s}]$, $A[\mathbb{M}_{r;s+1}]$, $A[\mathbb{G}_{a(r)}]$, $A[\mathbb{G}_a^-]$) of $A[\mathbb{M}_r]$, and from that observation the first two equalities in each statement follow immediately. The identification of all homomorphisms $\phi : A[\mathbb{M}_{r;1}] \to A[\mathbb{M}_r]$ follows from a word-for-word repetition of the argument in the first paragraph of the proof of [9, Lemma 3.3.2]. The proofs of statements (4) and (5) also follow via reasoning entirely similar to that in the proof of [9, Lemma 3.3.2], so for the remainder of this proof we focus on statements (2) and (3). We first prove (3) and then indicate how the argument should be modified for (2).

Suppose $r \geq 2$, let $s \geq 1$, and let $0 \neq \eta \in k$. Let $\phi : A[\mathbb{M}_{r;s,\eta}] \to A[\mathbb{M}_r]$ be a homomorphism of A-Hopf superalgebras, identifying the Hopf structure on $A[\mathbb{M}_{r;s,\eta}]$ as in Lemma 2.2.1. Our first goal is to describe the action of ϕ on the elements $\tau, \sigma_1, \sigma_p, \ldots, \sigma_{p^{s-1}} \in A[\mathbb{M}_{r;s,\eta}]$. Since ϕ is by definition an even map, it must map τ to an odd primitive element in $A[\mathbb{M}_r]$. Since A is purely even by assumption, this implies that $\phi(\tau) = \mu \cdot \tau$ for some $\mu \in A$. Next, by the assumption that $r \geq 2$, the element $\sigma_1 \in A[\mathbb{M}_{r;s,\eta}]$ is primitive and $(\sigma_1)^p = 0$.[2] Then $\phi(\sigma_1)$ must

[2] If $\eta \neq 0$ and $r = 1$, then σ_1 need not be primitive in $A[\mathbb{M}_{r;s,\eta}]$; cf. [9, Lemma 3.1.9].

be a primitive element in $A[\mathbb{M}_r]$ such that $\phi(\sigma_1)^p = 0$, so $\phi(\sigma_1) = a \cdot \sigma_1$ for some $a \in A$. Since $(\sigma_1)^n = n! \cdot \sigma_n$, this further implies for $0 \leq i < p$ that $\phi(\sigma_i) = a^i \cdot \sigma_i$.

Our goal is to show for $0 \leq i < p^{s-1}$ that $\phi(\sigma_i) = a^i \cdot \sigma_i$, and that $\phi(\sigma_{p^{s-1}}) = a^{p^{s-1}} \cdot \sigma_{p^{s-1}} + b_s \cdot \sigma_1$ for some $b_s \in A$. Taking $b_1 = 0$, the claim is true by the results of the preceding paragraph in the case $s = 1$, *so assume for the rest of this paragraph and the next that* $s \geq 2$. First consider $\phi(\sigma_p)$. Since ϕ is a Hopf superalgebra homomorphism, then

$$\begin{aligned}\Delta \circ \phi(\sigma_p) &= (\phi \otimes \phi) \circ \Delta(\sigma_p) \\ &= \phi(\sigma_p) \otimes 1 + 1 \otimes \phi(\sigma_p) + \left(\textstyle\sum_{i=1}^{p-1} \phi(\sigma_i) \otimes \phi(\sigma_{p-i})\right) + \phi(\tau) \otimes \phi(\tau) \\ &= \phi(\sigma_p) \otimes 1 + 1 \otimes \phi(\sigma_p) + a^p \cdot \left(\textstyle\sum_{i=1}^{p-1} \sigma_i \otimes \sigma_{p-i}\right) + \mu^2 \cdot (\tau \otimes \tau).\end{aligned} \tag{4}$$

By [9, Remark 3.1.3(3)], $A[\mathbb{M}_r]$ is a $\mathbb{Z}$-graded Hopf superalgebra with $\deg(\theta) = 2$, $\deg(\sigma_j) = 2jp^{r-1}$, and $\deg(\tau) = p^r$. In particular, the degree-$2p^r$ component of $A[\mathbb{M}_r]$ is spanned by σ_p. Since the coproduct on a Hopf (super)algebra is automatically injective, (4) then implies that the graded components of $\phi(\sigma_p)$ of $\mathbb{Z}$-degrees other than $2p^r$ must be primitive. Denoting the degree-$2p^r$ component of $\phi(\sigma_p)$ by $c \cdot \sigma_p$, (4) says that

$$a^p \cdot \left(\textstyle\sum_{i=1}^{p-1} \sigma_i \otimes \sigma_{p-i}\right) + \mu^2 \cdot (\tau \otimes \tau) = \Delta(c \cdot \sigma_p) = c \cdot \left(\textstyle\sum_{i=1}^{p-1} \sigma_i \otimes \sigma_{p-i} + \tau \otimes \tau\right),$$

and hence $a^p = c = \mu^2$. The primitive components of $\phi(\sigma_p)$ must be p-nilpotent because $(\sigma_p)^p = 0$. By the assumption that A is reduced, the only p-nilpotent primitive elements in $A[\mathbb{M}_r]$ are scalar multiples of σ_1. Then $\phi(\sigma_p) = a^p \cdot \sigma_p + c_1 \cdot \sigma_1$ for some $c_1 \in A$.

Now let $1 \leq n < s-1$. Suppose by way of induction that $\phi(\sigma_i) = a^i \cdot \sigma_i$ for $0 \leq i < p^n$, and that $\phi(\sigma_{p^n}) = a^{p^n} \cdot \sigma_{p^n} + c_n \cdot \sigma_1$ for some $c_n \in A$. We will then show that $c_n = 0$, that $\phi(\sigma_i) = a^i \cdot \sigma_i$ for $0 \leq i < p^{n+1}$, and that $\phi(\sigma_{p^{n+1}}) = a^{p^{n+1}} \cdot \sigma_{p^{n+1}} + c_{n+1} \cdot \sigma_1$ for some $c_{n+1} \in A$; by induction, this will complete the goal laid out at the beginning of the preceding paragraph. From the divided power relation $\sigma_i \sigma_j = \binom{i+j}{i} \sigma_{i+j}$ in $A[\mathbb{M}_r]$ and Lucas's Theorem for binomial coefficients mod p, it follows that if $0 \leq i < p^n$ and $0 \leq j < p$, then $\sigma_{i+jp^n} = \sigma_i \cdot \sigma_{jp^n} = \frac{1}{j!} \cdot \sigma_i \cdot (\sigma_{p^n})^j$, and hence

$$\begin{aligned}\phi(\sigma_{i+jp^n}) &= \frac{1}{j!}(a^i \cdot \sigma_i) \left(\sum_{\ell=0}^{j} \binom{j}{\ell} a^{\ell p^n} c_n^{j-\ell} \cdot (\sigma_{p^n})^{\ell} (\sigma_1)^{j-\ell} \right) \\ &= \sum_{\ell=0}^{j} a^{i+\ell p^n} c_n^{j-\ell} \cdot \sigma_{i+\ell p^n} \cdot \sigma_{j-\ell}.\end{aligned} \tag{5}$$

Since ϕ is a Hopf superalgebra homomorphism, then

$$\begin{aligned} \Delta \circ \phi(\sigma_{p^{n+1}}) &= (\phi \otimes \phi) \circ \Delta(\sigma_{p^{n+1}}) \\ &= \phi(\sigma_{p^{n+1}}) \otimes 1 + 1 \otimes \phi(\sigma_{p^{n+1}}) \\ &+ \sum_{i=1}^{p^{n+1}-1} \phi(\sigma_i) \otimes \phi(\sigma_{p^{n+1}-i}) + \sum_{i+j+p=p^{n+1}} \mu^2 \cdot \phi(\sigma_i)\tau \otimes \phi(\sigma_j)\tau. \end{aligned} \tag{6}$$

Using (5), and considering the $\mathbb{Z}$-grading on $A[\mathbb{M}_r]$, one can check that the homogeneous component of lowest $\mathbb{Z}$-degree in the sum $\sum_{i=1}^{p^{n+1}-1} \phi(\sigma_i) \otimes \phi(\sigma_{p^{n+1}-1})$ is

$$(c_n)^p \cdot \left(\sigma_1 \otimes \sigma_{p-1} + \sigma_2 \otimes \sigma_{p-2} + \cdots + \sigma_{p-1} \otimes \sigma_1\right).$$

By the injectivity of the coproduct, this implies that the degree-$2p^r$ component of $\phi(\sigma_{p^{n+1}})$ must be $(c_n)^p \cdot \sigma_p$. On the other hand, one can check that no scalar multiple of $\tau \otimes \tau$ occurs in the expansion of the sum $\sum_{i+j+p=p^{n+1}} \phi(\sigma_i)\tau \otimes \phi(\sigma_j)\tau$. This implies that the degree-$2p^r$ component of $\phi(\sigma_{p^{n+1}})$ must be 0, and hence because A is reduced implies that $c_n = 0$. So $\phi(\sigma_{p^n})$ is simply equal to $a^{p^n} \cdot \sigma_{p^n}$, and it follows from the algebra relations among $\sigma_1, \sigma_p, \ldots, \sigma_{p^n}$ (cf. [9, (3.1.1)]) that $\phi(\sigma_i) = a^i \cdot \sigma_i$ for $0 \leq i < p^{n+1}$. Now, considering (6) and arguing as in the previous paragraph for $\phi(\sigma_p)$, it follows that $\phi(\sigma_{p^{n+1}}) = a^{p^{n+1}} \cdot \sigma_{p^{n+1}} + c_{n+1} \cdot \sigma_1$ for some $c_{n+1} \in A$.

The inductive argument of the previous two paragraphs shows that $\phi(\sigma_i) = a^i \cdot \sigma_i$ for $0 \leq i < p^{s-1}$, and that $\phi(\sigma_{p^{s-1}}) = a^{p^{s-1}} \cdot \sigma_{p^{s-1}} + b_s \cdot \sigma_1$ for some $b_s \in A$, with $b_1 = 0$. To finish describing the Hopf superalgebra homomorphism $\phi : A[\mathbb{M}_{r;s,\eta}] \to A[\mathbb{M}_r]$, it remains to describe the action of ϕ on the generator $\theta \in A[\mathbb{M}_{r;s,\eta}]$. From the coproduct formula of Lemma 2.2.1 one gets

$$\begin{aligned} \Delta \circ \phi(\theta) &= (\phi \otimes \phi) \circ \Delta(\theta) \\ &= \phi(\theta) \otimes 1 + 1 \otimes \phi(\theta) \\ &- \eta \cdot \left(\sum_{i=1}^{p^s-1} \phi(\sigma_i) \otimes \phi(\sigma_{p^s-i}) + \sum_{i+j+p=p^s} \mu^2 \cdot \phi(\sigma_i)\tau \otimes \phi(\sigma_j)\tau \right). \end{aligned} \tag{7}$$

Applying (5), and considering the $\mathbb{Z}$-grading on $A[\mathbb{M}_r]$, it follows as in the third paragraph of the proof that the $\mathbb{Z}$-graded components of $\phi(\theta)$ of degrees less than $2p^r$ must be primitive, and hence must be an A-linear combination of the elements $\theta, \theta^p, \ldots, \theta^{p^{r-1}}$. One can further check, as in the previous paragraph, that the component in $\sum_{i=1}^{p^s-1} \phi(\sigma_i) \otimes \phi(\sigma_{p^s-i})$ of lowest $\mathbb{Z}$-degree is $(b_s)^p \cdot (\sum_{i=1}^{p-1} \sigma_i \otimes \sigma_{p-i})$, but that no scalar multiple of $\tau \otimes \tau$ occurs in $\sum_{i+j+p=p^s} \phi(\sigma_i)\tau \otimes \phi(\sigma_j)\tau$. Since $\eta \neq 0$ and since A is reduced, this implies as in the previous paragraph that $b_s = 0$. Then $\phi(\sigma_{p^{s-1}})$ is simply equal to $a^{p^{s-1}} \cdot \sigma_{p^{s-1}}$, so it follows for all $0 \leq i < p^s$ that $\phi(\sigma_i) = a^i \cdot \sigma_i$.

Now if $s \geq 2$, then $\mu^2 = a^p$ by the second paragraph of the proof, so we get

$$\Delta \circ \phi(\theta) =$$

$$\phi(\theta) \otimes 1 + 1 \otimes \phi(\theta) - (\eta \cdot a^{p^s}) \cdot \left(\sum_{i=1}^{p^s-1} \sigma_i \otimes \sigma_{p^s-i} + \sum_{i+j+p=p^s} \sigma_i \tau \otimes \sigma_j \tau \right),$$

and hence $\phi(\theta) = (\sum_{i=0}^{r-1} a_i \cdot \theta^{p^i}) - (\eta \cdot a^{p^s}) \cdot \sigma_{p^s}$ for some $a_0, a_1, \ldots, a_{r-1} \in A$ by the injectivity of the coproduct. But

$$a \cdot \sigma_1 = \phi(\sigma_1) = \phi(\theta^{p^{r-1}}) = \phi(\theta)^{p^{r-1}} = a_0^{p^{r-1}} \cdot \theta^{p^{r-1}} = a_0^{p^{r-1}} \cdot \sigma_1, \tag{8}$$

so $a = a_0^{p^{r-1}}$. Thus for $s \geq 2$, the homomorphism ϕ is completely determined by the elements $\mu, a_0, a_1, \ldots, a_{r-1} \in A$, and these elements satisfy the additional condition that $\mu^2 = a^p = a_0^{p^r}$. On the other hand, if $s = 1$ then we know from the second paragraph of the proof that $\phi(\sigma_i) = a^i \cdot \sigma_i$ for $0 \leq i < p$, and hence

$$\Delta \circ \phi(\theta) = \phi(\theta) \otimes 1 + 1 \otimes \phi(\theta) - \eta \cdot \left(\sum_{i=1}^{p-1} a^p \cdot (\sigma_i \otimes \sigma_{p-i}) + \mu^2 \cdot (\tau \otimes \tau) \right). \tag{9}$$

Now arguing as we did for $\phi(\sigma_p)$ in the third paragraph of the proof, and using the fact that $\eta \neq 0$, it follows that $a^p = \mu^2$ and $\phi(\theta) = (\sum_{i=0}^{r-1} a_i \cdot \theta^{p^i}) - (\eta \cdot a^p) \cdot \sigma_p$ for some scalars $a_0, \ldots, a_{r-1} \in A$. Then $a_0^{p^{r-1}} = a$ by (8), and hence $\mu^2 = a^p = a_0^{p^r}$. Thus for all $s \geq 1$, the Hopf superalgebra homomorphism $\phi : A[\mathbb{M}_{r;s,\eta}] \to A[\mathbb{M}_r]$ is completely specified by the elements $\mu, a_0, \ldots, a_{r-1} \in A$, which satisfy the additional condition that $\mu^2 = a_0^{p^r}$. Conversely, given any elements $\mu, a_0, \ldots, a_{r-1} \in A$ such that $\mu^2 = a_0^{p^r}$, the assignments

$$\tau \mapsto \mu \cdot \tau, \quad \theta \mapsto (\sum_{i=0}^{r-1} a_i \cdot \theta^{p^i}) - (\eta \cdot a_0^{p^{r+s-1}}) \cdot \sigma_{p^s}, \quad \text{and} \quad \sigma_i \mapsto a_0^{ip^{r-1}} \cdot \sigma_i$$

extend to a homomorphism of Hopf superalgebras $\phi : A[\mathbb{M}_{r;s,\eta}] \to A[\mathbb{M}_r]$. This completes the identification of the set $\mathbf{Hom}(\mathbb{M}_r, \mathbb{M}_{r;s,\eta})(A)$.

For the proof of (2), one first proceeds as in the second, third, and fourth paragraphs to show that $\phi(\tau) = \mu \cdot \tau$, $\phi(\sigma_i) = a^i \cdot \sigma_i$ for $0 \leq i < p^{s-1}$, and $\phi(\sigma_{p^{s-1}}) = a^{p^{s-1}} \cdot \sigma_{p^{s-1}} + b_s \cdot \sigma_1$, for some elements $\mu, a, b_s \in A$ with $\mu^2 = a^p$. Then one argues as in the fifth and sixth paragraphs (with $\eta = 0$) to show that $\phi(\theta) = \sum_{i=0}^{r-1} a_i \cdot \theta^{p^i}$ for some $a_0, \ldots, a_{r-1} \in A$ with $a_0^{p^{r-1}} = a$. So $\phi : A[\mathbb{M}_{r;s}] \to A[\mathbb{M}_r]$ is completely specified by the elements $\mu, a_0, \ldots, a_{r-1}, b_s \in A$, which satisfy the additional condition that $\mu^2 = a_0^{p^r}$. Conversely, given $s \geq 2$ and given any choice of elements $\mu, a_0, \ldots, a_{r-1}, b_s \in A$ such that $\mu^2 = a_0^{p^r}$, one can check that the assignments

$$\tau \mapsto \mu \cdot \tau, \qquad \sigma_i \mapsto a_0^{ip^{r-1}} \cdot \sigma_i \quad \text{for } 0 \le i < p^{s-1},$$
$$\theta \mapsto a_0 \cdot \theta + a_1 \cdot \theta^p + \cdots + a_{r-1} \cdot \theta^{p^{r-1}}, \quad \sigma_{p^{s-1}} \mapsto a_0^{p^{r+s-2}} \cdot \sigma_{p^{s-1}} + b_s \cdot \sigma_1,$$

uniquely specify a Hopf superalgebra homomorphism $\phi : A[\mathbb{M}_{r;s}] \to A[\mathbb{M}_r]$. □

Remark 2.2.3 1. Let $s \ge 1$. The canonical quotient homomorphisms $\pi_{r;s} : \mathbb{M}_r \to \mathbb{M}_{r;s}$, $q : \mathbb{M}_r \to \mathbb{G}_{a(r)}$, and $q^- : \mathbb{M}_r \to \mathbb{G}_a^-$ are labeled by $(1, 1, 0, \dots, 0)$, $(1, 0, \dots, 0)$, and $\mu = 1$, respectively. If $r \ge 2$ and $\eta \in k$, the canonical quotient map $\pi_{r;s,\eta} : \mathbb{M}_r \twoheadrightarrow \mathbb{M}_{r;s,\eta}$ is labeled by $(1, 1, 0, \dots, 0)$.

2. Suppose $s \ge 2$, and let $\rho = \rho_{(0,0,1)} : \mathbb{M}_1 \to \mathbb{M}_{1;s}$ be the homomorphism labeled by the tuple $(0, 0, 1) \in k^3$. The comorphism $\rho^* : k[\mathbb{M}_{1;s}] \to k[\mathbb{M}_1]$ satisfies $\rho^*(\tau) = 0$, $\rho^*(\sigma_i) = 0$ if i is not divisible by p^{s-1}, and $\rho^*(\sigma_{jp^{s-1}}) = \sigma_j$ for $0 \le j < p$. In particular, ρ^* factors through the subalgebra $k[\sigma_1]/\langle \sigma_1^p \rangle \cong k[\mathbb{G}_{a(1)}]$ of $k[\mathbb{M}_1]$ (note that $\theta = \theta^{p^{r-1}} = \sigma_1$ in $k[\mathbb{M}_1]$), so

$$\rho^* = q^* \circ \iota^* : k[\mathbb{M}_{1;s}] \twoheadrightarrow k[\mathbb{G}_{a(1)}] \hookrightarrow k[\mathbb{M}_1],$$
$$\text{and} \quad \rho = \iota \circ q : \mathbb{M}_1 \twoheadrightarrow \mathbb{G}_{a(1)} \hookrightarrow \mathbb{M}_{1;s},$$

where $\iota^* : k[\mathbb{M}_{1;s}] \twoheadrightarrow k[\mathbb{G}_{a(1)}]$ is the Hopf superalgebra homomorphism obtained by restricting the codomain of ρ^*, and $\iota : \mathbb{G}_{a(1)} \hookrightarrow \mathbb{M}_{1;s}$ is the corresponding map of supergroup schemes. The map ι is a closed embedding with $k[\mathrm{im}(\iota)] = k[\mathbb{M}_{1;s}]/\langle \tau, \sigma_1, \dots, \sigma_{p^{s-1}-1} \rangle$. Since τ is an element of the defining ideal of $\mathrm{im}(\iota)$, it follows (e.g., using the formula [9, (3.1.2)] for the group multiplication in $\mathbb{M}_{1;s}$) that $\mathrm{im}(\iota)$ is central and hence normal in $\mathbb{M}_{1;s}$, and that $\mathbb{M}_{1;s}/\mathrm{im}(\iota) \cong \mathbb{M}_{1;s-1}$.

3. Recall that $\mathbb{M}_{1;1} = \mathbb{G}_{a(1)} \times \mathbb{G}_a^-$. Then in a similar fashion to the previous item, the maps $\phi_{(0,1)} : \mathbb{M}_1 \to \mathbb{M}_{1;1}$ and $\phi_{(1,0)} : \mathbb{M}_1 \to \mathbb{M}_{1;1}$ labeled by $(0, 1)$ and $(1, 0)$, respectively, admit factorizations $\phi_{(0,1)} : \mathbb{M}_1 \twoheadrightarrow \mathbb{G}_{a(1)} \hookrightarrow \mathbb{M}_{1;1}$ and $\phi_{(1,0)} : \mathbb{M}_1 \twoheadrightarrow \mathbb{G}_a^- \hookrightarrow \mathbb{M}_{1;1}$, in which the first arrow in each composition is the canonical quotient map, and the second arrow is the obvious map arising from the direct product decomposition $\mathbb{M}_{1;1} = \mathbb{G}_{a(1)} \times \mathbb{G}_a^-$.

Corollary 2.2.4 *Let $A = A_{\overline{0}} \in \mathfrak{calg}_k$ be a purely even commutative k-algebra, and suppose A is reduced. Then there exists a natural identification*

$$\mathbf{Hom}(\mathbb{M}_r, \mathbb{M}_r)(A) = \left\{ (\mu, a_0, \dots, a_{r-1}) \in A^{r+1} : \mu^2 = a_0^{p^r} \right\}.$$

Under this identification, $(\mu, \underline{\alpha}) := (\mu, a_0, \dots, a_{r-1})$ corresponds to the supergroup homomorphism whose comorphism $\phi = \phi_{(\mu,\underline{\alpha})} : A[\mathbb{M}_r] \to A[\mathbb{M}_r]$ satisfies

$$\phi(\tau) = \mu \cdot \tau, \quad \phi(\theta) = \sum_{i=0}^{r-1} a_i \cdot \theta^{p^i}, \quad \textit{and} \quad \phi(\sigma_i) = a_0^{ip^{r-1}} \cdot \sigma_i \quad \textit{for } i \ge 0.$$

Proof As in the preceding proof, we first classify all Hopf superalgebra maps $\phi : A[\mathbb{M}_r] \to A[\mathbb{M}_r]$. Let ϕ be such a map. Then ϕ restricts for each $s \ge 1$ to a homomorphism $\phi_s : A[\mathbb{M}_{r;s}] \to A[\mathbb{M}_r]$. Since the restriction of ϕ_{s+1} to $A[\mathbb{M}_{r;s}]$ must be the same as ϕ_s, it follows that $\phi = \phi_{(\mu,\underline{\alpha})}$ for some tuple $(\mu, a_0, \ldots, a_{r-1}) \in A^{r+1}$ such that $\mu^2 = a_0^{p^r}$. Conversely, given any tuple $(\mu, \underline{\alpha}) \in A^{r+1}$ such that $\mu^2 = a_0^{p^r}$, one can check that the formulas

$$\phi(\tau) = \mu \cdot \tau, \quad \phi(\theta) = \textstyle\sum_{i=0}^{r-1} a_i \cdot \theta^{p^i}, \quad \text{and} \quad \phi(\sigma_i) = a_0^{ip^{r-1}} \cdot \sigma_i \quad \text{for } i \geq 0$$

uniquely specify a map of A-Hopf superalgebras $\phi_{(\mu,\underline{\alpha})} : A[\mathbb{M}_r] \to A[\mathbb{M}_r]$. □

Let $\mathbf{Aut}(G) \subset \mathbf{Hom}(G, G)$ be the subfunctor of automorphisms.

Lemma 2.2.5 *Under the identifications of Proposition 2.2.2 and Corollary 2.2.4,*

$$\begin{aligned}
\mathbf{Aut}(\mathbb{M}_{r;1})(k) &= \left\{(\mu, a_0, \ldots, a_{r-1}) \in k^{r+1} : \mu \neq 0 \textit{ and } a_0 \neq 0\right\}, \\
\mathbf{Aut}(\mathbb{M}_{r;s})(k) &= \left\{(\mu, a_0, \ldots, a_{r-1}, b_s) \in k^{r+2} : \mu^2 = a_0^{p^r} \neq 0\right\} \quad \textit{if } s \geq 2, \\
\mathbf{Aut}(\mathbb{G}_{a(r)})(k) &= \left\{(a_0, \ldots, a_{r-1}) \in k^r : a_0 \neq 0\right\}, \\
\mathbf{Aut}(\mathbb{G}_a^-)(k) &= k - \{0\}, \quad \textit{and} \\
\mathbf{Aut}(\mathbb{M}_r)(k) &= \left\{(\mu, a_0, \ldots, a_{r-1}) \in k^{r+1} : \mu^2 = a_0^{p^r} \neq 0\right\}.
\end{aligned}$$

Proof We sketch the details for the calculation of $\mathbf{Aut}(\mathbb{M}_{r;s})(k)$; the other calculations are similar and/or easier. First, the condition $\mu^2 = a_0^{p^r} \neq 0$ is evidently necessary in order for the corresponding homomorphism to be invertible, since otherwise the comorphism $\phi : k[\mathbb{M}_{r;s}] \to k[\mathbb{M}_{r;s}]$ has a nontrivial kernel. For the reverse set inclusion, let $(\mu, a_0, \ldots, a_{r-1}, b_s) \in k^{r+2}$ such that $\mu^2 = a_0^{p^r} \neq 0$, and let $\phi : k[\mathbb{M}_{r;s}] \to k[\mathbb{M}_{r;s}]$ be the corresponding comorphism. Since $k[\mathbb{M}_{r;s}]$ is a finite-dimensional k-vector space, to show that ϕ is invertible it suffices to show that ϕ admits a left inverse. For $1 \leq i \leq r-1$, let $\phi_i : k[\mathbb{M}_{r;s}] \to k[\mathbb{M}_{r;s}]$ be the homomorphism labeled by $(1, 1, 0, \ldots, -a_0^{-1}a_i, 0, \ldots, 0) \in k^{r+2}$, where $-a_0^{-1}a_i$ appears in the $(i+1)$-st coordinate. Now by induction on i, one can check that

$$\phi_i \circ \phi_{i-1} \circ \cdots \circ \phi_1 \circ \phi : k[\mathbb{M}_{r;s}] \to k[\mathbb{M}_{r;s}]$$

is labeled by $(\mu, a_0, 0, \ldots, 0, a'_{i+1}, \ldots, a'_{r-1}, b'_s)$ for some $a'_{i+1}, \ldots, a'_{r-1}, b' \in k$. In particular, $\phi_{r-1} \circ \cdots \circ \phi_1 \circ \phi$ is labeled by $(\mu, a_0, 0, \ldots, 0, b')$ for some $b' \in k$. Next let $\phi_0 : k[\mathbb{M}_{r;s}] \to k[\mathbb{M}_{r;s}]$ be the map labeled by $(\mu^{-1}, a_0^{-1}, 0, \ldots, 0)$. Then one can check that $\phi_0 \circ \phi_{r-1} \circ \cdots \circ \phi_1 \circ \phi$ is labeled by $(1, 1, 0, \ldots, 0, b'a_0^{-p^{r-1}})$. Finally, let $\psi : k[\mathbb{M}_{r;s}] \to k[\mathbb{M}_{r;s}]$ be the map labeled by $(1, 1, 0, \ldots, 0, -b'a_0^{-p^{r-1}})$. Then $\psi \circ \phi_0 \circ \phi_{r-1} \circ \cdots \circ \phi_1 \circ \phi$ is labeled by $(1, 1, 0, \ldots, 0)$, and hence is the identity map. Thus, $\psi \circ \phi_0 \circ \phi_{r-1} \circ \cdots \circ \phi_1$ is a left inverse for ϕ. □

2.3 Commuting Nilpotent Supermatrices

Given nonnegative integers m and n, the k-superfunctor $\mathbf{V}_r(GL_{m|n}) : \mathfrak{csalg}_k \to \mathfrak{sets}$ is defined by

$$\mathbf{V}_r(GL_{m|n})(A) = \Big\{(\alpha_0, \dots, \alpha_{r-1}, \beta) \in (\mathrm{Mat}_{m|n}(A)_{\overline{0}})^{\times r} \times \mathrm{Mat}_{m|n}(A)_{\overline{1}} : \\ [\alpha_i, \alpha_j] = [\alpha_i, \beta] = 0 \text{ for all } 0 \le i, j \le r-1, \\ \alpha_i^p = 0 \text{ for all } 0 \le i \le r-2, \text{ and } \alpha_{r-1}^p + \beta^2 = 0\Big\}.$$

Definition 2.3.1 For $m, n \in \mathbb{N}$, let $\mathbf{N}_r(GL_{m|n}) \subset \mathbf{V}_r(GL_{m|n})$ be defined by

$$\mathbf{N}_r(GL_{m|n})(A) = \left\{(\alpha_0, \dots, \alpha_{r-1}, \beta) \in \mathbf{V}_r(GL_{m|n})(A) : \alpha_{r-1} \text{ is nilpotent}\right\}.$$

Given a commutative superalgebra $A \in \mathfrak{csalg}_k$, let

$$A\mathbb{M}_r = \mathrm{Hom}_A(A[\mathbb{M}_r], A) = \mathrm{Hom}_A(k[\mathbb{M}_r] \otimes_k A, A) \cong \mathrm{Hom}_k(k[\mathbb{M}_r], A)$$

be the group algebra of the A-supergroup scheme $\mathbb{M}_r \otimes_k A$. Then

$$A\mathbb{M}_r = A[[u_0, \dots, u_{r-1}, v]]/\langle u_0^p, \dots, u_{r-2}^p, u_{r-1}^p + v^2\rangle.$$

The next result was stated, with a different proof, in [9, Proposition 3.3.5]. The proof given below is in the spirit of the proof of [19, Proposition 1.2]. The upshot of the new proof is that we get, in Remark 2.3.3, a description of homomorphisms in terms of exponential maps entirely analogous to the classical description given in [20, Remark 1.3].

Proposition 2.3.2 *Let $m, n \in \mathbb{N}$. Then for each $A \in \mathfrak{csalg}_k$, there exists a natural identification*

$$\mathbf{N}_r(GL_{m|n})(A) = \mathbf{Hom}(\mathbb{M}_r, GL_{m|n})(A), \quad (\underline{\alpha}|\beta) \mapsto \rho_{(\underline{\alpha}|\beta)}.$$

Given $(\underline{\alpha}|\beta) := (\alpha_0, \dots, \alpha_{r-1}, \beta) \in \mathbf{N}_r(GL_{m|n})(A)$, the corresponding supergroup homomorphism $\rho_{(\underline{\alpha}|\beta)} : \mathbb{M}_r \otimes_k A \to GL_{m|n} \otimes_k A$ is such that the induced action of the group algebra $A\mathbb{M}_r$ on $A^{m|n}$ is defined by having the generator $u_i \in k\mathbb{M}_r \subset A\mathbb{M}_r$ act via the matrix $\alpha_i \in Mat_{m|n}(A)_{\overline{0}}$, and by having $v \in k\mathbb{M}_r \subset A\mathbb{M}_r$ act via the matrix $\beta \in Mat_{m|n}(A)_{\overline{1}}$.

Proof Specifying an A-supergroup scheme homomorphism $\rho : \mathbb{M}_r \otimes A \to GL_{m|n} \otimes A$ is equivalent to defining a rational $\mathbb{M}_r \otimes A$-supermodule structure on the superspace $A^{m|n}$, which is in turn equivalent to defining a right $A[\mathbb{M}_r]$-supercomodule structure on $A^{m|n}$. So let

$$\Delta_\rho : A^{m|n} \to A^{m|n} \otimes_A A[\mathbb{M}_r] = A^{m|n} \otimes_k k[\mathbb{M}_r]$$

be the supercomodule structure map corresponding to ρ. Then for $w \in A^{m|n}$, we can write

$$\Delta_\rho(w) = \sum_{i=0}^{p^{r-1}-1} \sum_{j \geq 0} \left[\alpha_{ij}(w) \otimes \theta^i \sigma_j + \beta_{ij}(w) \otimes \tau\theta^i \sigma_j\right] \tag{10}$$

for some k-linear maps $\alpha_{ij}, \beta_{ij} : A^{m|n} \to A^{m|n}$. Define $\widetilde{\beta}_{ij} : A^{m|n} \to A^{m|n}$ by $\widetilde{\beta}_{ij}(w) = (-1)^{\overline{w}} \beta_{ij}(w)$. Then $\alpha_{ij} \in \mathrm{Mat}_{m|n}(A)_{\overline{0}}$ and $\widetilde{\beta}_{ij} \in \mathrm{Mat}_{m|n}(A)_{\overline{1}}$. (If A is a purely even k-superalgebra, then β_{ij} is already an element of $\mathrm{Mat}_{m|n}(A)_{\overline{1}}$.)

Write $\Delta_{A[\mathbb{M}_r]}$ and ε for the coproduct and counit on $A[\mathbb{M}_r]$, respectively. Then the comodule axiom $(1 \otimes \varepsilon) \circ \Delta_\rho = \mathrm{id}_{A^{m|n}}$ implies that $\alpha_{0,0} = \mathrm{id}_{A^{m|n}}$, while the axiom $(1 \otimes \Delta_{A[\mathbb{M}_r]}) \circ \Delta_\rho = (\Delta_\rho \otimes 1) \circ \Delta_\rho$ implies the following identities:

$$\alpha_{ij} \circ \alpha_{st} = \binom{i+s}{i} \alpha_{i+s,j+t} \text{ if } i+s < p^{r-1}, \qquad \alpha_{ij} \circ \alpha_{st} = 0 \text{ if } i+s \geq p^{r-1},$$

$$\beta_{ij} \circ \beta_{st} = \binom{i+s}{i} \alpha_{i+s,j+t+p}, \text{ if } i+s < p^{r-1}, \quad \beta_{ij} \circ \beta_{st} = 0 \text{ if } i+s \geq p^{r-1},$$

$$\alpha_{ij} \circ \beta_{st} = \binom{i+s}{i} \beta_{i+s,j+t} \text{ if } i+s < p^{r-1}, \qquad \alpha_{ij} \circ \beta_{st} = 0 \text{ if } i+s \geq p^{r-1},$$

$$\beta_{st} \circ \alpha_{ij} = \binom{i+s}{i} \beta_{i+s,j+t} \text{ if } i+s < p^{r-1}, \qquad \beta_{st} \circ \alpha_{ij} = 0 \text{ if } i+s \geq p^{r-1}.$$

In particular, the α_{ij} and β_{ij} all commute pairwise, $(\alpha_{ij})^p = 0$ if $i \neq 0$, $\alpha_{0,j} = (\alpha_{0,1})^j$, and $(\beta_{0,0})^2 = \alpha_{0,p}$. Also, for each fixed $w \in A^{m|n}$ the sum (10) is finite, so $\alpha_{ij}(w) = 0$ for all but finitely many j. Then varying w over an A-supermodule basis for $A^{m|n}$, this implies that $\alpha_{ij} = 0$ for all but finitely many j, and hence that $(\alpha_{0,1})^N = \alpha_{0,N} = 0$ for some $N \geq 0$.

Now set $\alpha_i = \alpha_{p^i,0}$ for $0 \leq i \leq r-2$, set $\alpha_{r-1} = \alpha_{0,1}$, and set $\beta = -\widetilde{\beta}_{0,0}$. Then $\beta^2 = (\widetilde{\beta}_{0,0})^2 = -(\beta_{0,0})^2$, and the observations of the previous paragraph imply that the tuple $(\alpha_0, \ldots, \alpha_{r-1}, \beta)$ is an element of $\mathbf{V}_r(GL_{m|n})(A)$ such that α_{r-1} is nilpotent. Conversely, let $(\alpha_0, \ldots, \alpha_{r-1}, \beta)$ be such a tuple. Define $\beta_{0,0} : A^{m|n} \to A^{m|n}$ by $\beta_{0,0}(w) = -(-1)^{\overline{w}}\beta(w)$, i.e., so that $\beta = -\widetilde{\beta}_{0,0}$, and given an integer $0 \leq i < p^{r-1}$ with p-adic decomposition $i = i_0 + i_1 p + \cdots + i_{r-2} p^{r-2}$, set

$$\alpha_{ij} = \frac{(\alpha_0)^{i_0}(\alpha_1)^{i_1} \cdots (\alpha_{r-2})^{i_{r-2}}}{(i_0)!(i_1)! \cdots (i_{r-2})!} (\alpha_{r-1})^j \quad \text{and} \quad \beta_{ij} = \alpha_{ij} \cdot \beta_{0,0}.$$

Then with these definitions, (10) defines an $A[\mathbb{M}_r]$-supercomodule structure on $A^{m|n}$. Thus, the elements of $\mathbf{Hom}(\mathbb{M}_r, GL_{m|n})(A)$ correspond bijectively to the elements of $\mathbf{V}_r(GL_{m|n})(A)$ such that α_{r-1} is nilpotent. Finally, the action of an element $\phi \in A\mathbb{M}_r$ on an element $w \in A^{m|n}$ is defined by $\phi \cdot w = (1 \otimes_A \phi)(\Delta_\rho(w))$. Then the description of the action of $A\mathbb{M}_r$ on $A^{m|n}$ immediately follows because $u_i \in k[\mathbb{M}_r]^\#$ is defined to be the functional that is linearly dual to the basis element $\theta^{p^i} \in k[\mathbb{M}_r]$, and $v \in k[\mathbb{M}_r]^\#$ is defined to be linearly dual to $\tau \in k[\mathbb{M}_r]$. □

Remark 2.3.3 Recall that the right action of A on $\mathrm{Mat}_{m|n}(A) = \mathrm{Hom}_A(A^{m|n}, A^{m|n})$ is defined by $(\phi \cdot a)(w) = (-1)^{\overline{a} \cdot \overline{w}} \phi(w) \cdot a$. In particular, if $\tau \in B_{\overline{1}}$ for some commutative A-superalgebra $B \in \mathfrak{csalg}_A$, and if $\beta = -\widetilde{\beta}_{0,0} \in \mathrm{Mat}_{m|n}(A)_{\overline{1}}$ as in the proof, then (via the homomorphism $A \to B$ making B into an A-superalgebra) one has $\beta \cdot \tau \in \mathrm{Mat}_{m|n}(B)_{\overline{0}}$. Note that while β^2 may not equal zero, $(\beta \cdot \tau)^2 = 0$ because $\tau^2 = 0$ (by the supercommutativity of B). For $w \in B^{m|n}$, one has

$$(-\beta \cdot \tau)(w) = (-1)^{\overline{w}}(-\beta)(w) \cdot \tau = \beta_{0,0}(w) \cdot \tau.$$

The preceding discussion and the proof of the proposition now imply the following explicit description for the homomorphism $\rho_{(\underline{\alpha}|\beta)} : \mathbb{M}_r \otimes_k A \to GL_{m|n} \otimes_k A$ corresponding to an element $(\underline{\alpha}|\beta) \in \mathbf{N}_r(GL_{m|n})(A)$. Let $B \in \mathfrak{csalg}_A$. For $\phi \in \mathrm{Mat}_{m|n}(B)_{\overline{0}}$, set

$$\exp(\phi) = 1 + \phi + \frac{1}{2!}\phi^2 + \cdots + \frac{1}{(p-1)!}\phi^{p-1} \in \mathrm{Mat}_{m|n}(B)_{\overline{0}}.$$

Then for $g = (\tau, \theta, \sigma_1, \sigma_2, \ldots,) \in \mathbb{M}_r(B) = (\mathbb{M}_r \otimes_k A)(B)$ (cf. the shorthand of [9, Sect. 3.1] for denoting an element of $\mathbb{M}_r(B)$), one has

$$\rho_{(\underline{\alpha}|\beta)}(g) = \Big(\prod_{i=0}^{r-1} \exp(\alpha_i \cdot \theta^{p^i})\Big) \cdot \exp(-\beta \cdot \tau) \cdot \Big(\prod_{i \geq 1} \exp(\alpha_{r-1}^{p^i} \cdot \sigma_{p^i})\Big) \in GL_{m|n}(B). \tag{11}$$

This expression is well-defined, independent of the ordering of the factors, by the assumption that the entries of $(\underline{\alpha}|\beta)$ commute pairwise and α_{r-1} is nilpotent. If $\alpha_{r-1}^{p^s} = 0$, then (11) also describes the element of $\mathbf{Hom}(\mathbb{M}_{r;s}, GL_{m|n})(A)$ corresponding to $(\underline{\alpha}|\beta)$ as in [9, Proposition 3.3.5].

Recall from [9, Proposition 3.3.5] that $\mathbf{Hom}(\mathbb{M}_{r;s}, GL_{m|n})$ identifies with the closed subfunctor $\mathbf{V}_{r;s}(GL_{m|n})$ of $\mathbf{V}_r(GL_{m|n})$, consisting of those $(\alpha_0, \ldots, \alpha_{r-1}, \beta) \in \mathbf{V}_r(GL_{m|n})$ such that $\alpha_{r-1}^{p^s} = 0$. More generally, if G is an algebraic k-supergroup scheme and if $G \hookrightarrow GL_{m|n}$ is a closed embedding, then the proof of [9, Theorem 3.3.7] shows that $\mathbf{Hom}(\mathbb{M}_{r;s}, G)$ identifies with a closed subfunctor of $\mathbf{Hom}(\mathbb{M}_{r;s}, GL_{m|n})$.

For each $s \geq 1$, the canonical quotient homomorphisms $\mathbb{M}_r \twoheadrightarrow \mathbb{M}_{r;s+1} \twoheadrightarrow \mathbb{M}_{r;s}$ arising from the subalgebra inclusions $k[\mathbb{M}_{r;s}] \hookrightarrow k[\mathbb{M}_{r;s+1}] \hookrightarrow k[\mathbb{M}_r]$ induce natural transformations

$$\mathbf{Hom}(\mathbb{M}_{r;s}, G) \hookrightarrow \mathbf{Hom}(\mathbb{M}_{r;s+1}, G) \hookrightarrow \mathbf{Hom}(\mathbb{M}_r, G).$$

In this way, the $\mathbf{Hom}(\mathbb{M}_{r;s}, G)$ for $s \geq 1$ form a directed system of subfunctors of $\mathbf{Hom}(\mathbb{M}_r, G)$. The first part of the next proposition is then another interpretation of [9, Proposition 3.3.5], while the second part follows from the observation [9,

Remark 3.1.3(4)] that if G is algebraic, then any A-supergroup scheme homomorphism $\rho : \mathbb{M}_r \otimes_k A \to G \otimes_k A$ necessarily factors for all sufficiently large s through the canonical quotient map $\mathbb{M}_r \otimes_k A \twoheadrightarrow \mathbb{M}_{r;s} \otimes_k A$.

Lemma 2.3.4 *The functor* $\mathbf{Hom}(\mathbb{M}_r, GL_{m|n})$ *is the union in* $\mathbf{V}_r(GL_{m|n})$ *of the closed subfunctors* $\mathbf{Hom}(\mathbb{M}_{r;s}, GL_{m|n})$ *for* $s \geq 1$. *That is, as subfunctors of* $\mathbf{V}_r(GL_{m|n})$,

$$\mathbf{Hom}(\mathbb{M}_r, GL_{m|n}) = \bigcup_{s\geq 1} \mathbf{Hom}(\mathbb{M}_{r;s}, GL_{m|n}) = \bigcup_{s\geq 1} \mathbf{V}_{r;s}(GL_{m|n}) = \mathbf{N}_r(GL_{m|n}).$$

More generally, let G be an algebraic k-supergroup scheme, and fix a closed embedding $G \hookrightarrow GL_{m|n}$ for some $m, n \in \mathbb{N}$. Then identifying $\mathbf{Hom}(\mathbb{M}_r, G)$ *and each* $\mathbf{Hom}(\mathbb{M}_{r;s}, G)$ *with a subfunctor of* $\mathbf{Hom}(\mathbb{M}_r, GL_{m|n})$ *via the embedding, one has* $\mathbf{Hom}(\mathbb{M}_r, G) = \bigcup_{s\geq 1} \mathbf{Hom}(\mathbb{M}_{r;s}, G)$.

The functor $\mathbf{N}_r(GL_{m|n})$ is not a closed subfunctor of $\mathbf{V}_r(GL_{m|n})$ because there is no set of polynomial equations that simultaneously captures, for all coefficient rings $A \in \mathfrak{csalg}_k$, the property that α_{r-1} is nilpotent. But by restricting attention to field coefficients we get:

Lemma 2.3.5 *Let G be an algebraic k-supergroup scheme, let $G \hookrightarrow GL_{m|n}$ be a closed embedding, and set* $N = \max(m, n)$. *Then for any field extension K of k, the unions*

$$\mathbf{Hom}(\mathbb{M}_r, GL_{m|n})(K) = \textstyle\bigcup_{s\geq 1} \mathbf{Hom}(\mathbb{M}_{r;s}, GL_{m|n})(K), \quad \textit{and}$$
$$\mathbf{Hom}(\mathbb{M}_r, G)(K) = \textstyle\bigcup_{s\geq 1} \mathbf{Hom}(\mathbb{M}_{r;s}, G)(K)$$

reach a stable value at $s = N$. In particular, if k is algebraically closed, then $\mathbf{N}_r(GL_{m|n})(k)$ *is a Zariski closed subset of the affine variety* $\mathbf{V}_r(GL_{m|n})(k)$, *and* $\mathbf{Hom}(\mathbb{M}_r, G)(k)$ *identifies via the embedding $G \hookrightarrow GL_{m|n}$ with a closed subvariety of* $\mathbf{N}_r(GL_{m|n})(k)$.

Proof By [9, Theorem 3.3.7], the functor $\mathbf{V}_r(GL_{m|n})$ admits the structure of an affine superscheme of finite type over k. Let $A = k[\mathbf{V}_r(GL_{m|n})]$ be the coordinate superalgebra of this affine superscheme, and let A_{red} be the largest (purely even) reduced quotient of A.[3] Then A_{red} is a finitely generated commutative (in the ordinary sense) k-algebra, and because the field k is a purely even reduced k-algebra,

$$\mathbf{V}_r(GL_{m|n})(k) = \mathrm{Hom}_{\mathfrak{salg}}(A, k) = \mathrm{Hom}_{\mathfrak{alg}}(A_{red}, k).$$

In particular, if k is algebraically closed then $\mathbf{V}_r(GL_{m|n})(k)$ admits the structure of an affine algebraic variety with coordinate algebra A_{red}.

[3]Since A is a commutative superalgebra, its odd elements are automatically nilpotent, and hence the largest reduced quotient of A will automatically be a purely even k-algebra.

Next consider the defining property of $\mathbf{N}_r(GL_{m|n})(K)$ that α_{r-1} is nilpotent. Since the matrix coefficients lie in the field K, it follows (e.g., from considering Jordan forms, perhaps over some algebraically closed extension field) that α_{r-1} is nilpotent if and only if $(\alpha_{r-1})^N = 0$. This implies that $\mathbf{V}_{r;s}(GL_{m|n})(K) = \mathbf{V}_{r;N}(GL_{m|n})(K)$ for all $s \geq N$, which in turn implies that $\mathbf{N}_r(GL_{m|n})(k) = \bigcup_{s \geq 1} \mathbf{Hom}(\mathbb{M}_{r;s}, GL_{m|n})(K)$ and $\bigcup_{s \geq 1} \mathbf{Hom}(\mathbb{M}_{r;s}, G)(K)$ reach stable values at $s = N$. Finally, by the proof of [9, Theorem 3.3.7], $\mathbf{Hom}(\mathbb{M}_{r;N}, GL_{m|n})$ identifies with a closed subsuperscheme of $\mathbf{V}_r(GL_{m|n})$, and $\mathbf{Hom}(\mathbb{M}_{r;N}, G)$ identifies via the embedding $G \hookrightarrow GL_{m|n}$ with a closed subsuperscheme of $\mathbf{Hom}(\mathbb{M}_{r;N}, GL_{m|n})$. Then taking k-points, it follows as in the first paragraph that if k is algebraically closed, then $\mathbf{N}_r(GL_{m|n})(k) = \mathbf{Hom}(\mathbb{M}_{r;N}, GL_{m|n})(k)$ identifies with a Zariski closed subset of the affine variety $\mathbf{V}(GL_{m|n})(k)$, and $\mathbf{Hom}(\mathbb{M}_{r;N}, G)(k)$ identifies with a closed subvariety of $\mathbf{N}_r(GL_{m|n})(k)$. □

Definition 2.3.6 Let G be an algebraic k-supergroup scheme. By Lemma 2.3.5, there exists a minimal integer $N \geq 1$ such that for any field extension K of k and any integer $N' \geq N$, the canonical quotient map $\mathbb{M}_r \twoheadrightarrow \mathbb{M}_{r;N'}$ induces an identification $\mathbf{Hom}(\mathbb{M}_{r;N'}, G)(K) = \mathbf{Hom}(\mathbb{M}_r, G)(K)$. (For $G = GL_{m|n}$, one has $N = \max(m, n)$.) Define the k-algebra $k[\mathcal{N}_r(G)]$ by

$$k[\mathcal{N}_r(G)] = k[\mathbf{Hom}(\mathbb{M}_{r;N}, G)]_{red},$$

the largest (purely even) reduced quotient of $k[\mathbf{Hom}(\mathbb{M}_{r;N}, G)]$. If k is algebraically closed, then $k[\mathcal{N}_r(G)]$ is the coordinate algebra of the affine algebraic variety

$$\mathcal{N}_r(G) := \mathbf{Hom}(\mathbb{M}_r, G)(k).$$

Remark 2.3.7 Let G be an algebraic k-supergroup scheme, and let N be as in Definition 2.3.6.

1. The affine k-superscheme structure on $\mathbf{Hom}(\mathbb{M}_{r;N}, G)$ is defined in [9] in terms of a choice of closed embedding $G \hookrightarrow GL_{m|n}$. However, Yoneda's Lemma then implies that the coordinate superalgebra $k[\mathbf{Hom}(\mathbb{M}_{r;N}, G)]$ is independent, up to isomorphism, of the choice of closed embedding. Consequently, $k[\mathcal{N}_r(G)]$ is also uniquely determined up to isomorphism.
2. Let $N' \geq N$. Then by the definition of N, the canonical quotient maps $\mathbb{M}_r \twoheadrightarrow \mathbb{M}_{r;N'} \twoheadrightarrow \mathbb{M}_{r;N}$ induce identifications

$$\mathbf{Hom}(\mathbb{M}_{r;N}, G)(k) = \mathbf{Hom}(\mathbb{M}_{r;N'}, G)(k) = \mathbf{Hom}(\mathbb{M}_r, G)(k). \tag{12}$$

The quotient map $\pi : \mathbb{M}_{r;N'} \twoheadrightarrow \mathbb{M}_{r;N}$ also induces a k-algebra homomorphism

$$\pi^* : k[\mathbf{Hom}(\mathbb{M}_{r;N'}, G)]_{red} \to k[\mathbf{Hom}(\mathbb{M}_{r;N}, G)]_{red} = k[\mathcal{N}_r(G)].$$

If k is algebraically closed, then (12) implies that π^* induces the identity map between maximal ideal spectra, and hence implies (by the anti-equivalence of categories between affine algebraic varieties over k, and finitely generated reduced commutative k-algebras) that π^* is an isomorphism.

3. Let K/k be a field extension. Base change to K defines a k-algebra map

$$k[\mathbf{Hom}(\mathbb{M}_{r;N}, G)] \to k[\mathbf{Hom}(\mathbb{M}_{r;N}, G)] \otimes_k K = K[\mathbf{Hom}(\mathbb{M}_{r;N} \otimes_k K, G \otimes_k K)].$$

Then passing to the reduced quotient rings, one gets a k-algebra map

$$k[\mathcal{N}_r(G)] \to K[\mathcal{N}_r(G \otimes_k K)].$$

Lemma 2.3.8 *Suppose k is algebraically closed, and let $\mu, \mathbf{a}_0, \ldots, \mathbf{a}_{r-1}, \mathbf{b}_s$ be indeterminates over k. Then*

$$\begin{aligned} k[\mathcal{N}_r(\mathbb{M}_{r;1})] &\cong k[\mu, \mathbf{a}_0, \ldots, \mathbf{a}_{r-1}], \\ k[\mathcal{N}_r(\mathbb{M}_{r;s})] &\cong k[\mu, \mathbf{a}_0, \ldots, \mathbf{a}_{r-1}, \mathbf{b}_s]/\langle \mu^2 - \mathbf{a}_0^{p^r} \rangle && \text{if } s \geq 2, \\ k[\mathcal{N}_r(\mathbb{M}_{r;s,\eta})] &\cong k[\mu, \mathbf{a}_0, \ldots, \mathbf{a}_{r-1}]/\langle \mu^2 - \mathbf{a}_0^{p^r} \rangle && \text{if } r \geq 2 \text{ and } 0 \neq \eta \in k, \\ k[\mathcal{N}_r(\mathbb{G}_{a(r)})] &\cong k[\mathbf{a}_0, \ldots, \mathbf{a}_{r-1}], && \text{and} \\ k[\mathcal{N}_r(\mathbb{G}_a^-)] &\cong k[\mu]. \end{aligned}$$

Proof These characterizations are immediate consequences of the calculation in Proposition 2.2.2 of the variety $\mathcal{N}_r(G) = \mathbf{Hom}(\mathbb{M}_r, G)(k)$. □

Lemma 2.3.9 *Suppose k is algebraically closed and let $\phi : G \to G'$ be a homomorphism of algebraic k-supergroup schemes. Then composition with ϕ defines a morphism of affine varieties*

$$\phi_* : \mathcal{N}_r(G) \to \mathcal{N}_r(G'), \quad \nu \mapsto \phi \circ \nu,$$

and hence also a homomorphism of k-algebras $\phi^* : k[\mathcal{N}_r(G')] \to k[\mathcal{N}_r(G)]$.

Proof Choose a positive integer N large enough so that $\mathcal{N}_r(G) = \mathbf{Hom}(\mathbb{M}_{r;N}, G)(k)$ and $\mathcal{N}_r(G') = \mathbf{Hom}(\mathbb{M}_{r;N}, G')(k)$. By [9, Theorem 3.3.7], composition with ϕ defines a morphism of affine k-superschemes $\mathbf{Hom}(\mathbb{M}_{r;N}, G) \to \mathbf{Hom}(\mathbb{M}_{r;N}, G')$. Then it follows that the induced map between the sets of k-points is a morphism of affine varieties. □

2.4 Universal Homomorphisms

Let G be an algebraic k-supergroup scheme. Recall from [9, Definition 3.3.9] the universal supergroup homomorphism

$$\rho : \mathbb{M}_{r;s} \otimes_k k[\mathbf{Hom}(\mathbb{M}_{r;s}, G)] \to G \otimes_k k[\mathbf{Hom}(\mathbb{M}_{r;s}, G)].$$

Then ρ is universal in the sense that if $B \in \mathfrak{csalg}_k$ and if $\rho' \in \mathbf{Hom}(\mathbb{M}_{r;s}, G)(B)$, then there exists a unique k-superalgebra homomorphism $\phi : k[\mathbf{Hom}(\mathbb{M}_{r;s}, G)] \to B$ such that $\rho' = \rho \otimes_\phi B$, i.e., such that $\rho' : \mathbb{M}_{r;s} \otimes_k B \to G \otimes_k B$ is obtained from ρ via base change along ϕ.

Definition 2.4.1 (*Universal homomorphism from* $\mathbb{M}_r$ *to* G) Let G be an algebraic k-supergroup scheme, and let $N \geq 1$ be as in Definition 2.3.6. Define the *universal supergroup homomorphism from* $\mathbb{M}_r$ *to* G to be the homomorphism of $k[\mathcal{N}_r(G)]$-supergroup schemes

$$u_G : \mathbb{M}_r \otimes_k k[\mathcal{N}_r(G)] \to G \otimes_k k[\mathcal{N}_r(G)]$$

that is obtained from the universal homomorphism

$$\rho : \mathbb{M}_{r;N} \otimes_k k[\mathbf{Hom}(\mathbb{M}_{r;N}, G)] \to G \otimes_k k[\mathbf{Hom}(\mathbb{M}_{r;N}, G)]$$

via base change along the canonical quotient map

$$k[\mathbf{Hom}(\mathbb{M}_{r;N}, G)] \twoheadrightarrow k[\mathbf{Hom}(\mathbb{M}_{r;N}, G)]_{red} = k[\mathcal{N}_r(G)].$$

Then u_G is universal in the sense that if K/k is a field extension and if $\rho' \in \mathbf{Hom}(\mathbb{M}_r, G)(K)$, then there is a unique k-algebra homomorphism $\phi : k[\mathcal{N}_r(G)] \to K$ such that $\rho' = u_G \otimes_\phi K$, i.e., such that $\rho' : \mathbb{M}_r \otimes_k K \to G \otimes_k K$ is obtained from u_G via base change along ϕ.

Let G be an algebraic k-supergroup scheme. As in [9, Sect. 6.1], set

$$H(G, k) := \mathrm{H}^{\mathrm{ev}}(G, k)_{\overline{0}} \oplus \mathrm{H}^{\mathrm{odd}}(G, k)_{\overline{1}}. \tag{13}$$

Then $H(G, k)$ inherits from $\mathrm{H}^\bullet(G, k)$ the structure of a $\mathbb{Z}$-graded commutative (in the ordinary sense) k-algebra. As in [9, §6.2], we use the homomorphism $u_G : \mathbb{M}_r \otimes_k k[\mathcal{N}_r(G)] \to G \otimes_k k[\mathcal{N}_r(G)]$ to define a k-algebra homomorphism

$$\psi_r : H(G, k) \to k[\mathcal{N}_r(G)].$$

Recall from [9, Proposition 3.2.1(2)] the calculation of $\mathrm{H}^\bullet(\mathbb{M}_r, k)$:

$$\mathrm{H}^\bullet(\mathbb{M}_r, k) \cong k[x_1, \dots, x_r, y]/\langle x_r - y^2 \rangle \ {}^g\!\otimes \Lambda(\lambda_1, \dots, \lambda_r).$$

Here $x_i \in \mathrm{H}^2(\mathbb{M}_r, k)_{\overline{0}}$ and $\lambda_i \in \mathrm{H}^1(\mathbb{M}_r, k)_{\overline{0}}$ for each i, $y \in \mathrm{H}^1(\mathbb{M}_r, k)_{\overline{1}}$, and ${}^g\!\otimes$ denotes the graded tensor product of graded superalgebras, as defined in [8, Definition 2.2.2]. The map $\mathrm{H}^\bullet(\mathbb{M}_r, k) \to k$ that sends x_r and y each to 1 but that sends the other algebra generators to 0 is a k-algebra homomorphism (though not a superalgebra homomorphism, since y is of odd superdegree). Extending scalars to $k[\mathcal{N}_r(G)]$, one gets a k-algebra homomorphism

$$\varepsilon : \mathrm{H}^\bullet(\mathbb{M}_r \otimes_k k[\mathcal{N}_r(G)], k[\mathcal{N}_r(G)]) = \mathrm{H}^\bullet(\mathbb{M}_r, k) \otimes_k k[\mathcal{N}_r(G)]$$
$$\to k \otimes_k k[\mathcal{N}_r(G)] = k[\mathcal{N}_r(G)].$$

Now define $\psi_r : H(G,k) \to k[\mathcal{N}_r(G)]$ to be the composite algebra homomorphism

$$\psi_r : H(G,k) \xrightarrow{\iota} \mathrm{H}^\bullet(G,k) \otimes_k k[\mathcal{N}_r(G)] = \mathrm{H}^\bullet(G \otimes_k k[\mathcal{N}_r(G)], k[\mathcal{N}_r(G)])$$
$$\xrightarrow{u_G^*} \mathrm{H}^\bullet(\mathbb{M}_r \otimes_k k[\mathcal{N}_r(G)], k[\mathcal{N}_r(G)]) \xrightarrow{\varepsilon} k[\mathcal{N}_r(G)], \tag{14}$$

where $\iota : H(G,k) \to \mathrm{H}^\bullet(G,k) \otimes_k k[\mathcal{N}_r(G)]$ is the base change map $z \mapsto z \otimes 1$. Thus, in the special case $r = 1$ and for $z \in H(G,k)$ homogeneous of degree n, one gets

$$(u_G^* \circ \iota)(z) = \psi(z) \cdot y^n \in H(\mathbb{M}_1, k) \otimes_k k[\mathcal{N}_1(G)] \cong k[y] \otimes_k k[\mathcal{N}_1(G)]. \tag{15}$$

The next proposition is an analogue of [20, Theorem 1.14], and is related to [9, Proposition 6.2.2].

Proposition 2.4.2 *Let G be an algebraic k-supergroup scheme. Then $k[\mathcal{N}_r(G)]$ admits the structure of a $\mathbb{Z}[\frac{p^r}{2}]$-graded connected k-algebra, and the homomorphism*

$$\psi_r : H(G,k) \to k[\mathcal{N}_r(G)]$$

is then a homomorphism of graded k-algebras that multiplies degrees by $\frac{p^r}{2}$. Moreover, ψ_r is natural with respect to homomorphisms $\phi : G \to G'$ of algebraic k-supergroup schemes.

Proof Let N be as in Definition 2.3.6. Recall from [9, Definition 3.3.8] that $k[V_{r;N}(G)]$ is the largest purely even quotient of $k[\mathbf{Hom}(\mathbb{M}_{r;N}, G)]$. Then $k[\mathcal{N}_r(G)] = k[V_{r;N}(G)]_{red}$. By [9, Corollary 3.4.3], $k[V_{r;N}(G)]$ admits the structure of a $\mathbb{Z}[\frac{p^r}{2}]$-graded connected k-algebra. Then the nilpotent elements in $k[V_{r;N}(G)]$ form a homogeneous ideal, and hence $k[\mathcal{N}_r(G)]$ inherits the structure of a $\mathbb{Z}[\frac{p^r}{2}]$-graded connected k-algebra. Next, recall from [9, Sect. 6.2] that the homomorphism

$$\psi_{r;N} : H(G,k) \to k[V_{r;N}(G)]$$

is defined as a composite exactly as in (14), except that $k[\mathcal{N}_r(G)]$ is replaced by $k[V_{r;N}(G)]$, and the homomorphism $u_G : \mathbb{M}_r \otimes_k k[\mathcal{N}_r(G)] \to G \otimes_k k[\mathcal{N}_r(G)]$ is replaced by the universal purely even supergroup homomorphism $u_{r;N} : \mathbb{M}_{r;N} \otimes_k k[V_{r;N}(G)] \to G \otimes_k k[V_{r;N}(G)]$ of [9, Definition 3.3.9(2)]. Since u_G can be obtained from $u_{r;N}$ via base change along the canonical quotient map $k[V_{r;N}(G)] \twoheadrightarrow k[V_{r;N}(G)]_{red} = k[\mathcal{N}_r(G)]$, it follows that there is a commutative diagram

$$
\begin{array}{ccccccc}
H(G,k) & \to & \mathrm{H}^\bullet(G \otimes_k B, B) & \to & \mathrm{H}^\bullet(\mathbb{M}_r \otimes_k B, B) & \to & B \\
\Big\| & & \Big\downarrow & & \Big\downarrow & & \Big\downarrow \\
H(G,k) & \to & \mathrm{H}^\bullet(G \otimes_k A, A) & \to & \mathrm{H}^\bullet(\mathbb{M}_r \otimes_k A, A) & \to & A
\end{array}
\tag{16}
$$

where $A = k[\mathscr{N}_r(G)]$, $B = k[V_{r;N}(G)]$, and in which the rows are the composites defining $\psi_{r;N}$ and ψ_r, respectively, and the vertical arrows are induced by the quotient map $B = k[V_{r;N}(G)] \twoheadrightarrow k[\mathscr{N}_r(G)] = A$. By [9, Proposition 6.6.2], $\psi_{r;N}$ is a homomorphism of graded k-algebras that multiplies degress by $\frac{p^r}{2}$. Then by the commutativity of the diagram it follows that ψ_r is as well. Finally, the last assertion concerning naturality in G follows from the same line of reasoning as the proof of [9, Lemma 6.2.1]. □

3 Cohomology and Support Varieties

3.1 Induced Maps in Cohomology

In this section we calculate the maps in cohomology induced by the homomorphisms described in Proposition 2.2.2. We identify the relevant cohomology rings as in [9, Proposition 3.2.1], using also the fact that base change induces identifications of the form $\mathrm{H}^\bullet(G \otimes_k A, A) = \mathrm{H}^\bullet(G, k) \otimes_k A$.

Lemma 3.1.1 *Let $A = A_{\overline{0}} \in \mathfrak{calg}_k$ be a purely even commutative k-algebra. In parts (2) and (3), assume that A is reduced.*

1. *Let $(\mu, a_0, \ldots, a_{r-1}) \in A^{r+1}$, and let $\rho : \mathbb{M}_{r;1} \otimes_k A \to \mathbb{M}_{r;1} \otimes_k A$ be the corresponding homomorphism as in Proposition 2.2.2(1). Then the induced map in cohomology $\rho^* : H^\bullet(\mathbb{M}_{r;1} \otimes_k A, A) \to H^\bullet(\mathbb{M}_{r;1} \otimes_k A, A)$ satisfies*

$$
\begin{aligned}
\rho^*(y) &= \mu \cdot y, \\
\rho^*(\lambda_i) &= a_0^{p^{i-1}} \lambda_i + a_1^{p^{i-1}} \lambda_{i+1} + \cdots + a_{r-i}^{p^{i-1}} \lambda_r && \textit{for } 1 \le i \le r, \textit{ and} \\
\rho^*(x_i) &= a_0^{p^i} x_i + a_1^{p^i} x_{i+1} + \cdots + a_{r-i}^{p^i} x_r && \textit{for } 1 \le i \le r.
\end{aligned}
$$

2. *Suppose $s \ge 2$. Let $(\mu, a_0, \ldots, a_{r-1}, b_s) \in A^{r+2}$ such that $\mu^2 = a_0^{p^r}$, and let $\rho : \mathbb{M}_{r;s} \otimes_k A \to \mathbb{M}_{r;s} \otimes_k A$ be the corresponding homomorphism as in Proposition 2.2.2(2). Then the induced map in cohomology $\rho^* : H^\bullet(\mathbb{M}_{r;s} \otimes_k A, A) \to H^\bullet(\mathbb{M}_{r;s} \otimes_k A, A)$ satisfies*

$$\rho^*(y) = \mu \cdot y,$$
$$\rho^*(\lambda_i) = a_0^{p^{i-1}} \lambda_i + a_1^{p^{i-1}} \lambda_{i+1} + \cdots + a_{r-i}^{p^{i-1}} \lambda_r \qquad \textit{for } 1 \leq i \leq r,$$
$$\rho^*(x_i) = a_0^{p^i} x_i + a_1^{p^i} x_{i+1} + \cdots + a_{r-i}^{p^i} x_r \qquad \textit{for } 1 \leq i \leq r, \textit{and}$$
$$\rho^*(w_s) = a_0^{p^{r+s-1}} w_s + (b_s)^p x_r.$$

3. *Suppose $r \geq 2$, and let $0 \neq \eta \in k$. Let $(\mu, a_0, \ldots, a_{r-1}) \in A^{r+1}$ such that $\mu^2 = a_0^{p^r}$, and let $\rho : \mathbb{M}_{r;s+1} \otimes_k A \to \mathbb{M}_{r;s,\eta} \otimes_k A$ be the corresponding homomorphism as in Proposition 2.2.2(3). Then identifying $H^\bullet(\mathbb{M}_{r;s,\eta}, k)$ as an algebra with*

$$H^\bullet(\mathbb{M}_{r-1;s+1}, k) = k[x_1, \ldots, x_{r-1}, y, w_{s+1}]/\langle x_{r-1} - y^2 \rangle \, {}^g\!\otimes \Lambda(\lambda_1, \ldots, \lambda_{r-1})$$

as in [9, Proposition 3.2.1(5)], the induced map in cohomology $\rho^* : H^\bullet(\mathbb{M}_{r;s,\eta} \otimes_k A, A) \to H^\bullet(\mathbb{M}_{r;s+1} \otimes_k A, A)$ satisfies

$$\rho^*(y) = \mu \cdot y,$$
$$\rho^*(\lambda_i) = a_0^{p^i} \lambda_{i+1} + a_1^{p^i} \lambda_{i+2} + \cdots + a_{r-i-1}^{p^i} \lambda_r \quad \textit{for } 1 \leq i \leq r-1,$$
$$\rho^*(x_i) = a_0^{p^{i+1}} x_{i+1} + a_1^{p^{i+1}} x_{i+2} + \cdots + a_{r-i-1}^{p^{i+1}} x_r \quad \textit{for } 1 \leq i \leq r-1, \textit{and}$$
$$\rho^*(w_{s+1}) = (-\eta^{-1})^p \cdot (a_0^p x_1 + a_1^p x_2 + \cdots + a_{r-1}^p x_r) + a_0^{p^{r+s}} w_{s+1}.$$

4. *Let $(a_0, \ldots, a_{r-1}) \in A^r$, and let $\rho : \mathbb{G}_{a(r)} \otimes_k A \to \mathbb{M}_{r;1} \otimes_k A$ be the corresponding homomorphism as in Proposition 2.2.2(4). Then the induced map in cohomology $\rho^* : H^\bullet(\mathbb{G}_{a(r)} \otimes_k A, A) \to H^\bullet(\mathbb{M}_{r;1} \otimes_k A, A)$ satisfies*

$$\rho^*(\lambda_i) = a_0^{p^{i-1}} \lambda_i + a_1^{p^{i-1}} \lambda_{i+1} + \cdots + a_{r-i}^{p^{i-1}} \lambda_r \qquad \textit{for } 1 \leq i \leq r, \textit{and}$$
$$\rho^*(x_i) = a_0^{p^i} x_i + a_1^{p^i} x_{i+1} + \cdots + a_{r-i}^{p^i} x_r \qquad \textit{for } 1 \leq i \leq r.$$

5. *Let $\mu \in A$, and let $\rho : \mathbb{G}_a^- \otimes_k A \to \mathbb{M}_{r;1} \otimes_k A$ be the corresponding homomorphism as in Proposition 2.2.2(5). Write $H^\bullet(\mathbb{G}_a^-, k) = k[y]$ as in [9, Proposition 3.2.1]. Then the induced map in cohomology $\rho^* : H^\bullet(\mathbb{G}_a^- \otimes_k A, A) \to H^\bullet(\mathbb{M}_{r;1} \otimes_k A, A)$ satisfies $\rho^*(y) = \mu \cdot y$.*

Proof We will prove in detail statements (2) and (3), but will leave the proofs of the other (easier) statements to the reader.

To begin, suppose $s \geq 2$, let $(\mu, a_0, \ldots, a_{r-1}, b_s) \in A^{r+2}$ such that $\mu^2 = a_0^{p^r}$, and let $\rho : \mathbb{M}_{r;s} \otimes_k A \to \mathbb{M}_{r;s} \otimes_k A$ be the corresponding homomorphism as in Proposition 2.2.2(2). To describe the map in cohomology induced by ρ, we'll consider how the comorphism $\rho^* : A[\mathbb{M}_{r;s}] \to A[\mathbb{M}_{r;s}]$ acts on the explicit cochain representatives described in [9, Proposition 3.2.1] for the generators of the cohomology ring $\mathrm{H}^\bullet(\mathbb{M}_{r;s} \otimes_k A, A) = \mathrm{H}^\bullet(\mathbb{M}_{r;s}, k) \otimes_k A$. First, since $y = [\tau]$ and $\rho^*(\tau) = \mu \cdot \tau$, then $\rho^*(y) = \mu \cdot y$. Next,

$$\rho^*(\theta^{p^{i-1}}) = \rho^*(\theta)^{p^{i-1}} = \left(\sum_{\ell=0}^{r-1} a_\ell \cdot \theta^{p^\ell}\right)^{p^{i-1}} = a_0^{p^{i-1}}\theta_i + a_1^{p^{i-1}}\theta_{i+1} + \cdots + a_{r-i}^{p^{i-1}}\theta_r.$$

Since λ_i is the cohomology class of $\theta^{p^{i-1}}$, this implies the description of $\rho^*(\lambda_i)$ in (2). Now recall from [6, Remark 3.4.4] the semilinear function

$$\overline{\beta} : \mathrm{H}^1(\mathbb{M}_{r;s}, k)_{\overline{0}} \to \mathrm{H}^2(\mathbb{M}_{r;s}, k)_{\overline{0}}.$$

Given a cocycle representative $f \in k[\mathbb{M}_{r;s}]_{\overline{0}}$ for an element of $\mathrm{H}^1(\mathbb{M}_{r;s}, k)_{\overline{0}}$, the action of $\overline{\beta}$ on the cohomology class $[f]$ is defined by $\overline{\beta}([f]) = [\beta(f)]$, where

$$\beta(f) = \sum_{\ell=1}^{p-1} \left\{ {p \atop \ell} \right\} f^\ell \otimes f^{p-\ell}, \tag{17}$$

and $\left\{ {p \atop \ell} \right\}$ denotes the integer $\frac{1}{p} \cdot \binom{p}{\ell}$. The map $\overline{\beta}$ is evidently natural with respect to Hopf superalgebra homomorphisms (equivalently, homomorphisms of supergroup schemes). Since $x_i = \overline{\beta}(\lambda_i)$ by the calculations in [9, Proposition 3.2.1] (cf. also the proof of [6, Proposition 3.4.2]), and since $\overline{\beta}$ is semilinear, the description of $\rho^*(x_i)$ in (2) follows from the description of $\rho^*(\lambda_i)$. Finally, recall from [9, Remark 3.1.3] that the superalgebra grading on $k[\mathbb{M}_{r;s}]$ lifts to a $\mathbb{Z}$-grading. This $\mathbb{Z}$-grading endows the cohomology ring $\mathrm{H}^\bullet(\mathbb{M}_{r;s}, k)$ with an internal $\mathbb{Z}$-grading in addition to the (external) cohomological grading. Then $\mathrm{H}^2(\mathbb{M}_{r;s}, k)_{\overline{0}}$ is spanned by

x_i	for $1 \leq i \leq r$,	of internal degree $2p^i$,
$\lambda_i\lambda_j$	for $1 \leq i < j \leq r$,	of internal degree $2p^{i-1} + 2p^{j-1} < 2p^r$, and
w_s		of internal degree $2p^{r+s-1}$.

In particular, $\mathrm{H}^2(\mathbb{M}_{r;s}, k)_{\overline{0}}$ has no nonzero elements of internal degrees strictly between $2p^r$ and $2p^{r+s-1}$. Now ρ^* maps w_s to the cohomology class of

$$\begin{aligned} -&\left[\sum_{j=1}^{p^s-1} \rho^*(\sigma_j) \otimes \rho^*(\sigma_{p^s-j}) + \sum_{i+j+p=p^s} \rho^*(\sigma_i\tau) \otimes \rho^*(\sigma_j\tau)\right] \\ &= -\left[\sum_{j=1}^{p^s-1} \rho^*(\sigma_j) \otimes \rho^*(\sigma_{p^s-j}) + \sum_{i+j+p=p^s} \mu^2 \cdot \rho^*(\sigma_i)\tau \otimes \rho^*(\sigma_j)\tau\right]. \end{aligned} \tag{18}$$

If $0 \leq i < p^{s-1}$ and $0 \leq j < p$, then $\rho^*(\sigma_{i+jp^{s-1}})$ is given by formula (5) with $n = s-1$ and $c_{s-1} = b_s$. Using this observation, one can check that the component of greatest internal $\mathbb{Z}$-degree in (18) is

$$-a_0^{p^{r+s-1}} \cdot \left[\sum_{j=1}^{p^s-1} \sigma_j \otimes \sigma_{p^s-j} + \sum_{i+j+p=p^s} \sigma_i \tau \otimes \sigma_j \tau \right],$$

which is of internal degree $2p^{r+s-1}$ and is a cochain representative for $a_0^{p^{r+s-1}} w_s$, and the component of smallest internal $\mathbb{Z}$-degree in (18) is

$$-(b_s)^p \cdot \left[\sum_{i=1}^{p-1} \sigma_i \otimes \sigma_{p-i} \right],$$

which is of internal degree $2p^r$ and is a cochain representative for $(b_s)^p x_r$. Since $\mathrm{H}^2(\mathbb{M}_{r;s}, k)_{\overline{0}}$ has no nonzero elements of internal degrees strictly between $2p^r$ and $2p^{r+s-1}$, the other homogeneous components of (18) must be cohomologous to zero. Then it follows that $\rho^*(w_s) = a_0^{p^{r+s-1}} w_s + (b_s)^p x_r$, as desired. This completes the proof of (2).

Now suppose $r \geq 2$ and $0 \neq \eta \in k$. The algebra identification $\mathrm{H}^\bullet(\mathbb{M}_{r-1;s+1}, k) \cong \mathrm{H}^\bullet(\mathbb{M}_{r;s,\eta}, k)$ arises from a coalgebra isomorphism $\pi^* : k[\mathbb{M}_{r-1;s+1}] \xrightarrow{\sim} k[\mathbb{M}_{r;s,\eta}]$, which in turn arises via duality from a group algebra isomorphism $\pi : k\mathbb{M}_{r;s,\eta} \xrightarrow{\sim} k\mathbb{M}_{r-1;s+1}$. The isomorphism π is defined by $\pi(v) = v$, $\pi(u_i) = u_{i-1}$ for $1 \leq i \leq r$, and $\pi(u_0) = -\eta^{-1} \cdot u_{r-1}^s$; see [9, Remark 3.1.8(4)]. Then given integers $0 \leq i < p^{r-2}$, $0 \leq j < p^s$, and $0 \leq i_0 < p$, one can check via duality that the coalgebra isomorphism $\pi^* : k[\mathbb{M}_{r-1;s+1}] \to k[\mathbb{M}_{r;s,\eta}]$ satisfies

$$\pi^*(\theta^i \sigma_{j+i_0 p^s}) = \frac{1}{(i_0)!}(-\eta^{-1})^{i_0} \cdot \theta^{i_0+pi} \sigma_j \qquad \text{and}$$
$$\pi^*(\theta^i \sigma_{j+i_0 p^s} \tau) = \frac{1}{(i_0)!}(-\eta^{-1})^{i_0} \cdot \theta^{i_0+pi} \sigma_j \tau.$$

Here as in [9, Lemma 3.1.7] we make the superalgebra identification $k[\mathbb{M}_{r;s,\eta}] \cong k[\mathbb{M}_{r;s}]$.

Now let $(\mu, a_0, \ldots, a_{r-1}) \in A^{r+1}$ such that $\mu^2 = a_0^{p^r}$, and let $\rho : \mathbb{M}_{r;s+1} \otimes_k A \to \mathbb{M}_{r;s,\eta} \otimes_k A$ be the corresponding homomorphism as in Proposition 2.2.2(3). Then to establish statement (3) of the lemma we must describe the map in cohomology induced by the composite coalgebra homomorphism $\rho^* \circ \pi^* : A[\mathbb{M}_{r-1;s+1}] \to A[\mathbb{M}_{r;s,\eta}] \to A[\mathbb{M}_{r;s+1}]$. First, since $\pi^*(\tau) = \tau$, one gets

$$(\rho^* \circ \pi^*)(y) = (\rho^* \circ \pi^*)([\tau]) = \rho^*([\tau]) = [\rho^*(\tau)] = [\mu \cdot \tau] = \mu \cdot y.$$

Next for $1 \leq i \leq r-2$ one has

$$\begin{aligned}(\rho^* \circ \pi^*)(\lambda_i) = (\rho^* \circ \pi^*)([\theta^{p^{i-1}}]) &= [\rho^*(\theta^{p^i})] \\ &= [a_0^{p^i} \theta^{p^i} + a_1^{p^i} \theta^{p^{i+1}} + \cdots + a_{r-i-1}^{p^i} \theta^{p^{r-1}}] \\ &= a_0^{p^i} \lambda_{i+1} + a_1^{p^i} \lambda_{i+2} + \cdots + a_{r-i-1}^{p^i} \lambda_r,\end{aligned}$$

while for $i = r-1$ one has, because $\theta^{p^{r-2}} = \sigma_1$ in $k[\mathbb{M}_{r-1;s+1}]$ and $\theta^{p^{r-1}} = \sigma_1$ in $k[\mathbb{M}_{r;s+1}]$,

$$(\rho^* \circ \pi^*)(\lambda_{r-1}) \\ = (\rho^* \circ \pi^*)([\sigma_1]) = [\rho^*(\sigma_1)] = [a_0^{p^{r-1}} \sigma_1] = [a_0^{p^{r-1}} \theta^{p^{r-1}}] = a_0^{p^{r-1}} \lambda_r.$$

Now for $1 \le i \le r-2$ one gets

$$\begin{aligned}(\rho^* \circ \pi^*)(x_i) &= (\rho^* \circ \pi^*)([\beta(\theta^{p^{i-1}})]) = \rho^*([\beta(\theta^{p^i})]) = \overline{\beta}([\rho^*(\theta^{p^i})]) \\ &= \overline{\beta}(a_0^{p^i} \lambda_{i+1} + a_1^{p^i} \lambda_{i+2} + \cdots + a_{r-i-1}^{p^i} \lambda_r) \\ &= a_0^{p^{i+1}} x_{i+1} + a_1^{p^{i+1}} x_{i+2} + \cdots + a_{r-i-1}^{p^{i+1}} x_r,\end{aligned}$$

while for $i = r-1$ one gets

$$(\rho^* \circ \pi^*)(x_{r-1}) \\ = (\rho^* \circ \pi^*)([\beta(\sigma_1)]) = \rho^*([\beta(\sigma_1)]) = \overline{\beta}([\rho^*(\sigma_1)]) = \overline{\beta}(a_0^{p^{r-1}} \lambda_r) = a_0^{p^r} x_r.$$

Finally,

$$\begin{aligned}(\rho^* \circ \pi^*)(w_{s+1}) &= (\rho^* \circ \pi^*) \left(- \left[\sum_{j=1}^{p^{s+1}-1} \sigma_j \otimes \sigma_{p^{s+1}-j} + \sum_{i+j+p=p^{s+1}} \sigma_i \tau \otimes \sigma_j \tau \right] \right) \\ &= (\rho^* \circ \pi^*) \left(- \left[\sum_{i=1}^{p-1} \sigma_{ip^s} \otimes \sigma_{(p-i)p^s} + \sum_{\substack{i_0+j_0=p-1 \\ i+j=p^s,\, i,j \ge 1}} \sigma_{i+i_0 p^s} \otimes \sigma_{j+j_0 p^s} \right.\right. \\ &\qquad\qquad \left.\left. + \sum_{i+j+p=p^{s+1}} \sigma_i \tau \otimes \sigma_j \tau \right] \right)\end{aligned}$$

The first summation inside the square brackets can be rewritten as

$$\sum_{i=1}^{p-1} \sigma_{ip^s} \otimes \sigma_{(p-i)p^s} = - \sum_{i=1}^{p-1} \left\{ {p \atop i} \right\} (\sigma_{p^s})^i \otimes (\sigma_{p^s})^{p-i} = -\beta(\sigma_{p^s}),$$

while the third summation inside the square brackets can be rewritten as

$$\begin{aligned}\sum_{i+j+p=p^{s+1}} \sigma_i \tau \otimes \sigma_j \tau &= \sum_{\substack{(i+i_0 p^s)+(j+j_0 p^s)+p=p^{s+1} \\ 0 \le i,j < p^s,\, 0 \le i_0, j_0 < p}} \sigma_{i+i_0 p^s} \tau \otimes \sigma_{j+j_0 p^s} \tau \\ &= \sum_{\substack{i_0+j_0=p-1 \\ i+j+p=p^s}} \sigma_{i+i_0 p^s} \tau \otimes \sigma_{j+j_0 p^s} \tau + \sum_{\substack{i_0+j_0=p-2 \\ i+j+p=2p^s,\, i,j \ge 1}} \sigma_{i+i_0 p^s} \tau \otimes \sigma_{j+j_0 p^s} \tau.\end{aligned}$$

Then

$$(\rho^* \circ \pi^*)(w_{s+1}) = \\ \rho^*\left(\left[(-\eta^{-1})^p \cdot \beta(\theta) - \sum_{\substack{i_0+j_0=p-1\\ i+j=p^s,\, i,j\geq 1}} \frac{(-\eta^{-1})^{p-1}}{(i_0!)(j_0!)}\theta^{i_0}\sigma_i \otimes \theta^{j_0}\sigma_j \right.\right. \\ - \sum_{\substack{i_0+j_0=p-1\\ i+j+p=p^s}} \frac{(-\eta^{-1})^{p-1}}{(i_0!)(j_0!)}\theta^{i_0}\sigma_i\tau \otimes \theta^{j_0}\sigma_j\tau \\ \left.\left. - \sum_{\substack{i_0+j_0=p-2\\ i+j+p=2p^s,\, i,j\geq 1}} \frac{(-\eta^{-1})^{p-2}}{(i_0!)(j_0!)}\theta^{i_0}\sigma_i\tau \otimes \theta^{j_0}\sigma_j\tau\right]\right).$$

The space $\mathrm{H}^2(\mathbb{M}_{r;s+1}, k)_{\bar{0}}$ has no nonzero elements of internal $\mathbb{Z}$-degrees strictly between $2p^r$ and $2p^{r+(s+1)-1} = 2p^{r+s}$, so when considering the image in $\mathrm{H}^2(\mathbb{M}_{r;s+1} \otimes_k A, A)$ of the expression above, we can ignore all terms whose internal $\mathbb{Z}$-degrees fall in that range (since the sum of those terms must be cohomologous to zero). In particular, from the three summations in the square brackets we only need to consider contributions of the maximum $\mathbb{Z}$-degree $2p^{r+s}$. Then it follows that

$$(\rho^* \circ \pi^*)(w_{s+1}) = \\ \left[(-\eta^{-1})^p \cdot \beta(\textstyle\sum_{i=0}^{r-1} a_i\theta^{p^i}) + (-\eta^{-1})^p \cdot \beta(-\eta \cdot a_0^{p^{r+s-1}}\sigma_{p^s})\right. \\ - \sum_{\substack{i_0+j_0=p-1\\ i+j+p=p^s}} \frac{(-\eta^{-1})^{p-1}}{(i_0!)(j_0!)}(-\eta a_0^{p^{r+s-1}}\sigma_{p^s})^{i_0}(a_0^{ip^{r-1}}\sigma_i)(\mu\tau) \\ \otimes(-\eta a_0^{p^{r+s-1}}\sigma_{p^s})^{j_0}(a_0^{jp^{r-1}}\sigma_j)(\mu\tau) \\ - \sum_{\substack{i_0+j_0=p-2\\ i+j+p=2p^s,\, i,j\geq 1}} \frac{(-\eta^{-1})^{p-2}}{(i_0!)(j_0!)}(-\eta a_0^{p^{r+s-1}}\sigma_{p^s})^{i_0}(a_0^{ip^{r-1}}\sigma_i)(\mu\tau) \\ \left.\otimes(-\eta a_0^{p^{r+s-1}}\sigma_{p^s})^{j_0}(a_0^{jp^{r-1}}\sigma_j)(\mu\tau)\right] \\ = \left[(-\eta^{-1})^p\beta(\textstyle\sum_{i=0}^{r-1} a_i\theta^{p^i}) + a_0^{p^{r+s}}\beta(\sigma_{p^s}) - \displaystyle\sum_{\substack{i_0+j_0=p-1\\ i+j+p=p^s}} a_0^{p^{r+s}}\sigma_{i_0p^s}\sigma_i\tau \otimes \sigma_{j_0p^s}\sigma_j\tau\right.$$

$$- \sum_{\substack{i_0+j_0=p-2 \\ i+j+p=2p^s,\, i,j\geq 1}} a_0^{p^{r+s}} \sigma_{i_0 p^s}\sigma_i\tau \otimes \sigma_{j_0 p^s}\sigma_j\tau \Bigg]$$

$$= (-\eta^{-1})^p \overline{\beta}([\textstyle\sum_{i=0}^{r-1} a_i \theta^{p^i}]) - a_0^{p^{r+s}} \left[\sum_{j=1}^{p^{s+1}-1} \sigma_j \otimes \sigma_{p^{s+1}-j} + \sum_{i+j+p=p^{s+1}} \sigma_i\tau \otimes \sigma_j\tau \right],$$

and hence

$$(\rho^* \circ \pi^*)(w_{s+1}) = (-\eta^{-1})^p (a_0^p x_1 + \cdots + a_{r-1}^p x_r) + a_0^{p^{r+s}} w_{s+1}.$$

□

Corollary 3.1.2 *Under the identifications and assumptions of Lemma 2.3.8:*

1. *The homomorphism* $\psi_r : H(\mathbb{M}_{r;1}, k) \to k[\mathcal{N}_r(\mathbb{M}_{r;1})]$ *satisfies*

$$\psi_r(y) = \mu, \quad \text{and} \quad \psi_r(x_i) = \mathbf{a}_{r-i}^{p^i} \quad \text{for } 1 \leq i \leq r.$$

2. *Suppose* $s \geq 2$*. Then* $\psi_r : H(\mathbb{M}_{r;s}, k) \to k[\mathcal{N}_r(\mathbb{M}_{r;s})]$ *satisfies*

$$\psi_r(y) = \mu, \quad \psi_r(x_i) = \mathbf{a}_{r-i}^{p^i} \quad \text{for } 1 \leq i \leq r, \quad \text{and} \quad \psi_r(w_s) = (\mathbf{b}_s)^p.$$

3. *Suppose* $r \geq 2$*, let* $0 \neq \eta \in k$*, and identify* $H^\bullet(\mathbb{M}_{r;s,\eta}, k)$ *as an algebra with* $H^\bullet(\mathbb{M}_{r-1;s+1}, k)$*. Then* $\psi_r : H(\mathbb{M}_{r;s,\eta}, k) \to k[\mathcal{N}_r(\mathbb{M}_{r;s,\eta})]$ *satisfies* $\psi_r(y) = \mu$*,*

$$\psi_r(x_i) = \mathbf{a}_{r-i-1}^{p^{i+1}} \quad \text{for } 1 \leq i \leq r-1, \quad \text{and} \quad \psi_r(w_{s+1}) = (-\eta^{-1})^p \cdot \mathbf{a}_{r-1}^p.$$

4. $\psi_r : H(\mathbb{G}_{a(r)}, k) \to k[\mathcal{N}_r(\mathbb{G}_{a(r)})]$ *satisfies* $\psi_r(x_i) = \mathbf{a}_{r-i}^{p^i}$ *for* $1 \leq i \leq r$*.*
5. $\psi_r : H(\mathbb{G}_a^-, k) \to k[\mathcal{N}_r(\mathbb{G}_a^-)]$ *satisfies* $\psi_r(y) = \mu$*.*

In general, if k is algebraically closed and if G is a height-r infinitesimal elementary k-supergroup scheme, then $\psi_r : H(G, k) \to k[\mathcal{N}_r(G)]$ *is injective modulo nilpotents and its image contains the* p^r*-th powers of all elements in* $k[\mathcal{N}_r(G)]$*, so the induced map between maximal ideal spectra*

$$\Psi : \mathcal{N}_r(G) \to |G| := Max(H(G, k)).$$

is a homeomorphism of algebraic varieties.

Proof The five numbered statements follow from the explicit calculations in Lemma 3.1.1, the definition of the map ψ_r, and the descriptions in [9, Proposition 3.2.1] and [9, Lemma 3.2.4] of the inflation maps in cohomology arising from the quotient maps $\pi_{r;s} : \mathbb{M}_r \twoheadrightarrow \mathbb{M}_{r;s}$, $q : \mathbb{M}_r \twoheadrightarrow \mathbb{G}_{a(r)}$, and $q^- : \mathbb{M}_r \twoheadrightarrow \mathbb{G}_a^-$. The last statement is then an immediate consequence. □

3.2 Lyndon–Hochschild–Serre Spectral Sequences

Let $s \geq 1$. In this section we make some observations concerning various Lyndon–Hochschild–Serre (LHS) spectral sequences that arise from the canonical quotient homomorphisms $\pi_{r;s} : \mathbb{M}_r \twoheadrightarrow \mathbb{M}_{r;s}$, $q : \mathbb{M}_r \twoheadrightarrow \mathbb{G}_{a(r)}$, and $q^- : \mathbb{M}_r \twoheadrightarrow \mathbb{G}_a^-$.

3.2.1 LHS Spectral Sequence for $\pi_{r;s}$

As in the proof of [9, Proposition 3.2.1], let $N_{r;s}$ be the closed subsupergroup scheme of $\mathbb{M}_r$ defined by $k[N_{r;s}] = k[\mathbb{M}_r]/\langle \tau, \theta, \sigma_1, \ldots, \sigma_{p^s-1}\rangle$. Then $N_{r;s}$ is the kernel of the canonical quotient homomorphism $\pi_{r;s} : \mathbb{M}_r \twoheadrightarrow \mathbb{M}_{r;s}$. In particular, $N_{r;s}$ is normal in $\mathbb{M}_r$. As in [14, I.6.6] one gets for each rational $\mathbb{M}_r$-supermodule V an LHS spectral sequence

$$E(V) : \quad E_2^{i,j}(V) = \mathrm{H}^i(\mathbb{M}_{r;s}, \mathrm{H}^j(N_{r;s}, V)) \Rightarrow \mathrm{H}^{i+j}(\mathbb{M}_r, V). \tag{19}$$

Moreover, $E(k)$ is a spectral sequence of algebras, and $E(V)$ is a spectral sequence of right modules over $E(k)$ such that the product on the E_2-page, $E_2^{i,j}(V) \otimes E_2^{m,n}(k) \to E_2^{i+m,j+n}(V)$, identifies up to the sign $(-1)^{j \cdot m}$ with the usual cup product

$$\mathrm{H}^i(\mathbb{M}_{r;s}, \mathrm{H}^j(N_{r;s}, V)) \otimes \mathrm{H}^m(\mathbb{M}_{r;s}, \mathrm{H}^n(N_{r;s}, k)) \to \mathrm{H}^{i+m}(\mathbb{M}_{r;s}, \mathrm{H}^{j+n}(N_{r;s}, V)).$$

Since τ is an element of the defining ideal of $N_{r;s}$, it follows that $N_{r;s}$ is central in $\mathbb{M}_r$. This implies that $\mathbb{M}_{r;s}$ acts trivially on $\mathrm{H}^\bullet(N_{r;s}, k)$, and hence

$$E_2^{i,j}(k) = \mathrm{H}^i(\mathbb{M}_{r;s}, \mathrm{H}^j(N_{r;s}, k)) \cong \mathrm{H}^i(\mathbb{M}_{r;s}, k) \otimes \mathrm{H}^j(N_{r;s}, k).$$

As discussed in the proof of [9, Proposition 3.2.1], $\mathrm{H}^\bullet(N_{r;s}, k)$ is an exterior algebra generated by a cohomology class $[\widetilde{\sigma}_1] \in \mathrm{H}^1(N_{r;s}, k)_{\overline{0}}$, and the differential $d_2 : E_2^{0,1}(k) \to E_2^{2,0}(k)$ maps $[\widetilde{\sigma}_1]$ to $w_s \in \mathrm{H}^2(\mathbb{M}_{r;s}, k)$, where by convention $w_1 = x_1 - y^2$ (cf. [9, Remark 3.2.2]). By the derivation property of the differential, the map $d_2 : E_2^{\bullet,1}(k) \to E_2^{\bullet+2,0}(k)$ then identifies with multiplication by w_s, and this is the only nontrivial differential in (19).

Now suppose V is an $\mathbb{M}_{r;s}$-supermodule, considered as an $\mathbb{M}_r$-supermodule via $\pi_{r;s}$. Then $N_{r;s}$ acts trivially on V, and it follows that one has identifications

$$\mathrm{H}^\bullet(\mathbb{M}_{r;s}, \mathrm{H}^\bullet(N_{r;s}, V)) \cong \mathrm{H}^\bullet(\mathbb{M}_{r;s}, V \otimes \mathrm{H}^\bullet(N_{r;s}, k)) \cong \mathrm{H}^\bullet(\mathbb{M}_{r;s}, V) \otimes \mathrm{H}^\bullet(N_{r;s}, k),$$

which are compatible with the product structure on the spectral sequence. In particular, $E_2^{i,j}(V) = 0$ for $j \geq 2$, and it follows from the module structure of $E_2(V)$ over $E_2(k)$ that the differential

$$d_2 : \mathrm{H}^\bullet(\mathbb{M}_{r;s}, V) \otimes (k \cdot [\widetilde{\sigma}_1]) \cong E_2^{\bullet,1}(V) \to E_2^{\bullet+2,0}(V) = \mathrm{H}^{\bullet+2}(\mathbb{M}_{r;s}, V)$$

identifies with right multiplication on $\mathrm{H}^\bullet(\mathbb{M}_{r;s}, V)$ by $w_s \in \mathrm{H}^2(\mathbb{M}_{r;s}, k)$, and this is the only nontrivial differential of the spectral sequence $E(V)$. This description of the differential implies that the kernel of the inflation map $\mathrm{H}^\bullet(\mathbb{M}_{r;s}, V) \to \mathrm{H}^\bullet(\mathbb{M}_r, V)$ (which identifies with the horizontal edge map of (19)) is $\mathrm{H}^\bullet(\mathbb{M}_{r;s}, V).w_s$, the image of $\mathrm{H}^\bullet(\mathbb{M}_{r;s}, , V)$ under right multiplication by w_s.

Lemma 3.2.1 *Let V be a finite-dimensional rational $\mathbb{M}_r$-supermodule. Then the cohomology module $H^\bullet(\mathbb{M}_r, V)$ is finitely generated as a right $H^\bullet(\mathbb{M}_r, k)$-supermodule (or equivalently, as an $H(\mathbb{M}_r, k)$-supermodule).*

Proof First, let $\rho : \mathbb{M}_r \to GL(V)$ be the homomorphism that affords V the structure of a rational $\mathbb{M}_r$-supermodule. Since V is finite-dimensional, then $GL(V) \cong GL_{m|n}$ for some nonnegative integers m and n. In particular, $GL(V)$ is an algebraic k-supergroup scheme, so ρ factors for $s \gg 0$ through the quotient map $\pi_{r;s} : \mathbb{M}_r \twoheadrightarrow \mathbb{M}_{r;s}$ by [9, Remark 3.1.3(4)]. Then without loss of generality we may assume that V is an $\mathbb{M}_{r;s}$-supermodule for some fixed integer $s \geq 1$, and we may assume that the $\mathbb{M}_r$-supermodule structure on V is obtained by pulling back along $\pi_{r;s}$.

Now consider the LHS spectral sequence $E_2(V) \Rightarrow \mathrm{H}^\bullet(\mathbb{M}_r, V)$ as in the paragraph preceding the lemma, and recall from [7, Theorem 1.1] that $\mathrm{H}^\bullet(\mathbb{M}_{r;s}, V)$ is finite as a module under the cup product action of $\mathrm{H}^\bullet(\mathbb{M}_{r;s}, k)$. Since $\mathrm{H}^\bullet(N_{r;s}, k)$ is finite-dimensional, this implies that $E_2(V)$ is finite under the action of $E_2^{\bullet,0}(k) = \mathrm{H}^\bullet(\mathbb{M}_{r;s}, k)$, and hence implies by [13, Lemma 1.6] and [4, III.2.9 Corollary 1] that the inflation map $\mathrm{H}^\bullet(\mathbb{M}_{r;s}, k) \to \mathrm{H}^\bullet(\mathbb{M}_r, k)$ (which identifies with the horizontal edge map of the spectral sequence) makes $\mathrm{H}^\bullet(\mathbb{M}_r, V)$ into a finitely generated $\mathrm{H}^\bullet(\mathbb{M}_{r;s}, k)$-supermodule. In particular, $\mathrm{H}^\bullet(\mathbb{M}_r, V)$ is finite under the cup product action of $\mathrm{H}^\bullet(\mathbb{M}_r, k)$. Since $\mathrm{H}^\bullet(\mathbb{M}_r, k)$ and $H(\mathbb{M}_r, k)$ differ only by nilpotent elements, the last statement is equivalent to saying that $\mathrm{H}^\bullet(\mathbb{M}_r, V)$ is finite as a $H(\mathbb{M}_r, k)$-supermodule. □

3.2.2 LHS Spectral Sequence for q

The kernel of the quotient homomorphism $q : \mathbb{M}_r \twoheadrightarrow \mathbb{G}_{a(r)}$ is the closed subsupergroup scheme Q of $\mathbb{M}_r$ defined by $k[Q] = k[\mathbb{M}_r]/\langle\theta\rangle$. Let $\widetilde{\tau}$ and $\widetilde{\sigma}_i$ denote the images in $k[Q]$ of the generators $\tau, \sigma_{ip} \in k[\mathbb{M}_r]$, respectively. Then the set $\{\widetilde{\sigma}_i, \widetilde{\sigma}_i\widetilde{\tau} : i \in \mathbb{N}\}$ is a homogeneous basis for $k[Q]$. Now let $T(v) \cong k[v]$ be the tensor (super)algebra over k generated by the element v of odd superdegree. We consider the elements of $k[Q]$ as linear functionals on $T(v)$ such that $\widetilde{\sigma}_i(v^j) = (-1)^i \cdot \delta_{2i,j}$ and $(\widetilde{\sigma}_i\widetilde{\tau})(v^j) = \delta_{2i+1,j}$, where $\delta_{i,j}$ denotes the usual Kronecker delta. In this way, one can then check that the Hochschild complex $C^\bullet(Q, k) = k[Q]^{\otimes\bullet}$ for Q identifies with the cobar complex for $T(v)$ as defined in [17]. Then by [17, Theorem 2.5], the cohomology ring $\mathrm{H}^\bullet(Q, k)$ is the exterior algebra over k generated by the cohomology class $[\widetilde{\tau}] \in \mathrm{H}^1(Q, k)_{\overline{1}}$, i.e., $\mathrm{H}^\bullet(Q, k) \cong \Lambda([\widetilde{\tau}])$. Since the generator $y \in \mathrm{H}^\bullet(\mathbb{M}_r, k)$ is by definition the cohomology class of the cochain $\tau \in k[\mathbb{M}_r]$, it immediately follows that the restriction map in cohomology $\mathrm{H}^\bullet(\mathbb{M}_r, k) \to \mathrm{H}^\bullet(Q, k)$ is surjective and sends y to $[\widetilde{\tau}]$.

Now consider the LHS spectral sequence for the extension $Q \hookrightarrow \mathbb{M}_r \overset{q}{\twoheadrightarrow} \mathbb{G}_{a(r)}$:

$$E_2^{i,j}(k) = \mathrm{H}^i(\mathbb{G}_{a(r)}, \mathrm{H}^j(Q, k)) \Rightarrow \mathrm{H}^{i+j}(\mathbb{M}_r, k). \tag{20}$$

Since $\mathrm{H}^0(Q, k)$ and $\mathrm{H}^1(Q, k)$ are each one-dimensional, and since $\mathbb{G}_{a(r)}$ is unipotent, it follows that $\mathbb{G}_{a(r)}$ acts trivially on $\mathrm{H}^\bullet(Q, k)$. Then $E_2^{i,j}(k) \cong \mathrm{H}^i(\mathbb{G}_{a(r)}, k) \otimes \mathrm{H}^j(Q, k)$, and the only nontrivial differential in (20) is $d_2 : E_2^{\bullet,1} \to E_2^{\bullet+2,0}$. The horizontal edge map of (20) identifies with the inflation map $\mathrm{H}^\bullet(\mathbb{G}_{a(r)}, k) \to \mathrm{H}^\bullet(\mathbb{M}_r, k)$, and so is injective by [9, Lemma 2.3.4]. Then $d_2 \equiv 0$.

Finally, let V be a finite-dimensional $\mathbb{G}_{a(r)}$-supermodule, considered as a rational $\mathbb{M}_r$-supermodule via the quotient $q : \mathbb{M}_r \twoheadrightarrow \mathbb{G}_{a(r)}$. Then as in Sect. 3.2.1 above, the E_2-page of the LHS spectral sequence $E_2^{i,j}(V) = \mathrm{H}^i(\mathbb{G}_{a(r)}, \mathrm{H}^j(Q, V)) \Rightarrow \mathrm{H}^{i+j}(\mathbb{M}_r, V)$ admits a factorization

$$E_2^{i,j}(V) \cong \mathrm{H}^i(\mathbb{G}_{a(r)}, V) \otimes \mathrm{H}^j(Q, k).$$

Using this, the right module structure of $E_2(V)$ over $E_2(k)$, the derivation property of the differential, and the fact that the differential on $E_2(k)$ is trivial, it follows that the differential on $E_2(V)$ is trivial as well. Consequently, the inflation map $\mathrm{H}^\bullet(\mathbb{G}_{a(r)}, V) \to \mathrm{H}^\bullet(\mathbb{M}_r, V)$ is an injection.

3.2.3 LHS Spectral Sequence for q^-

The kernel of the quotient homomorphism $q^- : \mathbb{M}_r \twoheadrightarrow \mathbb{G}_a^-$ is the closed subsupergroup scheme Q^- of $\mathbb{M}_r$ defined by $k[Q^-] = k[\mathbb{M}_r]/\langle \tau \rangle$. In particular, Q^- is abelian because $k[Q^-]$ is cocommutative in the sense of Hopf superalgebras. Let $\widetilde{\theta}$ and $\widetilde{\sigma}_i$ denote the images in $k[Q^-]$ of the generators $\theta, \sigma_i \in k[\mathbb{M}_r]$, respectively. The supergroup $N_{r;1}$ defined in Sect. 3.2.1 is a (normal) subsupergroup of Q^- because $k[Q^-]/\langle \widetilde{\theta} \rangle \cong k[\mathbb{M}_r]/\langle \theta, \tau \rangle = k[N_{r;1}]$, and $Q^-/N_{r;1} \cong \mathbb{G}_{a(r)}$. Consider the LHS spectral sequence for the extension $N_{r;1} \hookrightarrow Q^- \twoheadrightarrow \mathbb{G}_{a(r)}$:

$$E_2^{i,j} = \mathrm{H}^i(\mathbb{G}_{a(r)}, \mathrm{H}^j(N_{r;1}, k)) \Rightarrow \mathrm{H}^{i+j}(Q^-, k). \tag{21}$$

Since Q^- is abelian, the action of $Q^-/N_{r;1} \cong \mathbb{G}_{a(r)}$ on $\mathrm{H}^\bullet(N_{r;1}, k)$ (induced by conjugation) is trivial, and hence $E_2^{i,j} \cong \mathrm{H}^i(\mathbb{G}_{a(r)}, k) \otimes \mathrm{H}^j(N_{r;1}, k)$. From the discussion in Sect. 3.2.1 above, the cohomology ring $\mathrm{H}^\bullet(N_{r;1}, k)$ is an exterior algebra generated by the cohomology class $[\widehat{\sigma}] \in \mathrm{H}^1(N_{r;1}, k)$, where $\widehat{\sigma}$ denotes the image of $\widetilde{\sigma}_p \in k[Q^-]$ under the quotient map $k[Q^-] \twoheadrightarrow k[Q^-]/\langle \widehat{\theta} \rangle = k[N_{r;1}]$. The differential $d_2 : E_2^{0,1} \to E_2^{2,0}$ is induced by the differential ∂ on the Hochschild complex for Q^-. Then in terms of the map β of (17),

$$d_2([\widehat{\sigma}]) = [\partial(\widetilde{\sigma}_p)] = [-\textstyle\sum_{i=1}^{p-1} \widetilde{\sigma}_i \otimes \widetilde{\sigma}_{p-i}] = [\beta(\widetilde{\theta}^{p^{r-1}})] = x_r \in \mathrm{H}^2(\mathbb{G}_{a(r)}, k).$$

Since $d_2 : E_2^{\bullet,1} \to E_2^{\bullet+2,0}$ is the only nontrivial differential in (21), and since x_r is not a zero divisor in $\mathrm{H}^\bullet(\mathbb{G}_{a(r)}, k)$, this implies that the inflation map $\mathrm{H}^\bullet(\mathbb{G}_{a(r)}, k) \to \mathrm{H}^\bullet(Q^-, k)$, which identifies with the horizontal edge map of (21), is surjective with kernel generated by x_r, i.e.,

$$\mathrm{H}^\bullet(Q^-, k) \cong k[x_1, x_2, \dots, x_{r-1}] \,{}^{g}\!\otimes \Lambda(\lambda, \dots, \lambda_r).$$

Moreover, replacing θ by $\widetilde{\theta}$, cochain representatives for the generators of $\mathrm{H}^\bullet(Q^-, k)$ are given by the same formulas as in [9, Proposition 3.2.1].

Now consider the LHS spectral sequence (with trivial coefficients) for the supergroup extension $Q^- \hookrightarrow \mathbb{M}_r \twoheadrightarrow \mathbb{G}_a^-$. Since τ is an element of the defining ideal of Q^-, it follows that Q^- is central in $\mathbb{M}_r$, and hence the spectral sequence takes the form

$$E_2^{i,j} \cong \mathrm{H}^i(\mathbb{G}_a^-, k) \otimes \mathrm{H}^j(Q^-, k) \Rightarrow \mathrm{H}^{i+j}(\mathbb{M}_r, k). \tag{22}$$

This spectral sequence is not confined simply to two rows, but it follows from inspecting the explicit formulas for the cochain representatives of the generators that the restriction map $\mathrm{H}^\bullet(\mathbb{M}_r, k) \to \mathrm{H}^\bullet(Q^-, k)$, which identifies with the vertical edge map of (22), is surjective with kernel generated by $y \in \mathrm{H}^\bullet(\mathbb{M}_r, k)$. Then the column $E_2^{0,\bullet}$ in (22) consists of permanent cycles, which implies by the derivation property of the differential that all differentials in (22) are zero.

Finally, let V be a finite-dimensional $\mathbb{G}_a^-$-supermodule, considered as a rational $\mathbb{M}_r$-supermodule via the quotient $q^- : \mathbb{M}_r \to \mathbb{G}_a^-$. Then by reasoning precisely analogous to that in Sect. 3.2.2 above, it follows that the LHS spectral sequence

$$E_2^{i,j}(V) = \mathrm{H}^i(\mathbb{G}_a^-, \mathrm{H}^j(Q^-, V)) \Rightarrow \mathrm{H}^{i+j}(\mathbb{M}_r, V)$$

admits the factorization $E_2^{i,j}(V) \cong \mathrm{H}^i(\mathbb{G}_a^-, V) \otimes \mathrm{H}^j(Q^-, k)$ on the E_2-page, the differentials are all trivial, and the inflation map $\mathrm{H}^\bullet(\mathbb{G}_a^-, V) \to \mathrm{H}^\bullet(\mathbb{M}_r, V)$ is an injection.

3.3 Support Sets for Modules

Definition 3.3.1 Let G be an affine k-supergroup scheme. Given a rational G-supermodule V, define the injective dimension of V (in the category of rational G-supermodules) by

$$\mathrm{id}_G(V) = \sup \left\{ n \in \mathbb{N} : \mathrm{Ext}_G^n(-, V) \neq 0 \right\}.$$

Lemma 3.3.2 *Let G be an affine k-supergroup scheme, and let V be a rational G-supermodule. Then the following are equivalent:*

1. $id_G(V) \leq n < \infty$

2. *$\mathrm{Ext}^i_G(L, V) = 0$ for all $i > n$ and all irreducible rational G-supermodules L.*
3. *There exists an injective resolution of V,*

$$0 \to V \to Q_0 \to Q_1 \to Q_2 \to \cdots,$$

such that $Q_i = 0$ for all $i > n$.

Proof The first condition evidently implies the second, and the third condition evidently implies the first, so suppose that $\mathrm{Ext}^i_G(L, V) = 0$ for all $i > n$ and all irreducible rational G-supermodules L. Let $Q_\bullet$ be a minimal injective resolution of V. The minimality of $Q_\bullet$ implies for each irreducible G-supermodule L that $\mathrm{Ext}^i_G(L, V) = \mathrm{Hom}_G(L, Q_i)$, and hence $\mathrm{Hom}_G(L, Q_i) = 0$ for all $i > n$ and each irreducible L. Then $\mathrm{soc}_G(Q_i) = 0$ for $i > n$, which implies that $Q_i = 0$ for $i > n$. □

Lemma 2.1.4 and the local finiteness of rational representations imply that $\mathbb{M}_r$ is unipotent, and hence that up to isomorphism and parity flip, the only irreducible $\mathbb{M}_r$-supermodule is the trivial module k. Thus for $G = \mathbb{M}_r$, it suffices to consider just $L = k$ in Lemma 3.3.2(2).

Lemma 3.3.3 *Let G be an affine k-supergroup scheme, and let V be a finite-dimensional rational G-supermodule. Then*

$$id_G(V) = id_G(V \otimes V^\#) = id_G(Hom_k(V, V)).$$

Proof First observe that since V is finite-dimensional, the natural map $\alpha : V \otimes V^\# \to \mathrm{Hom}_k(V, V)$, defined for $v, v' \in V$ and $\phi \in V^\#$ by $\alpha(v \otimes \phi)(v') = \phi(v') \cdot v$, is a G-supermodule isomorphism. Then it suffices to show that $\mathrm{id}_G(V \otimes V^\#) \leq \mathrm{id}_G(V)$ and $\mathrm{id}_G(V) \leq \mathrm{id}_G(V \otimes V^\#)$. To prove the first inequality, suppose $\mathrm{id}_G(V) = n < \infty$. Then by Lemma 3.3.2, there exists an injective resolution $Q_\bullet$ of V with the property that $Q_i = 0$ for $i > n$. Then $Q_\bullet \otimes V^\#$ is an injective resolution of $V \otimes V^\#$ (cf. [14, I.3.10]), so we deduce that there exists an injective resolution $Q'_\bullet$ of $V \otimes V^\#$ such that $Q'_i = 0$ for $i > n$. Then $\mathrm{id}_G(V \otimes V^\#) \leq n = \mathrm{id}_G(V)$.

For the other inequality, suppose $\mathrm{id}_G(V \otimes V^\#) = n < \infty$. Then by the same type of reasoning as in the previous paragraph, $\mathrm{id}_G((V \otimes V^\#) \otimes V) \leq n < \infty$, and hence $\mathrm{Ext}^i_G(L, V \otimes V^\# \otimes V) = 0$ for all $i > n$ and all irreducible G-supermodules L. Define $\tau : k \to \mathrm{Hom}_k(V, V)$ by $\tau(\lambda) = \lambda \cdot 1_V$, and let $c : V^\# \otimes V \to k$ be the contraction map defined by $c(\phi \otimes v) = \phi(v)$. It is straightforward to check that τ and c are both G-supermodule homomorphisms. If $\{v_1, \dots, v_m\}$ is a fixed homogeneous basis for V, and if $\{v_1^*, \dots, v_m^*\}$ is the corresponding dual basis such that $v_i^*(v_j) = \delta_{i,j}$, then under the identification $\mathrm{Hom}_k(V, V) \cong V \otimes V^\#$ the map τ satisfies $\tau(\lambda) = \lambda \cdot (\sum_{i=1}^m v_i \otimes v_i^*)$. Now consider the composite G-supermodule homomorphism

$$V \cong k \otimes V \xrightarrow{\tau \otimes 1} V \otimes V^\# \otimes V \xrightarrow{1 \otimes c} V \otimes k \cong V.$$

It is straightforward to check (e.g., by first considering the effect on basis vectors and then extending linearly) that this composite is the identity. Consequently, V is isomorphic to a G-supermodule direct summand of $V \otimes V^{\#} \otimes V$. Then the equality $\mathrm{Ext}^i_G(L, V \otimes V^{\#} \otimes V) = 0$ for $i > n$ implies that $\mathrm{Ext}^i_G(L, V) = 0$ for $i > n$, and hence implies that $\mathrm{id}_G(V) \leq n = \mathrm{id}_G(V \otimes V^{\#})$. □

Given a supergroup scheme homomorphism $\phi : G \to H$ and a rational H-supermodule V, let $\phi^* V$ denote the rational G-supermodule obtained from V by pulling back along ϕ.

Definition 3.3.4 Let G be an algebraic k-supergroup scheme. Given a rational G-supermodule M, define the support set $\mathcal{N}_1(G)_M \subset \mathcal{N}_1(G)$ by

$$\mathcal{N}_1(G)_M = \left\{\phi \in \mathcal{N}_1(G) = \mathbf{Hom}(\mathbb{M}_1, G)(k) : \mathrm{id}_{\mathbb{M}_1}(\phi^* M) = \infty\right\}.$$

Applying Lemma 3.3.2 and the observation immediately following it, Lemma 3.3.3, and the adjoint associativity isomorphism, it follows that if M is finite-dimensional, then

$$\mathcal{N}_1(G)_M = \left\{\phi \in \mathcal{N}_1(G) : \mathrm{Ext}^i_{\mathbb{M}_1}(\phi^* M, \phi^* M) \neq 0 \text{ for infinitely many } i \geq 1\right\}.$$

In Sect. 3.4 we will show that if G is a height-one infinitesimal elementary supergroup scheme and if M is a finite-dimensional G-supermodule, then $\mathcal{N}_1(G)_M$ is closed in $\mathcal{N}_1(G)$ and identifies with the cohomological support variety $|G|_M$. More generally, in Sect. 3.5 we will discuss how we expect that results recently announced by Benson, Iyengar, Krause, and Pevtsova (BIKP) can be applied to obtain more general results concerning the cohomological spectrum $|G|$ and the support varieties $|G|_M$ when G is a more general infinitesimal unipotent supergroup scheme.

Question 3.3.5 Suppose k is algebraically closed. Let G be an arbitrary height-1 infinitesimal k-supergroup scheme, and let M be a finite-dimensional G-supermodule. Is $\mathcal{N}_1(G)_M$ Zariski closed in $\mathcal{N}_1(G)$?

As mentioned in the introduction to the paper, our definition of the support set $\mathcal{N}_1(G)_M$ is inspired by similar definitions appearing in the literature in the context of commutative local rings [1, 2, 15], and which were brought to our attention by way of a talk by Srikanth Iyengar. At present we do not understand how the existing support theory can be related to the rational cohomology of $\mathbb{M}_1$, but some obvious intermediate actors are the group algebra $k\mathbb{M}_1 = k[[u, v]]/\langle u^p + v^2\rangle$ and its polynomial subalgebra $\mathbb{P}_1 = k[u, v]/\langle u^p + v^2\rangle$. In the rest of this section we will consider the more general question of the relationship between rational cohomology for $\mathbb{M}_r$ and cohomology for its polynomial subalgebra $\mathbb{P}_r$.

Let $_{\mathbb{M}_r}\mathfrak{smod}$ denote the category of rational left $\mathbb{M}_r$-supermodules, and let $_{\mathbb{P}_r}\mathfrak{smod}$ denote the category of left $\mathbb{P}_r$-supermodules. It follows from Proposition 2.3.2 that $_{\mathbb{M}_r}\mathfrak{smod}$ identifies with the full subcategory of $_{\mathbb{P}_r}\mathfrak{smod}$ consisting of the objects on which the generator $u_{r-1} \in \mathbb{P}_r$ acts via a locally nilpotent endomorphism. The exact inclusion of categories ${}_{\mathbb{M}_r}\mathfrak{smod} \hookrightarrow {}_{\mathbb{P}_r}\mathfrak{smod}$ then induces for each pair of rational $\mathbb{M}_r$-supermodules M and N a linear map on cohomology groups

$$\operatorname{Ext}^\bullet_{\mathbb{M}_r}(M, N) \to \operatorname{Ext}^\bullet_{\mathbb{P}_r}(M, N). \tag{23}$$

We will show that, if M and N are finite-dimensional then (23) is an isomorphism.

Proposition 3.3.6 *Let M and N be finite-dimensional rational $\mathbb{M}_r$-supermodules. Then the map in cohomology* (23) *induced by the inclusion ${}_{\mathbb{M}_r}\mathfrak{smod} \hookrightarrow {}_{\mathbb{P}_r}\mathfrak{smod}$ is an isomorphism.*

Proof Set $V = N \otimes M^\#$. Then by the usual adjoint associativity isomorphisms (cf. [14, I.4.4] and [8, Lemma 2.3.4]), it suffices to show that the map $\mathrm{H}^\bullet(\mathbb{M}_r, V) \to \mathrm{H}^\bullet(\mathbb{P}_r, V)$ is an isomorphism. Let $\Delta_V : V \to V \otimes k[\mathbb{M}_r]$ be the comodule map that defines the rational $\mathbb{M}_r$-supermodule structure on V. Since V is finite-dimensional, there exists an integer $s \geq 1$ such that the image of Δ_V is contained in $V \otimes k[\mathbb{M}_{r;s}]$, and hence such that the action of $\mathbb{M}_r$ on V factors through the quotient $\mathbb{M}_r \twoheadrightarrow \mathbb{M}_r/N_{r;s} = \mathbb{M}_{r;s}$. Set $A = \mathbb{P}_r$, and let $B = k[x]$ be the Hopf subalgebra of A generated by $x := u_{r-1}^{p^s}$. Then B is a central, primitively generated Hopf subalgebra of A, and the Hopf quotient $A /\!/ B = A/\langle x\rangle$ is just $\mathbb{P}_r/\langle u_{r-1}^{p^s}\rangle = k\mathbb{M}_{r;s}$. Since $N_{r;s}$ acts trivially on V, it follows that B does as well, and hence that the action of $\mathbb{P}_r$ on V factors through the quotient $A /\!/ B$. Now our strategy for showing that the map $\mathrm{H}^\bullet(\mathbb{M}_r, V) \to \mathrm{H}^\bullet(\mathbb{P}_r, V)$ is an isomorphism will be to show that the inclusion of categories ${}_{\mathbb{M}_r}\mathfrak{smod} \hookrightarrow {}_{\mathbb{P}_r}\mathfrak{smod}$ induces a morphism of spectral sequences between (i) the LHS spectral sequence of rational cohomology groups for the group extension

$$1 \to N_{r;s} \to \mathbb{M}_r \to \mathbb{M}_{r;s} \to 1, \tag{24}$$

and (ii) the LHS spectral sequence for the extension of Hopf superalgebras

$$0 \to B \to A \to A /\!/ B \to 0. \tag{25}$$

We'll show that this morphism is an isomorphism on the E_2-page, and hence induces a morphism between the abutments $\mathrm{H}^\bullet(\mathbb{M}_r, V)$ and $\mathrm{H}^\bullet(\mathbb{P}_r, V)$.

Let $V \to I^\bullet$ be a resolution of V by injective rational $\mathbb{M}_r$-supermodules, and let $V \to Q^\bullet$ be a resolution of V by injective $\mathbb{P}_r$-supermodules. Then $V \to I^\bullet$ remains exact as a complex of $\mathbb{P}_r$-supermodules, so there exists a $\mathbb{P}_r$-supermodule chain homomorphism $I^\bullet \to Q^\bullet$ that lifts the identity on V. For W a rational $\mathbb{M}_r$-supermodule, considered also as a $\mathbb{P}_r$-supermodule, one has $W^B = W^{N_{r;s}}$. Then the map $I^\bullet \to Q^\bullet$ restricts to a chain map $(I^\bullet)^{N_{r;s}} \to (Q^\bullet)^B$. Since $\mathbb{M}_r/N_{r;s} = \mathbb{M}_{r;s}$ and $A /\!/ B = k\mathbb{M}_{r;s}$, and since $\mathbb{M}_{r;s}$ is a finite k-supergroup scheme, we can consider $(I^\bullet)^{N_{r;s}}$ and $(Q^\bullet)^B$ both either as complexes of rational $\mathbb{M}_{r;s}$-supermodules or as complexes of $k\mathbb{M}_{r;s}$-supermodules. So let $(I^\bullet)^{N_{r;s}} \to J^{\bullet,\bullet}$ and $(Q^\bullet)^B \to K^{\bullet,\bullet}$ be Cartan-Eilenberg resolutions of $(I^\bullet)^{N_{r;s}}$ and $(Q^\bullet)^B$, respectively, in the category of $k\mathbb{M}_{r;s}$-supermodules. Then the chain map $(I^\bullet)^{N_{r;s}} \to (Q^\bullet)^B$ lifts to a double complex chain map $J^{\bullet,\bullet} \to K^{\bullet,\bullet}$ of $k\mathbb{M}_{r;s}$-supermodules. Now the LHS spectral sequence of rational cohomology groups for the group extension (24),

$$E_2^{i,j}(\mathbb{M}_r) = \mathrm{H}^i(\mathbb{M}_{r;s}, \mathrm{H}^j(N_{r;s}, V)) \Rightarrow \mathrm{H}^{i+j}(\mathbb{M}_r, V),$$

arises from the double complex $(J^{\bullet,\bullet})^{\mathbb{M}_{r;s}}$ by first computing cohomology along columns and then along rows, while the LHS spectral sequence for the extension of Hopf superalgebras (25),

$$E_2^{i,j}(\mathbb{P}_r) = \mathrm{H}^i(A /\!\!/ B, \mathrm{H}^j(B, V)) \Rightarrow \mathrm{H}^{i+j}(A, V),$$

arises from the double complex $(K^{\bullet,\bullet})^B$ in the same manner. The chain map $J^{\bullet,\bullet} \to K^{\bullet,\bullet}$ restricts to a double complex chain map $(J^{\bullet,\bullet})^{\mathbb{M}_{r;s}} \to (K^{\bullet,\bullet})^{k\mathbb{M}_{r;s}}$, and hence we get a morphism of spectral sequences $E(\mathbb{M}_r) \to E(\mathbb{P}_r)$. On the E_2-page, the map

$$\mathrm{H}^i(\mathbb{M}_{r;s}, \mathrm{H}^j(N_{r;s}, V)) \to \mathrm{H}^i(k\mathbb{M}_{r;s}, \mathrm{H}^j(B, V))$$

is induced by the equivalence of categories ${}_{\mathbb{M}_{r;s}}\mathfrak{smod} \cong {}_{k\mathbb{M}_{r;s}}\mathfrak{smod}$ and by the inclusion of categories ${}_{N_{r;s}}\mathfrak{smod} \hookrightarrow {}_B\mathfrak{smod}$, the latter of which gives rise to a homomorphism of $\mathbb{M}_{r;s}$-supermodules (equivalently, of $k\mathbb{M}_{r;s}$-supermodules) $\mathrm{H}^j(N_{r;s}, V) \to \mathrm{H}^j(B, V)$. Thus to show that the morphism of spectral sequences $E(\mathbb{M}_r) \to E(\mathbb{P}_r)$ is an isomorphism from the E_2-page onwards, and hence show that the abutments $\mathrm{H}^\bullet(\mathbb{M}_r, V)$ and $\mathrm{H}^\bullet(\mathbb{P}_r, V)$ are isomorphic, it suffices to show that the inclusion of categories ${}_{N_{r;s}}\mathfrak{smod} \hookrightarrow {}_B\mathfrak{smod}$ induces an isomorphism $\mathrm{H}^\bullet(N_{r;s}, V) \cong \mathrm{H}^\bullet(B, V)$. Since $N_{r;s}$ and B each act trivially on V, we have $\mathrm{H}^\bullet(N_{r;s}, V) \cong \mathrm{H}^\bullet(N_{r;s}, k) \otimes V$ and $\mathrm{H}^\bullet(B, V) \cong \mathrm{H}^\bullet(B, k) \otimes V$, so it suffices even to assume that $V = k$. In this case we know that $\mathrm{H}^\bullet(N_{r;s}, k)$ and $\mathrm{H}^\bullet(B, k) = \mathrm{H}^\bullet(k[x], k)$ are both exterior algebras generated in cohomological degree one, and one can check that the generator $[\widetilde{\sigma}_1] \in \mathrm{H}^1(N_{r;s}, k)$ gets mapped to the generator in $\mathrm{H}^1(k[x], k)$. Thus $\mathrm{H}^\bullet(N_{r;s}, k) \cong \mathrm{H}^\bullet(B, k)$. □

3.4 Support Varieties in Height One

In this section, we describe the support variety $|G|_M$ of a finite-dimensional rational G-module M in the case when G is a height-one infinitesimal elementary supergroup scheme, i.e., when G is one of $\mathbb{G}_{a(1)}$, $\mathbb{G}_a^-$, or $\mathbb{M}_{1;s}$ for some $s \geq 1$. *Throughout this section assume that k is an algebraically closed field.*

Let G be a finite supergroup scheme. Recall that the cohomological spectrum $|G|$ of G is the maximal ideal spectrum of the cohomology ring $\mathrm{H}^\bullet(G, k)$, or equivalently (cf. [8, Corollary 2.2.5]) the maximal ideal spectrum of the subring $H(G, k)$ defined in (13):

$$|G| := \mathrm{Max}\,(\mathrm{H}^\bullet(G, k)) = \mathrm{Max}\,(H(G, k))\,.$$

Now let M be a finite-dimensional rational G-supermodule. Set $\Lambda = \mathrm{Hom}_k(M, M)$. Then Λ is a finite-dimensional unital rational G-algebra, and the unit map $1_\Lambda : k \to \Lambda$ is a G-supermodule homomorphism. Let $\rho_\Lambda : \mathrm{H}^\bullet(G, k) \to \mathrm{H}^\bullet(G, \Lambda)$ be the map in cohomology induced by 1_Λ, and set $I_M = \ker(\rho_\Lambda)$. By abuse of notation, we also denote the intersection $I_M \cap H(G, k)$ by I_M. Then the cohomological support variety $|G|_M$ is the Zariski closed subvariety of $|G|$ defined by I_M:

$$|G|_M := \mathrm{Max}\,(\mathrm{H}^\bullet(G, k)/I_M) = \mathrm{Max}\,(H(G, k)/I_M) = \mathrm{Max}\left(H(G, k)/\sqrt{I_M}\right).$$

Using the adjoint associativity isomorphism

$$\mathrm{H}^\bullet(G, \Lambda) = \mathrm{Ext}^\bullet_G(k, \mathrm{Hom}_k(M, M)) \cong \mathrm{Ext}^\bullet_G(k, M \otimes M^\#) \cong \mathrm{Ext}^\bullet_G(M, M),$$

one can check that this definition for $|G|_M$ agrees with [8, Definition 2.3.8]. Letting $1_M \in \mathrm{Hom}_G(M, M) \cong \mathrm{H}^0(G, \Lambda)$ denote the identity map on M, one can check that $\rho_\Lambda(z) = 1_M \cdot z$, the image of 1_M under the right cup product action of z.

Now let G be a height-one infinitesimal elementary k-supergroup scheme, and let M be a finite-dimensional rational G-supermodule. By Corollary 3.1.2, the map $\psi_1 : H(G, k) \to k[\mathcal{N}_1(G)]$ induces a homeomorphism of varieties $\Psi : \mathcal{N}_1(G) \to |G|$. The main result of this section is to relate the cohomological support variety $|G|_M$ to the support set $\mathcal{N}_1(G)_M$ defined in Definition 3.3.4.

Theorem 3.4.1 *Suppose k is algebraically closed. Let G be a height-one infinitesimal elementary k-supergroup scheme, and let M be a finite-dimensional rational G-supermodule. Then*

$$\Psi^{-1}(|G|_M) = \mathcal{N}_1(G)_M = \left\{\phi \in \mathcal{N}_1(G) : id_{\mathbb{M}_1}(\phi^* M) = \infty\right\}$$

In particular, $\mathcal{N}_1(G)_M$ is closed in $\mathcal{N}_1(G)$.

We break the proof up into cases, depending on the isomorphism type of G.

3.4.1 Proof of Theorem 3.4.1 for $G = \mathbb{M}_{1;s}$ and $s \geq 2$

In this case

$$H(\mathbb{M}_{1;s}, k) \cong k[x_1, y, w_s]/\langle x_1 - y^2\rangle \cong k[y, w_s].$$

The points of the cohomological spectrum $|\mathbb{M}_{1;s}|$ may be identified with k-algebra homomorphisms $H(\mathbb{M}_{1;s}, k) \to k$. Given scalars $\mu, a, b \in k$ such that $\mu^2 = a$, let $\widehat{\phi}_{(\mu,a,b)} : H(\mathbb{M}_{1;s}, k) \to k$ denote the homomorphism such that $\widehat{\phi}_{(\mu,a,b)}(y) = \mu$, $\widehat{\phi}_{(\mu,a,b)}(x_1) = a$, and $\widehat{\phi}_{(\mu,a,b)}(w_s) = b$. Similarly, given scalars $\mu, a, b \in k$ such that $\mu^2 = a^p$, let $\phi_{(\mu,a,b)} : \mathbb{M}_1 \to \mathbb{M}_{1;s}$ denote the homomorphism labeled by (μ, a, b) as in Proposition 2.2.2(2). Then Corollary 3.1.2 implies that $\Psi(\phi_{(\mu,a,b)}) = \widehat{\phi}_{(\mu,a^p,b^p)}$.

Observe that $\mathrm{Aut}(\mathbb{M}_{1;s}) = \mathbf{Aut}(\mathbb{M}_{1;s})(k)$ acts on $\mathcal{N}_1(\mathbb{M}_{1;s})$ via composition, $\nu_* : \phi \mapsto \nu \circ \phi$, and acts on $|\mathbb{M}_{1;s}|$ by the $\nu_*(\widehat{\phi}) = \widehat{\phi} \circ \nu^* : H(\mathbb{M}_{1;s}, k) \to H(\mathbb{M}_{1;s}, k) \to k$. Moreover, it follows from the last assertion of Proposition 2.4.2 that Ψ is $\mathrm{Aut}(\mathbb{M}_{1;s})$-equivariant. Given scalars $\lambda, c, d \in k$ such that $\lambda^2 = c^p \neq 0$, let $\nu = \nu_{(\lambda,c,d)} : \mathbb{M}_{1;s} \to \mathbb{M}_{1;s}$ be the corresponding automorphism as in Lemma 2.2.5. Then an elementary computation (noting that comorphisms compose in the reverse order from homomorphisms) and Lemma 3.1.1(2) imply that

$$\nu_*(\phi_{(\mu,a,b)}) = \phi_{(\mu\lambda,ac,bc^{p^{s-1}}+ad)} \quad \text{and} \quad \nu_*(\widehat{\phi}_{(\mu,a,b)}) = \widehat{\phi}_{(\mu\lambda,ac^p,bc^{p^s}+ad^p)}.$$

Furthermore, the action of the automorphism ν on $|G|$ maps $|G|_{\nu^* M}$ isomorphically onto $|G|_M$. Now, there are precisely two orbits of $\mathrm{Aut}(\mathbb{M}_{1;s})$ in $\mathcal{N}_1(\mathbb{M}_{1;s})$: the orbit of the canonical quotient map $\pi_{1;s} = \phi_{(1,1,0)} : \mathbb{M}_1 \to \mathbb{M}_{1;s}$ and the orbit of $\phi_{(0,0,1)} : \mathbb{M}_1 \to \mathbb{M}_{1;s}$. Then to finish the proof of the theorem in the case $G = \mathbb{M}_{1;s}$ with $s \geq 2$, it suffices to show that

1. $\widehat{\phi}_{(1,1,0)} \in |\mathbb{M}_{1;s}|_M$ if and only if $\phi_{(1,1,0)} \in \mathcal{N}_1(\mathbb{M}_{1;s})_M$, and
2. $\widehat{\phi}_{(0,0,1)} \in |\mathbb{M}_{1;s}|_M$ if and only if $\phi_{(0,0,1)} \in \mathcal{N}_1(\mathbb{M}_{1;s})_M$.

First we prove statement (1). Set $\Lambda = \mathrm{Hom}_k(M, M)$, and set $\pi = \pi_{1;s} = \phi_{(1,1,0)}$. Then there exists a commutative diagram

$$\begin{array}{ccc} H(\mathbb{M}_{1;s}, k) & \xrightarrow{\rho_\Lambda} & \mathrm{H}^\bullet(\mathbb{M}_{1;s}, \Lambda) \\ \downarrow{\scriptstyle \pi^*} & & \downarrow{\scriptstyle \pi^*} \\ H(\mathbb{M}_1, k) & \xrightarrow{\rho_{\pi^*\Lambda}} & \mathrm{H}^\bullet(\mathbb{M}_1, \pi^*\Lambda) \end{array}$$

in which the horizontal arrows are induced by the unit map $k \to \Lambda$, and the vertical arrows are the inflation maps induced by π. First suppose $\pi \in \mathcal{N}_1(\mathbb{M}_{1;s})_M$, so that $\mathrm{H}^i(\mathbb{M}_1, \pi^*\Lambda) \neq 0$ for infinitely many $i \geq 0$. Since $\mathrm{H}^\bullet(\mathbb{M}_1, \pi^*\Lambda)$ is finitely generated over $H(\mathbb{M}_1, k) \cong k[y]$ by Lemma 3.2.1, this implies that $y^n \notin \ker(\rho_{\pi^*\Lambda})$ for all $n \geq 1$. Then the commutativity of the diagram implies that $\ker(\rho_\Lambda)$ is contained in the kernel of the inflation map $\pi^* : H(\mathbb{M}_{1;s}, k) \to H(\mathbb{M}_1, k)$, i.e., $I_M \subseteq \langle w_s \rangle$, which in turn implies that $\widehat{\pi} = \widehat{\phi}_{(1,1,0)} \in |G|_M$.

For the converse of (1), suppose $\pi \notin \mathcal{N}_1(\mathbb{M}_{1;s})_M$. We want to show that $\widehat{\pi} \notin |G|_M$, or equivalently that $\mathrm{H}^\bullet(\mathbb{M}_{1;s}, \Lambda)_\mathfrak{m} = 0$. Here $\mathrm{H}^\bullet(\mathbb{M}_{1;s}, \Lambda)_\mathfrak{m}$ denotes the localization of $\mathrm{H}^\bullet(\mathbb{M}_{1;s}, \Lambda)$ at the maximal homogeneous ideal $\mathfrak{m} = \langle w_s \rangle$ that defines the line in $|\mathbb{M}_{1;s}|$ through $\widehat{\pi}$. Since $\pi \notin \mathcal{N}_1(\mathbb{M}_{1;s})_M$, there exists an integer $N \geq 1$ such that $\mathrm{H}^i(\mathbb{M}_1, \pi^*\Lambda) = 0$ for $i \geq N$. Then $(\pi^* \circ \rho_\Lambda)(y^N) = 0$. By the discussion in Sect. 3.2.1, this implies that $\rho_\Lambda(y^N) = z.w_s$ for some $z \in \mathrm{H}^{N-2}(\mathbb{M}_{1;s}, \Lambda)$. In other words, $1_M.y^N = z.w_s$ in $\mathrm{H}^\bullet(\mathbb{M}_{1;s}, \Lambda)$. Now consider the homogeneous localizations of $\mathrm{H}^\bullet(\mathbb{M}_{1;s}, k)$ and $\mathrm{H}^\bullet(\mathbb{M}_{1;s}, \Lambda)$ at the maximal homogeneous ideal $\mathfrak{m} = \langle w_s \rangle$. Since y is a unit in $\mathrm{H}^\bullet(\mathbb{M}_{1;s}, k)_\mathfrak{m}$, the identity $1_M \cdot y^N = z.w_s$ implies that $\mathrm{H}^\bullet(\mathbb{M}_{1;s}, \Lambda)_\mathfrak{m} = (\mathrm{H}^\bullet(\mathbb{M}_{1;s}, \Lambda)_\mathfrak{m}).\mathfrak{m}$. Then the graded analogue of Nakayama's Lemma [5, Ex. 1.5.24] implies that $\mathrm{H}^\bullet(\mathbb{M}_{1;s}, \Lambda)_\mathfrak{m} = 0$, and hence that $\widehat{\pi} \notin |G|_M$.

Now we prove (2). Set $\phi = \phi_{(0,0,1)}$, and write $\phi = \rho \circ \pi : \mathbb{M}_1 \twoheadrightarrow \mathbb{M}_{1;s} \to \mathbb{M}_{1;s}$, where $\rho = \rho_{(0,0,1)}$ as in Remark 2.2.3(2). First suppose $\phi \in \mathscr{N}_1(\mathbb{M}_{1;s})_M$. Then $\mathrm{H}^i(\mathbb{M}_1, \phi^*\Lambda) = \mathrm{H}^i(\mathbb{M}_1, \pi^*(\rho^*\Lambda))$ is nonzero for infinitely many $i \geq 0$, which means that $\pi \in \mathscr{N}_1(\mathbb{M}_{1;s})_{\rho^*M}$, and hence $\widehat{\pi} \in |\mathbb{M}_{1;s}|_{\rho^*M}$ by (1). The homomorphism $\rho : \mathbb{M}_{1;s} \to \mathbb{M}_{1;s}$ induces a morphism of varieties $\rho_* : |\mathbb{M}_{1;s}|_{\rho^*M} \to |\mathbb{M}_{1;s}|_M$, and one can check using Lemma 3.1.1(2) that $\rho_*(\widehat{\pi}) = \widehat{\phi}$. Now for the converse of (2), suppose that $\phi \notin \mathscr{N}_1(\mathbb{M}_{1;s})_M$, and let $N \geq 1$ such that $\mathrm{H}^i(\mathbb{M}_1, \phi^*\Lambda) = 0$ for all $i \geq N$. Let $\mathfrak{m} = \langle y \rangle$ be the maximal homogeneous ideal in $H(\mathbb{M}_{1;s}, k)$ that defines the line in $|\mathbb{M}_{1;s}|$ through $\widehat{\phi}$. To prove that $\widehat{\phi} \notin |\mathbb{M}_{1;s}|_M$, we will prove the equivalent condition that $\mathrm{H}^\bullet(\mathbb{M}_{1;s}, \Lambda)_\mathfrak{m} = 0$.

The homomorphism ϕ admits the factorization $\phi = \iota \circ q : \mathbb{M}_1 \twoheadrightarrow \mathbb{G}_{a(1)} \hookrightarrow \mathbb{M}_{1;s}$, where ι is the homomorphism discussed in Remark 2.2.3(2). By the discussion of Sect. 3.2.2, the map in cohomology induced by q, $\mathrm{H}^\bullet(\mathbb{G}_{a(1)}, \iota^*\Lambda) \to \mathrm{H}^\bullet(\mathbb{M}_1, \phi^*\Lambda)$, is an injection. Then $\mathrm{H}^j(\mathbb{G}_{a(1)}, \iota^*\Lambda) = 0$ for $j \geq N$. In particular, $\mathrm{H}^\bullet(\mathbb{G}_{a(1)}, \iota^*\Lambda)$ is finite-dimensional. Now consider the LHS spectral sequence for the extension $\mathbb{G}_{a(1)} \overset{\iota}{\to} \mathbb{M}_{1;s} \overset{\pi}{\twoheadrightarrow} \mathbb{M}_{1;s-1}$, where now $\pi : \mathbb{M}_{1;s} \twoheadrightarrow \mathbb{M}_{1;s-1}$ denotes the quotient homomorphism whose comorphism is the subalgebra inclusion $k[\mathbb{M}_{1;s-1}] \hookrightarrow k[\mathbb{M}_{1;s}]$:

$$E_2^{i,j} = \mathrm{H}^i(\mathbb{M}_{1;s-1}, \mathrm{H}^j(\mathbb{G}_{a(1)}, \iota^*\Lambda)) \Rightarrow \mathrm{H}^{i+j}(\mathbb{M}_{1;s}, \Lambda). \tag{26}$$

Since $\mathrm{H}^\bullet(\mathbb{G}_{a(1)}, \iota^*\Lambda)$ is finite-dimensional, one gets by [7, Theorem 1.1] that the E_2-page is finite as a module over the cohomology ring $\mathrm{H}^\bullet(\mathbb{M}_{1;s}, k)$, and hence gets by [13, Lemma 1.6] and [4, III.2.9 Corollary 1] that the inflation map $\mathrm{H}^\bullet(\mathbb{M}_{1;s-1}, k) \to \mathrm{H}^\bullet(\mathbb{M}_{1;s}, k)$ makes $\mathrm{H}^\bullet(\mathbb{M}_{1;s}, \Lambda)$ into a finite module over $\mathrm{H}^\bullet(\mathbb{M}_{1;s-1}, k)$. By [9, Lemma 3.2.4(3)], the inflation map $\mathrm{H}^\bullet(\mathbb{M}_{1;s-1}, k) \to \mathrm{H}^\bullet(\mathbb{M}_{1;s}, k)$ satisfies $y \mapsto y$ and $w_{s-1} \mapsto 0$, so $\mathrm{H}^\bullet(\mathbb{M}_{1;s}, \Lambda)$ is finite as a module over the subalgebra $k[y]$ of $H(\mathbb{M}_{1;s}, k)$. Fix a homogeneous generating set for $\mathrm{H}^\bullet(\mathbb{M}_{1;s}, \Lambda)$ as a $k[y]$-module, and let m be the maximum of the cohomological degrees of the generators. Then $1_M.(w_s)^m \in \mathrm{H}^{2m}(\mathbb{M}_{1;s}, \Lambda)$. Writing $1_M.(w_s)^m$ as a $k[y]$-linear combination of the homogeneous generators, it follows by degree consideration that each term in the sum is a multiple of y, and hence $1_M.(w_s)^m = z.y$ for some $z \in \mathrm{H}^{2m-1}(\mathbb{M}_{1;s}, \Lambda)$. Now localizing at the maximal homogeneous ideal $\mathfrak{m} = \langle y \rangle$, and using the fact that w_s is a unit in $\mathrm{H}^\bullet(\mathbb{M}_{1;s}, k)_\mathfrak{m}$, it follows as in the proof of (1) that $\mathrm{H}^\bullet(\mathbb{M}_{1;s}, \Lambda)_\mathfrak{m} = 0$. □

3.4.2 Proof of Theorem 3.4.1 for $G = \mathbb{M}_{1;1} = \mathbb{G}_{a(1)} \times \mathbb{G}_a^-$

The proof in this case is similar to that for $G = \mathbb{M}_{1;s}$ when $s \geq 2$, except that now there are three orbits of $\mathrm{Aut}(\mathbb{M}_{1;1})$ in $\mathscr{N}_1(\mathbb{M}_{1;1})$ to consider, namely, the orbits of the canonical quotient map $\phi_{(1,1)} : \mathbb{M}_1 \twoheadrightarrow \mathbb{M}_{1;1}$, and the orbits of the homomorphisms $\phi_{(0,1)}$ and $\phi_{(1,0)}$ discussed in Remark 2.2.3(3). Then it suffices to show that

1. $\widehat{\phi}_{(1,1)} \in |\mathbb{M}_{1;1}|_M$ if and only if $\phi_{(1,1)} \in \mathcal{N}_1(\mathbb{M}_{1;1})_M$,
2. $\widehat{\phi}_{(0,1)} \in |\mathbb{M}_{1;1}|_M$ if and only if $\phi_{(0,1)} \in \mathcal{N}_1(\mathbb{M}_{1;1})_M$, and
3. $\widehat{\phi}_{(1,0)} \in |\mathbb{M}_{1;1}|_M$ if and only if $\phi_{(1,0)} \in \mathcal{N}_1(\mathbb{M}_{1;1})_M$.

The proofs of (1) and (2) are entirely similar to the arguments in Sect. 3.4.1 above, with (26) replaced by the LHS spectral sequence for the extension $\mathbb{G}_{a(1)} \hookrightarrow \mathbb{M}_{1;1} \twoheadrightarrow \mathbb{G}_a^-$. The proof of (3) is also similar to that of (2) in Sect. 3.4.1, except that one now applies the observation at the end of Sect. 3.2.3 when considering the map in cohomology induced by the quotient $\mathbb{M}_1 \twoheadrightarrow \mathbb{G}_a^-$, the spectral sequence (26) is replaced by the LHS spectral sequence for the extension $\mathbb{G}_a^- \hookrightarrow \mathbb{M}_{1;1} \twoheadrightarrow \mathbb{G}_{a(1)}$, and the maximal homogeneous ideal $\mathfrak{m} = \langle y \rangle \subset H(\mathbb{M}_{1;s}, k)$ is replaced by the maximal homogeneous ideal $\langle x_1 \rangle \subset H(\mathbb{M}_{1;1}, k)$. □

3.4.3 Proof of Theorem 3.4.1 for $G = \mathbb{G}_{a(1)}$

In this case $H(\mathbb{G}_{a(1)}, k) \cong k[x_1]$, and hence $|\mathbb{G}_{a(1)}|$ and $\mathcal{N}_1(\mathbb{G}_{a(1)})$ are both one-dimensional affine spaces, and the following are equivalent:

- $|\mathbb{G}_{a(1)}|_M = |\mathbb{G}_{a(1)}|$
- $\mathrm{H}^i(\mathbb{G}_{a(1)}, \Lambda) = \mathrm{Ext}^i_{\mathbb{G}_{a(1)}}(M, M) \neq 0$ for infinitely many $i \geq 0$.

Since $\mathrm{H}^\bullet(\mathbb{G}_{a(1)}, \Lambda)$ embeds into $\mathrm{H}^\bullet(\mathbb{M}_1, \Lambda)$ by the observation at the end of Sect. 3.2.2, it follows that if $\mathcal{N}_1(\mathbb{G}_{a(1)})_M = \{0\}$, then $\mathrm{H}^i(\mathbb{G}_{a(1)}, \Lambda) = 0$ for $i \gg 0$, and hence $|\mathbb{G}_{a(1)}|_M = \{0\}$. Conversely, suppose $\mathcal{N}_1(\mathbb{G}_{a(1)})_M = \mathcal{N}_1(\mathbb{G}_{a(1)})$, so that $\mathrm{H}^i(\mathbb{M}_1, q^*\Lambda) \neq 0$ for infinitely many $i \geq 0$. Consider the commutative diagram

$$\begin{array}{ccc} H(\mathbb{G}_{a(1)}, k) & \xrightarrow{\rho_\Lambda} & \mathrm{H}^\bullet(\mathbb{G}_{a(1)}, \Lambda) \\ \downarrow{\scriptstyle q^*} & & \downarrow{\scriptstyle q^*} \\ H(\mathbb{M}_1, k) & \xrightarrow{\rho_{q^*\Lambda}} & \mathrm{H}^\bullet(\mathbb{M}_1, q^*\Lambda). \end{array}$$

As in the proof of (1) in Sect. 3.4.1 above, it follows that $\rho_{q^*\Lambda}(y^n) \neq 0$ for $n \geq 1$. Since $q^*(x_1) = x_1$ by [9, Lemma 3.2.4], and since $x_1 = y^2$ in $H(\mathbb{M}_1, k)$, this implies that $\ker(\rho_\Lambda) = \{0\}$, and hence implies that $|\mathbb{G}_{a(1)}|_M = |\mathbb{G}_{a(1)}|$. □

3.4.4 Proof of Theorem 3.4.1 for $G = \mathbb{G}_a^-$

The reasoning is entirely similar to that for $G = \mathbb{G}_{a(1)}$. □

3.5 Applications of the BIKP Detection Theorem

The following theorem of Benson, Iyengar, Krause, and Pevtsova (BIKP) was announced by Julia Pevtsova at the Mathematical Congress of the Americas in Montreal in July 2017.

Theorem 3.5.1 (BIKP Detection Theorem) *Let G be a finite unipotent k-supergroup scheme. Given $z \in H^n(G,k)$ and a field extension K/k, set $z_K = z \otimes_k 1 \in H^n(G,k) \otimes_k K = H^n(G \otimes_k K, K)$. Given a rational G-supermodule M, set $M_K = M \otimes_k K$. Then the following hold:*

1. *A cohomology class $z \in H^\bullet(G,k)$ is nilpotent if and only if for every field extension K of k and every elementary subsupergroup scheme E of $G \otimes_k K$, the restriction of $z_K \in H^\bullet(G \otimes_k K, K)$ to $H^\bullet(E,K)$ is nilpotent.*
2. *Let M be a rational G-supermodule. Then M is projective if and only if for every field extension K of k and every elementary subsupergroup scheme E of $G \otimes_k K$, the restriction of M_K to E is projective.*

In the remainder of the paper we will discuss a few of the ways we expect the BIKP detection theorem may be applied (à la Suslin, Friedlander, and Bendel [19, 20]) to generalize our results in Corollary 3.1.2 and Theorem 3.4.1. Since at the time of writing this article the BIKP detection theorem has not yet appeared in preprint form, we will state these expected applications only as conjectures.

Conjecture 3.5.2 *Suppose k is algebraically closed, and let G be an infinitesimal unipotent k-supergroup scheme of height $\leq r$. Then the kernel of the algebra homomorphism*

$$\psi_r : H(G,k) \to k[\mathcal{N}_r(G)]$$

consists of nilpotent elements, and its image contains the p^r-th power of each element of $k[\mathcal{N}_r(G)]$. Consequently, the associated morphism of varieties

$$\Psi = \Psi_r : \mathcal{N}_r(G) \to |G| = Max(H(G,k))$$

is a homeomorphism.

We can already show (for arbitrary infinitesimal G) that ψ_r is surjective onto p^r-th powers.

Lemma 3.5.3 *Suppose k is algebraically closed, and let G be an infinitesimal k-supergroup scheme of height $\leq r$. Then the image of the homomorphism*

$$\psi_r : H(G,k) \to k[\mathcal{N}_r(G)]$$

contains the p^r-th power of each element of $k[\mathcal{N}_r(G)]$.

Proof Let $\iota : G \hookrightarrow GL_{m|n}$ be a closed embedding, and set $N = \max(m, n)$. Then by Remark 2.3.7(2) and the hypothesis that k is algebraically closed, $k[\mathcal{N}_r(G)] = k[\mathbf{Hom}(\mathbb{M}_{r;N}, G)]_{red}$. The embedding of G into $GL_{m|n}$ defines a closed embedding of affine k-superschemes $\mathbf{Hom}(\mathbb{M}_{r;N}, G) \subset \mathbf{Hom}(\mathbb{M}_{r;N}, GL_{m|n})$, and hence a surjection of coordinate superalgebras $k[\mathbf{Hom}(\mathbb{M}_{r;N}, GL_{m|n})] \twoheadrightarrow k[\mathbf{Hom}(\mathbb{M}_{r;N}, G)]$. Passing to the largest purely even quotients, the embedding induces surjective k-algebra homomorphism $\iota^* : k[V_{r;N}(GL_{m|n})] \twoheadrightarrow k[V_{r;N}(G)]$. Next, by [6, Lemma 4.4.1], the image of $\iota : G \hookrightarrow GL_{m|n}$ is contained in $GL_{m|n(r)}$, the r-th Frobenius kernel of $GL_{m|n}$. Then by [9, Lemma 6.2.1], there exists a commutative diagram

$$\begin{array}{ccc} H(GL_{m|n(r)}, k) & \xrightarrow{\psi_{r;N}} & k[V_{r;N}(GL_{m|n})] \\ \downarrow{\scriptstyle \iota^*} & & \downarrow{\scriptstyle \iota^*} \\ H(G, k) & \xrightarrow{\psi_{r;N}} & k[V_{r;N}(G)]. \end{array}$$

The top horizontal map is surjective onto p^r-th powers by [9, Theorem 6.2.3], so the commutativity of the diagram implies that the bottom horizontal map is surjective onto p^r-th powers as well. Finally, the commutative diagram (16) implies that ψ_r is surjective onto p^r-th powers. $\square$

To argue that $\ker(\psi_r)$ is nilpotent, we first need a strengthening of Proposition 2.2.2 and of Corollary 3.1.2. Recall from [9, Remark 3.1.3(5)] that the super Frobenius morphism $\mathbf{F}^\ell : \mathbb{M}_{r+\ell} \to \mathbb{M}_r$ is the supergroup homomorphism whose comorphism $(\mathbf{F}^\ell)^* : k[\mathbb{M}_r] \to k[\mathbb{M}_{r+\ell}]$ is defined by $\theta \mapsto \theta^{p^\ell}$, $\tau \mapsto \tau$, and $\sigma_i \mapsto \sigma_i$ for $i \in \mathbb{N}$. Note that $\theta^{p^{r-1}} = \sigma_1$ in $k[\mathbb{M}_r]$, but $\theta^{p^{r+\ell-1}} = \sigma_1$ in $k[\mathbb{M}_{r+\ell}]$, so the assignment $(\mathbf{F}^\ell)^*(\theta) = \theta^{p^\ell}$ does make sense. We could have incorporated the first statement of the following lemma into Proposition 2.2.2, but to do so would have made the proof more notationally cumbersome than it already is.

Lemma 3.5.4 *Let G be a height-r infinitesimal elementary k-supergroup scheme, and let $A = A_{\overline{0}}$ be a reduced purely even commutative k-algebra. Then for all $\ell \geq 1$, composition with the super Frobenius morphism $\mathbf{F}^\ell : \mathbb{M}_{r+\ell} \to \mathbb{M}_r$ defines a bijection*

$$\mathbf{Hom}(\mathbb{M}_r, G)(A) \xrightarrow{\sim} \mathbf{Hom}(\mathbb{M}_{r+\ell}, G)(A).$$

In particular, if k is algebraically closed then $\mathcal{N}_r(G) \cong \mathcal{N}_{r+\ell}(G)$.

Proof The coordinate algebra $A[\mathbb{M}_r]$ identifies via the comorphism $(\mathbf{F}^\ell)^* : A[\mathbb{M}_r] \to A[\mathbb{M}_{r+\ell}]$ with a Hopf subsuperalgebra of $A[\mathbb{M}_{r+\ell}]$. Then the proof proceeds by the same reasoning as the proof of Proposition 2.2.2: One begins by classifying all Hopf superalgebra homomorphisms $\phi : A[G] \to A[\mathbb{M}_{r+\ell}]$, and then one observes from the classification that the image of each homomorphism is contained in the subalgebra $A[\mathbb{M}_r]$ of $A[\mathbb{M}_{r+\ell}]$, and hence each homomorphism $\phi : A[G] \to A[\mathbb{M}_{r+\ell}]$ uniquely factors through the comorphism $(\mathbf{F}^\ell)^* : A[\mathbb{M}_r] \to A[\mathbb{M}_{r+\ell}]$. The classification of all Hopf superalgebra homomorphisms $\phi : A[G] \to A[\mathbb{M}_{r+\ell}]$ follows from essentially a repetition of the argument in the proof Proposition 2.2.2. The first difference, which is really just an issue of bookkeeping, occurs when considering the $\mathbb{Z}$-degree of elements in $A[\mathbb{M}_{r+\ell}]$, as when analyzing the formulas (4), (6), (7), and (9). The $\mathbb{Z}$-grading on $k[\mathbb{M}_{r+\ell}]$ is defined instead by $\deg(\theta) = 2$, $\deg(\sigma_j) = 2jp^{r+\ell-1}$, and $\deg(\tau) = p^{r+\ell}$ (so $(\mathbf{F}^{\ell})^*$ is a morphism of graded algebras that multiplies degrees by p^{ℓ}), but this difference in $\mathbb{Z}$-gradings between $k[\mathbb{M}_r]$ and $k[\mathbb{M}_{r+\ell}]$ does not alter the substance of any of the arguments in the proof of Proposition 2.2.2. The second difference comes in considering the space of primitive elements in $A[\mathbb{M}_{r+\ell}]$. An element $z \in A[\mathbb{M}_{r+\ell}]$ is primitive if and only if it is an A-linear combination of $\theta, \theta^p, \ldots, \theta^{p^{r+\ell-1}}$. Then the only p-nilpotent primitive elements in $A[\mathbb{M}_{r+\ell}]$ are multiples of $\theta^{p^{r+\ell-1}} = \sigma_1$, and the only p^r-nilpotent primitive elements in $A[\mathbb{M}_{r+\ell}]$ are A-linear combinations of $\theta^{p^{\ell}}, \theta^{p^{\ell+1}}, \ldots, \theta^{p^{r+\ell-1}}$. Thus when considering the primitive components of $\phi(\sigma_p)$, $\phi(\sigma_{p^{n+1}})$, or $\phi(\theta)$, as in (4), (6), (7), and (9), one only gets primitive elements in the image of $(\mathbf{F}^{\ell})^*$. □

Corollary 3.5.5 *Suppose k is algebraically closed, and let G be an infinitesimal elementary k-supergroup scheme of height $\leq r$. Then the algebra homomorphism*

$$\psi_r : H(G, k) \to k[\mathcal{N}_r(G)]$$

is injective modulo nilpotents and its image contains all p^r-th powers in $k[\mathcal{N}_r(G)]$.

Proof Suppose G is infinitesimal of height $r' \leq r$. Set $A = k[\mathcal{N}_r(G)]$, and set $B = k[\mathcal{N}_{r'}(G)]$. By Lemma 3.5.4, the super Frobenius morphism defines an identification of coordinate algebras $A = B$. Lemma 3.5.4 also implies that the universal supergroup homomorphism from $\mathbb{M}_r$ to G factors through the super Frobenius morphism $\mathbb{M}_r \otimes_k A \to \mathbb{M}_{r'} \otimes_k A$. Then there exists a commutative diagram

$$\begin{array}{ccccccc} H(G,k) & \to & \mathrm{H}^\bullet(G \otimes_k B, B) & \to & \mathrm{H}^\bullet(\mathbb{M}_{r'} \otimes_k B, B) & \to & B \\ \| & & \| & & \downarrow & & \| \\ H(G,k) & \to & \mathrm{H}^\bullet(G \otimes_k A, A) & \to & \mathrm{H}^\bullet(\mathbb{M}_r \otimes_k A, A) & \to & A \end{array}$$

in which the rows are the composites defining $\psi_{r'}$ and ψ_r, respectively, and the third vertical arrow is the map in cohomology induced by the super Frobenius morphism. The top row of the diagram is injective modulo nilpotents and surjective onto p^r-th powers, by Corollary 3.1.2, so the commutativity of the diagram implies the corresponding conclusion for the bottom row. □

Now we explain how the previous results, when combined with the BIKP detection theorem, should imply the validity of Conjecture 3.5.2.

Proof (Justification for Conjecture 3.5.2) Suppose k is algebraically closed, and let G be an infinitesimal unipotent k-supergroup scheme of height $\leq r$. We have already shown in Lemma 3.5.3 that the image of ψ_r contains the p^r-th power of each element of $k[\mathcal{N}_r(G)]$, so let $z \in H(G, k)$ be a homogeneous element, and suppose $\psi_r(z) = 0$.

Then for each (algebraically closed) field extension K/k and each (infinitesimal) elementary subsupergroup scheme E of $G \otimes_k K$, there exists by Proposition 2.4.2 a commutative diagram

$$\begin{array}{ccc} H(G \otimes_k K, K) & \xrightarrow{\psi_r \otimes_k K} & K[\mathcal{N}_r(G)] \\ \big\downarrow \iota^* & & \big\downarrow \iota^* \\ H(E, K) & \xrightarrow{\psi_r} & K[\mathcal{N}_r(E)]. \end{array}$$

Here $\iota : E \hookrightarrow G \otimes_k K$ denotes the closed embedding of E into $G \otimes_k K$. Since $\psi_r(z) = 0$, the commutativity of the diagram implies that $(\psi_r \circ \iota^*)(z_K) = 0$. But the bottom arrow of the diagram is injective modulo nilpotents by Corollary 3.5.5, so $\iota^*(z_K)$ must be nilpotent in $H(E, K)$. Since K and E were arbitrary, this implies by the BIKP detection theorem that z is nilpotent. □

Conjecture 3.5.6 *Suppose k is algebraically closed. Let G be a height-*1 *infinitesimal unipotent k-supergroup scheme, and let M be a finite-dimensional rational G-supermodule. Then under the homeomorphism* $\Psi = \Psi_1 : \mathcal{N}_1(G) \to |G|$ *of Conjecture 3.5.2,*

$$\Psi^{-1}(|G|_M) = \mathcal{N}_1(G)_M = \left\{\phi \in \mathcal{N}_1(G) : id_{\mathbb{M}_1}(\phi^* M) = \infty\right\}.$$

To justify this conjecture we must assume a stronger version of Theorem 3.5.1(1)

- (BIKP 1′) Let Λ be a unital associative rational G-algebra, and let $z \in \mathrm{H}^\bullet(G, \Lambda)$. Then z is nilpotent if and only if for every field extension K of k, and every elementary subsupergroup scheme E of $G \otimes_k K$, the restriction of $z_K \in \mathrm{H}^\bullet(G \otimes_k K, \Lambda_K)$ to $\mathrm{H}^\bullet(E, \Lambda_K)$ is nilpotent.

We also need to assume an affirmative answer to Question 3.3.5, i.e., we must assume that $\mathcal{N}_1(G)_M$ is a Zariski closed subset of $\mathcal{N}_1(G)$, and we need to assume that the ideal of functions $\mathscr{I}_M = \mathscr{I}(G)_M \subset k[\mathcal{N}_1(G)]$ vanishing on $\mathcal{N}_1(G)_M$ is compatible with field extensions. Specifically, let K be an algebraically closed field extension of k. Then $M_K := M \otimes_k K$ is a rational $G \otimes_k K$-supermodule, and one can consider the set

$$\mathcal{N}_1(G \otimes_k K)_{M_K} = \left\{\phi \in \mathbf{Hom}(\mathbb{M}_1 \otimes_k K, G \otimes_k K)(K) : \mathrm{id}_{\mathbb{M}_1 \otimes_k K}(\phi^* M_K) = \infty\right\}.$$

Then we want to assume that the image of $\mathscr{I}(G)_M$ in $K[\mathcal{N}_1(G \otimes_k K)]$ under base change is contained in the ideal of functions $\mathscr{I}(G \otimes_k K)_{M_K} \subset K[\mathcal{N}_1(G \otimes_k K)]$ vanishing on $\mathcal{N}_1(G \otimes_k K)_{M_K}$.

Proof (Justification (modulo assumptions) for Conjecture 3.5.6) Given the preceding assumptions, set $\Lambda = \mathrm{Hom}_k(M, M)$, and let $I_M = \ker(\rho_\Lambda)$ be the defining ideal of $|G|_M$. By the assumption that $\mathcal{N}_1(G)_M$ is closed in $\mathcal{N}_1(G)$, the ideal $\mathscr{I}_M$ is a radical ideal and is the kernel of the canonical quotient map $k[\mathcal{N}_1(G)] \twoheadrightarrow k[\mathcal{N}_1(G)_M]$. Now to prove Conjecture 3.5.6, it suffices as in the proof of [20, Corollary 6.3.1] to show that the homomorphism $\psi = \psi_1 : H(G,k) \to k[\mathcal{N}_1(G)]$ satisfies

$$\psi^{-1}(\mathcal{I}_M) = \sqrt{I_M}. \tag{27}$$

It follows from Theorem 3.4.1 that (27) is true in the special case that G is a height-one infinitesimal elementary supergroup scheme.

Let $\phi \in \mathcal{N}_1(G) = \mathbf{Hom}(\mathbb{M}_1, G)(k)$, and let $z \in H(G,k)$ be homogeneous of degree n. Then by the definition of ψ, $\phi^*(z) = \psi(z)(\phi) \cdot y^n \in H(\mathbb{M}_1, k)$; cf. (15). Now suppose $\phi \in \mathcal{N}_1(G)_M$ but that $\psi(z)(\phi) \neq 0$, so that $\psi(z) \notin \mathcal{I}_M$. Then $\phi^*(z)$ is a nonzero multiple of y^n in $H(\mathbb{M}_1, k)$. Since $\phi \in \mathcal{N}_1(G)_M$, then $\mathrm{H}^i(\mathbb{M}_1, \phi^*\Lambda) \neq 0$ for infinitely many $i \geq 1$. But $\mathrm{H}^\bullet(\mathbb{M}_1, \phi^*\Lambda)$ is finite over $H(\mathbb{M}_1, k) \cong k[y]$ by Lemma 3.2.1, so this implies that $\rho_{\phi^*\Lambda}(y) \in H(\mathbb{M}_1, \phi^*\Lambda)$ is not nilpotent, and hence $\rho_{\phi^*\Lambda}(\phi^*(z))$ is not nilpotent. Since $\rho_{\phi^*\Lambda}(\phi^*(z)) = \phi^*(\rho_\Lambda(z))$, we deduce that $\rho_\Lambda(z)$ is not nilpotent, and hence $z \notin \sqrt{I_M}$. Thus, if $\psi(z) \notin \mathcal{I}_M$, then $z \notin \sqrt{I_M}$.

For the converse, suppose $\psi(z) \in \mathcal{I}_M$. Let K/k be an extension of (algebraically closed) fields, and let E be an elementary (height-one infinitesimal) subsupergroup scheme of $G \otimes_k K$. Write $\nu : E \hookrightarrow G \otimes_k K$ for the inclusion of E into $G \otimes_k K$. Then we want to show that the restricted cohomology class $\nu^*(\rho_{\Lambda_K}(z_K)) = \rho_{\nu^*\Lambda_K}(\nu^*(z_K)) \in \mathrm{H}^\bullet(E, \nu^*\Lambda_K)$ is nilpotent. Write $\psi(z)_K$ for the image of $\psi(z)$ under the base change map $k[\mathcal{N}_1(G)] \to K[\mathcal{N}_1(G \otimes_k K)]$ of Remark 2.3.7(3). By the assumption that the ideal $\mathcal{I}_M = \mathcal{I}(G)_M$ is compatible with field extensions, we get $\psi(z)_K \in \mathcal{I}(G \otimes_k K)_{M_K}$. Next, it is evident from the definitions that composition with ν maps $\mathcal{N}_1(E)_{\nu^*M_K}$ into the support set $\mathcal{N}_1(G \otimes_k K)_{M_K}$, and hence the restriction map $\nu^* : K[\mathcal{N}_1(G \otimes_k K)] \to K[\mathcal{N}_1(E)]$ maps $\mathcal{I}(G \otimes_k K)_{M_K}$ into the ideal $\mathcal{I}(E)_{\nu^*M_K}$. Then by the naturality of ψ,

$$\psi(\nu^*(z_K)) = \nu^*(\psi(z_K)) = \nu^*(\psi(z)_K) \in \mathcal{I}(E)_{\nu^*M_K}.$$

Since (27) is true for height-one infinitesimal elementary supergroup schemes, this means that the cohomology class $\rho_{\nu^*\Lambda_K}(\nu^*(z_K)) \in \mathrm{H}^\bullet(E, \nu^*\Lambda_K)$ is nilpotent. Thus, since K and E were arbitrary, we deduce by (BIKP 1′) that if $\psi(z) \in \mathcal{I}_M$, then $z \in \sqrt{I_M}$. □

Acknowledgements The authors are pleased to thank David Benson, Srikanth Iyengar, and Julia Pevtsova for enlightening conversations that helped contribute to this work. In particular, the authors thank Julia Pevtsova for explaining the BIKP detection theorem (Theorem 3.5.1), and they thank Srikanth Iyengar for generously sharing a preprint copy of his work with Luchezar Avramov on restricting homology to hypersurfaces. The first author thanks the organizers of the Southern Regional Algebra Conference, held March 17–19, 2017, at the University of South Alabama, for giving him the opportunity to speak on the work in [9].

References

1. Avramov, L.L.: Modules of finite virtual projective dimension. Invent. Math. **96**(1), 71–101 (1989)
2. Avramov, L.L., Buchweitz, R.O.: Support varieties and cohomology over complete intersections. Invent. Math. **142**(2), 285–318 (2000)
3. Benson, D., Iyengar, S.B., Krause, H., Pevtsova, J.: Detecting nilpotence and projectivity over finite unipotent supergroup schemes (2018)
4. Bourbaki, N.: Commutative algebra. Chapters 1–7. Elements of Mathematics (Berlin). Springer, Berlin (1998). Translated from the French, Reprint of the 1989 English translation
5. Bruns, W., Herzog, J.: Cohen-Macaulay rings, Cambridge Studies in Advanced Mathematics, vol. 39. Cambridge University Press, Cambridge (1993)
6. Drupieski, C.M.: Cohomological finite generation for restricted Lie superalgebras and finite supergroup schemes. Represent. Theory **17**, 469–507 (2013). https://doi.org/10.1090/S1088-4165-2013-00440-5
7. Drupieski, C.M.: Cohomological finite generation for finite supergroup schemes. Adv. Math. **288**, 1360–1432 (2016). https://doi.org/10.1016/j.aim.2015.11.017
8. Drupieski, C.M., Kujawa, J.R.: On Support Varieties for Lie Superalgebras and Finite Supergroup Schemes (2016)
9. Drupieski, C.M., Kujawa, J.R.: Graded analogues of one-parameter subgroups and applications to the cohomology of $GL_{m|n(r)}$ (2017)
10. Friedlander, E.M., Parshall, B.J.: Support varieties for restricted Lie algebras. Invent. Math. **86**(3), 553–562 (1986). https://doi.org/10.1007/BF01389268
11. Friedlander, E.M., Parshall, B.J.: Geometry of p-unipotent Lie algebras. J. Algebra **109**(1), 25–45 (1987). https://doi.org/10.1016/0021-8693(87)90161-X
12. Friedlander, E.M., Pevtsova, J.: Π-supports for modules for finite group schemes. Duke Math. J. **139**(2), 317–368 (2007). https://doi.org/10.1215/S0012-7094-07-13923-1
13. Friedlander, E.M., Suslin, A.: Cohomology of finite group schemes over a field. Invent. Math. **127**(2), 209–270 (1997). https://doi.org/10.1007/s002220050119
14. Jantzen, J.C.: Representations of Algebraic Groups, Mathematical Surveys and Monographs, vol. 107, 2nd edn. American Mathematical Society, Providence, RI (2003)
15. Jorgensen, D.A.: Support sets of pairs of modules. Pacific J. Math. **207**(2), 393–409 (2002)
16. Milnor, J.W., Moore, J.C.: On the structure of Hopf algebras. Ann. Math. **2**(81), 211–264 (1965)
17. Priddy, S.B.: Koszul resolutions. Trans. Amer. Math. Soc. **152**, 39–60 (1970)
18. Quillen, D.: The spectrum of an equivariant cohomology ring I, II. Ann. Math. **94**(2), 549–572; ibid. **94**(2), 573–602 (1971)
19. Suslin, A., Friedlander, E.M., Bendel, C.P.: Infinitesimal 1-parameter subgroups and cohomology. J. Am. Math. Soc. **10**(3), 693–728 (1997). https://doi.org/10.1090/S0894-0347-97-00240-3
20. Suslin, A., Friedlander, E.M., Bendel, C.P.: Support varieties for infinitesimal group schemes. J. Am. Math. Soc. **10**(3), 729–759 (1997). https://doi.org/10.1090/S0894-0347-97-00239-7

A Survey of the Marcus–de Oliveira Conjecture

Huajun Huang

Abstract We classify and survey the progress on the famous Marcus–de Oliveira determinantal conjecture (MOC) and related problems. The MOC claims that for two normal matrices A and B with eigenvalues $a_1, \ldots, a_n$ and $b_1, \ldots, b_n$, respectively, the set $\Delta(A, B) = \{\det(A + UBU^*) : U \text{ is unitary}\}$ is in the convex hull of the set $\{\prod_{i=1}^{n}(a_i + b_{\sigma(i)}) : \sigma \in S_n\}$. We review the origin and the motivations of this conjecture from M. Fiedler's work in the Hermitian case to Marcus and de Oliveira's independent questions. Then, we survey the major positive cases of the MOC in terms of matrix degree, eigenvalues, and other things. We also describe some known properties of the set $\Delta(A, B)$, including compactness, connectivity, simply connectivity, convexity, star shapedness, and boundary and corner properties. Finally, we list some extended results that are related to the MOC. The main goal of this article is to provide a fairly comprehensive and brief overview of the progress of the MOC to interested readers and researchers for further explorations of the subject.

Keywords Marcus-de Oliveira conjecture · Normal matrix · Determinant · Eigenvalues · Unitary matrix

1 Introduction

The Marcus–de Oliveira determinantal conjecture (MOC), independently presented by Marcus in 1973 [33] and de Oliveira in 1982 [13], is one of the well-known open problems in matrix analysis. Let M_n, N_n, and U_n denote the sets of $n \times n$ complex matrices, normal matrices, and unitary matrices of degree n, respectively. The MOC claims that for two normal matrices $A, B \in \mathrm{N}_n$ with eigenvalues $a_1, \ldots, a_n$ and $b_1, \ldots, b_n$, respectively,

$$\Delta(A, B) \subset \mathrm{co}\,\{z_\sigma : \sigma \in S_n\}, \tag{1}$$

H. Huang (✉)
Department of Mathematics and Statistics, Auburn University,
Auburn, AL 36849, USA
e-mail: huanghu@auburn.edu

J. Feldvoss et al. (eds.), *Advances in Algebra*, Springer Proceedings
in Mathematics & Statistics 277, https://doi.org/10.1007/978-3-030-11521-0_8

where $\mathrm{co}\{\cdot\}$ is the convex hull, S_n is the symmetric group of degree n, $z_\sigma = \prod_{i=1}^n (a_i + b_{\sigma(i)})$, and

$$\Delta(A, B) = \{\det(A + UBU^*) : U \in \mathrm{U}_n\}. \tag{2}$$

Since every normal matrix is unitarily similar to a diagonal matrix, it suffices to prove the MOC for diagonal matrices A and B. The z_σ's are called *σ-points*. Let $P_\sigma \in \mathrm{M}_n$ be the permutation matrix with 1's in the $(\sigma(i), i)$ positions for $i = 1, \ldots, n$. Then,

$$z_\sigma = \det(A_0 + P_\sigma B_0 P_\sigma^*) \in \Delta(A, B) \tag{3}$$

for $A_0 = \mathrm{diag}\,(a_1, \ldots, a_n)$ and $B_0 = \mathrm{diag}\,(b_1, \ldots, b_n)$.

The origin of the MOC was inspired by Fiedler's result published in 1971 [22] that $\Delta(A, B) = [\min_{\sigma \in S_n} z_\sigma, \max_{\sigma \in S_n} z_\sigma]$ when A and B are Hermitian matrices. Besides, the motivations of Marcus and de Oliveira in posing the MOC involve several extensions of *numerical range* of $A \in \mathrm{M}_n$:

$$W(A) = \{x^* A x : x \in \mathbb{C}^n, \|x\| = 1\}. \tag{4}$$

The Hausdorff–Toeplitz theorem shows that $W(A)$ is always convex and compact. Moreover, when A is normal, $W(A)$ is the convex hull of the eigenvalues of A [24]. In [33], Marcus studied the convexity of a set $W_{r,m}^n$ which consists of all function values

$$f_{r,m}(x_1, \ldots, x_m) = \sum_{\omega \in Q_{r,m}} \det[(Ax_{\omega(i)}, x_{\omega(j)})], \tag{5}$$

where $1 \le r \le m \le n$, $x_1, \ldots, x_m \in \mathbb{C}^n$ is an orthonormal set, $Q_{r,m}$ is the set of all sequences ω of length r chosen from $1, \ldots, m$ such that $\omega(1) < \cdots < \omega(r)$. Obviously, $W(A) = W_{1,1}^n(A)$. Marcus proved that $W_{r,n-1}^n(A)$ is convex when $1 \le r \le n-1$, and $W_{r,r}^n(A)$ is not convex for some normal A when $2 \le r < n-1$. He further studied the convexity of the set $W(A, \delta)$ of all $\delta(X)$ for a given function δ and $X = \left[\,(Ax_i, x_j)\,\right]_{m \times m}$. When $m = n$, $A, B \in \mathrm{N}_n$ and $\delta(X) = \det(B + X)$, it leads to the MOC. De Oliveira presented the MOC as a research problem [13]. He used the Binet–Cauchy determinant formula to prove the MOC when one of A and B has rank one. He remarked that "there are some true relations such that when we replace sums with products and products with sums, we obtain other true relations", and then gave a short proof to the analog statement of the MOC: if $A, B \in \mathrm{N}_n$, then

$$W(A, B) = \{\mathrm{tr} AUBU^* : U \in \mathrm{U}_n\} \subset \mathrm{co}\{\sum_{i=1}^n a_i b_{\sigma(i)} : \sigma \in S_n\}.$$

Replacing B by C, the set $W_C(A) = W(A, C)$ is called the *C-numerical range of A* and $\Delta_C(A) = \Delta(A, C)$ is called the *C-determinantal range of A*.

The MOC has attracted attensions of many researchers since 1980s. Significant progress has been made by several groups of mathematicians, including Bebiano et al. [3–5, 7–9, 11, 12], Drury [14–21, 27], Kovačec [11, 25–27, 39], Merikoski and Virtanen [9, 34–37], and Li et al. [28, 29, 31, 32]. In 1992, Bebiano gave a talk "New developments on the Marcus-Oliveira Conjecture" in the Second Conference of the International Linear Algebra Society, where she summarized the works on the MOC up to that time [30].

This article tries to give an up-to-date survey on the progress of the MOC. The covered results here are far from complete. We hope that a brief overview of the history and major breakthroughs on this open problem could motivate interested readers and probably foster new ideas to the related research.

2 Confirmed Cases of the MOC

The MOC has been proved affirmatively in many situations. We will classify and enumerate known results in this section.

2.1 Degree

An easy calculation shows that the MOC holds for normal matrices of degree 2, where $\Delta(A, B)$ is exactly the line segment with the two σ-points $(a_1 + b_1)(a_2 + b_2)$ and $(a_1 + b_2)(a_2 + b_1)$ as endpoints. More generally, the set $\Delta(A, B)$ for $A, B \in \mathrm{M}_2$ (not assuming normal) has been determined in [2, Theorem 1], [8, Theorem 2.1], and rephrased in [32, Theorem 3.1] as follows.

Theorem 1 (elliptical range theorem) *Let $A, B \in \mathrm{M}_2$. Suppose $A - (\mathrm{tr} A/2)I_2$ and $B - (\mathrm{tr} B/2)I_2$ have singular values $a \geq b \geq 0$ and $c \geq d \geq 0$, respectively. Then, $\Delta(A, B)$ is an elliptical disk with the two σ-points as the foci and with length of minor semi-axis equal to $(ac - bd)$.*

The MOC is equivalent to that for $A, B \in \mathrm{N}_n$ and $U \in \mathrm{U}_n$, there exists $\{t_\sigma : \sigma \in S_n\} \subset \mathbb{R}$ such that $0 \leq t_\sigma \leq 1$, $\sum_{\sigma \in S_n} t_\sigma = 1$, and

$$\det(A + UBU^*) = \sum_{\sigma \in S_n} t_\sigma z_\sigma. \tag{6}$$

The term $\sum_{\sigma \in S_n} t_\sigma z_\sigma$ is called a *convex combination* of z_σ's. The degree 3 case for the MOC was proved by Bebiano, Merikoski, and da Providência in 1987 [9] using direct computation to express $\det(A + UBU^*)$ as a convex combination of σ-points.

Assuming $A = \mathrm{diag}\,(a_1, \ldots, a_n)$ and $B = \mathrm{diag}\,(b_1, \ldots, b_n)$, by the Binet–Cauchy formula (see [13]) , the equality in (6) becomes

$$\sum_{|I|=|J|} |\det U[I|J]|^2 \prod_{I^c} a_i \prod_J b_j = \sum_{|I|=|J|} \left(\sum_{\sigma:\sigma(I)=J} t_\sigma \right) \prod_{I^c} a_i \prod_J b_j, \tag{7}$$

where $I, J \subset \{1, \ldots, n\}$, and $U[I|J]$ is the submatrix of U with rows in I and columns in J. We can rewrite

$$\sum_{\sigma:\sigma(I)=J} t_\sigma = \sum_{\sigma \in S_n} t_\sigma |\det P_\sigma[I|J]|.$$

Note that $\det U[I|J]$ for $|I| = |J| = m$ are entries of the m-th compound $U^{(m)}$ of U. Naturally, J. K. Merikoski and A. Virtanen raised the following conjecture (MVC) [34]: for $A, B \in \mathrm{N}_n$ and $U \in \mathrm{U}_n$, do there exist real numbers t_σ, $\sigma \in S_n$, satisfying $0 \le t_\sigma \le 1$, $\sum_{\sigma \in S_n} t_\sigma = 1$, and

$$|U^{(m)}|^2 = \sum_{\sigma \in S_n} t_\sigma |P_\sigma^{(m)}| \qquad \text{for any } m = 1, \ldots, n? \tag{8}$$

Here, $|\cdot|$ and $|\cdot|^2$ are understood entrywise. Obviously, the MVC is stronger than the MOC.

Unfortunately, the MVC holds only for $n \le 3$ and fails for $n \ge 4$. When $n = 4$, a counterexample was constructed by Drury [14]. However, the further studies on MVC were fruitful in developing results and new tools. In 1994, Kovačec proved in [25] that (7) holds when at least $(n-2)$ of the eigenvalues of A or B are zero. Thus, the MOC holds if $(n-2)$ of the eigenvalues of A or B are equal. In 1999, Drury proved in [17] that the MVC holds when $n = 4$ and U is a Householder reflection. Thus, the MOC holds in this case. However, when $n = 5$, there exists a Householder reflection U that makes the MVC fail.

2.2 *Eigenvalues*

The MOC has been proved when the eigenvalues of the normal matrices A and B display certain special patterns.

Let $\sigma(A) = \{a_1, \ldots, a_n\}$ denote the *spectrum* of A. Obviously, M. Fiedler's result in 1971 [22] says that the MOC holds for Hermitian matrices A and B, which corresponds to the case $\sigma(A) \cup \sigma(B) \subset \mathbb{R}$. It could be easily extended to the case that $\sigma(A) \cup \sigma(B)$ is on a line through the origin. Moreover, the following progress has been made:

1. The MOC holds if A or B has at most one nonzero eigenvalue (de Oliveira 1982 [13]). Using the fact $A + UBU^* = (A + \lambda I_n) + U(B - \lambda I_n)U^*$ for $U \in \mathrm{U}_n$, we see that the MOC holds if all but possibly one of a_i's or of b_i's is equal (Bebiano-Queiró 1985 [7]).

2. The MOC holds if one of A and B is positive definite Hermitian, and the other one is skew-Hermitian, that is, $\sigma(A) \subset \mathbb{R}^+$ and $\sigma(B) \subset i\mathbb{R}$, or $\sigma(B) \subset \mathbb{R}^+$ and $\sigma(A) \subset i\mathbb{R}$ (da Providência-Bebiano 1986 [12]).
3. If all of a_i's and b_j's have the same modulus, then $\Delta(A, B)$ is contained in a line through the origin of argument $\frac{1}{2}\sum_{i=1}^{n}(\arg a_i + \arg b_i)$, and the MOC holds in this case (Bebiano-da Providência 1988 [5]).
4. The MOC holds if the σ-points lie all on a straight line, not necessarily passing through the origin (Merikoski–Virtanen 1989 [34]). This extends Fiedler's result on Hermitian matrices A and B. At the first glance, the conditions are about the σ-points instead of eigenvalues. However, [32, Theorems 3.3] shows that the σ-points are collinear if and only if the eigenvalues of A and $-B$ satisfy certain conditions. See 10 in this section for the details.
5. The MOC holds if A and B differ from Hermitian matrices by scalar matrices (Drury [15]). In 1994, Bebiano, Kovačec, and da Providência [11] extended this result and the result 2 in [12] to show that the MOC holds if $\sigma(A) \subset \mathbb{R}^+$ and $\sigma(B)$ lies on a line through the origin, or if $\sigma(A)$ lies on a line ℓ and $\sigma(B)$ lie on a line parallel to ℓ. The original statement required that the eigenvalues of A or B are pairwise distinct, but this constraint could be removed by small perturbations. The result was further extended by Drury in [19, Theorem 1.2, Corollary 1.3] and [20, Corollary 3] using Möbius transforms, the MOC holds if $\sigma(A)$ is in a line or circle and $\sigma(B)$ is in a line or circle.
6. The MOC holds if A or B has at most two distinct eigenvalues, namely, $|\sigma(A)| \leq 2$ or $|\sigma(B)| \leq 2$ (Merikoski–Virtanen 1993 [36]).
7. The MOC holds if $(n-2)$ of the elements in $\sigma(A)$ or $\sigma(B)$ is equal (Kovačec 1994 [25, Theorem 1]).
8. The MOC holds if at most one of the elements in $\sigma(A) \cup \sigma(B)$ is unreal (Kovačec 1999 [26, Corollary 4]).

Fractional linear (Möbius) transforms play an important role in several research papers like [19, 20, 32]. Let $a, b, c, d \in \mathbb{C}$, let $X = \begin{bmatrix} a & b \\ c & d \end{bmatrix}$, and denote the Möbius transform on $\mathbb{C} \setminus \{-\frac{d}{c}\}$ (or on $\mathbb{C}$ if $c = 0$):

$$\mu_X(z) := (az + b)(cz + d)^{-1}. \tag{9}$$

The $\mu_X(A)$ for $A \in \mathrm{M}_n$ can be similarly defined when $\det(cA + dI_n) \neq 0$. Suppose $\det[(cA + dI_n)(cB + dI_n)] \neq 0$. We can verify that (Li-Poon-Sze [32, Lemma 3.4]) for $A, B \in \mathrm{N}_n$ and $U \in \mathrm{U}_n$:

$$\det(\mu_X(A) - U\mu_X(B)U^*) = \frac{(\det X)^n}{\det[(cA + dI_n)(cB + dI_n)]}\det(A - UBU^*). \tag{10}$$

It immediately leads to the following result.

Theorem 2 *Suppose* $\det X \neq 0$ *and* $\det[(cA + dI_n)(cB + dI_n)] \neq 0$. *The pair of matrices* $(A, -B)$ *satisfies the MOC if and only if* $(\mu_X(A), -\mu_X(B))$ *satisfies the MOC.*

In 1988, Bebiano and da Providência showed in [5, Theorem 2] that if the MOC holds for (A, B), then the MOC holds for $((A - tI_n)^{-1}, (B + tI_n)^{-1}))$ for any $t \in \mathbb{C} \setminus \sigma(A \oplus (-B))$. This could be viewed as a special case of Theorem 2. A Möbius transform on the complex plane will send any circle or line to another circle or line (except probably for a pole on a circle). One may derive new results from the existing ones by Möbius transforms. For example, Kovačec's result 8 implies that

8′. The MOC holds if all but possibly one element of $\sigma(A) \cup \sigma(B)$ is located on a line or a circle.

In 2008, applying Möbius transform and other techniques, C. K. Li, Y. T. Poon, and N. S. Sze obtained sufficient and necessary conditions for $\Delta(A, B)$ being a singleton or a line segment for arbitrary matrices $A, B \in \mathrm{M}_n$ [32]. When $A, B \in \mathrm{N}_n$, the MOC certainly holds in these cases.

9. ([32, Theorem 3.2]) Suppose $A, B \in \mathrm{M}_n$. Then, $\Delta(A, B) = \{\delta\}$ if and only if one of the following holds:

   a. $\delta = 0$, and there is $\mu \in \mathbb{C}$ such that rank $(A - \mu I_n)$ + rank $(B + \mu I_n) < n$. Equivalently, $A \oplus (-B)$ has an eigenvalue with multiplicity greater than n.
   b. $\delta \neq 0$, one of A and B is a scalar matrix, and $\delta = \det(A + B)$.

   This extends a result of Bebiano, Poon, and da Providência in [10] which says that for $A, B \in \mathrm{N}_n$, $\Delta(A, B)$ is a point of $\mathbb{C} \setminus \{0\}$ if and only if either A or B is a scalar matrix and $\prod_{i=1}^{n}(a_i + b_i) \neq 0$.
10. ([32, Theorem 3.3]) Suppose $A, B \in \mathrm{M}_n$. Then, $\Delta(A, B)$ has empty interior if and only if it is either a singleton or a line segment. Moreover, $\Delta(A, B)$ is a line segment if and only if it is not a singleton and one of the following conditions holds:

    a. $A, B \in \mathrm{N}_n$, and $\sigma(A \oplus (-B))$ lie on a straight line or on a circle.
    b. $A, B \in \mathrm{N}_n$, and there is $\mu \in \mathbb{C}$ such that one of $(A - \mu I_n)$ and $(-B - \mu I_n)$ has rank 1 and the other one is the inverse of an invertible normal matrix whose eigenvalues are collinear, that is, it has the form $(cH + dI_n)^{-1}$, where $c, d \in \mathbb{C}$ and H is a Hermitian matrix so that $(cH + dI_n)$ is invertible.
    c. $A \oplus (-B)$ has an eigenvalue of multiplicity exactly n.

    This extends a result of Bebiano and Soares in [8, Theorem 3.1] about the sufficient and necessary conditions for $\Delta(A, B)$ lying on a line through the origin.

2.3 Other Cases

There are some other results confirming the MOC in situations not directly linked to the degree or the eigenvalues of normal matrices A and B.

1. If $\det(A + UBU^*) = 0$ for some $U \in \mathrm{U}_n$, then $0 \in \mathrm{co}\{z_\sigma : \sigma \in S_n\}$ (Drury-Cload 1992 [21]).
2. Assume $A = \mathrm{diag}\,(a_1, \ldots, a_n)$ and $B = \mathrm{diag}\,(b_1, \ldots, b_n)$. Then, for $U \in \mathrm{U}_n$,

$$\det(A + UBU^*) = \frac{\det(AU + UB)}{\det U} = \frac{\det(Z \circ U)}{\det U}, \quad Z = \left[(a_i + b_j) \right]_{i,j}, \tag{11}$$

where $\circ$ denotes the Hadamard (entrywise) product. Let S_n^0 be the subset of S_n consisting of the identity permutation, the transpositions, and the three cycles. When rank $Z = 2$ and U is a Householder reflection, Drury proved that (1999 [17, Corollary 3])

$$\det(A + UBU^*) = \frac{\det(Z \circ U)}{\det U} \in \mathrm{co}\{z_\sigma : \sigma \in S_n^0\}. \tag{12}$$

3. When $n = 4$, Drury proved a weaker form of MOC (2004 [18, Corollary 1.2]): suppose $A, B \in \mathrm{N}_4$, and $0 \notin \mathrm{co}\{z_\sigma : \sigma \in S_4\}$. Then,

$$\Delta(A, B) \subset \mathrm{co}\{r z_\sigma : \sigma \in S_4, r \geq 1\}. \tag{13}$$

4. In 2013, Guterman, Lemos, and Soares studied when the equality of the MOC holds for arbitrary $A, B \in \mathrm{M}_n$, namely,

$$\Delta(A, B) = \mathrm{co}\{z_\sigma : \sigma \in S_n\}. \tag{14}$$

Using the result of [32] on corner points (defined in the next section), they proved that (14) holds if and only if the boundary of $\Delta(A, B)$ is a convex polygon [23, Theorem 3.4]. They further used the elliptical range theorem to show that if B has pairwise distinct eigenvalues and (14) holds, then A must be a normal matrix [23, Theorem 3.5].

3 Properties of $\Delta(A, B)$

The geometry properties of $\Delta(A, B)$ has been extensively studied due to the nature of the MOC. Bebiano's synopses in [30] of her talk "New developments on the Marcus-Oliveira Conjecture" gave a fairly comprehensive review of the known properties of $\Delta(A, B)$ up to the year 1994.

The continuity of the map $U \mapsto \det(A + UBU^*)$ implies that $\Delta(A, B)$ is connected and compact. When $n = 2$, the elliptical range theorem (Theorem 1) shows that $\Delta(A, B)$ is a line segment for normal matrices A and B, and it is an elliptical disk for general $A, B \in \mathrm{M}_n$. One may hope to know whether or not $\Delta(A, B)$ is convex, star shaped, or simply connected for $A, B \in \mathrm{N}_n$. When $n = 3$, Bebiano, Poon, and da Providência showed in 1988 [10] that $\Delta(A, B)$ for $A, B \in \mathrm{N}_3$ is star shaped, and they also gave sufficient and necessary conditions for this $\Delta(A, B)$ being convex. However, Bebiano constructed an example for $n = 12$ (suggested by da Providência) to show that $\Delta(A, B)$ may not be simply connected [1], and thus $\Delta(A, B)$ may not be convex or star shaped. The example is $A = (A')^{\oplus 6}$ and $B = (B')^{\oplus 6}$, where $A' = \mathrm{diag}\,(i, -i)$ and $B' = \mathrm{diag}\,(1, \frac{1+\sqrt{3}}{-1+\sqrt{3}})$. By computation,

$$\Delta(A', B') = \left\{ \frac{2\sqrt{3}}{-1+\sqrt{3}} + i\frac{2}{-1+\sqrt{3}}t : -1 \le t \le 1 \right\}.$$

For $U = (U')^{\oplus 6}$ where $U' \in \mathrm{U}_2$, we have

$$\det(A + UBU^*) = \left(\frac{2\sqrt{3}}{-1+\sqrt{3}} + i\frac{2}{-1+\sqrt{3}}t \right)^6 = z(t).$$

When t varies from -1 to 1, $z(t)$ describes a continous curve surrounding the origin. Obviously, $\{z(t) : -1 \le t \le 1\} \subset \Delta(A, B)$. We show that $0 \notin \Delta(A, B)$ so that $\Delta(A, B)$ is not simply connected. Write $A = iA_1$ for $A_1 = \mathrm{diag}\,(1, -1)^{\oplus 6}$. Then, for any $U \in \mathrm{U}_{12}$,

$$\begin{aligned} |\det(A + UBU^*)|^2 &= |\det A|^2 |\det(I_{12} - iA_1^{-1}UBU^*)|^2 \\ &= |\det A|^2 |\det[I_{12} + (A_1^{-1}UBU^*)^2]|. \end{aligned}$$

Since A_1^{-1} is Hermitian and UBU^* is positive definite, all eigenvalues of $A_1^{-1}UBU^*$ are real. Hence, $\det(A + UBU^*) \neq 0$.

As mentioned earlier, the cases $\Delta(A, B)$ having empty interior have been resolved in [32, Theorems 3.2, 3.3], where $\Delta(A, B)$ must be a point or a line segment. Moreover, there are some situations where points or regions were proved to be inside or outside of $\Delta(A, B)$.

1. If σ and τ differ by a transposition, then the segment $[z_\sigma, z_\tau]$ is contained in $\Delta(A, B)$ (Bebiano-Queiró 1985 [7, Theorem 1]).
2. If τ and ϕ both differ from σ by a transposition, then the region $[z_\sigma, z_\tau, z_\phi, z_{\phi\sigma^{-1}\tau}]$ is contained in $\Delta(A, B)$ (Bebiano-Queiró 1985 [7, Theorem 2]).
3. Given z_σ, consider the points z_τ generated by permutations τ which differ from σ by a transposition. If these points do not all lie in a half plane determined by a straight line through z_σ, then z_σ is an interior point of $\Delta(A, B)$ (Bebiano-Queiró 1985 [7, Theorem 3]).
4. $0 \notin \Delta(A, B)$ if and only if $\sigma(A)$ and $\sigma(-B)$ are separable by a circle (Wielandt 1955 [42]). Let $\mathscr{E}(A, B)$ be the set of eigenvalues of matrices of the form

$VAV^* + UBU^*$ for $V, U \in \mathrm{U}_n$. The condition $0 \notin \Delta(A, B)$ is equivalent to that $0 \notin \mathscr{E}(A, B)$. Similarly, for any $\mu \in \mathbb{C}$, $\mu \notin \mathscr{E}(A, B)$ if and only if $\sigma(A)$ and $\sigma(\mu I_n - B)$ are separable by a circle. The result was extended by Li et al. in [31] to arbitrary matrices $A, B \in \mathrm{M}_n$.

The boundary $\partial\Delta$ of $\Delta(A, B)$ was also explored in quite a few papers. A *corner point of* $\Delta(A, B)$ is a point $z \in \mathbb{C}$ such that, in a neighborhood of z, $\Delta(A, B)$ is contained in an angle with vertex at z and measuring less than π. Here are some results about $\partial\Delta$.

1. Bebiano and Queiró proved in 1985 [7, Theorem 4] that every nonzero corner point must be a σ-point. Later Drury covered the case of zero corner point [21]. The results were further extended to matrices $A, B \in \mathrm{M}_n$ in (Li-Poon-Sze [32, Theorem 3.10]).
2. Bebiano, Kovačec, and da Providência showed in 1994 [11] that for certain classes of essentially Hermitian matrices A and B:
   a. through any of the regular boundary point $z \neq 0$ of $\partial\Delta$ there passes a regular curve contained in $\Delta(A, B)$ and of curvature 0 in z;
   b. each support point $z \neq 0$ of $\Delta(A, B)$ is either a σ-point, or there passes a curve through z contained in $\Delta(A, B)$ and of curvature 0 in z.
3. In the MOC, it suffices to consider only $\det(A + UBU^*)$ for $A = \mathrm{diag}\,(a_1, \ldots, a_n)$, $B = \mathrm{diag}\,(b_1, \ldots, b_n)$, and $U \in \mathrm{SU}_n$. Then by (11), $\det(A + UBU^*)$ can be expressed as the Hadamard product

$$\det(A + UBU^*) = \det(Z \circ U), \qquad Z = \left[\, (a_i + b_j)\,\right]_{i,j}. \tag{15}$$

For $M \in \mathrm{M}_n$ and $\sigma \in S_n$, let

$$d_\sigma(M) = \prod_{i=1}^{n} m_{i\sigma(i)} \tag{16}$$

be the diagonal product of M associated to permutation σ. In particular, $d_\sigma(Z) = z_\sigma$. Let $\mathscr{T} = \{(i, j) : 1 \leq i < j \leq n\}$ be the set of transpositions in S_n. In 2007, Kovacec, Bebiano, and da Providência provided a corner description of $\Delta_0(M) = \{\det(M \circ Q) : Q \in \mathrm{SO}_n\}$ [27, Theorem 11], for $M \in \mathrm{M}_n$ and $\sigma \in S_n$, under certain constraints, $\Delta_0(M)$ has in $d_\sigma(M)$ the corner

$$d_\sigma(M) + \mathrm{cone}\{d_{\sigma\tau}(M) - d_\sigma(M) : \tau \in \mathscr{T}\}.$$

4 Extensions

The studies of questions related to the MOC have raised many interesting new results and new questions. One common approach is to ask similar questions in boarder or in different settings. We list some major progress here.

1. In the paper [32], Li, Poon, and Sze studied the following rank set of matrices $A, B \in \mathrm{M}_n$:

$$R(A, B) = \{\text{rank } (VAV^* + UBU^*) : U, V \in \mathrm{U}_n\}. \tag{17}$$

They present the best upper bound and a lower bound for this set.

a. Let

$$m = \min\{\text{rank } (A - \mu I_n) + \text{rank } (-B - \mu I_n) : \mu \in \sigma(A \oplus -B)\}.$$

Then, $\max R(A, B) = \min\{m, n\}$. In particular, if $A, B \in \mathrm{N}_n$, and ℓ is the maximal multiplicity of an eigenvalue of $A \oplus -B$, then $m = 2n - \ell$ and $\max R(A, B) = \min\{2n - \ell, n\}$. The set $\Delta(A, B) = \{0\}$ if and only if there is an eigenvalue of $A \oplus -B$ of multiplicity greater than n.

b. Let $\mu \in \mathbb{C}$ such that rank $(A - \mu I_n) = p$ and rank $(-B - \mu I_n) = q$. Then, $\min R(A, B) \leq \max\{p, q\}$.

2. Let $A, B \in \mathrm{M}_n$ be such that all the eigenvalues of $A \oplus B$ have multiplicity less than n. Assume A or B is non-normal. The $\Delta(A, B)$ contains a nondegenerated elliptical disk [23, Theorem 2.4]. It is natural to be expected that many properties of $\Delta(A, B)$ for normal matrices A and B could be extended to non-normal case.
3. (Drury [15]) Let A and B be Hermitian matrices with eigenvalues $(a_1, \ldots, a_n)$ and $(b_1, \ldots, b_n)$, respectively. Then,

$$\det(xI_n + A + B) \in \text{co}\left\{\prod_{i=1}^{n}(x + a_i + b_{\sigma(i)}) : \sigma \in S_n\right\}, \tag{18}$$

where co $\{\cdot\}$ denotes the convex hull in $\mathbb{C}[x]$, the space of polynomials. In fact, Drury investigated what symmetric function f could satisfy that

$$f(t_1, \ldots, t_n) \in \text{co}\{f(a_1 + b_{\sigma(1)}, \ldots, a_n + b_{\sigma(n)}) : \sigma \in S_n\},$$

where $t_1, \ldots, t_n$ are eigenvalues of $A + B$. He proved affirmatively for the functions $f(t_1, \ldots, t_n) = \prod_{i=1}^{n}(\lambda + t_i)$ and $f(t_1, \ldots, t_n) = \sum_{i=1}^{n} t_i^m$ when $m \in \mathbb{Z}^+$. However, if A and B are only normal matrices, then (18) is not true when $n \geq 4$ [21].

4. The determinantal inequalities involving sums of matrices were explored by researchers during a long period. Queiró and Kovačec in [39] proved the following inequality for $A, B \in \mathrm{N}_n$ and $U \in \mathrm{U}_n$:

$$|\det(A + UBU^*)| \leq \min \left\{ \prod_{i=1}^{n} \max_j |a_i + b_j|, \prod_{j=1}^{n} \max_i |a_i + b_j| \right\}. \tag{19}$$

The proof uses the expression (11) and then applies the classical Hadamard's inequality: $|\det M| \leq \prod_{i=1}^n \left(\prod_{j=1}^n |m_{ij}|^2 \right)^{1/2}$. In 1994, Bebiano and Miranda [3] examined the equality case and proved that the equality in (19) holds if and only if there is $\sigma \in S_n$ such that

$$\max_j |a_i + b_j| = |a_i + b_{\sigma(i)}|, \qquad i = 1, \ldots, n, \tag{20a}$$

or

$$\max_i |a_i + b_j| = |a_{\sigma(j)} + b_j|, \qquad j = 1, \ldots, n. \tag{20b}$$

As a direct consequence,

$$\max\{|\det(A + UBU^*)| : U \in \mathrm{U}_n\} \leq \max\{|z_\sigma| : \sigma \in S_n\}. \tag{21}$$

5. The following result was independently proved in [38], [29, Theorem 1], [6, Theorem 2], for arbitrary $A, B \in \mathrm{M}_n$ with singular values $\alpha_1 \geq \cdots \geq \alpha_n \geq 0$ and $\beta_1 \geq \cdots \geq \beta_n \geq 0$, respectively,

$$\prod_{i=1}^{n} (\alpha_i + \beta_{n-i+1}) \geq |\det(A + B)| \tag{22a}$$

$$\geq \begin{cases} 0 & \text{if } [\alpha_n, \alpha_1] \cap [\beta_n, \beta_1] \neq \emptyset, \\ \prod_{i=1}^n (\alpha_i - \beta_{n-i+1}) & \text{otherwise.} \end{cases} \tag{22b}$$

The lower bound inequality (22b) was extended to certain Schur concave functions [29, Theorem 5]. The equality case was studied by Bebiano and Perdigão and they got the following result in 1998 [4, Theorem 2.1]. The set $S(A, B) = \{|\det(A + UBV)| : U, V \in \mathrm{U}_n\}$ satisfies that

$$\max S(A, B) = \prod_{i=1}^{n} (\alpha_i + \beta_{n-i+1}), \tag{23a}$$

$$\min S(A, B) = \begin{cases} 0 & [\alpha_n, \alpha_1] \cap [\beta_n, \beta_1] \neq \emptyset, \\ \prod_{i=1}^n (\alpha_i - \beta_{n-i+1}) & \text{otherwise.} \end{cases} \tag{23b}$$

The inequalities (22a) and (22b) were further extended by Tam and Thompson to real matrices and some other real simple Lie groups [41]. They also extended the Hermitian case proved by Fiedler [22] to the Lie algebra of a compact connected subgroup in U_n [40].

References

1. Bebiano, N.: Some variations on the concept of the c-numerical range. Linear algebra and its applications (Coimbra 1984). Portugal. Math. **43**(1), 189–200 (1985/86)
2. Bebiano, N.: Some analogies between the c-numerical range and a certain variation on this concept. Linear Algebra Appl. **81**, 47–54 (1986)
3. Bebiano, N., Miranda, M.E.: On a recent determinantal inequality. Linear Algebra Appl. **201**, 99–102 (1994)
4. Bebiano, N., Perdigão, C.: Extremal matrices in certain determinantal inequalities. Linear Multilinear Algebra **44**(3), 261–276 (1998)
5. Bebiano, N., da Providência, J.: Some remarks on a conjecture of de Oliveira. Linear Algebra Appl. **102**, 241–246 (1988)
6. Bebiano, N., da Providência, J.: On the dual of a result of Miranda and Thompson. Linear Algebra Appl. 262, 119–129 (1997)
7. Bebiano, N., Queiró, J.F.: The determinant of the sum of two normal matrices with prescribed eigenvalues. Linear Algebra Appl. **71**, 23–28 (1985)
8. Bebiano, N., Soares, G.: Three observations on the determinantal range. Linear Algebra Appl. **401**, 211–220 (2005)
9. Bebiano, N., Merikoski, J.K., da Providência, J.: On a conjecture of G. N. de Oliveira on determinants. Linear Multilinear Algebra **20**(2), 167–170 (1987)
10. Bebiano, N., Poon, Y.T., da Providência, J.: On c-det spectral and c-det-convex matrices. Linear Multilinear Algebra **23**(4), 343–351 (1988)
11. Bebiano, N., Kovačec, A., da Providência, J.: The validity of the Marcus-de Oliveira conjecture for essentially Hermitian matrices. Linear Algebra Appl. **197/198**, 411–427 (1994). In: Second Conference of the International Linear Algebra Society (ILAS). Lisbon (1992)
12. da Providência, J., Bebiano, N.: Matrices satisfying a conjecture of G. N. de Oliveira on determinants. Linear Algebra Appl. **78**, 187–198 (1986)
13. de Oliveira, G.: Normal matrices (research problem). Linear Multilinear Algebra **12**, 153–154 (1982)
14. Drury, S.W.: A counterexample to a question of Merikoski and Virtanen on the compounds of unitary matrices. Linear Algebra Appl. **168**, 251–257 (1992)
15. Drury, S.W.: On symmetric functions of the eigenvalues of the sum of two Hermitian matrices. Linear Algebra Appl. **176**, 211–222 (1992)
16. Drury, S.W.: The determinantal conjecture and Hadamard type inequalities. Linear Algebra Appl. **246**, 279–297 (1996)
17. Drury, S.W.: OMC for householder reflections. Linear Algebra Appl. **298**(1–3), 159–169 (1999)
18. Drury, S.W.: The external vertices conjecture in case $n = 4$. Electron. J. Linear Algebra **11**, 180–191 (2004)
19. Drury, S.W.: Essentially Hermitian matrices revisited. Electron. J. Linear Algebra **15**, 285–296 (2006)
20. Drury, S.W.: OMC for scalar multiples of unitaries. Linear Algebra Appl. **422**(1), 318–325 (2007)
21. Drury, S.W., Cload, B.: On the determinantal conjecture of Marcus and de Oliveira. Linear Algebra Appl. **177**, 105–109 (1992)

22. Fiedler, M.: Bounds for the determinant of the sum of Hermitian matrices. Proc. Am. Math. Soc. **30**, 27–31 (1971)
23. Guterman, A., Lemos, R., Soares, G.: Extremal case in Marcus-Oliveira conjecture and beyond. Linear Multilinear Algebra **61**(9), 1206–1222 (2013)
24. Kippenhahn, R.: über den Wertevorrat einer matrix. Math. Nachr. **6**, 193–228 (1951)
25. Kovačec, A.: On a conjecture of Marcus and de Oliveira. Linear Algebra Appl. **201**, 91–97 (1994)
26. Kovačec, A.: The Marcus-de Oliveira conjecture, bilinear forms, and cones. Linear Algebra Appl. **289**(1–3), 243–259 (1999). In: Linear Algebra and Statistics. Istanbul (1997)
27. Kovacec, A., Bebiano, N., da Providência, J.: On the corners of certain determinantal ranges. Linear Algebra Appl. **426**(1), 96–108 (2007)
28. Li, C.-K.: Some convexity theorems for the generalized numerical ranges. Linear Multilinear Algebra **40**(3), 235–240 (1996)
29. Li, C.-K., Mathias, R.: The determinant of the sum of two matrices. Bull. Austral. Math. Soc. **52**(3), 425–429 (1995)
30. Li, C.-K, da Silva, J.D., de Oliveira, G., (Eds.): Second Conference of the International Linear Algebra Society (ILAS)
31. Li, C.-K., Poon, Y.-T., Sze, N.-S.: Eigenvalues of the sum of matrices from unitary similarity orbits. SIAM J. Matrix Anal. Appl. **30**(2), 560–581 (2008)
32. Li, C.-K., Poon, Y.-T., Sze, N.-S.: Ranks and determinants of the sum of matrices from unitary orbits. Linear Multilinear Algebra **56**(1–2), 105–130 (2008)
33. Marcus, M.: Derivations, Plücker relations, and the numerical range. Indiana Univ. Math. J. **22**, 1137–1149 (1972/73)
34. Merikoski, J.K., Virtanen, A.: Some notes on de Oliveira's determinantal conjecture. Linear Algebra Appl. **121**, 345–352 (1989). In: Linear Algebra and Applications. Valencia (1987)
35. Merikoski, J.K., Virtanen, A.: On elementary symmetric functions of the eigenvalues of the sum and product of normal matrices. Czechoslovak Math. J. **42**(117)(2), 193–198 (1992)
36. Merikoski, J.K., Virtanen, A.: Some further notes on the Marcus-de Oliveira determinantal conjecture. Erratum to: "Some notes on de Oliveira's determinantal conjecture" [Linear Algebra Appl. 121, 345–352; MR1011745 (90g:15014)]. Linear Algebra Appl. **187**(259–262), 1993 (1989)
37. Merikoski, J.K., Virtanen, A., Bebiano, N., da Providência, J.: Interdependence of some problems arising in generalizing the Marcus-de Oliveira determinantal conjecture. Portugal. Math. **50**(2), 227–236 (1993)
38. Miranda, M.E.F.: On the trace of the product and the determinant of the sum of complex matrices with prescribed singular values. In: Linear Algebra and Applications. Vitoria-Gasteiz (1983)
39. Queiró, J.F., Kovačec, A.: A bound for the determinant of the sum of two normal matrices. Linear Multilinear Algebra **33**(3–4), 171–173 (1993)
40. Tam, T.-Y. Thompson, M.C.: Determinants of sum of orbits under compact Lie group. Linear Algebra Appl. **436**(6), 1644–1650 (2012)
41. Tam, T.-Y., Thompson, M.C.: Determinants of sums of two real matrices and their extensions. Linear Multilinear Algebra **60**(11–12), 1409–1431 (2012)
42. Wielandt, H.: On eigenvalues of sums of normal matrices. Pacific J. Math. **5**, 633–638 (1955)

Schubert Polynomial Analogues for Degenerate Involutions

Michael Joyce

Abstract We survey the recent study of involution Schubert polynomials and a modest generalization that we call degenerate involution Schubert polynomials. We cite several conditions when (degenerate) involution Schubert polynomials have simple factorization formulae. Such polynomials can be computed by traversing through chains in certain weak order posets, and we provide explicit descriptions of such chains in weak order for involutions and degenerate involutions. As an application, we give several examples of how certain multiplicity-free sums of Schubert polynomials factor completely into very simple linear factors.

Keywords Schubert polynomial · Involution · Degenerate involution · Weak order · Symmetric subgroup

1 Introduction

The study of the complete flag variety $\mathscr{F}$, the algebraic variety that parameterizes the complete flags of an n-dimensional (complex) vector space, leads to many rich links between geometry, algebra, representation theory, and combinatorics. The general linear group $G = GL_n(\mathbb{C})$ acts transitively on $\mathscr{F}$ and of particular interest are actions of subgroups $H \subset G$ that act with finitely many orbits in $\mathscr{F}$; such subgroups are called spherical.

In this survey, we focus on the case when $H = O_n(\mathbb{C})$, the orthogonal group, and on variations H_μ associated to compositions μ of n. When μ is the composition with a single part equal to n, $H_\mu = H = O_n(\mathbb{C})$, while the composition μ with n parts equal to 1 corresponds to $H_\mu = B$, the Borel subgroup of G consisting of upper triangular matrices. In the general case, when $\mu = (\mu_1, \mu_2, \ldots, \mu_k)$, H_μ is a semidirect product of a Levi factor $L_\mu = O_{\mu_1} \times \cdots \times O_{\mu_k}$ (embedded into G diagonally) and the unipotent radical of a parabolic subgroup $P_\mu \subset G$. The H-orbits on $\mathscr{F}$ are parameterized by involutions, while the H_μ-orbits on $\mathscr{F}$ are parameterized

M. Joyce (✉)
Mathematics Department, Tulane University, New Orleans, LA 70118, USA
e-mail: mjoyce3@tulane.edu

J. Feldvoss et al. (eds.), *Advances in Algebra*, Springer Proceedings in Mathematics & Statistics 277, https://doi.org/10.1007/978-3-030-11521-0_9

by combinatorial objects called μ-involutions. We refer to a μ-involution for some composition μ of n as a degenerate involution of rank n.

The B-orbits on $\mathscr{F}$ are the classical Schubert cells of $\mathscr{F}$ and are parameterized by permutations in S_n, the symmetric group on n letters. For $w \in S_n$, the Schubert polynomial $\mathfrak{S}_w$ is a natural representative of the class of the closure of the Schubert cell associated to w in $H^*(\mathscr{F})$, the integral cohomology ring of $\mathscr{F}$. Letting $\mathscr{I}_n$ denote the set of involutions in S_n, for $\pi \in \mathscr{I}_n$, the involution Schubert polynomial $\hat{\mathfrak{S}}_\pi$ is a natural representative of the class of the closure of the H-orbit of $\mathscr{F}$ associated to π in $H^*(\mathscr{F})$. Likewise, for a fixed composition μ of n, we let $\mathscr{I}_\mu$ denote the set of μ-involutions and for $\pi \in \mathscr{I}_\mu$, the μ-involution Schubert polynomial $\hat{\mathfrak{S}}^\mu_\pi$ is a natural representative of the class of the closure of the H_μ-orbit of $\mathscr{F}$ associated to π in $H^*(\mathscr{F})$.

Let us give a brief history of the study of involution Schubert polynomials and degenerate involution Schubert polynomials. The first detailed investigation of H-orbits on $\mathscr{F}$ was carried out in the more general setting of symmetric varieties by Richardson and Springer [21, 23]. One of their key results is that the inclusion order of H-orbit closures in $\mathscr{F}$ is given by the restriction of the Bruhat order on S_n to $\mathscr{I}_n$ [22]. In a combinatorial framework, Can and the author [6] decomposed the (degenerate) involution Schubert polynomial associated to the longest permutation (which can be viewed as a μ-involution for any composition μ) as a sum of ordinary Schubert polynomials, by studying maximal chains in the associated weak order poset using a result of Brion [4]. Can, Wyser, and the author [7] then decomposed an arbitrary involution Schubert polynomial as a multiplicity-free sum of ordinary Schubert polynomials. Hamaker et al. [14] gave the first detailed account of involution Schubert polynomials, creating a uniform combinatorial language for the study and connecting the combinatorics explicitly to the geometry of H-orbit closures in $\mathscr{F}$. Hamaker et al. [13] then developed the theory of involution words, describing the maximal chains for any interval in the weak order poset associated to H-orbits on $\mathscr{F}$ and providing evidence for their explicit conjecture that such chains are closely linked to the ordinary Bruhat order on S_n. Can, Wyser, and the author [8] return to the study of the degenerate involution Schubert polynomial associated to the longest permutation viewed as a μ-involution and use geometric considerations to show that certain multiplicity-free sums of ordinary Schubert polynomials have very simple factorizations. Hamaker et al. [15] have given a transition formula for involution Schubert polynomials, generalizing the transition formula for ordinary Schubert polynomials established by Lascoux and Schützenberger [19]. Hamaker et al. [16] have also initiated a study of involution Stanley symmetric functions, a natural limit of involution Schubert polynomials, and shown that they can be expanded positively in the Schur P-basis; they have also established a similar result for fixed-point-free involution Stanley symmetric functions [17].

We now describe the contents of this survey. In Sect. 2, we establish our notation and conventions. Then, in Sect. 3, we explain how to define K-Schubert polynomials for a spherical variety K and give three important characterizations in the case $K = H_\mu$. We then discuss factorization results for (degenerate) involution Schubert polynomials in Sect. 4. In Sect. 5, we describe chains in the weak order for μ-involutions. Finally, in Sect. 6, we apply our results to give new identities expressing multiplicity-free sums of ordinary Schubert polynomials as a product of simple linear factors.

2 Notation and Conventions

In this paper, G denotes a connected, reductive algebraic group over $\mathbb{C}$ (though our results apply for any algebraically closed field of characteristic not equal to 2). We let B denote a Borel subgroup of G and K a *spherical subgroup* of G, i.e., K contains a dense orbit on G/B under the left multiplication action. Most of our results are specialized to the case where $G = GL_n(\mathbb{C})$; in that case, we let B be the Borel subgroup of upper triangular matrices for definiteness. In this case, G/B is isomorphic to the complete flag variety $\mathscr{F}$.

Our main choices of spherical subgroup K will be $H = O_n(\mathbb{C})$ and a subgroup H_μ associated to a composition μ of n that we now describe. Recall that a composition μ is a sequence of positive integers $(\mu_1, \ldots, \mu_k)$ such that $\mu_1 + \cdots + \mu_k = n$. The subgroup $H_\mu \subseteq GL_n(\mathbb{C})$ is the semidirect product $L_\mu \ltimes R_\mu$, where

$$L_\mu = O_{\mu_1}(\mathbb{C}) \times \cdots \times O_{\mu_k}(\mathbb{C})$$

embedded diagonally in $GL_n(\mathbb{C})$, and R_μ is the unipotent radical of the parabolic subgroup P_μ containing B whose associated Levi factor is $GL_{\mu_1}(\mathbb{C}) \times \cdots \times GL_{\mu_k}(\mathbb{C})$.

We will make use of some elementary combinatorics of the symmetric group S_n. The simple transpositions of S_n are denoted s_i, $i = 1, \ldots, n-1$, where s_i is the permutation that interchanges i and $i+1$ and fixes all other numbers. Given a permutation $w \in S_n$, a *reduced decomposition* of w is a sequence $(s_{i_1}, \ldots, s_{i_\ell})$ of simple transpositions such that $w = s_{i_1}, \ldots, s_{i_\ell}$ with ℓ minimal. The number $\ell = \ell(w)$ is the *length* of w and is equal to the number of pairs (i, j) with $1 \leq i < j \leq n$ such that $w(i) > w(j)$. As a slight abuse of notation, we identify a reduced decomposition with the corresponding product $s_{i_1}, \ldots, s_{i_\ell}$. The longest permutation of S_n is w_0 defined by $w_0(i) = n + 1 - i$.

The group S_n acts on $f \in \mathbb{Z}[x_1, \ldots, x_n]$ as follows. For $1 \leq i \leq n-1$, let $s_i \in S_n$ denote the simple transposition that interchanges i and $i+1$ while fixing all other values. Then, $s_i \cdot f$ is the polynomial obtained by interchanging the variables x_i and x_{i+1}. Note that $f - s_i \cdot f$ is always divisible by $x_i - x_{i+1}$. Thus, we can define divided difference operators $\partial_i : \mathbb{Z}[x_1, \ldots, x_n] \to \mathbb{Z}[x_1, \ldots, x_n]$ by

$$\partial_i(f) = \frac{f - s_i \cdot f}{x_i - x_{i+1}}.$$

It can be easily seen that if I denotes the ideal in $\mathbb{Z}[x_1, \ldots, x_n]$ generated by all homogeneous symmetric polynomials in $x_1, \ldots, x_n$ of positive degree, then the divided difference operator descends to an operator $\partial_i : \mathbb{Z}[x_1, \ldots, x_n]/I \to \mathbb{Z}[x_1, \ldots, x_n]/I$, which by abuse of notation we also refer to as ∂_i.

We now recall the definition of the Richardson–Springer monoid [21] of the symmetric group S_n. Recall that S_n is generated by the $n-1$ simple transpositions s_i, $1 \le i \le n-1$. The Richardson–Springer monoid, denoted $M(S_n)$, is generated by the elements $m(s_1), \ldots, m(s_{n-1})$ subject to the relations

$$m(s_i)^2 = m(s_i)$$

for $1 \le i \le n-1$,

$$m(s_i)m(s_j) = m(s_j)m(s_i)$$

whenever $|i-j| > 1$ and

$$m(s_i)m(s_{i+1})m(s_i) = m(s_{i+1})m(s_i)m(s_{i+1})$$

for $1 \le i \le n-2$. Then, $M(S_n)$ is a finite monoid and every element of $M(S_n)$ has the form $m(w)$ for $w \in S_n$, where $m(w) := m(s_{i_1}), \ldots, m(s_{i_\ell})$ for any reduced decomposition $s_{i_1}, \ldots, s_{i_\ell}$ of w.

An *involution* is a permutation w such that w^2 is the identity permutation. In particular, we consider the identity permutation itself to be an involution. We let $\mathscr{I}_n$ denote the set of all involutions in S_n. To define degenerate involutions, we need a convention for interpreting certain strings as permutations. Here, we use the convention that $[n] = \{1, 2, \ldots, n\}$.

Convention 2.1 *Given any string α containing each element of an alphabet $\mathbb{A} \subset [n]$ in exactly one position, we interpret α as the one-line notation of a permutation of $\mathbb{A}$, where we order $\mathbb{A}$ in increasing order. For example, the string $\alpha = [5264]$ should be interpreted as the permutation of the alphabet $\mathbb{A} = \{2, 4, 5, 6\}$ given by* $2 \mapsto 5, 4 \mapsto 2, 5 \mapsto 6, 6 \mapsto 4$.

Let $\mu = (\mu_1, \ldots, \mu_k)$ be a composition of n. Then a *μ-involution* is a permutation in S_n such that, when the one-line notation is partitioned into strings of length $\mu_1, \ldots, \mu_k$, when each string is viewed as permutation of its alphabet according to Convention 2.1, the corresponding permutation is an involution. We let $\mathscr{I}_\mu$ denote the set of all μ-involutions. A *degenerate involution* (of rank n) is a μ-involution for some composition μ (of n).

We adopt a notational convention that τ, τ', etc. denote involutions, while π, π', etc. denote μ-involutions.

Finally, we introduce a standard notation from the theory of posets. If X is any partially ordered set and $a, b \in X$, then $[a, b] := \{x \in X : a \le x \le b\}$.

3 K-Schubert Polynomials

Throughout this section, we only consider the group $G = GL_n(\mathbb{C})$.

The integral singular cohomology ring of $\mathcal{F}$ can be viewed from two complementary perspectives. First, an additive basis for $H^*(\mathcal{F})$ is given by the *Schubert classes*, which are the classes of the *Schubert varieties*. The Schubert varieties are the closures of the orbits of the B-action on $\mathcal{F}$, and are parameterized by the permutations $w \in S_n$; the corresponding Schubert variety is denoted by X_w and its Schubert class by $\sigma_w \in H^*(\mathcal{F})$. Schubert varieties can also be described concretely by rank conditions (c.f. [12, 20]).

Second, there is the Borel presentation which describes the ring structure,

$$H^*(\mathcal{F}) \cong \mathbb{Z}[x_1, \dots, x_n]/I,$$

where I is the ideal generated by all homogeneous symmetric polynomials in $x_1, \dots, x_n$ of positive degree, or equivalently, I is generated by the elementary symmetric polynomials in $x_1, \dots, x_n$. Under this isomorphism, x_i is mapped to the Chern class of the line bundle L_i/L_{i-1} constructed from the tautological sequence of bundles

$$0 = L_0 \subset L_1 \subset \cdots \subset L_{n-1} \subset L_n = \mathcal{F} \times \mathbb{C}^n.$$

Here, the fiber of L_i over the flag $F \in \mathcal{F}$ is V_i if F is given by

$$0 = V_0 \subset V_1 \subset \cdots \subset V_{n-1} \subset V_n = \mathbb{C}^n.$$

A natural question emerges, how are Schubert classes represented in the Borel presentation? Of course since $H^*(\mathcal{F})$ is a quotient of $\mathbb{Z}[x_1, \dots, x_n]$, there are many different choices of representatives for σ_w. Lascoux and Schützenberger found a choice of natural representatives with rich combinatorial structure [18], which they called *Schubert polynomials*. The Schubert polynomial $\mathfrak{S}_w$, which represents the Schubert class σ_w, has several characterizations. First, Schubert polynomials can be defined recursively. The Schubert polynomial for the longest permutation w_0 is given by $\mathfrak{S}_{w_0} := x_1^{n-1}x_2^{n-2}, \dots, x_{n-1}$. Then, for any $w \in S_n$, $\mathfrak{S}_w := \partial_i(\mathfrak{S}_{ws_i})$ for any i such that $\ell(ws_i) = \ell(w) + 1$.

Second, $\mathfrak{S}_w$ can be characterized as the unique polynomial in the $\mathbb{Z}$-span of the Artin basis, $\{x_1^{a_1}, \dots, x_n^{a_n} : 0 \le a_i \le n - i\}$, that represents σ_w in $\mathbb{Z}[x_1, \dots, x_n]/I$. We adapt the latter point of view in defining the K-Schubert polynomial of a K-orbit closure Y for any spherical subgroup K. There is one technical complication, due to the fact that the greatest common divisor of the coefficients of the polynomial representing a K-orbit closure may be greater than 1. In fact, the gcd will always be of the form $2^{\kappa(Y)}$ for some nonnegative integer $\kappa(Y)$ [5]. Therefore, we will define the K-Schubert polynomial to be the representative polynomial in the Artin basis divided by this common factor.

Definition 3.1 Let K be a spherical subgroup of $GL_n(\mathbb{C})$ and let Y be a K-orbit closure in $\mathscr{F}$. The *K-Schubert polynomial* of Y, denoted $\mathfrak{S}_Y$, is the unique polynomial in the $\mathbb{Z}$-span of $\{x_1^{a_1}, \ldots, x_n^{a_n} : 0 \leq a_i \leq n-i\}$ such that $2^{\kappa(Y)}\mathfrak{S}_Y$ represents the class of Y in $H^*(\mathscr{F})$. If $K = H$ is the orthogonal group and Y is the closure of $H\tau B$ for some involution τ, then we write $\hat{\mathfrak{S}}_\tau$ for $\mathfrak{S}_Y$. More generally, if $K = H_\mu$ and Y is the K-orbit closure associated to the degenerate involution π, then we write $\hat{\mathfrak{S}}_\pi$ for $\mathfrak{S}_Y$.

While Definition 3.1 connects K-Schubert polynomials to geometry, it does not give an explicit algebraic or combinatorial description of them. In the cases of interest, we can give a recursive description using divided difference operators, analogous to the case of ordinary Schubert polynomials.

If $K = H = O_n(\mathbb{C})$, then there is a unique closed K-orbit on $\mathscr{F}$, parameterized by the longest permutation τ_0 viewed as an involution. Then, as a corollary to Theorem 4.5, we have that

$$\hat{\mathfrak{S}}_{\tau_0} = x_1, \ldots, x_{\lfloor n/2 \rfloor} \prod_{0<i<j\leq n-i} (x_i + x_j).$$

When $K = H_\mu$, there is a unique closed K-orbit on $\mathscr{F}$, parameterized by the longest permutation π_0 viewed as a μ-involution. Then, Theorem 4.7 gives a similar factorization formula for $\hat{\mathfrak{S}}_{\pi_0}$, and we defer the result until Sect. 4 where the necessary notation is introduced.

Then, we can compute $\hat{\mathfrak{S}}_\tau$ (resp., $\hat{\mathfrak{S}}_\pi$) for an involution τ (resp., μ-involution π) recursively via $\hat{\mathfrak{S}}_\tau = \partial_i(\hat{\mathfrak{S}}_{\tau'})$ (resp., $\hat{\mathfrak{S}}^\mu_\pi = \partial_i(\hat{\mathfrak{S}}^\mu_{\pi'})$) when $m(s_i) \cdot \tau = \tau'$ (resp., $m(s_i) \cdot \pi = \pi'$), where the latter notation refers to the action of the Richardson–Springer monoid which is described explicitly in Sect. 5.

There is another approach to understanding K-Schubert polynomials, and that is to express them in the basis of ordinary Schubert polynomials. For a K-orbit closure Y, we may write

$$\hat{\mathfrak{S}}_Y = \sum_{w \in S_n} c_{Y,w} \mathfrak{S}_w,$$

with $c_{Y,w} \in \mathbb{Z}$. In our cases of interest, we will see in Sect. 6 that each $c_{Y,w}$ is a nonnegative integer (this is true, in general, for geometric reasons [4]) and that, in fact, $c_{Y,w}$ is equal to 0 or 1.

4 Factorization Results for Involution Schubert Polynomials

4.1 Permutations

We recall the definition of the *Rothe diagram* of a permutation $w \in S_n$. It is defined to be the set

$$D(w) := \{(i, j) \in [n] \times [n] : j < w(i) \text{ and } i < w^{-1}(j)\}.$$

More concretely, the diagram can be obtained as follows. Begin with the full set $[n] \times [n]$ and then eliminate every entry $(i, w(i))$ as well as every entry directly to the right or directly below that entry. The length of w is equal to the cardinality of $D(w)$. The *code* of w is the sequence $c(w) = (c_1(w), \ldots, c_n(w))$ where $c_i(w)$ is equal to the number of $j \in [n]$ such that $(i, j) \in D(w)$. We also recall that $D(w^{-1})$ is the transpose of $D(w)$.

We next recall the notion of dominant permutations. We begin by recalling a well-known theorem.

Theorem 4.1 *The following are equivalent for a permutation $w \in S_n$:*

1. *The diagram of w is the diagram of a partition, i.e., a left-arrayed collection of rows of weakly decreasing length.*
2. *The code of w is a partition, i.e., $c_1(w) \geq c_2(w) \geq \cdots \geq c_n(w)$.*
3. *The permutation w is 132-avoiding, i.e., there is no $i < j < k$ with $w(i) < w(k) < w(j)$.*

Definition 4.2 A permutation satisfying any of the equivalent conditions of Theorem 4.1 is said to be *dominant*.

Theorem 4.3 *If $w \in S_n$ is a dominant permutation, then*

$$\mathfrak{S}_w = \prod_{(i,j)\in D(w)} x_i = \prod_{i=1}^{n} x_i^{c_i}.$$

Example 4.4 The permutation $w = [6435721] \in S_7$ is dominant, with code $c(w) = (5, 3, 2, 2, 2, 1, 0)$. Thus, $\mathfrak{S}_w = x_1^5 x_2^3 x_3^2 x_4^2 x_5^2 x_6$.

4.2 Involutions

Following [14], we define the *involution diagram* of $\tau \in \mathscr{I}_n$. It is important to note that the involution diagram of $\tau \in \mathscr{I}_n$ is different than its diagram when viewed as a permutation in S_n. We have

$$\hat{D}(\tau) := \{(i, j) \in [n] \times [n] : j < \tau(i) \text{ and } i < \tau(j) \text{ and } i \leq j\}.$$

Thus, $\hat{D}(\tau)$ is equal to the lower left half of $D(\tau)$ including the diagonal. The *involution length* of τ, denoted $\hat{\ell}(\tau)$, is equal to the cardinality of $\hat{D}(\tau)$. The *involution code* of τ is the sequence $\hat{c}(\tau) = (\hat{c}_1(\tau), \ldots, \hat{c}_n(\tau))$ where $\hat{c}_i(\tau)$ is equal to the number of $j \in [n]$ such that $(i, j) \in \hat{D}(\tau)$.

We introduce some definitions needed to state our next result. Let $\kappa(\tau)$ denote the number of disjoint 2-cycles of $\tau \in \mathscr{I}_n$. Let $\hat{D}_1(\tau) = \{(i, j) \in \hat{D}(\tau) : i = j\}$.

We note that $(a, a) \in \hat{D}_1(\tau)$ if and only if τ contains a 2-cycle (a, b) with $a < b$. In particular, $\kappa(\tau) = \#\hat{D}_1(\tau)$. Let $\hat{D}_2(\tau) = \{(i, j) \in \hat{D}(\tau) : i < j\}$. An involution $\tau \in \mathscr{I}_n$ is said to be dominant if it is dominant as a permutation.

Theorem 4.5 ([14, Theorem 3.26]) *If $\tau \in \mathscr{I}_n$ is a dominant involution, then*

$$\hat{\mathfrak{S}}_\tau = \prod_{(i,i)\in\hat{D}_1(\tau)} x_i \prod_{(i,j)\in\hat{D}_2(\tau)} (x_i + x_j) = \frac{1}{2^{\kappa(\tau)}} \prod_{(i,j)\in\hat{D}(\tau)} (x_i + x_j).$$

Example 4.6 The involution $\tau = (1, 6)(2, 5)(3, 7) \in \mathscr{I}_7$ is dominant and

$$\hat{\mathfrak{S}}_\tau = x_1x_2x_3(x_1 + x_2)(x_1 + x_3)(x_1 + x_4)(x_1 + x_5)(x_2 + x_3)(x_2 + x_4)(x_3 + x_4).$$

4.3 Degenerate Involutions

There is not yet any general theory of diagrams and dominance for degenerate involutions. However, the degenerate involution Schubert polynomial associated to the longest permutation, viewed as a μ-involution for any composition μ, has a very simple factorization.

Let $\mu = (\mu_1, \ldots, \mu_k)$ be a composition of n and let $\nu_i = \mu_1 + \cdots + \mu_i$, with the convention $\nu_0 = 0$. Let $\tau_n \in \mathscr{I}_n$ denote the longest permutation on n letters viewed as an involution. We define the degenerate involution diagram associated to the longest permutation $\pi_\mu \in \mathscr{I}_\mu$ viewed as a μ-involution. Let

$$\hat{D}_0^\mu = \{(i, j) \in [n] \times [n] : \nu_a + 1 \leq i \leq \nu_{a+1} \text{ and } \nu_b + 1 \leq j \leq \nu_{b+1} \text{ with } a < b\},$$

$$\hat{D}_1^\mu = \{(i, j) \in [n] \times [n] : \nu_a + 1 \leq i, j \leq \nu_{a+1} \text{ and } (i - \nu_a, j - \nu_a) \in \hat{D}_1(\pi_{\mu_a})\},$$

$$\hat{D}_2^\mu = \{(i, j) \in [n] \times [n] : \nu_a + 1 \leq i, j \leq \nu_{a+1} \text{ and } (i - \nu_a, j - \nu_a) \in \hat{D}_2(\pi_{\mu_a})\}.$$

Then $\hat{D}^\mu(\pi_\mu) := \hat{D}_0^\mu \cup \hat{D}_1^\mu \cup \hat{D}_2^\mu$, the union being disjoint.

Theorem 4.7 ([8, Corollary 3.6]) *Let μ be a composition of n and let π_μ be the longest permutation viewed as a μ-involution. Then*

$$\hat{\mathfrak{S}}_{\pi_\mu} = \prod_{(i,j)\in\hat{D}_0^\mu} x_i \prod_{(i,i)\in\hat{D}_1^\mu} x_i \prod_{(i,j)\in\hat{D}_2^\mu} (x_i + x_j) = \frac{1}{2^{\hat{\ell}_\mu(\pi_\mu)}} \prod_{(i,j)\in\hat{D}^\mu(\pi_\mu)} (x_i + x_j).$$

The integer $\hat{\ell}_\mu(\pi_\mu)$ is the involution length of π_μ and is defined in Sect. 5.2. In the case of the longest permutation μ-involution π_μ, we have $\hat{\ell}^\mu(\pi_\mu) = \#(\hat{D}_0^\mu \cup \hat{D}_1^\mu)$.

We remark that the proof of this result in [8] is geometric in nature, relying on the localization theorem in equivariant cohomology. It would be interesting to give a purely combinatorial proof of this result.

5 Weak Order Chains

There are well-known left and right weak orders on S_n that have a geometric interpretation in terms of minimal parabolic subgroups of $GL_n(\mathbb{C})$ acting on Schubert varieties. In this section, we define the analogues of weak order for involutions and μ-involutions. We will discuss the geometric meaning of these weak orders in Sect. 6.

5.1 Involutions

We recall the definition of weak order on involutions. Given $\tau \in \mathscr{I}_n$, we begin by defining an action of the generators of $M(S_n)$ via

$$m(s_i) \cdot \tau = \begin{cases} \tau & \text{if } \tau(i+1) < \tau(i) \\ s_i\tau & \text{if } \tau(i) = i \text{ and } \tau(i+1) = i+1 \\ s_i\tau s_i & \text{otherwise} \end{cases} .$$

It is straightforward to check that this defines an action of $M(S_n)$. For $\tau, \tau' \in \mathscr{I}_n$, say that $\tau \to \tau'$ if $\tau' = m(s_i) \cdot \tau$ for some s_i. The weak order on $\mathscr{I}_n$ is the transitive closure of the relation $\to$.

The weak order poset on $\mathscr{I}_n$ is a ranked poset with rank function $\text{rank}(\tau) = \hat{\ell}(\tau)$ for $\tau \in \mathscr{I}_n$. The poset has both a unique minimal element, the identity involution, and a unique maximal element, the longest permutation, which is an involution because it interchanges 1 and n, 2 and $n-1$, etc.

We now introduce a central combinatorial problem, to describe maximal chains of intervals in the weak order poset of involutions, using the language of [13]. Let $\tau, \tau' \in \mathscr{I}_n$ and suppose that $\tau \leq \tau'$. An *involution word* from τ to τ' is a sequence $(s_{i_1}, \ldots, s_{i_k})$ such that $\tau' = m(s_{i_1}), \ldots, m(s_{i_k}) \cdot \tau$ and $k = \hat{\ell}(\tau') - \hat{\ell}(\tau)$. In this case, if we let $w = s_{i_1}, \ldots, s_{i_k}$, we may write $\tau' = m(w) \cdot \tau$. An *atom* of τ' relative to τ is any $w \in S_n$ such that $\tau' = m(w) \cdot \tau$ and $\ell(w) = \hat{\ell}(\tau') - \hat{\ell}(\tau)$, and the set of all atoms of τ' relative to τ is denoted $\mathscr{A}_*(\tau, \tau')$. When $\tau \nleq \tau'$, $\mathscr{A}_*(\tau, \tau') = \emptyset$. When τ is the identity involution, we define the atoms of τ' to be $\mathscr{A}(\tau') := \mathscr{A}_*(\text{identity}, \tau')$.

Remark 5.1 There is a natural surjective map from reduced decompositions of elements in $\mathscr{A}_*(\tau, \tau')$ to maximal chains in the weak order poset of the interval $[\tau, \tau']$ in $\mathscr{I}_n$, in which a reduced decomposition $s_{i_1}, \ldots, s_{i_k}$ of some $w \in \mathscr{A}_*(\tau, \tau')$ maps to the maximal chain consisting of $\tau, m(s_{i_k}) \cdot \tau, \ldots, m(s_{i_1}), \ldots, m(s_{i_k}) \cdot \tau = \tau'$. The map may fail to be one to one because an edge in the Hasse diagram of $\mathcal{I}_n$ may have more than one s_j labeling it. To get a bijection, one should instead consider the directed graph $\mathcal{G}$ with vertices $\mathcal{I}_n$ and with an edge from $\tau \in \mathcal{I}_n$ to $\tau' \in \mathcal{I}_n$ labeled j if $\tau' = m(s_j) \cdot \tau$. By labeling edges, we allow for the possibility of more than one edge from τ to τ'. Then, the reduced decompositions of elements in $\mathcal{A}_*(\tau, \tau')$ correspond bijectively to the maximal paths in the induced subgraph of $\mathcal{G}$ on the vertex set $[\tau, \tau']$.

A similar remark applies to the case of weak order for degenerate involutions considered in Sect. 5.2.

For $\tau \in \mathcal{I}_n$, define $\mathrm{Fix}(\tau) := \{i \in [n] : \tau(i) = i\}$ and

$$\mathrm{Cyc}(\tau) := \{(i, j) \in [n] \times [n] : j = \tau(i) \text{ and } i \leq j\}.$$

The set $\mathrm{Cyc}(\tau)$ describes the 1- and 2-cycles of τ, while $\mathrm{Fix}(\tau)$ consists of the fixed points of τ.

We now state a main result describing the atoms of an arbitrary involution.

Theorem 5.2 ([7, Theorem 2.6], [13, Corollary 5.13]) *Let $\tau \in \mathcal{I}_n$. Then, $\mathcal{A}(\tau)$ consists of all $w \in S_n$ such that*

1. *If $(i, j) \in Cyc(\tau)$, then $w(i) \geq w(j)$.*
2. *If $(i, j) \in Cyc(\tau)$, then there does not exist $i < k < j$ such that $w(i) > w(k) > w(j)$.*
3. *If $(i, j), (k, l) \in Cyc(\tau)$ with $i < k$ and $j < l$, then $w(k) \geq w(l) > w(i) \geq w(j)$.*

Hamaker, Marberg, and Pawlowski extended this result to describe the relative atoms of an arbitrary pair of involutions (Fig. 1).

Theorem 5.3 ([13, Theorem 5.11]) *Let $\pi, \tau \in \mathcal{I}_n$. Then, the set $\mathcal{A}_*(\tau, \tau')$ consists of all $w \in S_n$ such that for all $(i, j), (k, l) \in Cyc(\tau')$:*

1. *If $w(i) < w(j)$, then $(w(i), w(j) \in Cyc(\tau)$ and otherwise, $w(i), w(j) \in Fix(\tau)$.*
2. *If $i \leq j < k \leq l$, then $w(i) < w(k)$, $w(i) < w(l)$, $w(j) < w(k)$ and $w(j) < w(l)$.*
3. *If $i < k < j < l$, then $w(i) < w(k)$, $w(i) < w(l)$ and $w(j) < w(l)$.*
4. *If $i < k < l < j$, then none of the following inequalities holds:*
   a. *$w(j) < w(k) < w(i)$*
   b. *$w(j) < w(l) < w(i)$*
   c. *$w(k) < w(i) < w(j) < w(l)$*
   d. *$w(k) < w(j) \leq w(i) < w(l)$*
5. *If $i < k = l < j$, then it is not the case that $w(j) < w(k) = w(l) < w(i)$.*

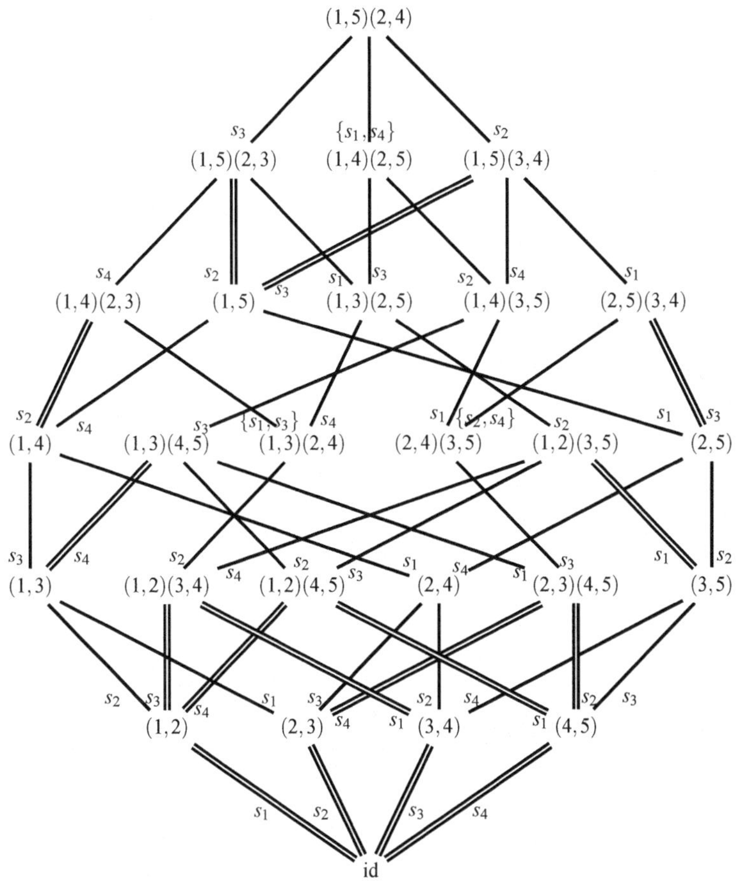

Fig. 1 Weak order on $\mathscr{I}_5$

5.2 Degenerate Involutions

We now recall the definition of weak order on μ-involutions. We again define an action of the generators of $M(S_n)$, but this time the definition is more involved.

Definition 5.4 Let $\pi = [\alpha_1|\alpha_2|\cdots|\alpha_k] \in \mathscr{I}_\mu$. Then, $s_i \cdot \pi$ is defined according to the following exhaustive list of mutually disjoint cases.

1. If i occurs after $i+1$ in π, then $s_i \cdot \pi = \pi$.
2. If i occurs before $i+1$ in π and i and $i+1$ occur in different μ-strings of π, then $s_i \cdot \pi = s_i\pi$, where $s_i\pi$ is the μ-involution obtained from π by interchanging the values of i and $i+1$.
3. If i occurs before $i+1$ in π and i and $i+1$ occur in the same μ-string α_j, then there are two subcases to consider.

a. If α_j fixes both i and $i+1$ (when the string α_j is viewed as a permutation as in Convention 2.1), then $s_i \cdot \pi = [\alpha_1|\cdots|s_i\alpha_j|\cdots|\alpha_k]$.
b. Otherwise, $s_i \cdot \pi = [\alpha_1|\cdots|s_i\alpha_j s_i|\cdots|\alpha_k]$.

It is a more tedious, but still straightforward, exercise to see that this defines an action of $M(S_n)$. (It also follows from geometry, using arguments from [5, 21].) For $\pi, \pi' \in \mathscr{I}_\mu$, say that $\pi \to \pi'$ if $\pi' = m(s_i) \cdot \pi$ for some s_i. The weak order on $\mathscr{I}_\mu$ is the transitive closure of the relation $\to$.

The weak order poset on $\mathscr{I}_\mu$ is a ranked poset, with $\text{rank}(\pi) = \hat{\ell}_\mu(\pi)$ for $\pi \in \mathscr{I}_\mu$, where $\hat{\ell}_\mu$ is defined as follows. Let $\pi = [\alpha_1|\cdots|\alpha_k]$ be decomposed into its μ-strings. Then, using Convention 2.1, each α_i can be viewed as an involution of its alphabet, and hence has an involution length $\hat{\ell}(\alpha_i)$. Let $\text{sort}(\pi)$ be the permutation whose one-line notation is obtained by concatenating the increasing rearrangements of each μ-string α_i. Then

$$\hat{\ell}_\mu(\pi) := \sum_{i=1}^{k} \hat{\ell}(\alpha_i) + \ell(\text{sort}(\pi)).$$

For example, if $\pi = [586|21|743]$, then $\hat{\ell}(\alpha_1) = 1$, $\hat{\ell}(\alpha_2) = 1$, $\hat{\ell}(\alpha_3) = 2$, while $\text{sort}(\pi) = [56812347]$ and $\ell(\text{sort}(\pi)) = 13$, so $\hat{\ell}_\mu(\pi) = 17$.

We now define the notion of atoms for μ-involutions, which is a straightforward generalization of the notion of atoms for involutions. Let $\pi, \pi' \in \mathscr{I}_\mu$ and suppose that $\pi \leq \pi'$. An involution word from π to π' is a sequence $(s_{i_1}, \ldots, s_{i_k})$ such that $\pi' = m(s_{i_1}), \ldots, m(s_{i_k}) \cdot \pi$ and $k = \hat{\ell}_\mu(\pi') - \hat{\ell}_\mu(\pi)$. Letting $w = s_{i_1}, \ldots, s_{i_k}$, we may write $\pi' = m(w) \cdot \pi$. An atom of π' relative to π is any $w \in S_n$ such that $\pi' = m(w) \cdot \pi$ and $\ell(w) = \hat{\ell}_\mu(\pi') - \hat{\ell}_\mu(\pi)$, and the set of all atoms of π' relative to π is denoted $\mathscr{A}_*(\pi, \pi')$. When $\pi \not\leq \pi'$, $\mathscr{A}_*(\pi, \pi') = \emptyset$. When π is the identity μ-involution, we define the atoms of π' to be $\mathscr{A}(\pi') := \mathscr{A}_*(\text{identity}, \pi')$.

Let $w \in S_n$ and fix a composition $\mu = (\mu_1, \ldots, \mu_k)$ of n. Via its one-line notation, we can view w as a string of length n. We subdivide w into k strings, where the i-th string is $\text{str}_i(w)$ consists of the string of length μ_i consisting of the elements in positions $\nu_i + 1$ through ν_{i+1}. For example, if $w = [37184265]$ and $\mu = (4, 1, 3)$, then $\text{str}_1(w) = 3718$, $\text{str}_2(w) = 4$, and $\text{str}_3(w) = 265$.

For any integer n, let τ_n denote the longest permutation of S_n, viewed as an involution in $\mathscr{I}_n$ and define $\mathscr{A}_n := \mathscr{A}(\tau_n)$. For any composition μ, let π_μ denote the longest permutation of S_n viewed as a μ-involution in $\mathscr{I}_\mu$ and define $\mathscr{A}_\mu := \mathscr{A}(\pi_\mu)$.

Theorem 5.5 ([8, Proposition 2.5]) *The set $\mathscr{A}_\mu$ consists of all $w \in S_n$ such that the letters of $\text{str}_i(w)$ are $\nu_{k-i} + 1$ through ν_{k+1-i} and, viewed as a permutation of its alphabet via Convention 2.1, $\text{str}_i(w)$ belongs to $\mathscr{A}_{\mu_i}$.*

For example, if $\mu = (4, 1, 3)$ (Fig. 2),

$$\mathscr{A}_\mu = \{76854231, 76854312, 78564231, 78564312, 85764231, 85764312\}.$$

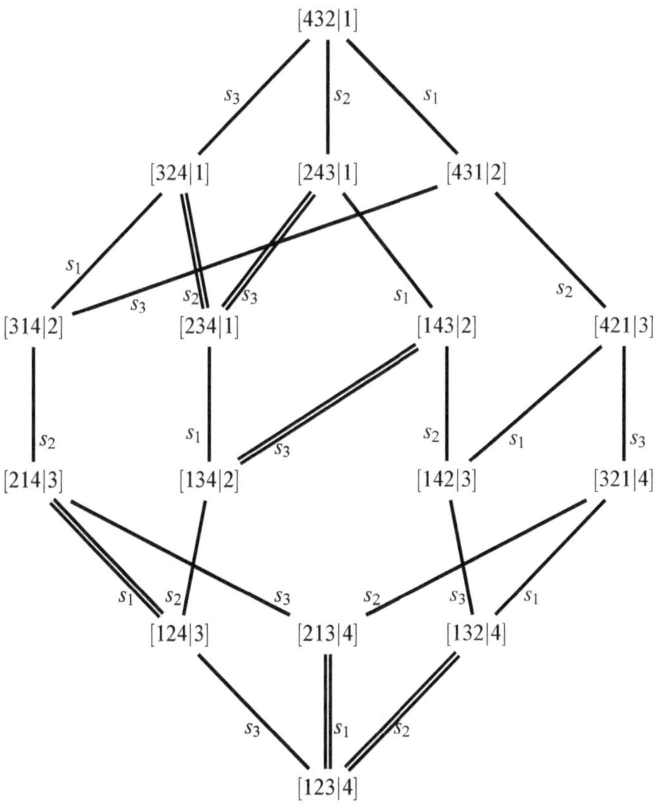

Fig. 2 Weak order on $\mathscr{I}_{3,1}$

6 Schubert Polynomial Identities

We now give results for how certain sums of Schubert polynomials factor completely. The primary tool is a result of Brion which relates the expansion of cohomology classes of K-orbit closures for spherical subgroups K in the Schubert basis to chains in a weak order poset associated to K. We give a brief description of this result, referring to [4, 5] for more details.

To begin, let G be an arbitrary connected reductive group, B a Borel subgroup of G and K a spherical subgroup of G. By definition, K has a dense orbit in G/B, which we denote Y_0 in the sequel. The set $\mathscr{O}_K$ of all K-orbits Y in G/B acquires a partial order, called weak order, in which Y_0 is the unique maximal element.

The minimal parabolic subgroups of G containing B are denoted P_s, where s runs over all of the simple reflections associated to the simple roots of the root system for (G, B). For any s, let $p_s : G/B \to G/P_s$ be the natural projection map. For two distinct K-orbits Y_1 and Y_2 write $Y_1 \to Y_2$ via s or $Y_1 \xrightarrow{s} Y_2$, if the following occurs:

Y_2 is the dense orbit of $p_s^{-1}(p_s(\overline{Y_1}))$ for some s. The weak order on the set of K-orbits in G/B is the transitive closure of the relation $\rightarrow$.

For our applications, we now specialize to $G = GL_n(\mathbb{C})$. In this case, the above construction allows us to define an action of $M(S_n)$ on $\mathscr{O}_K$. In this case, the s parameterizing minimal parabolic subgroups are precisely the simple transpositions s_i of S_n and $G/B \cong \mathscr{F}$. If Y is a K-orbit in $\mathscr{F}$, then $m(s_i) \cdot Y$ is defined to be the dense K-orbit in $p_{s_i}^{-1}(p_{s_i}(\overline{Y}))$. This action on generators yields a well-defined action of $M(S_n)$ on $\mathscr{O}_K$.

The weak order is intimately related to the theory of K-Schubert polynomials. If $Y_1 \xrightarrow{s_i} Y_2$, then $\mathfrak{S}_{Y_2} = \partial_i \mathfrak{S}_{Y_1}$. (In the general setup, one can define more general divided difference operators ∂_s, as in [2, 11], but making a choice of polynomial representative for the Schubert classes is a more subtle problem.)

Definition 6.1 Let Y be a K-orbit on $\mathscr{F}$. Define a set $\mathscr{W}(Y) \subseteq S_n$ as follows. The set $\mathscr{W}(y)$ consists of all permutations $w \in S_n$ that admit a reduced decomposition $w = s_{i_1}, \ldots, s_{i_\ell}$ such that there are K-orbits $Y_1, \ldots, Y_\ell$ such that

$$Y = Y_\ell \xrightarrow{s_{i_\ell}} Y_{\ell-1} \xrightarrow{s_{i_{\ell-1}}} \cdots \xrightarrow{s_{i_2}} Y_1 \xrightarrow{s_{i_1}} Y_0$$

and $\dim Y_{i-1} = \dim Y_i + 1$ for $1 \leq \imath \leq \ell$. (In particular, the codimension of Y in $\mathscr{F}$ must be ℓ.)

Brion [4] proved an important theorem expressing the class $[\overline{Y}] \in H^*(\mathscr{F})$ positively in the Schubert basis. Reinterpreted in terms of K-Schubert polynomials, it says

Theorem 6.2 ([4, Theorem 1.5(ii)]) *Let Y be a K-orbit closure in $\mathscr{F}$. Then*

$$\mathfrak{S}_Y = \sum_{w \in \mathscr{W}(Y)} \mathfrak{S}_w,$$

where w_0 denotes the longest permutation in S_n, i.e., $w_0(i) = n + 1 - i$.

In the case, where $K = H = O_n(\mathbb{C})$ (resp., $K = H_\mu$), the weak order on involutions (resp., μ-involutions) is the *opposite* of the weak order for the K-orbits on $\mathscr{F}$. In particular, if $Y(\tau)$ is the K-orbit on $\mathscr{F}$ corresponding to $\tau \in \mathscr{I}_n$, then $w \in \mathscr{W}(Y(\tau))$ if and only if $w^{-1} \in \mathscr{A}(\tau)$. Similarly, if $Y(\pi)$ is the K-orbit on $\mathscr{F}$ corresponding to $\pi \in \mathscr{I}_\mu$, then $w \in \mathscr{W}(Y(\pi))$ if and only if $w^{-1} \in \mathscr{A}(\pi)$.

Theorem 6.3 *Let $\pi \in \mathscr{I}_n$ be a dominant involution in S_n, and let $Y(\tau)$ denote the K-orbit of τ in $\mathscr{F} \cong GL_n(\mathbb{C})/B$. Then*

$$\sum_{w \in \mathscr{A}(\tau)} \mathfrak{S}_{w^{-1}} = \prod_{(i,i) \in \hat{D}_1(\tau)} x_i \prod_{(i,j) \in \hat{D}_2(\tau)} (x_i + x_j).$$

Proof The left-hand side is equal to $\mathfrak{S}_{Y(\tau)}$ by Theorem 6.2, while the right-hand side is equal to $\mathfrak{S}_{Y(\tau)}$ by Theorem 4.7.

Example 6.4 Let $\tau = (1, 5)(2, 3) \in \mathscr{I}_5$. Then

$$\mathscr{A}(\tau) = \{32451, 32514, 35124, 51324\}$$

and

$$\mathfrak{S}_{52134} + \mathfrak{S}_{42153} + \mathfrak{S}_{34152} + \mathfrak{S}_{24351} = x_1x_2(x_1 + x_2)(x_1 + x_3)(x_1 + x_4).$$

Theorem 6.5 *Let $\pi \in \mathscr{I}_n$ be the longest permutation in S_n viewed as a μ-involution, and let $Y(\pi)$ denote the corresponding (closed) K-orbit of π in $\mathscr{F} \cong GL_n(\mathbb{C})/B$. Then*

$$\sum_{w\in\mathscr{A}_\mu} \mathfrak{S}_{w^{-1}} = \prod_{(i,j)\in\hat{D}_0^\mu} x_i \prod_{(i,i)\in\hat{D}_1^\mu} x_i \prod_{(i,j)\in\hat{D}_2^\mu} (x_i + x_j).$$

Proof The left-hand side is equal to $\mathfrak{S}_{Y(\pi)}$ by Theorem 6.2, while the right-hand side is equal to $\mathfrak{S}_{Y(\pi)}$ by Theorem 4.5.

Example 6.6 Let $\mu = (3, 1)$ and let $\pi_\mu = [432|1]$, the longest permutation considered as μ-involution. Then, $\mathscr{A}_\mu = \{4231, 4312\}$ and

$$\mathfrak{S}_{4231} + \mathfrak{S}_{3421} = x_1^2x_2x_3(x_1 + x_2).$$

7 Other Directions

We close with a list of several open problems in the field.

(1) It is possible to define a natural notion of diagrams for μ-involutions. However, there does not appear to be an obvious notion of dominance for μ-involutions. In particular, a μ-involution whose underlying permutation is dominant may not factor into linear factors. Can one define such a notion? In particular, can one find a large class of μ-involutions whose degenerate involution Schubert polynomials factor completely into simple linear factors described in terms of the combinatorics of the associated diagrams?

(2) There is an obvious bijection between K-orbits on G/B and B-orbits on G/K. One can transport the weak order structure for B-orbits on G/K. In general, G/K will not be a complete variety, so one is naturally led to study completions of G/K. In many cases, including when $G = GL_n(\mathbb{C})$ and $K = H = O_n(\mathbb{C})$, there is a natural completion to use called the wonderful compactification of G/K [10]. The wonderful compactification of G/H is the classical variety X of complete quadrics, and the G-orbits are in bijection with compositions μ of n. Moreover, the stabilizer subgroup of a point in the G-orbit associated to μ is

conjugate to the subgroup H_μ. Thus, a geometric study of X can unite the various cases considered here. In particular, the B-orbits on X are parameterized by the degenerate involutions of rank n, and the weak order on X is the disjoint union of the weak orders on μ-involutions for all compositions μ of n.

The geometry of X is intricate. The cohomology ring of X has been studied directly [9] and as an example of a complete symmetric variety [3]. Recently, several combinatorial models related to the geometry of X have been introduced [1]. Still, there is much to be learned about X. Are there are any hidden symmetries in $H^*(X)$? Can one give a combinatorial description of the analogue of the Bruhat order, the inclusion order of B-orbit closures on X?

(3) The combinatorics studied here relates to the symmetric subgroup $O_n(\mathbb{C})$ of $GL_n(\mathbb{C})$. There are similar results for the symmetric subgroups $Sp_n(\mathbb{C})$ [7, 13, 14, 26] and $GL_p(\mathbb{C}) \times GL_q(\mathbb{C})$, $p+q=n$, [7, 25] of $GL_n(\mathbb{C})$. But there are symmetric subgroups associated to other reductive algebraic groups as well. There are four more classical families of symmetric subgroups and 12 exceptional symmetric subgroups [24, Table 26.3], and it would be interesting to extend the combinatorial and geometric results from the "Type A" cases to the other Lie types.

Acknowledgements The author is grateful to Mahir Can for many helpful discussions. The author thanks the referee for their thorough reading of the paper and their helpful suggestions for improvement.

References

1. Banerjee, S., Can, M.B., Joyce, M.: Combinatorial Models for Complete Quadrics (2016)
2. Bernšteĭn, I.N., Gel'fand, I.M., Gel'fand, S.I.: Schubert cells, and the cohomology of the spaces G/P. Usphei Mat. Nauk, **28**(3(171)), 3–26 (1973)
3. Bifet, E., De Concini, C., Procesi, C.: Cohomology of regular embeddings. Adv. Math. **82**(1), 1–34 (1990)
4. Brion, M.: The behaviour at infinity of the Bruhat decomposition. Comment. Math. Helv. **73**(1), 137–174 (1998)
5. Brion, M.: On orbit closures of spherical subgroups in flag varities. Comment. Math. Helv. **76**(2), 263–299 (2001)
6. Can, M.B., Joyce, M.: Weak order on complete quadrics. Trans. Amer. Math. Soc. **365**(12), 6269–6282 (2013)
7. Can, M.B., Joyce, M., Wyser, B.: Chains in weak order posets associated to involutions. J. Combin. Theory Ser. A **137**, 207–225 (2016)
8. Can, M.B., Joyce, M., Wyser, B.: Wonderful symmetric varieties and Schubert polynomials. To appear in ARS Math. Contemp
9. De Concini, C., Goresky, M., MacPherson R., Procesi C.: On the geometry of quadrics and their degenerations.Comment. Math. Helv. **63**(3), 337–413 (1988)
10. De Concini, C., Procesi, C.: Complete symmetric varieties. In: Invariant Theory (Montecatini 1982). Lecture Notes in Mathematics, vol. 996, pp. 1–44. Springer, Berlin (1983)
11. Demazure, M.: Désingularisation des variétés de Schubert généralisées. Ann. Sci. École Norm. Sup. **4**(7), 53–88 (1974)
12. Fulton, W.: Young Tableaux. Number 35 in London Mathematical Society Student Texts. Cambridge University Press (1997)

13. Hamaker, Z., Marberg, E., Pawlowski, B.: Involution words II: Braid relations and atomic structures. J. Algebr. Combin. **45**(3), 701–743 (2017)
14. Hamaker, Z., Marberg, E., Pawlowski, B.: Involution Words: Counting Problems and Connections to Schubert Calculus for Symmetric Orbit Closures (2015)
15. Hamaker, Z., Marberg, E., Pawlowski, B.: Transition Formulas for Involution Schubert Polynomials (2016)
16. Hamaker, Z., Marberg, E., Pawlowski, B.: Schur P-Positivity and Involution Stanley Symmetric Functions (2017)
17. Hamaker, Z., Marberg, E., Pawlowski, B.: Fixed-Point-Free Involutions and Schur P-Positivity (2017)
18. Lascoux, A., Schützenberger, M.-P.: Polynômes de Schubert. C. R. Acad. Sci. Paris Sér I Math. **294**(13), 447–450 (1982)
19. Lascoux, A., Schützenberger, M.-P.: Schubert polynomials and the Littlewood-Richardson rule. Lett. Math. Phys. **10**(2–3), 111–124 (1985)
20. Manivel, L.: Symmetric Functions, Schubert Polynomials and Degeneracy Loci. SMF/AMS Texts and Monographs, vol. 6. American Mathematical Society, Providence, RI (2001) (Translated from the 1998 French original by John R. Swallow)
21. Richardson, R.W., Springer, T.A.: The Bruhat order on symmetric varieties. Geom. Dedicata **35**(1–3), 389–436 (1990)
22. Richardson, R.W., Springer, T.A., Combinatorics and geometry of K-orbits on the flag manifold. In: Linear Algebraic Groups and Their Representations, Los Angeles, CA (1992). Contemporary Mathematics, vol. 153. American Mathematical Society, Providence, RI (1992)
23. Richardson, R.W., Springer, T.A.: Complements to: "The Bruhat order on symmetric varieties". Geom. Dedicata **49**(2), 231–238 (1994)
24. Timashev, D.: Homogeneous Spaces and Equivariant Embeddings. Encyclopaedia of Mathematical Sciences, vol. 138. Springer, Heidelberg (2011) (Invariant Theory and Algebraic Transformation Groups, 8)
25. Wyser, B.J., Yong, A.: Polynomials for $\mathrm{GL}_p \times \mathrm{GL}_q$ orbit closures in the flag variety. Selecta Math. (N.S.) **20**(4), 1083–1110 (2014)
26. Wyser, B., Yong, A.: Polynomials for symmetric orbit closures in the flag variety. Transform. Groups **22**(1), 267–290 (2017)

The Structure of Hopf Algebras Acting on Dihedral Extensions

Alan Koch, Timothy Kohl, Paul J. Truman and Robert Underwood

Abstract We discuss isomorphism questions concerning the Hopf algebras that yield Hopf–Galois structures for a fixed separable field extension L/K. We study in detail the case where L/K is Galois with dihedral group D_p, $p \geq 3$ prime and give explicit descriptions of the Hopf algebras which act on L/K. We also determine when two such Hopf algebras are isomorphic, either as Hopf algebras or as algebras. For the case $p = 3$ and a chosen L/K, we give the Wedderburn–Artin decompositions of the Hopf algebras.

Keywords Hopf–Galois extension · Dihedral extension · Wedderburn–Artin decomposition

1 Introduction

Galois theory for purely inseparable field extensions was first considered by Jacobson [15]. More broadly, Chase and Sweedler defined the notion of a Hopf algebra acting on a purely inseparable extension of fields to obtain a weak analog to the Fundamental Theorem of Galois theory [6]. The construction of these "Hopf–Galois" structures,

A. Koch
Department of Mathematics, Agnes Scott College, 141 E. College Ave.,
Decatur, GA 30030, USA
e-mail: akoch@agnesscott.edu

T. Kohl
Department of Mathematics and Statistics, Boston University, 111 Cummington Mall, Boston, MA 02215, USA
e-mail: tkohl@math.bu.edu

P. J. Truman
School of Computing and Mathematics, Keele University, Staffordshire ST5 5BG, UK
e-mail: P.J.Truman@Keele.ac.uk

R. Underwood (✉)
Department of Mathematics and Computer Science, Auburn University at Montgomery, Montgomery, AL 36124, USA
e-mail: runderwo@aum.edu

J. Feldvoss et al. (eds.), *Advances in Algebra*, Springer Proceedings in Mathematics & Statistics 277, https://doi.org/10.1007/978-3-030-11521-0_10

however, applies not only to purely inseparable field extensions but also to separable field extensions, as well as extensions of commutative rings (which will not be considered here).

In the groundbreaking paper [13], Greither and Pareigis obtain a classification of Hopf–Galois structures on separable field extensions L/K. The most remarkable aspect of their classification is that it is entirely group theoretic, depending on $Gal(E/K)$ and $Gal(E/L)$, where E is the normal closure of L/K, the structure of L/K is irrelevant (aside from these Galois groups). Much of the work to date has focused on counting the number of Hopf–Galois extensions, either directly (see, e.g., [4, 5, 8, 12, 18–21]) or through results which facilitate computations (e.g., [2]).

In the three decades since the publication of [13], what has been lacking is a thorough investigation into the structure of the Hopf algebras which produce these Hopf–Galois structures. It is unclear how much the structure of the Hopf algebras also depends on group theory. To this end, we introduce some questions for study, including:

1. Can a single K-Hopf algebra determine more than one Hopf–Galois structure on L/K?
2. Can two non-isomorphic K-Hopf algebras, each of which giving a Hopf–Galois structure on L/K, become isomorphic upon base change to some intermediate field $K \subset F \subset L$?
3. Can a single K-algebra be endowed with multiple coalgebra structures, resulting in multiple (non-isomorphic) Hopf algebras, giving different Hopf–Galois structures on L/K?
4. Can two non-isomorphic K-algebras, each of which giving a Hopf–Galois structure on L/K (after being endowed a coalgebra structure), become isomorphic as algebras upon base change to some intermediate field $K \subset F \subset L$?

It is not known whether the answers to these questions depend on knowledge of the fields, aside from their automorphism groups. Some specific cases—most notably the cases where $Gal(L/K)$ is cyclic or elementary abelian—have been investigated by the authors in [17]. We note that Childs [10, Theorem 5] has shown that abelian fixed-point-free endomorphisms of $Gal(L/K)$ determine Hopf–Galois structures on L/K whose Hopf algebras are isomorphic to H_λ (see Example 3 for the definition of H_λ). Childs applies his result to the cases where $Gal(L/K)$ is the symmetric group S_n, $n \geq 5$, and the dihedral group of order $4n$, $n \geq 2$. Thus, in these cases, Childs has obtained an affirmative answer to Question (1). In this paper, we will focus on the case where L/K is Galois with $Gal(L/K) = D_p$, the dihedral group of order $2p$ for some prime $p \geq 3$. In this instance, the questions above do in fact have satisfying group theoretic answers.

We start by reviewing Hopf–Galois theory and the Greither–Pareigis theory that yields the classification of Hopf–Galois structures in the separable case. We then apply Greither–Pareigis theory to describe the Hopf–Galois structures in the case that L/K is Galois with group D_p; there are $p + 2$ such structures, and while this was known previously (see [4]) we provide a simpler description. We will show that there are three Hopf algebra isomorphism classes, and that base changing to a proper intermediate field still results in three distinct Hopf algebra classes. On the other hand, there are two K-algebra isomorphism classes, and base changing to any intermediate field (even L itself) does not change these isomorphism classes. We then find explicit bases for each of our Hopf algebras, and specializing to the case $p = 3$, we give an even more detailed description of the algebra structure.

In contrast, we point out that the Hopf–Galois theory in the purely inseparable case differs greatly to what is presented here. For example, if $L = K(x),\ x^p \in K,\ x \notin K$ is a purely inseparable extension (p prime), then $H = K[t]/(t^p)$, t primitive, can act in an infinite number of ways (see, e.g., [16]), allowing for an infinite number of Hopf–Galois structures.

2 Hopf–Galois Theory

In this section, we recall the notion of a Hopf algebra, a Hopf–Galois extension, and the Greither–Pareigis classification.

A *bialgebra* over a field K is a K-algebra B together with K-algebra maps $\Delta : B \to B \otimes_K B$ (comultiplication) and $\varepsilon : B \to K$ (counit) which satisfy the conditions

$$(I \otimes \Delta)\Delta = (\Delta \otimes I)\Delta,$$

$$mult(I \otimes \varepsilon)\Delta = I = mult(\varepsilon \otimes I)\Delta,$$

where $mult : B \otimes_K B \to B$ is the multiplication map of B and I is the identity map on B. A *Hopf algebra* over K is a K-bialgebra H with a K-linear map $\sigma : H \to H$ which satisfies

$$mult(I \otimes \sigma)\Delta(h) = \varepsilon(h)1_H = mult(\sigma \otimes I)\Delta(h),$$

for all $h \in H$. A K-Hopf algebra H is *cocommutative* if $\Delta = \tau \circ \Delta$, where $\tau : H \otimes_K H \to H \otimes_K H, a \otimes b \mapsto b \otimes a$ is the twist map.

Let L be a finite extension of K and let $\mathrm{m} : L \otimes_K L \to L$ denote multiplication in L. Let H be a finite dimensional, cocommutative K-Hopf algebra and suppose there is a K-linear action of H on L which satisfies

$$h \cdot (xy) = (\mathrm{m} \circ \Delta)(h)(x \otimes y)$$
$$h \cdot 1 = \varepsilon(h)1$$

for all $h \in H,\ x, y \in L$, and that the K-linear map $j : L \otimes_K H \to End_K(L),\ j(x \otimes h)(y) = x(h \cdot y)$ is an isomorphism of vector spaces over K. Then, we say H provides a *Hopf–Galois structure* on L/K.

Example 1 Suppose L/K is Galois with Galois group G. Let $H = K[G]$ be the group algebra, which is a Hopf algebra via $\Delta(g) = g \otimes g,\ \varepsilon(g) = 1, \sigma(g) = g^{-1}$, for all $g \in G$. The action

$$\left(\sum r_g g\right) \cdot x = \sum r_g(g(x))$$

provides the "usual" Hopf–Galois structure on L/K which we call the *classical* Hopf–Galois structure.

In general, the process of finding a Hopf algebra and constructing an action may seem daunting, but in the separable case Greither and Pareigis [13] have provided a complete classification of such structures. Let L/K be separable with normal closure E. Let $G = Gal(E/K)$, $G' = Gal(E/L)$, and $X = G/G'$. Denote by $Perm(X)$ the group of permutations of X. A subgroup $N \leq Perm(X)$ is *regular* if $|N| = |X|$ and $\eta[xG'] \neq xG'$ for all $\eta \neq 1_N$, $xG' \in X$. Let $\lambda : G \to Perm(X)$, $\lambda(g)(xG') = gxG'$ denote the left translation map. A subgroup $N \leq Perm(X)$ is *normalized* by $\lambda(G) \leq Perm(X)$ if $\lambda(G)$ is contained in the normalizer of N in $Perm(X)$.

Theorem 1 (Greither–Pareigis) *Let L/K be a finite separable extension. There is a one-to-one correspondence between Hopf–Galois structures on L/K and regular subgroups of* $\mathrm{Perm}(X)$ *that are normalized by $\lambda(G)$.*

One direction of this correspondence works by Galois descent: Let N be a regular subgroup normalized by $\lambda(G)$. Then, G acts on the group algebra $E[N]$ through the Galois action on E and conjugation by $\lambda(G)$ on N, i.e.,

$$g(x\eta) = g(x)(\lambda(g)\eta\lambda(g^{-1})), \ g \in G, \ x \in E, \ \eta \in N.$$

For simplicity, we will denote the conjugation action of $\lambda(g) \in \lambda(G)$ on $\eta \in N$ by ${}^{g}\eta$. We then define

$$H = (E[N])^G = \{x \in E[N] : \ g(x) = x, \forall g \in G\}.$$

The action of H on L/K is thus

$$\left(\sum_{\eta \in N} r_\eta \eta\right) \cdot x = \sum_{\eta \in N} r_\eta \eta^{-1}[1_G](x),$$

see [9, Proposition 1].

The fixed ring H is an n-dimensional K-Hopf algebra, $n = [L : K]$, and L/K has a Hopf–Galois structure via H [13, p. 248, proof of 3.1 (b) $\Rightarrow$ (a)], [7, Theorem 6.8, pp. 52–54]. By [13, p. 249, proof of 3.1, (a) $\Rightarrow$ (b)],

$$E \otimes_K H \cong E \otimes_K K[N] \cong E[N],$$

as E-Hopf algebras, that is, H is an *E-form* of $K[N]$.

Theorem 1 can be applied to the case where L/K is Galois with group G (thus, $E = L$, $G' = 1_G$, $G/G' = G$). In this case, the Hopf–Galois structures on L/K

correspond to regular subgroups of $\mathrm{Perm}(G)$ normalized by $\lambda(G)$, where $\lambda : G \to \mathrm{Perm}(G)$, $\lambda(g)(h) = gh$, is the left regular representation.

Example 2 Suppose L/K is a Galois extension, $G = Gal(L/K)$. Let $\rho : G \to \mathrm{Perm}(G)$ be the right regular representation defined as $\rho(g)(h) = hg^{-1}$ for $g, h \in G$. Then, $N = \rho(G)$ is a regular subgroup normalized by $\lambda(G)$, since $\lambda(g)\rho(h)\lambda(g^{-1}) = \rho(h)$ for all $g, h \in G$; N corresponds to a Hopf–Galois structure with K-Hopf algebra $H = L[\rho(G)]^G = K[G]$, the usual group ring Hopf algebra with its usual action on L. Consequently, $\rho(G)$ corresponds to the classical Hopf–Galois structure.

Example 3 Again, suppose L/K is Galois with group G. Let $N = \lambda(G)$. Then, N is a regular subgroup of $Perm(G)$ which is normalized by $\lambda(G)$, and $N = \rho(G)$ if and only if G is abelian. We denote the corresponding Hopf algebra by H_λ. If G is non-abelian, then $\lambda(G)$ corresponds to the *canonical nonclassical* Hopf–Galois structure.

Thus, for G non-abelian there are at least two Hopf–Galois structures on L/K. We remark that if L/K is Galois with G simple and non-abelian, then these are the only Hopf–Galois structures on L/K [3].

Note that with the classical and the canonical nonclassical structures, the regular subgroup $N \leq Perm(G)$ is isomorphic to G. The following example shows that this need not be the case in general.

Example 4 Let $L = \mathbb{Q}(\sqrt{2}, \sqrt{3})$. Then, $L/\mathbb{Q}$ is Galois with elementary abelian Galois group $G = \langle r, s\rangle$ with

$$r(\sqrt{2}+\sqrt{3}) = \sqrt{2}-\sqrt{3},\ s(\sqrt{2}+\sqrt{3}) = -\sqrt{2}+\sqrt{3}.$$

Let $\eta \in Perm(G)$ be defined by $\eta(r^i s^j) = r^{i-1}s^{i+j-1}$, and let $N = \langle \eta \rangle$. It is routine to verify that N is a regular subgroup of $Perm(G)$ which is normalized by $\lambda(G)$. Since N is cyclic of order 4, $N \not\cong G$.

3 The Group D_p

Throughout this section, we let D_p denote the dihedral group of order $2p$ for p an odd prime. Explicitly, we write

$$D_p = \langle r, s : r^p = s^2 = rsrs = 1\rangle.$$

Let L/K be a Galois extension with group D_p. We shall describe all of the regular subgroups of $Perm(D_p)$ normalized by $\lambda(D_p)$, and then address the isomorphism questions given in the introduction. By Examples 2 and 3, we have regular subgroups $\rho(D_p)$, $\lambda(D_p)$ normalized by $\lambda(D_p)$. We construct other regular subgroups of $Perm(D_p)$.

Lemma 1 *Pick* $0 \leq c \leq p-1$. *Let* $\eta_c = \lambda(r)\rho(r^c s) \in Perm(D_p)$, *and let* $N_c = \langle \eta_c \rangle$. *Then,* $N_c \cong C_{2p}$, *the cyclic group of order* $2p$, *and the* N_c *are distinct as sets. Moreover,* N_c *is a regular subgroup of* $Perm(D_p)$ *normalized by* $\lambda(D_p)$.

Proof Suppose $0 \leq c \leq p-1$. Because left and right representations commute with each other, $\eta_c^i = \lambda(r^i)\rho((r^c s)^i)$. As $|r| = p$ and $|r^c s| = 2$ it follows that $|N_c| = 2p$, thus $N_c \cong C_{2p}$. Now, suppose $0 \leq d \leq p-1, c \neq d$. Since $N_c \cong C_{2p}$, it contains a unique element of order 2, which is $\eta_c^p = \rho(r^c s)$. Similarly, the unique element of order 2 in N_d is $\eta_d^p = \rho(r^d s)$. Since $c \neq d$, this shows that $N_c \neq N_d$.

It remains to show that the stabilizer in N_c of any element in D_p is trivial, and that N_c is normalized by $\lambda(D_p)$. For the remainder of the proof, we write η for η_c and N for N_c. Let $x \in D_p$ and suppose $\eta^i[x] = x$. Then

$$x = \eta^i[x] = \lambda(r^i)\rho((r^c s)^i)[x] = r^i x (r^c s)^{-i},$$

and so

$$1_{D_p} = x^{-1} r^i x (r^c s)^{-i} = r^{\pm i}(r^c s)^{-i},$$

which cannot happen unless $i = 0$. Hence, $\eta^i = 1_N$ and $N \leq Perm(D_p)$ is regular.

We now show that N is normalized by $\lambda(D_p)$. Of course, it suffices to show that ${}^r\eta \in N$ and ${}^s\eta \in N$. We have, for $x \in D_p$,

$$\begin{aligned}
{}^r\eta[x] &= \lambda(r)\big(\lambda(r)\rho(r^c s)\big)\lambda(r^{-1})[x] = r x r^c s = \eta[x] \\
{}^s\eta[x] &= \lambda(s)\big(\lambda(r)\rho(r^c s)\big)\lambda(s)[x] = s r s x r^c s = r^{-1} x r^c s = r^{-1} x (r^c s)^{-1} = \eta^{-1}[x].
\end{aligned}$$

Thus, ${}^r\eta = \eta$, ${}^s\eta = \eta^{-1}$, and N is normalized by $\lambda(D_p)$.

By [4, Corollary 6.5], the collection $\{\rho(D_p), \lambda(D_p), N_0, \ldots, N_{p-1}\}$ is the complete set of all regular subgroups of $Perm(D_p)$ normalized by $\lambda(D_p)$, and hence the corresponding Hopf algebras give all of the Hopf–Galois structures on L/K. We will denote the Hopf algebra corresponding to N_c by H_c for all c.

4 The Hopf Algebra Isomorphism Questions

Let L/K be Galois with group D_p. We have seen that the Hopf algebras which give Hopf–Galois structures on L/K are

$$\{K[D_p], H_\lambda, H_0, \ldots, H_{p-1}.\}$$

Here, we will investigate when two of these Hopf algebras are isomorphic. Note that throughout this section, when working with Hopf algebras, "isomorphic" refers to isomorphic as Hopf algebras; considering isomorphisms as algebras will be discussed in the next section.

Clearly, H_c cannot be isomorphic to either $K[D_p]$ or H_λ since it is commutative. It remains to determine whether $K[D_p] \cong H_\lambda$ or whether $H_c \cong H_d$ for some $0 \leq c < d \leq p-1$.

Generally, Hopf isomorphism questions reduce to group isomorphism questions.

Proposition 1 *Let L/K be a finite separable extension, with Galois closure E, and let $G = Gal(E/K)$, $G' = Gal(E/L)$. Let $X = G/G'$. Let N, N' be regular subgroups of $Perm(X)$ which are normalized by $\lambda(G)$, and let H, H' be their corresponding Hopf algebras. If $H \cong H'$, then $N \cong N'$.*

Proof If $H \cong H'$, then $(E[N])^G \cong (E[N'])^G$. Therefore, $E \otimes_K (E[N])^G \cong E \otimes_K (E[N'])^G$. However, $E \otimes_K (E[N])^G \cong E[N]$ and similarly for N', and hence $E[N] \cong E[N']$. Since group algebras (over the same field) are isomorphic as Hopf algebras if and only if their groups are isomorphic, the result follows.

The converse to Proposition 1 is not true as we shall see below in Proposition 4. However, we have

Proposition 2 *Using the notation as in Proposition 1, $H \cong H'$ if and only if there exists an isomorphism $\phi : N \to N'$ which respects the actions of G.*

Proof See [17, Corollary 2.3].

We can use this proposition to show that the H_c are all isomorphic.

Proposition 3 *For $0 \leq c, d \leq p-1$ we have $H_c \cong H_d$.*

Proof Define $\phi : N_c \to N_d$ by $\phi(\eta_c) = \eta_d$. This is clearly an isomorphism; it remains to show that it respects the D_p-actions. But since the D_p-actions are identical with respect to the generators of the groups this is immediate:

$$\begin{aligned} \phi({}^r\eta_c) &= \phi(\eta_c) = \eta_d = {}^r\eta_d = {}^r\phi(\eta_c) \\ \phi({}^s\eta_c) &= \phi(\eta_c^{-1}) = \eta_d^{-1} = {}^s\eta_d = {}^s\phi(\eta_c). \end{aligned}$$

On the other hand, we have

Proposition 4 $K[D_p] \ncong H_\lambda$.

Proof Suppose $K[D_p] \cong H_\lambda$. Then, there exists an isomorphism $\phi : \rho(D_p) \to \lambda(D_p)$ which respects the D_p-actions. Note that D_p acts trivially on $\rho(D_p)$. Pick $1 \leq i \leq p-1$ such that $\phi(\rho(r)) = \lambda(r^i)$. Then

$$\phi({}^s\rho(r)) = \phi(\rho(r)) = \lambda(r^i)$$

while

$${}^s\phi(\rho(r)) = {}^s(\lambda(r^i)) = \lambda(s)\lambda(r^i)\lambda(s) = \lambda(sr^is) = \lambda(r^{-i}),$$

which is a contradiction since $r^i = r^{-i}$ if and only if $i = 0$. Thus, $K[D_p] \ncong H_\lambda$.

Remark 1 One could also prove Proposition 4 as follows. Over L, $K[D_p]$ and H_λ are isomorphic to $L[D_p]$ as Hopf algebras, thus their duals $K[D_p]^*$ and H_λ^* are finite dimensional as algebras over K and separable (as defined in [23, 6.4, p. 47]). Using the classification of such K-algebras [23, 6.4, Theorem], we conclude that $K[D_p]^*$ and H_λ^* are not isomorphic as K-Hopf algebras, and so neither are $K[D_p]$ and H_λ. In fact, by [23, 6.3, Theorem], $K[D_p]^*$ and H_λ^* are not isomorphic as K-algebras, and consequently, $K[D_p]$ and H_λ are not isomorphic as K-coalgebras. As we will show in Sect. 5, however, $K[D_p] \cong H_\lambda$ as K-algebras.

Picking $c = 0$, we obtain the following.

Theorem 2 *There are three K-Hopf algebras which provide Hopf–Galois structures on a dihedral extension L/K of degree $2p$, namely:*

1. *The group algebra $K[D_p]$, which provides the classical structure.*
2. *The Hopf algebra H_λ, which provides the canonical nonclassical structure.*
3. *A commutative K-Hopf algebra H_0 which provides p different structures.*

We now wish to consider whether any two of the Hopf algebras above become isomorphic after base change to an intermediate field $K \subset F \subset L$. This question is relatively easy to answer in the dihedral case. Since a K-Hopf algebra H is commutative if and only if $F \otimes_K H$ is commutative the Hopf algebra H_0 is not isomorphic to either $K[D_p]$ or H_λ after base change. What remains is to determine whether we can have $F \otimes_K K[D_p] \cong F \otimes_K H_\lambda$. But this is also easy since the center $Z(D_p)$ is trivial, the subgroup of D_p defined as $\{g \in D_p : {}^g\lambda(h) = \lambda(h), \forall h \in D_p\}$ is trivial. Hence, by [13, Corollary 3.2], $F = L$ is the smallest field extension of L for which $F \otimes_K H_\lambda \cong F[D_p]$. Thus, $F = L$ is minimal so that $F \otimes_K K[D_p] \cong F \otimes_K H_\lambda$.

5 The Algebra Structure

Let L/K be Galois with group D_p and $\mathbb{Q} \subseteq K$. In this section, we investigate the question of when two Hopf algebras providing Hopf–Galois structures on L/K are isomorphic as algebras. Throughout this section, when working with Hopf algebras, "isomorphic" refers to isomorphic as algebras. Since $\mathrm{char}(K)$ does not divide $[L : K]$, Maschke's theorem and a result of Amitsur [1, Theorem 1] show that the Hopf algebras are left semisimple.

Of course, $H_c \cong H_d$ for all $0 \le c, d \le p - 1$, and these (Hopf) algebras are not isomorphic to either $K[D_p]$ or H_λ since they are commutative; they remain non-isomorphic after base change for the same reason.

It remains to consider the classical and the canonical nonclassical structures, $K[D_p]$ and H_λ, respectively. Our main tool will be the Wedderburn–Artin decomposition.

The decomposition of $\mathbb{Q}[D_p]$ is given in [11, Example (7.39)].

Theorem 3 *Let D_p, $p \geq 3$ prime, be the dihedral group of order $2p$. Then*

$$\mathbb{Q}[D_p] \cong \mathbb{Q} \times \mathbb{Q} \times \mathrm{Mat}_2(\mathbb{Q}(\zeta_p + \zeta_p^{-1})),$$

where ζ_p denotes a primitive pth root of unity.

Let $E = K \cap \mathbb{Q}(\zeta_p + \zeta_p^{-1})$ and $l = [\mathbb{Q}(\zeta_p + \zeta_p^{-1}) : E]$. Then

$$K[D_p] \cong \begin{cases} K \times K \times \left(\mathrm{Mat}_2\big(K(\zeta_p + \zeta_p^{-1})\big)\right)^{(p-1)/(2l)} & \zeta_p + \zeta_p^{-1} \notin K \\ K \times K \times \left(\mathrm{Mat}_2(K)\right)^{(p-1)/2} & \zeta_p + \zeta_p^{-1} \in K. \end{cases} \tag{1}$$

The former case follows by noting that $E = \mathbb{Q}(\alpha)$ for $\alpha \in E$ satisfying an irreducible monic polynomial of degree $(p-1)/(2l)$ over $\mathbb{Q}$. The latter case follows from observing that if $\zeta_p + \zeta_p^{-1} \in K$, then the $(p-1)/2$ two-dimensional irreducible representations of D_p over $\mathbb{C}$ correspond to $(p-1)/2$ characters with values in K [22, Chap. 5, 5.3].

What can be said about the decomposition of H_λ? Since H_λ is left semisimple, the K-algebra H_λ decomposes into a product of matrix rings over division rings,

$$H_\lambda \cong \mathrm{Mat}_{q_1}(R_1) \times \mathrm{Mat}_{q_2}(R_2) \times \cdots \times \mathrm{Mat}_{q_t}(R_t).$$

The division rings R_i are finite dimensional K-algebras.

Now, there are exactly two one-dimensional irreducible representations of D_p, with characters χ_1 and χ_2, corresponding to mutually orthogonal idempotents

$$e_1 = \frac{1}{2p} \sum_{g \in D_p} \chi_1(g^{-1})g, \quad e_2 = \frac{1}{2p} \sum_{g \in D_p} \chi_2(g^{-1})g$$

in $L[D_p]$. Both e_1 and e_2 are fixed by the action of D_p, and hence $e_1, e_2 \in H_\lambda$. It follows that

$$H_\lambda \cong K \times K \times \prod_{j=1}^{m} \mathrm{Mat}_{q_j}(R_j), \tag{2}$$

where $q_j \geq 1$ are integers and R_j are division rings. For later use, we set $S_j = \mathrm{Mat}_{q_j}(R_j)$ for $1 \leq j \leq m$. Observe that

$$\sum_{j=1}^{m} \dim_K(R_j) \cdot q_j^2 = 2(p-1).$$

For the moment, we assume $p = 3$, so that $\sum_{j=1}^{m} \dim_K(R_j) \cdot q_j^2 = 4$. Since the dimension of a division algebra over its center is a perfect square, we conclude that

$$H_\lambda \cong K \times K \times \mathrm{Mat}_q(R), \tag{3}$$

where $1 \le q \le 2$ and R is a division ring. If $q = 1$, then the corresponding division ring R is noncommutative. If $q = 2$, then $R = K$.

Assume that the base field $K = \mathbb{Q}$, and let L be the splitting field of $x^3 - v$ over $\mathbb{Q}$, where v is not a 3rd power in $\mathbb{Q}$. Let ζ_3 denote a primitive third root of unity and let $\alpha = \sqrt[3]{v}$. Then, $L = \mathbb{Q}(\alpha, \zeta_3)$ is Galois with group D_3. The Galois action is given as $r(\alpha) = \zeta_3\alpha$, $r(\zeta_3) = \zeta_3$, $s(\alpha) = \alpha$, $s(\zeta_3) = \zeta_3^2$. Let

$$b = \alpha s + \alpha\zeta_3 sr + \alpha\zeta_3^2 sr^2.$$

Then $b \in L[D_3]$, and since b is fixed by all elements of D_3, $b \in H_\lambda$. Moreover, direct computation yields $b^2 = 0$. Thus, the only possibility is that $q = 2$ in (3), and so,

$$H_\lambda \cong \mathbb{Q} \times \mathbb{Q} \times \mathrm{Mat}_2(\mathbb{Q}).$$

Since $\mathbb{Q}[D_3] \cong \mathbb{Q} \times \mathbb{Q} \times \mathrm{Mat}_2(\mathbb{Q})$ by Theorem 3, here we have an instance of $H_\lambda \cong \mathbb{Q}[D_3]$ as $\mathbb{Q}$-algebras. Surprisingly, this is true for *any* L/K Galois with group D_p, $p \ge 3$. In fact, it holds even more generally:

Theorem 4 (Greither) *Let $\mathbb{Q} \subseteq K$, and let L/K be a Galois extension with group G. Then*

$$H_\lambda \cong K[G]$$

as K-algebras.

Proof We prove the special case where $G = D_p$. We thank C. Greither for the method of proof.

From (1), we obtain

$$L \otimes_K H_\lambda \cong L \otimes_K K[D_p] \cong L[D_p] \cong \begin{cases} L \times L \times \left(\mathrm{Mat}_2(L \otimes_K K(\zeta_p + \zeta_p^{-1}))\right)^{(p-1)/(2l)} \\ \text{if } \zeta_p + \zeta_p^{-1} \notin K \\ L \times L \times \left(\mathrm{Mat}_2(L)\right)^{(p-1)/2} \\ \text{if } \zeta_p + \zeta_p^{-1} \in K, \end{cases}$$

with $l = [\mathbb{Q}(\zeta_p + \zeta_p^{-1}) : E] = [K(\zeta_p + \zeta_p^{-1}) : K]$, $E = K \cap \mathbb{Q}(\zeta_p + \zeta_p^{-1})$. Thus, the decomposition of $L[D_p]$ contains components of the form L and $\mathrm{Mat}_2\big(L \otimes_K F\big)$, with $F = K(\zeta_p + \zeta_p^{-1})$.

The Hopf algebra H_λ descends from $L[D_p]$ via the action of D_p as usual, an element of D_p acts by conjugation on D_p and by the Galois action on L. A character of D_p has the same value on conjugate elements in D_p [22, Chap. 2, 2.1, Proposition 1(iii)], and so the central indecomposable idempotents of $\mathbb{C}[D_p]$ as constructed in [22, Chap. 6, 6.3, Exercise 6.4] are fixed by conjugation by elements of D_p. Let e be a central indecomposable idempotent in $K[D_p]$. Since e is in the center of $\mathbb{C}[D_p]$, e

is a sum of central indecomposable idempotents of $\mathbb{C}[D_p]$ [22, Chap. 6, 6.3, Exercise 6.4]. Thus, conjugation by elements of D_p fixes e.

Now, the central indecomposable idempotents of $K[D_p]$ correspond to the components in the decomposition (1); let M be the component of $K[D_p]$ corresponding to e. For $\alpha \in M$, $g \in D_p$,

$$g\alpha g^{-1} = g\alpha e g^{-1} = g\alpha g^{-1} g e g^{-1} = g\alpha g^{-1} e \in M,$$

and so, conjugation by g, which is an automorphism of $K[D_p]$, restricts to an automorphism of each component of $K[D_p]$. Hence, the action of D_p preserves the components in the decomposition of $L \otimes_K K[D_p] \cong L[D_p]$. So D_p can be thought of as acting on these components. The two copies of L in the decomposition of $L[D_p]$ descend to the two copies of K in the decomposition (2) of H_λ, and each copy of $\mathrm{Mat}_2\big(L \otimes_K F\big)$ descends to a component $S = S_j$ in the decomposition (2) of H_λ; the K-algebra S is an L-form of $\mathrm{Mat}_2\big(F\big)$. We want to show that $S \cong \mathrm{Mat}_2\big(F\big)$ as K-algebras, and so $H_\lambda \cong K[D_p]$.

Let $\mathrm{Aut}(\mathrm{Mat}_2(F))$ denote the automorphism group scheme of $\mathrm{Mat}_2(F)$ in the sense of [23, Sect. 7.6]. By [23, Theorem, p. 137], the isomorphism classes of L-forms of $\mathrm{Mat}_2(F)$ correspond to the cohomology set $\mathrm{H}^1(D_p, \mathrm{Aut}(\mathrm{Mat}_2(L \otimes_K F))$. In this setting, a 1-cocycle (crossed homomorphism) is a function $f : D_p \to \mathrm{Aut}(\mathrm{Mat}_2(L \otimes_K F))$ which satisfies $f(gh) = f(g) \circ (g \cdot f(h))$, for $g, h \in D_p$. The action $g \cdot f(h)$ of the element $g \in D_p$ on the automorphism $f(h)$ in $\mathrm{Aut}(\mathrm{Mat}_2(L \otimes_K F))$ is induced by the Galois action on L we have

$$g \cdot f(h) = (g \otimes I_F) f(h) (g^{-1} \otimes I_F),$$

where $g, h \in D_p$, and I_F is the identity map on F. The trivial element in $\mathrm{H}^1(D_p, \mathrm{Aut}(\mathrm{Mat}_2(L \otimes_K F))$ is represented by the 1-cocycle

$$g \mapsto (g \otimes I_F)\phi(g^{-1} \otimes I_F)\phi^{-1},$$

where ϕ is any element of $\mathrm{Aut}(\mathrm{Mat}_2(L \otimes_K F))$. The L-form S comes from a particularly simple 1-cocycle $\hat{f} : D_p \to \mathrm{Aut}(\mathrm{Mat}_2(L \otimes_K F))$. For $g \in D_p$, $\hat{f}(g)$ is conjugation by g on $\mathrm{Mat}_2(F) \subseteq \mathrm{Mat}_2(L \otimes_K F)$. Let $[\hat{f}]$ denote the class of $\hat{f}$ in $\mathrm{H}^1(D_p, \mathrm{Aut}(\mathrm{Mat}_2(L \otimes_K F))$.

Let $\mathrm{Inn}(\mathrm{Mat}_2(L \otimes_K F))$ denote the group of inner automorphisms. Since every element of the group $\mathrm{Inn}(\mathrm{Mat}_2(L \otimes_K F))$ is given as conjugation by some element of $\mathrm{GL}_2(L \otimes_K F)$, there is a surjection of groups $\psi : \mathrm{GL}_2(L \otimes_K F) \to \mathrm{Inn}(\mathrm{Mat}_2(L \otimes_K F))$ with $\ker(\psi) = (L \otimes_K F)^\times$. Thus, there is an induced map in cohomology

$$\mathrm{H}^1(D_p, \mathrm{GL}_2(L \otimes_K F)) \xrightarrow{\psi} \mathrm{H}^1(D_p, \mathrm{Inn}(\mathrm{Mat}_2(L \otimes_K F))).$$

A 1-cocycle class $[q] \in \mathrm{H}^1(D_p, \mathrm{GL}_2(L \otimes_K F))$ is represented by a function $q : D_p \to \mathrm{GL}_2(L \otimes_K F)$ which satisfies $q(gh) = q(g)(g \cdot q(h))$. The action of $g \in D_p$ on $q(h) \in \mathrm{GL}_2(L \otimes_K F)$ is through the Galois action on L as above.

There is a special cocycle class $[\hat{q}] \in \mathrm{H}^1(D_p, \mathrm{GL}_2(L \otimes_K F))$ represented by the function $\hat{q}$ in which each $g \in D_p$ is identified with its image in $\mathrm{Mat}_2(F) \subseteq \mathrm{Mat}_2(L \otimes_K F)$ under the map $K[D_p] \to \mathrm{Mat}_2(F)$. So conjugation by g in $\mathrm{Mat}_2(F)$ is precisely the conjugation action of g on $K[D_p]$ restricted to the component $\mathrm{Mat}_2(F)$. It follows that

$$\psi([\hat{q}]) = [\hat{f}].$$

The class $[\hat{f}]$ corresponds to the isomorphism class of the L-form S.

Now, if $\zeta_p + \zeta_p^{-1} \in K$, then $F = K$. Thus, $L = L \otimes_K F$ and

$$\mathrm{H}^1(D_p, \mathrm{GL}_2(L \otimes_K F)) = \mathrm{H}^1(D_p, \mathrm{GL}_2(L)).$$

By Hilbert's Theorem 90, $\mathrm{H}^1(D_p, \mathrm{GL}_2(L))$ is trivial, and so, $[\hat{q}]$ is trivial, and consequently, $[\hat{f}]$ is trivial. It follows that $S \cong \mathrm{Mat}_2(K)$ as K-algebras, and so $H_\lambda \cong K[D_p]$.

If $\zeta_p + \zeta_p^{-1} \notin K$, then C. Greither has provided a generalization of Hilbert's Theorem 90 to yield $\mathrm{H}^1(D_p, \mathrm{GL}_2(L \otimes_K F)$ being trivial. As above, $[\hat{f}]$ is trivial, and so $S \cong \mathrm{Mat}_2(F)$ as K-algebras. It follows that $H_\lambda \cong K[D_p]$.

We summarize our findings in this section. The Hopf algebras that provide Hopf–Galois structures in the case that L/K is Galois with group D_p fall into two K-algebra isomorphism classes represented by $K[D_p]$ and H_0. So a single K-algebra (e.g., $K[D_p]$) can be endowed with multiple coalgebra structures, resulting in multiple (non-isomorphic) Hopf algebras (e.g., $K[D_p]$, H_λ) giving different Hopf–Galois structures on L/K (e.g., classical and canonical nonclassical).

6 Explicit Structure Computations

Let L/K be Galois with group D_p and $\mathbb{Q} \subseteq K$. We find generators over K for the Hopf algebras $K[D_p]$, H_λ, H_0, $H_1, \ldots, H_{p-1}$ constructed above. Let $L^{\langle r \rangle}$ be the unique quadratic extension of K contained in L. Pick $d \in L$ such that $L^{\langle r \rangle} = K(\sqrt{d})$. Note that $s(\sqrt{d}) = -\sqrt{d}$. Additionally, let $y \in L$ be so that $K(y) = L^{\langle s \rangle}$.

The simplest case, of course, is $K[D_p]$, it has a K-basis D_p, and $\{r, s\}$ generates $K[D_p]$ as a K-algebra.

We next turn to H_λ. Suppose $h \in H_\lambda$. Identifying D_p with $\lambda(D_p)$, we have $h \in L[D_p]$ with h fixed by D_p. Let

$$h = \sum_{i=0}^{p-1} a_i r^i + \sum_{i=0}^{p-1} b_i r^i s, \; a_i, b_i \in L.$$

Then

$$ {}^{r}h = \sum_{i=0}^{p-1} r(a_i)r^i + \sum_{i=0}^{p-1} r(b_i)r^{i+1}sr^{-1} = \sum_{i=0}^{p-1} r(a_i)r^i + \sum_{i=0}^{p-1} r(b_i)r^{i+2}s, $$

and since ${}^{r}h = h$ we have $a_i \in L^{\langle r\rangle}$ and $r(b_i) = b_{i+2}$ for all i (where $i+2$ is considered mod p). Thus, $b_i = r^{i(p+1)/2}(b_0)$ for all i.

Furthermore,

$$ {}^{s}h = \sum_{i=0}^{p-1} s(a_i)sr^is + \sum_{i=0}^{p-1} s(b_i)sr^is^2 = \sum_{i=0}^{p-1} s(a_i)r^{-i} + \sum_{i=0}^{p-1} s(b_i)r^{-i}s, $$

which after interchanging i with $p-i$ for all $i \neq 0$ gives

$$ {}^{s}h = s(a_0) + s(b_0)s + \sum_{i=1}^{p-1} s(a_{p-i})r^i + \sum_{i=1}^{p-1} s(b_{p-i})r^is $$

and so $a_0 \in K$, $s(a_i) = a_{p-i}$, $s(b_0) = b_0$. Note that $s(b_i) = b_{p-i}$ as well, but this followed previously since

$$ s(b_i) = sr^{i(p+1)/2}(b_0) = r^{-i(p+1)/2}s(b_0) = r^{-i(p+1)/2}(b_0) = r^{(p-i)(p+1)/2}(b_0) = b_{p-i}. $$

Since $r^{(p-i)(p+1)/2} = r^{i(p-1)/2}$, it follows that

$$ H_\lambda = \left\{ a_0 + \sum_{i=1}^{(p-1)/2} (a_ir^i + s(a_i)r^{-i}) + b_0s + \sum_{i=1}^{p-1} r^{i(p-1)/2}(b_0)r^{-i}s : \right. \\ \left. a_0 \in K, a_i \in L^{\langle r\rangle}, b_0 \in L^{\langle s\rangle} \right\}. $$

Example 5 Suppose $p = 3$. Then

$$ H_\lambda = \left\{ a_0 + a_1r + s(a_1)r^2 + b_0s + r(b_0)sr + r^2(b_0)sr^2 : a_0 \in K, a_1 \in L^{\langle r\rangle}, b_0 \in L^{\langle s\rangle} \right\}. $$

A K-basis for H_λ is

$$ \{1, r+r^2, \sqrt{d}(r-r^2), s+rs+r^2s, ys+r(y)rs+r^2(y)r^2s, y^2s \\ + r(y^2)rs + r^2(y^2)r^2s\}. $$

We next consider the structure of H_c, $0 \le c \le p-1$. If we write

$$ h = \sum_{i=0}^{2p-1} a_i\eta^i, \quad \eta = \eta_c, $$

where $a_i \in L$ for all i, then, since r acts trivially on η,

$$ {}^r h = \sum_{i=0}^{2p-1} r(a_i)\eta^i = \sum_{i=0}^{2p-1} a_i \eta^i. $$

Thus, $r(a_i) = a_i$ for all i, and hence $a_i \in L^{\langle r \rangle}$. Also, since ${}^s\eta = \eta^{-1}$,

$$ {}^s h = \sum_{i=0}^{2p-1} s(a_i)\eta^{-i} = \sum_{i=0}^{2p-1} a_i \eta^i, $$

from which it follows that $s(a_0) = a_0$ and $s(a_i) = a_{2p-i}$ for all $i > 0$. In particular, $s(a_p) = a_p$, so $a_0, a_p \in K$. Thus,

$$ H_c = \left\{ a_0 + a_p \eta^p + \sum_{i=1}^{p-1} \left(a_i \eta^i + s(a_i)\eta^{-i} \right) : \ a_0, a_p \in K, a_i \in L^{\langle r \rangle}, 1 \le i \le p-1 \right\}, $$

and H_c has K-basis

$$ \big\{ 1, \eta^p, \eta + \eta^{-1}, \eta^2 + \eta^{-2}, \ldots, \eta^{p-1} + \eta^{-(p-1)}, \sqrt{d}(\eta - \eta^{-1}), \sqrt{d}(\eta^2 - \eta^{-2}), \\ \ldots, \sqrt{d}(\eta^{p-1} - \eta^{-(p-1)}) \big\}. $$

Example 6 Suppose $p = 3$. Then

$$ H_c = \left\{ a_0 + a_3\eta^3 + a_1\eta + s(a_1)\eta^5 + a_2\eta^2 + s(a_2)\eta^4 : \ a_0, a_3 \in K, a_1, a_2 \in L^{\langle r \rangle} \right\}, $$

and H_c has K-basis

$$ \{1, \eta^3, \eta + \eta^5, \eta^2 + \eta^4, \sqrt{d}(\eta - \eta^5), \sqrt{d}(\eta^2 - \eta^4)\}. $$

7 Example: Hopf–Galois Structures in the Case D_3

We close with an analysis of the Hopf–Galois structures in the case $p = 3$. Let L/K be any Galois extension with group D_3 and $\mathbb{Q} \subseteq K$. As we have seen, there are two regular subgroups normalized by $\lambda(D_3)$ and isomorphic to D_3, namely, $\rho(D_3)$ and $\lambda(D_3)$, and three regular subgroups normalized by $\lambda(D_3)$ and isomorphic to C_6, the cyclic group of order 6, namely, N_0, N_1, and N_2.

By Proposition 4, $K[D_3] \not\cong H_\lambda$, as K-Hopf algebras, and by Theorem 4, $K[D_3] \cong H_\lambda$, as K-algebras, with Wedderburn–Artin decomposition

$$ K[D_3] \cong H_\lambda \cong K \times K \times \mathrm{Mat}_2(K). $$

By Proposition 3, $H_0 \cong H_1 \cong H_2$ as Hopf algebras, and hence as K-algebras.

We seek the Wedderburn–Artin decomposition and the Hopf algebra structure of H_0 (hence of H_1 and H_2). In contrast to the situation with H_λ, the structure of H_0 seems to depend on the extension L/K, specifically on the fixed field $L^{\langle r\rangle}$.

Here is how we can compute the structure of H_0. By [20, Corollary 3.6],

$$\{g \in \lambda(D_3) :\ {}^g\eta = \eta, \forall \eta \in N_0\}$$

is precisely the 3-Sylow subgroup $\lambda(\langle r\rangle) \leq \lambda(D_3)$, which we identify with $\langle r\rangle$. There is an induced action of $D_3/\langle r\rangle$ on $L[N_0]$. Note that $D_3/\langle r\rangle \cong C_2$, the cyclic group of order 2. By the Fundamental Theorem of Galois Theory, $D_3/\langle r\rangle \cong C_2$ is the group of the Galois extension F/K, where $F = L^{\langle r\rangle}$; F is a quadratic extension of K. We write $F = K[z]/(z^2 - b)$ for $b \in K$, z indeterminate.

Now, $D_3/\langle r\rangle \cong C_2$ can be viewed as the group of automorphisms of $N_0 \cong C_6$. We have

$$H_0 = (L[N_0])^{D_3} = (F[C_6])^{C_2},$$

where the action of C_2 on $F[C_6]$ is by the Galois group on F and as automorphisms on C_6.

Now, F is a C_2-Galois extension of K, [14, p. 130]. So by [14, Theorem 5], F corresponds to an F-Hopf algebra form of $K[C_6]$, namely, $(F[C_6])^{C_2}$, which must of course be H_0.

And so, H_0 is the fixed ring of $F[C_6]$ under the action of C_2, and H_0 is an F-form of $K[C_6]$. Under these conditions, H_0 can be characterized. Let x, y be indeterminates and recall $F = K[z]/(z^2 - b)$. The method of Haggenmüller and Pareigis in [14, Theorem 6, p. 134] applies to yield $H_0 \cong K[x, y]/I$,

$$I = (y^2 - bx^2 + u, (x-2)(x-1)(x+1)(x+2), (x-1)(x+1)(xy)),$$

with $u \in K^\times$, $u = 4b$. The Hopf algebra structure of H_0 is defined by

$$\Delta(\bar{x}) = \frac{1}{2}\bar{x} \otimes \bar{x} + \frac{1}{2b}\bar{y} \otimes \bar{y},$$

$$\Delta(\bar{y}) = \frac{1}{2}\bar{x} \otimes \bar{y} + \frac{1}{2}\bar{y} \otimes \bar{x},$$

$$\varepsilon(\bar{x}) = 2, \quad \varepsilon(\bar{y}) = 0, \quad \sigma(\bar{x}) = \bar{x}, \quad \sigma(\bar{y}) = -\bar{y},$$

where $\bar{x} = x \bmod I$, $\bar{y} = y \bmod I$.

Remark 2 The Hopf algebra structure of H_0 does not depend on the choice of generator for the quadratic extension F. Indeed, suppose that $F = F'$ where $F' = K[z]/((z^2 - a'z - b')$. Then, the induced isomorphism $\phi : F[C_6] \to F'[C_6]$ respects the action of C_2, and hence the fixed rings H_0, H_0' are isomorphic as K-Hopf algebras.

Example 7 We assume that the base field $K = \mathbb{Q}$ and compute the structure of H_0 in the case that L is the splitting field of $x^3 - v$, irreducible over $\mathbb{Q}$. In this case, $F = L^{\langle r \rangle} = \mathbb{Q}(\zeta_3)$, and hence $F = \mathbb{Q}[z]/(z^2 + 3)$, and so, $b = -3$, $u = -12$. We then have $H_0 = \mathbb{Q}[x, y]/I$, with

$$I = (y^2 + 3x^2 - 12, (x-2)(x-1)(x+1)(x+2), (x-1)(x+1)(xy)).$$

The Hopf algebra structure of H_0 is given as

$$\Delta(\bar{x}) = \frac{1}{2}\bar{x} \otimes \bar{x} - \frac{1}{6}\bar{y} \otimes \bar{y},$$

$$\Delta(\bar{y}) = \frac{1}{2}\bar{x} \otimes \bar{y} + \frac{1}{2}\bar{y} \otimes \bar{x},$$

$$\varepsilon(\bar{x}) = 2, \quad \varepsilon(\bar{y}) = 0, \quad \sigma(\bar{x}) = \bar{x}, \quad \sigma(\bar{y}) = -\bar{y}.$$

We can also obtain the Wedderburn–Artin decomposition of H_0 in Example 7.

Proposition 5 *Assume the conditions of Example 7. Then*

$$H_0 \cong \mathbb{Q} \times \mathbb{Q} \times \mathbb{Q} \times \mathbb{Q} \times \mathbb{Q} \times \mathbb{Q},$$

as $\mathbb{Q}$*-algebras.*

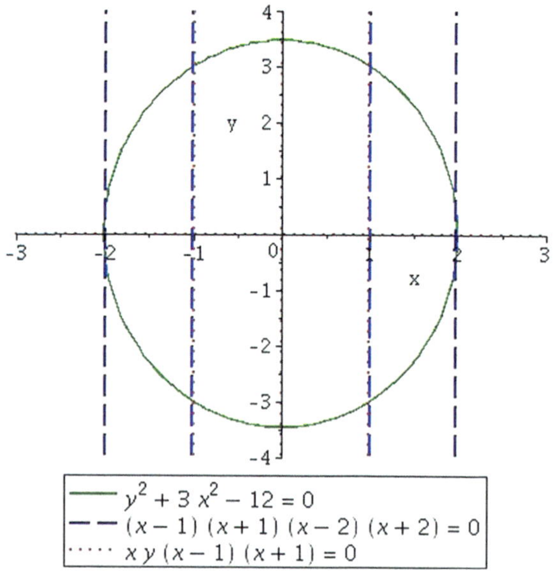

Fig. 1 Graph of variety determined by I

Proof The ideal I determines an affine variety in $X \subseteq \mathbb{Q}^2$ consisting of exactly six points:

$$P_1 = (-2, 0), P_2 = (-1, 3), P_3 = (1, 3), P_4 = (2, 0), P_5 = (1, -3), P_6 = (-1, -3).$$

This is the set of common zeros of the polynomials in I, see Fig. 1.

As a commutative $\mathbb{Q}$-algebra, $H_0 \cong \mathbb{Q}[x, y]/I$ is a product of fields with $\dim_{\mathbb{Q}}(H_0) = 6$. For $1 \leq j \leq 6$, let (x_j, y_j) be the coordinates of the point P_j and let $\Psi_j : H_0 \to \mathbb{Q}$ denote the ring homomorphism defined as $\bar{x} \mapsto x_j$, $\bar{y} \mapsto y_j$. Then, Ψ_j is surjective and $\ker(\Psi_j)$ is an ideal I_j of H_0 of dimension five over $\mathbb{Q}$. Now, $\bar{y} - y_j - 6\bar{x} + 6x_j \in I_j$, yet $\bar{y} - y_j - 6\bar{x} + 6x_j \notin I_k$ whenever $j \neq k$. Thus, the ideals I_j, $1 \leq j \leq 6$ are distinct, and therefore must arise by omitting one factor isomorphic to $\mathbb{Q}$ from the Wedderburn–Artin decomposition of H_0. It follows that the decomposition of H_0 must contain at least 6 factors isomorphic to $\mathbb{Q}$, and hence $H_0 \cong \mathbb{Q}^6$.

Acknowledgements The authors would like to thank the referee for comments and suggestions which improved the exposition and content of this paper.

References

1. Amitsur, S.A.: The radical of field extensions. Bull. Res. Council Israel Sect. F **7F**(1), 1–10 (1957/1958)
2. Byott, N.P.: Uniqueness of Hopf Galois structure of separable field extensions. Comm. Algebra **24**(10), 3217–3228 (1996)
3. Byott, N.P.: Hopf Galois structures on field extensions with simple Galois groups. Bull. London. Math. Soc. **36**, 23–29 (2004)
4. Byott, N.P.: Hopf–Galois structures on Galois field extensions of degree pq. J. Pure Appl. Algebra **188**(1–3), 45–57 (2004)
5. Carnahan, S., Childs, L.: Counting Hopf Galois structures on non-abelian Galois field extensions. J. Algebra **218**(1), 81–92 (1999)
6. Chase, S.U., Sweedler, M.E.: Hopf Algebras and Galois Theory. Lecture Notes in Mathematics, vol. 97. Springer-Verlag, Berlin (1969)
7. Childs, L.N.: Taming Wild Extensions: Hopf Algebras and Local Galois Module Theory. AMS: Mathematical Surveys and Monographs, vol. 80 (2000)
8. Childs, L.N.: Some Hopf Galois structures arising from elementary abelian p-groups. Proc. Amer. Math. Soc. **135**(11), 3453–3460 (2007)
9. Childs, L.N.: Hopf Galois structures on Kummer extensions of prime power degree. New York J. Math. **17**, 51–74 (2011)
10. Childs, L.N.: Fixed-Point free endomorphisms and Hopf Galois structures. Proc. Amer. Math. Soc. **141**(4), 1255–1265 (2013)
11. Curtis, C.W., Reiner, I.: Methods of Representation Theory, vol. I. Wiley, New York (1981)
12. Featherstonhaugh, S.C., Caranti, A., Childs, L.N.: Abelian Hopf Galois structures on prime-power Galois field extensions. Trans. Amer. Math. Soc. **364**(7), 3675–3684 (2012)
13. Greither, C., Pareigis, B.: Hopf Galois theory for separable field extensions. J. Algebra **106**(1), 239–258 (1987)
14. Haggenmüller, R., Pareigis, B.: Hopf algebra forms on the multiplicative group and other groups. Manuscripta Math. **55**, 121–135 (1986)

15. Jacobson, N.: Galois theory of purely inseparable fields of exponent one. Amer. J. Math. **66**, 645–648 (1944)
16. Koch, A.: Scaffolds and integral Hopf Galois module structure on purely inseparable extensions. New York J. Math. **21**, 73–91 (2015)
17. Koch, A., Kohl, T., Truman, P., Underwood, R.: Isomorphism problems for Hopf–Galois structures on separable field extensions. J. Pure Appl. Algebra **223**(5), 2230–2245 (2019). https://doi.org/10.1016/j.jpaa.2018.07.014
18. Kohl, T.: Classification of the Hopf Galois structures on prime power radical extensions. J. Algebra **207**(2), 525–546 (1998)
19. Kohl, T.: Groups of order $4p$, twisted wreath products and Hopf-Galois theory. J. Algebra **314**(1), 42–74 (2007)
20. Kohl, T.: Regular permutation groups of order mp and Hopf Galois structures. Algebra Number Theory **7**(9), 2203–2240 (2013)
21. Kohl, T.: Hopf-Galois structures arising from groups with unique subgroup of order p. Algebra Number Theory **10**(1), 37–59 (2016)
22. Serre, J.-P.: Linear Representations of Finite Groups. Springer-Verlag, New York (1977)
23. Waterhouse, W.C.: Introduction to Affine Group Schemes. Springer-Verlag, New York (1979)

Semi-extraspecial Groups

Mark L. Lewis

Abstract We survey the results regarding semi-extraspecial p-groups. Semi-extraspecial groups can be viewed as generalizations of extraspecial groups. We present the connections between semi-extraspecial groups, Camina groups, and VZ-groups, and give upper bounds on the order of the center and the orders of abelian normal subgroups. We define ultraspecial groups to be semi-extraspecial groups where the center is as large as possible, and demonstrate a connection between ultraspecial groups that have at least two abelian subgroups whose order is the maximum and semifields.

Keywords p-group · Extraspecial group · Semifields

Mathematics Subject Classification 2010: 20D15

1 Introduction

We have two main goals for this paper. The first is to give an expository account of the known results regarding semi-extraspecial groups. The second is to connect semifields with a certain class of ultraspecial groups and then show how a number of results from finite geometry apply to these groups.

A non-abelian p-group G is *special* if $G' = Z(G) = \Phi(G)$. Furthermore, a group G is *extraspecial* if G is a special p-group and $|G'| = |Z(G)| = p$. Extraspecial groups are central extensions of Z_p by Z_p^e, and as their name suggests, they are very special and come in just two types. We say that a p-group G is *semi-extraspecial*, if G satisfies the property for every maximal subgroup N of $Z(G)$ that G/N is an extraspecial group.

As far as we can tell, semi-extraspecial groups were first studied by Beisiegel in 1977 [4]. The paper [4] is in German. One of the other primary places where results

M. L. Lewis (✉)
Department of Mathematical Sciences, Kent State University,
Kent, OH 44242, USA
e-mail: lewis@math.kent.edu

J. Feldvoss et al. (eds.), *Advances in Algebra*, Springer Proceedings
in Mathematics & Statistics 277, https://doi.org/10.1007/978-3-030-11521-0_11

regarding semi-extraspecial groups appear is the paper [50] by Verardi and concerns a number of the geometric aspects of these groups. Many of the results regarding semi-extraspecial groups have not appeared in English. It is important to place these groups into the larger context of Camina groups and VZ-groups.

A group G is a *Camina group* if for every element $g \in G\backslash G'$, the conjugacy class of g is gG'. It is in the context of Camina groups that other results about semi-extraspecial groups have been proved. The main contributions have been by MacDonald, but also by Chillag et al. (see [6, 11, 38–41]). A group G is a VZ-group if every nonlinear irreducible character vanishes off of the center of G. I. e., $\chi(g) = 0$ for every nonlinear character $\chi \in \mathrm{Irr}(G)$ and for every element $g \in G\backslash Z(G)$. An important source regarding VZ-groups is the paper [16] by Fernández-Alcober and Moretó, which looks at generalizing several of the key properties of semi-extraspecial groups.

In Sect. 5, we will present a number of equivalent conditions for a group to be a Camina group and a similar looking set of equivalent conditions for when a group is a VZ-group. A group G is a semi-extraspecial group if and only if it is a Camina group and a VZ-group. Combining the list of conditions, we obtain a list of equivalent conditions for a special group to be a semi-extraspecial group.

Verardi has proved that there exists an upper bound on the order of abelian subgroups of semi-extraspecial groups (see Theorem 9). We see that a particular class of semi-extraspecial groups can be better understood in terms of the number of abelian subgroups whose order equals this upper bound.

The particular class of semi-extraspecial groups that we focus on are the ultraspecial groups. We will define ultraspecial groups in Sect. 7. In particular, we are interested in the ultraspecial groups that have at least two abelian subgroups of the largest possible order. We note that these are not all of the ultraspecial groups. In particular, Verardi has shown that there exist examples of semi-extraspecial groups with no abelian subgroups of this maximal order and one abelian subgroup of the maximal order. Verardi also notes that there are many examples known of semi-extraspecial groups with at least two abelian subgroups of maximal order.

We will show that there is a way of constructing a group from the algebraic object called a semifield. We will give the definition of semifields in Sect. 8, and we then give the definition of the group associated with each semifield which we will call a *semifield group*. The groups produced by this method from semifields are ultraspecial groups with at least two abelian subgroups. Semifields were initially studied algebraically as division algebras by Albert and Dickson. For Albert see: [1–3] and for Dickson see: [13, 14].

The term semifield seems to have originated with Knuth with the papers [28, 29]. In fact, [29] contains the results from Knuth's dissertation. Knuth studied semifields in the context of finite geometries. Semifields have continued to be an active topic in finite geometries. In this paper, we review the results regarding semifields that we apply to obtain results regarding semifield groups. There are many more results known about semifields. We suggest consulting the expository papers [9, 25] for a more thorough introduction to semifields. We believe that there are many other results regarding semifields that can be used to obtain information about semifield groups.

We will see that the construction of these semifield groups has appeared in the finite geometry literature a number of times (see [10, 20, 27, 45]). Our second goal is to present the results from the finite geometry literature regarding these groups.

We would like to thank Professor Alireza Moghaddamfar for his assistance in transcribing the original notes that were the starting point for this paper and for several useful comments while writing this paper. We would also like to thank Professor James Wilson for helpful discussions and Josh Maglione for computer computations with these groups and many helpful comments.

2 Overview

In this paper, all groups are finite p-groups for some prime p. This is an expository paper where we present results regarding semi-extraspecial groups and ultraspecial groups, but we will not present proofs.

The first time we seriously came to grips with semi-extraspecial groups was when we were writing the paper [30] where we determined when two semi-extraspecial groups have the same character table. We had encountered these groups earlier when we were working on the paper [34], although their involvement in that paper is peripheral as quotients of Camina p-groups of nilpotence class 3.

The genesis of this current paper is several folds. When presenting the paper [30] at a conference, we were asked how many semi-extraspecial groups there are, and then at a later conference where we were presenting the paper [31], someone commented that there were many semi-extraspecial groups. However, no one seems to be able to quantify how many semi-extraspecial groups exist in "many." In our paper with Lewis and Wilson [35], we give an estimate on the number of quotients of a Heisenberg group, but that paper does not touch the number of ultraspecial groups or the other semi-extraspecial groups that are not quotients of a Heisenberg group. We will see that this paper gives a way to count one class of ultraspecial groups, but further work is needed to count all ultraspecial groups and, in fact, to count other semi-extraspecial groups.

Perhaps a more immediate motivation for this paper is to better understand Camina groups of nilpotence class 3. It is known that the quotient of a Camina group of nilpotence class 3 by its center is an ultraspecial group with at least one abelian subgroup of maximal possible order (see Theorem 5.2 of [38] and Theorem 1.3 of [41]). However, until our recent paper [33], all of the examples of Camina p-groups that have been constructed have had the Heisenberg group as the quotient modulo the center. In [33], we construct two Camina groups of nilpotence class 3 whose quotients modulo their centers are not isomorphic to the Heisenberg group. We note that these two groups have quotients modulo the center that are not isomorphic. This raises the question of which ultraspecial groups can occur as the quotient modulo the center of a Camina group of nilpotence class 3. Also, in Corollary 6.5 of [33], we show that a certain subgroup of Camina groups of nilpotence class 3 are semi-extraspecial. With this in mind, we believe that understanding the structure of Camina groups of nilpotence class 3 will be linked to understanding the structure of semi-extraspecial groups.

We also note that semi-extraspecial groups and ultraspecial groups have appeared recently in a number of classifications or examples of groups with certain properties on their characters and/or their conjugacy classes. Some of the papers we have in mind are [15, 17, 49]. It seems to us that the understanding of the groups arising in these situations would be enhanced by increasing the knowledge that we have regarding semi-extraspecial groups and ultraspecial groups.

In particular, we hope to encourage new research on these groups. In this paper, we will focus on abelian subgroups of maximal possible order of ultraspecial groups. We believe that results regarding abelian subgroups of maximal possible order for semi-extraspecial groups that are not ultraspecial will deepen our understanding of these groups. We note that for ultraspecial groups, these abelian subgroups of maximal possible order are centralizers. We also believe that there would be much benefit to understanding the structure of centralizers in ultraspecial groups that are not abelian. For semi-extraspecial groups that are not ultraspecial, the centralizers are not abelian. However, there seems to be a close connection between centralizers in these groups and the abelian subgroups.

We also would like to see a better understanding of the quotients of semi-extraspecial groups, and we believe it would be fruitful to determine the automorphism groups of semi-extraspecial and ultraspecial groups.

3 Extraspecial Groups

Semi-extraspecial groups can be thought of as a particular type of extension or a generalization of extraspecial groups. There are many extensions of extraspecial groups that are not semi-extraspecial. In this section, we review the properties of extraspecial groups. We highlight the properties that are generalized by semi-extraspecial groups.

Recall that a non-abelian p-group G is *special* if $G' = Z(G) = \Phi(G)$. Furthermore, a group G is *extraspecial* if G is a special p-group and $|G'| = |Z(G)| = p$.

Here are some of the important facts about extraspecial groups. These facts are proved in group theory texts such as [21] or [23].

(1) Because $\Phi(G) = G'$, we know that G/G' is elementary abelian; that is, a vector space over $\mathrm{GF}(p)$.
(2) $|G : G'|$ is a square, so $|G : G'| = p^{2a}$ for some positive integer a. (Satz III.13.7 (c) of [21]).
(3) For every prime p and every positive integer a, there exists, up to isomorphism, exactly two extraspecial groups of order p^{2a+1}.
(4) For every element $x \in G \backslash G'$, the conjugacy class $\mathrm{cl}(x) = xG'$, so $|C_G(x)| = p^{2a} = |G : G'|$.

(5) If $|G| = p^{2a+1}$, then every noncentral element of G lies in an abelian subgroup of order p^{a+1}.
(6) Every irreducible character of G has degree p^a and "vanishes" (i.e., is 0) on $G \backslash Z(G)$ (Theorem 7.5 and Example 7.6 (b) of [22]).
(7) Every extraspecial group is a central product of extraspecial groups of order p^3 (Satz III.13.7 (d) of [21]).

When p is odd, the pair of extraspecial groups of order p^{2a+1} can be distinguished by the fact that one has exponent p and the other has exponent p^2. When $p = 2$, it is more complicated to distinguish these groups. It is well known that a group of exponent 2 must be abelian, so we cannot have extraspecial groups of exponent 2. It is not difficult to see that every extraspecial 2-group will have exponent 4. The two extraspecial groups of order 8 are Q_8, the quaternion group of order 8, and D_8, the dihedral group of order 8. For any positive integer a, the pair of extraspecial groups of order 2^{2a+1} can be distinguished by the number of involutions.

Applying (7) above, every extraspecial 2-group will be a central product of Q_8s and D_8s. One can show that the central product of two Q_8s is isomorphic to the central product of two D_8s (see the proof of Satz III.13.8 of [21]). Hence, a deep theorem shows that the pair of extraspecial groups of order 2^{2a+1} can also be distinguished by whether the number of quaternion (or dihedral) factors in the central product is even or odd.

4 Semi-extraspecial Groups

Recall that a p-group G is *semi-extraspecial*, if G satisfies the property for every maximal subgroup N of $Z(G)$ that G/N is an extraspecial group. We will often abbreviate semi-extraspecial by s.e.s. Beisiegel showed that every semi-extraspecial group is a special group (Lemma 1 of [4]). So if G is s.e.s., then $G' = Z(G) = \Phi(G)$. He also proved that $|G : G'|$ is a square and that $|G'| \leqslant \sqrt{|G : G'|}$ (Satz 1 of [4]). In particular, this gives the lemma as follows:

Lemma 1 *Let G be a s.e.s. p-group for a prime p. If $|G : G'| = p^{2a}$ and $|G'| = p^b$ for positive integers a and b, then $b \leqslant a$.*

In addition, Beisiegel gave a number of examples of semi-extraspecial groups that appear as Sylow subgroups. In particular, he showed that a Sylow p-subgroup of $\mathrm{SL}_3(p^a)$ or $\mathrm{SU}_3(p^{2a})$ will be s.e.s. of order p^{3a} (see Lemmas 4 and 5 of [4]). He also shows that the Suzuki 2-groups of Types B, C, and D are semi-extraspecial groups (Satz 2 of [4]). We say that the group G is the *Heisenberg p-group of degree a* if G is isomorphic to a Sylow p-subgroup of $\mathrm{GL}_3(p^a)$. The existence of the Heisenberg groups shows that there exist s.e.s. groups G with $|G : G'| = p^{2a}$ and $|G'| = p^a$ for every prime p and every positive integer a.

It is not difficult to prove the following fact.

Lemma 2 *If G is an s.e.s. p-group for some prime p and $N < G'$, then G/N is an s.e.s. group.*

When G is the Heisenberg p-group of degree a and b is an integer satisfying $1 \leqslant b \leqslant a$, we can find a subgroup N of G' so that $|G' : N| = p^b$. Thus, for every prime p and for all pairs of integers a, b with $1 \leqslant b \leqslant a$, we can find an s.e.s. group G with $|G : G'| = p^a$ and $|G'| = p^b$.

The next result follows from Lemma 2.2 of [38]. It is also noted at the beginning of Sect. 2 of [50].

Lemma 3 *If G is an s.e.s. p-group, then G/G' and G' are elementary abelian p-groups.*

5 Generalizations

Before we go further, we mention two generalizations of s.e.s. groups. First, we say that a group G is a *Camina group* if for every element $g \in G \backslash G'$, the conjugacy class of g is gG'. Camina groups have been studied in a number of places, namely, [6, 11, 38, 39, 41]. It has been shown by Dark and Scoppola that if G is a Camina group, then either G is a Frobenius group whose Frobenius complement is abelian, G is a Frobenius group whose Frobenius complement is isomorphic to the quaternions, or G is a p-group. (An alternate proof of this fact is presented in [24, 32].) Furthermore, MacDonald proved in Theorem 3.1 of [39] that a Camina 2-group will have nilpotence class 2, and Dark and Scoppola have proved in [11] when p is odd that a Camina p-group will have nilpotence class 2 or 3.

There are a number of equivalent conditions to being a Camina group. The following has been proved as Proposition 3.1 of [6]:

Theorem 1 *Let G be a group. Then, the following are equivalent:*

1. *G is a Camina group.*
2. *For every element $g \in G \backslash G'$, $|C_G(g)| = |G : G'|$.*
3. *For every element $g \in G \backslash G'$ and for every element $z \in G'$, there is an element $y \in G$ so that $[g, y] = z$.*
4. *Every character $\chi \in \mathrm{Irr}(G)$ vanishes on $G \backslash G'$.*

The key fact is the following theorem proved by Verardi in Theorem 1.2 of [50].

Theorem 2 *A group G is a s.e.s. p-group for some prime p if and only if G is Camina group of nilpotence class 2.*

Second, we say that a group G is a VZ-group if every nonlinear irreducible character vanishes off of the center of G. I. e., $\chi(g) = 0$ for every nonlinear character $\chi \in \mathrm{Irr}(G)$ and for every element $g \in G \backslash Z(G)$. If G is a VZ-group, then G is nilpotent of class 2. We defined the term VZ-group in [30], although groups satisfying this hypothesis had been earlier studied in a number of places including [16, 49].

To characterize VZ-groups, it is helpful to introduce the concept of isoclinism. Isoclinism was first introduced by Hall [18]. Two groups G and H are *isoclinic* if there exist isomorphisms $\alpha : \frac{G}{Z(G)} \to \frac{H}{Z(H)}$ and $\beta : G' \to H'$ such that

$$[\alpha(\overline{g_1}), \alpha(\overline{g_2})] = \beta(\overline{[g_1, g_2]}), \quad \forall g_1, g_2 \in G,$$

where $\overline{g} = gZ(G)$ for all $g \in G$. It is not difficult to show that isoclinism is an equivalence relation. If G and H are isomorphic, then G and H are isoclinic. On the other hand, if G and H are any two extraspecial groups of the same order, then G and H are isoclinic. Thus, isoclinism is weaker than isomorphism.

In fact, G and H being isoclinic do not even imply that $|G| = |H|$. On the other hand, a number of properties such as solvability, nilpotence, and nilpotence class are preserved by isoclinism. Hall showed that G is always isoclinic to a group H such that $Z(H) \leqslant H'$. For groups of nilpotence class 2, this implies that every group of nilpotence class 2 is isoclinic to a special group.

Most of the following result was proved by van der Waals and Kuisch as Theorem 2.4 of [49]. The result as stated is Theorem A of [16] by Fernández-Alcober and A. Moretó.

Theorem 3 *Let G be a group. The following conditions are equivalent:*

1. *G is a VZ-group.*
2. *G is isoclinic to a semi-extraspecial p-group for some prime p.*
3. $\mathrm{cd}(G) = \{1, \sqrt{|G : Z(G)|}\}$.
4. *For every element $x \in G \setminus Z(G)$, the conjugacy class of x is xG'.*
5. *$Z(G/N) = Z(G)/N$ for every normal subgroup N of G such that $G' \nleq N$.*

In Theorem B of [16], they prove.

Theorem 4 *Let G be a group where $|G : Z(G)|$ is a square. Then, G is a VZ-group if and only if every normal subgroup contains G' or is contained in $Z(G)$.*

Also, it is not difficult to observe that if G is a VZ-group, then G is a semi-extraspecial if and only if G is special. With this in mind, we combine the equivalent conditions in Theorems 1, 3 and 4 specialized to special groups, and we obtain the following equivalent conditions for being semi-extraspecial.

Theorem 5 *Let G be a special group. Then, the following are equivalent:*

1. *G is semi-extraspecial.*
2. *For every element $g \in G \setminus G'$, the conjugacy class of g is gG'.*
3. *For every element $g \in G \setminus G'$, we have $|C_G(g)| = |G : G'|$.*
4. *For every element $g \in G \setminus G'$ and every element $z \in Z(G)$, there exists an element $y \in G$ such that $[g, y] = z$.*
5. *Every nonlinear character $\chi \in \mathrm{Irr}(G)$ vanishes on $G \setminus G'$.*

6. $\operatorname{cd}(G) = \{1, \sqrt{|G : G'|}\}$.
7. *For normal subgroup N in G, either $G' \leqslant N$ or $N \leqslant G'$.*

Notice that many of the conditions of this generalize a number of the facts listed about extraspecial groups in Sect. 3.

6 Character Tables

We came to these groups from character theory. Recall that the character table of a group is a $k(G) \times k(G)$ complex-valued matrix, where $k(G)$ is the number of conjugacy classes of G which also equals the number of irreducible characters. A pair of groups is said to have *isomorphic character tables* if the rows and columns of their character tables can be permuted to give identical matrices. In [30], we determined which VZ-groups have isomorphic character tables. The condition we found there can be simplified to the following condition for deciding when two s.e.s. groups have isomorphic character tables.

Theorem 6 *If G and H are s.e.s. groups, then G and H have isomorphic character tables if and only if $|G : G'| = |H : H'|$ and $|G'| = |H'|$.*

If G is a p-group and C is a conjugacy class of G, then $C^p = \{x^p \mid x \in C\}$ is a conjugacy class of G. The map on the conjugacy classes of G given by $C \mapsto C^p$ is called the *power map* of G. Note that G has exponent p if and only if the power map on G maps every conjugacy class to the conjugacy class of the identity. We say that G and H form a *Brauer pair* if G is not isomorphic to H and G and H has the same character table so that the power maps on the corresponding classes match up.

Let G be a p-group. Set $\Omega_1(G) = \langle x \in G \mid x^p = 1\rangle$ and $\mho_1(G) = \langle x^p \mid x \in G\rangle$. In [43], Nenciu determined which pairs of VZ-groups form Brauer pairs. We then used Nenciu's result to determine which pairs of s.e.s. groups yield Brauer pairs (see [31]).

Theorem 7 *Let P and Q be non-isomorphic s.e.s. p-groups for an odd prime p. Then, P and Q form a Brauer pair if and only if $|P : P'| = |Q : Q'|$, $|P'| = |Q'|$, and $|\mho_1(P)| = |\mho_1(Q)|$. In particular, if P and Q have exponent p, $|P : P'| = |Q : Q'|$, and $|P'| = |Q'|$, then P and Q form a Brauer pair.*

Thus, the previous theorem says that s.e.s. groups with exponent p form Brauer pairs if and only if they have the same character tables. The following result shows that this gives many different Brauer pairs. Recall that the Heisenberg groups have exponent p, so in light of the previous two theorems, these quotients have the same character tables if they have the same size, and so, they form Brauer pairs if they have same size. This next theorem appeared in [35].

Theorem 8 *For every odd prime p and every integer $n \geqslant 12$, there is a Heisenberg group of order $p^{5n/4+O(1)}$ that has $p^{n^2/24+O(n)}$ pairwise non-isomorphic quotients of order p^n.*

In particular, these groups are not distinguishable by their character tables, and thus, it seems that character theory does not have much to say about these groups. Before we leave character tables, we want to mention the recent result by Casey Wynn in his dissertation. He studies the super-character theories of semi-extraspecial groups. He shows that determining the super-character theories of a semi-extraspecial group G can be reduced to determining the super-character theories of the subgroups of G/G' and G', both of which are elementary abelian groups. See [53] which is being published as [36].

7 Abelian Subgroups

We now turn to looking at abelian subgroups of these groups. We will see that some of these groups can be distinguished by their abelian subgroups. We begin by stating the following theorem which Verardi proved as Proposition 1.7 and Theorem 1.8 of [50]. This theorem gives an upper bound for the order of abelian subgroups of an s.e.s. group.

Theorem 9 *If G is an s.e.s. p-group with $|G : G'| = p^{2a}$ and $|G'| = p^b$, then the following are true:*

1. *Every abelian subgroup of G has order at most p^{a+b}.*
2. *For every element $g \in G \backslash G'$, we have $p^{b+1} \leqslant |Z(C_G(g))| \leqslant p^{2b}$.*

Recall from Theorem 5 that $|C_G(g)| = p^{2a}$ for all $g \in G \backslash G'$, where we use the notation of Theorem 9. It follows from part (2) that the only way that $C_G(g)$ can be abelian is if $a = b$. Thus, we follow Beisiegel in [4], and we say that G is *ultraspecial* if G is semi-extraspecial and $|G'| = \sqrt{|G : G'|}$. The next theorem of Verardi is Corollary 5.11 of [50].

Theorem 10 *For each odd prime p, the Heisenberg group of degree p^2 is the unique ultraspecial group of order p^6 and exponent p up to isomorphism.*

Using the classification of groups of order p^6, one can see that there are $p + 3$ ultraspecial groups of order p^6. The classification of groups of order p^6 can be found in [44]. Section 2 of that paper has a nice history of the problem of classifying groups of order p^6. Using the classification, one can also see for each prime p that all of the ultraspecial groups of order p^6 are isoclinic (including for $p = 2$).

Next, we consider the intersections of the abelian subgroups that have maximal order. Notice that if A is an abelian subgroup of G, then $AZ(G)$ is also an abelian subgroup. It follows that if A is an abelian subgroup of G of maximal order, then $Z(G) \leqslant A$. When G is an s.e.s. group, we have that $Z(G) = G'$, and we deduce that every abelian subgroup of G having maximal order contains G' which implies that abelian subgroups of maximal order are all normal in G. This next theorem of Verardi's was proved as Theorem 1.9 of [50].

Theorem 11 *If G is an s.e.s. group with $|G : G'| = p^{2a}$ and $|G'| = p^b$ and A and B are distinct abelian subgroups of G of order p^{a+b} with $|A \cap B| > p^b$, then $p^{2b} \leqslant |A \cap B| \leqslant p^a$. In particular, if $b > \frac{a}{2}$, then $A \cap B = G'$.*

In particular, as noted in Corollary 1.10 of [50], this theorem implies for ultraspecial groups of order p^{3a} that distinct abelian subgroups of order p^{2a} intersect in the center.

Next, we consider the question of how many abelian subgroups of maximal possible order an s.e.s. can have. Verardi provides an example of an ultraspecial group of order 3^9 with no abelian subgroups of order 3^6 on pages 148–149 of [50]. Using the computer algebra system Magma [5], we have found 4162 ultraspecial groups of order 2^9 with no abelian subgroups of order 2^6. One of which is SmallGroups(512, 10477021). Also, using Magma, we have found ultraspecial groups of orders 5^9 and 7^9 with no abelian subgroups of order 5^6 and 7^6, respectively.

We believe it is likely that there exist s.e.s groups G with $|G : G'| = p^{2a}$ and $|G'| = p^b$ having no abelian subgroup of order p^{a+b} for every prime p and for all pairs of integers a and b satisfying $a \geqslant b \geqslant 3$. However, at this time, we do not see any way to prove this fact. In fact, we do not see any uniform way of producing these groups. At this point, finding such groups has been a matter of trial and error.

There are many open questions that one can ask about these groups. One obvious question is what is the smallest order possible for maximal abelian subgroups. We note that it is easy to find abelian subgroups of order p^{b+1}, but it seems unlikely that these will be maximal abelian subgroups.

On pages 149–150 of [50], Verardi produces an example of an ultraspecial group of order 7^9 with one abelian subgroup of order 7^6. Using Magma, we have found many examples of ultraspecial groups of order 3^{12} with one abelian subgroup of order 3^8. We also found similar examples for many other orders. In a future paper, we plan to describe how to construct ultraspecial groups of order p^{3a} with one abelian subgroup of order p^{2a} for any prime p and every integer $a \geqslant 3$. We also will show that this construction can be generalized to other s.e.s. groups.

Finally, in Example 3.9 of [50], Verardi gives examples of various s.e.s. groups G with $|G : G'| = p^{2a}$ and $|G'| = p^b$ with at least two abelian subgroups of order p^{a+b} for all primes p and various choices of a and b. Heineken in [19] gave examples of s.e.s. groups with exactly two abelian subgroups of order p^{a+b}. Verardi has shown that the set of possible values for the number of abelian subgroups of order p^{a+b} is very limited when $b > \frac{a}{2}$. The following theorem is a combination of Theorems 3.8 and Corollary 5.9 of [50].

Theorem 12 *Let G be an s.e.s. group with $|G : G'| = p^{2a}$ and $|G'| = p^b$. If $b > \frac{a}{2}$ and G has at least two abelian subgroups of order p^{a+b}, then the number of abelian subgroups of G of order p^{a+b} has the form $1 + p^h$ where h is an integer satisfying $0 \leqslant h \leqslant a$. Furthermore, if G is an ultraspecial group and $h > 0$, then h divides a.*

We will use $h(G)$ to denote the quantity h in Theorem 12. As noted above, Heineken has produced examples where $h = 0$. On the other extreme, the Heisenberg groups have $h = a$. In fact, the following theorem which is Theorem 5.10 in [50] due to Verardi shows that the number of s.e.s. groups with $h = a$ is very limited.

Theorem 13 *If p is an odd prime, G is ultraspecial group of order, $h(G) = a$, and G has exponent p, then G is isomorphic to the Heisenberg group for p of degree a.*

Now, it seems reasonable to ask for ultraspecial groups of order p^{3a} which values are possible for $h(G)$. This seems to depend on p and a. We will see when $p = 3$ and $a = 3$ that the possible values for $h(G)$ are 1 and 3, and when $p = 3$ and $a = 4$, the possible values for $h(G)$ are 0, 2, and 4. On the other hand, for $p = 2$ and $a = 3$, the only possible value for $h(G)$ is 3; for $p = 2$ and $a = 4$, the possible values for $h(G)$ are 0 and 4, and for $p = 2$ and $a = 5$, the possible values for $h(G)$ are 0, 1, and 5 (all that are allowed). Note that the condition that $h(G)$ must divide a is only proved for ultraspecial groups. It is an open question as to whether other values of $h(G)$ can occur for the other s.e.s. groups covered by that theorem. We note that Theorem 12 is a consequence of Corollary 3.6 of [50] which is the following.

Theorem 14 *Let G be an s.e.s. group with $|G : G'| = p^{2a}$ and $|G'| = p^b$ abelian subgroups A and B so that $G = AB$. Then, the number of abelian subgroups C of G having order p^{a+b} so that $C \cap A = G'$ is p^h.*

In Example 3.9 (c) of [50], Verardi gives an example of an s.e.s. group G with $|G : G'| = 3^8$ and $|G'| = 3^2$ with $G = AB$ for abelian subgroups A and B, where $h = 6$ which is greater than $a = 4$. So this shows that the value for h in Theorem 14 does not obey the upper bound given in Theorem 12. A natural open question is to determine what values of h are possible in this case.

8 Semifields

For the rest of this paper, we will focus on ultraspecial groups of order p^{3a} with at least two abelian subgroups of order p^{2a}. We begin by looking at semifields. As we have stated in the introduction, semifields were initially studied under the name finite division algebras. This is the name that Verardi uses in [50]. Since he does not refer to any of the papers on semifield, we feel it is likely that Verardi was not aware of the literature regarding semifields. Also, as we mentioned in the introduction, there are two expository articles regarding semifields [9, 25]. Most of these basic definitions and results can be found there, and we do not offer any more specific reference.

We have already seen ultraspecial groups of order p^{3a} with at least two abelian subgroups of order p^{2a}. Recall that the Heisenberg p-group of degree a is a Sylow p-subgroup of $GL_3(p^a)$. It is known that the upper triangular matrices with 1's on the diagonal form a Sylow p-subgroup of $GL_3(p^a)$. Thus, we may represent the Heisenberg group by

$$\left\{ \begin{bmatrix} 1 & a & c \\ 0 & 1 & b \\ 0 & 0 & 1 \end{bmatrix} \mid a, b, c \in \mathrm{GF}(p^a) \right\},$$

where $\mathrm{GF}(p^a)$ is the finite field of order p^a. It is not difficult to see that this group is isomorphic to the group whose set is $\{(a, b, c) \mid a, b, c \in \mathrm{GF}(p^a)\}$ with multiplication given by

$$(a_1, b_1, c_1) \cdot (a_2, b_2, c_2) = (a_1 + a_2, b_1 + b_2, c_1 + c_2 + a_1 b_2).$$

As we mentioned before, this group is an ultraspecial group. It is not difficult to see that $A_1 = \{(a, 0, c) \mid a, c \in F\}$ and $A_2 = \{(0, b, c) \mid b, c \in F\}$ are abelian subgroups of order p^{2a}.

Following the literature, we say $(F, +, *)$ is a *pre-semifield* if $(F, +)$ is an abelian group with at least two elements whose identity is 0 and $*$ is a multiplication that satisfies the distributive laws and $a * c = 0$ for $a, c \in F$ implies $a = 0$ or $c = 0$. (Note we are *not* assuming $*$ is associative.) We say F is a *semifield* if in addition F has an identity 1. When F is finite, one can show that F has a vector space structure over some finite field. This implies that $|F|$ is a power of a prime p. For the remainder of this paper, all (pre-)semifields will be finite. Note that if $*$ is associative, then F will, in fact, itself be a field. If $|F|$ is p, p^2, or 8, then $*$ must be associative and $(F, +, *)$ will be a field. We say that $(F, +, *)$ is a *proper* semifield if $*$ is nonassociative. It has been shown that there exist proper semifields for every prime power p^n such that $n \geqslant 3$ and $p^n \geqslant 16$.

Let $(F, +, *)$ be a (finite) pre-semifield. Fix the set $G(F) = \{(a, b, c) \mid a, b, c \in F\}$. We define the multiplication $\cdot$ on $G(F)$ by

$$(a_1, b_1, c_1) \cdot (a_2, b_2, c_2) = (a_1 + a_2, b_1 + b_2, c_1 + c_2 + a_1 * b_2).$$

The following theorem has been proved in a number of places. See Lemma 3 in [4], Lemma 2.1 of [20], Proposition 2.2 of [27], the Theorem in [45], and page 139 of [50] when p is odd.

Theorem 15 *If F is a (pre-)semifield, then $G(F)$ is an ultraspecial group of order $|F|^3$ with at least two abelian subgroups of order $|F|^2$.*

As far as we can tell, semifield groups were first studied by Cronheim in Sect. 5 of [10]. Cronheim calls these groups T-groups. Verardi calls these groups B-groups in honor of Beisiegel. Note when F is a finite field, then $G(F)$ is the Heisenberg group for F. Because of this analogy, Knarr and Stroppel call these groups Heisenberg groups, but with wish to reserve the name Heisenberg groups for only the case when F is a field. Finally, Hiranme calls these groups semifield groups, and that is the name that we shall adopt for these groups. In particular, we say that $G(F)$ is the *semifield group* associated with F.

We can identify two of abelian subgroups of $|F|^2$ in $G(F)$. It is not difficult to see that $A_1 = \{(a, 0, c) \mid a, c \in F\}$ and $A_2 = \{(0, b, c) \mid b, c \in F\}$ are abelian subgroups of order $|F|^2$. If p is odd, $G(F)$ has exponent p (see Proposition 2.3 (5) of [27] or page 139 of [50]). For $p = 2$, the elements in $A_1 \cup A_2$ have order 2 and every element of $G(F)$ outside of $A_1 \cup A_2$ has order 4 (see Proposition 2.3 (6) of [27]). The following theorem by Verardi (Proposition 3.1 of [50]) yields the connection between semifields and ultraspecial groups with two abelian subgroups of maximal possible order. Note that Proposition 3.5 of [20] gives a similar result.

Theorem 16 *If G is an ultraspecial p-group for an odd prime p with at least two abelian subgroups of order $|G : G'|$ and G has exponent p, then there is a semifield F so that $G \cong G(F)$.*

When $p = 2$, we know that a non-abelian 2-group cannot have exponent 2, the hypotheses of the previous theorem could not be met with $p = 2$. However, if we only require the two specified abelian subgroups to have exponent 2, then a similar result is obtained. This follows from Proposition 3.5 of [20].

Theorem 17 *If G is an ultraspecial 2-group with two abelian subgroups, each having order $|G : G'|$ and each having exponent 2, then there is a semifield F so that $G \cong G(F)$.*

We note that Hiranime includes a condition that is equivalent to being a semifield group (Lemma 2.2 and Proposition 3.5 of [20]).

Theorem 18 *A p-group G of order p^{3a} is a semifield group if and only if there exist elementary abelian groups A and B of order p^{2a} so that $a'b' = b'a'$ implies $a' \in A \cap B$ or $b' \in A \cap B$ for all $a' \in A$ and $b' \in B$.*

Using the Universal Coefficients Theorem (see Chap. 5 of [52]), one can show that every s.e.s. p-group is isoclinic to a unique s.e.s p-group of the same order with exponent p when p is odd. Thus, when p is odd, one can determine all of the ultraspecial groups G with at least two abelian subgroups of order $|G : G'|$ by determining all the semifield groups and then determining all the possible isoclinisms for each semifield group.

One can also use the Universal Coefficients Theorem to see that every ultraspecial 2-groups with at least two abelian subgroups of order $|G : G'|$ is isoclinic to an ultraspecial group with two abelian subgroups of order $|G : G'|$ and exponent 2. If this is true, again we can find all of the ultraspecial 2-groups having at least two abelian subgroups of order $|G : G'|$ by computing the semifield groups and then determining all possible isoclinisms. We note that for ultraspecial 2-groups that do not have at least two abelian subgroups of order $|G : G'|$ that it is not so clear how one should determine representatives from each isoclinism class.

9 More on Semifield Groups

We have seen that every ultraspecial group G with at least two abelian subgroups is isoclinic to a unique (up to isomorphism) semifield group for some semifield. We now want to determine when two semifield groups are isomorphic.

We say two (pre-)semifields $(F_1, +, *_1)$ and $(F_2, +, *_2)$ are *isotopic* if there exist additive isomorphisms $\alpha, \beta, \gamma : F_1 \to F_2$ that satisfy

$$\gamma(a *_1 b) = \alpha(a) *_2 \beta(b), \quad \forall a, b \in F_1.$$

One can show that isotopism is an equivalence relation on (pre-)semifields. It is well known in the semifield literature that every pre-semifield is isotopic to some semifields. We note that it is usually easier to describe a pre-semifield for each isotopism class of semifields; so often it is pre-semifields that are described. For example, Albert's "twisted semifields" are defined as follows. Let F be a finite field. For an element $j \in F$ and nontrivial automorphisms α, β define $*$ by $x * y = xy + j\alpha(x)\beta(y)$ (see (3.3) of [25]). Then, $(F, +, *)$ will be a pre-semifield.

We now begin to see how isotopism of the semifields yields isomorphic semifield groups. The following result has been proved by a number of authors (see Lemma 2.4 of [20], Proposition 3.2 (1) of [27], Proposition A of [45], and Lemma 3.13 of [50]).

Theorem 19 *If F_1 and F_2 are isotopic (pre-)semifields, then $G(F_1)$ and $G(F_2)$ are isomorphic groups.*

If $|F|$ is p, p^2, or 8, then F must be isotopic to the field of that order. (This has been proved by Menichetti.) Thus, for these orders, there is only one isotopism class of (pre)-semifield. Equivalently, this says that there is a unique semifield group of order 8^3. On the other hand, it is known for every prime power p^a with either p odd and $a \geqslant 3$ or $p = 2$ and $a \geqslant 4$ that there exists some semifield that is *not* isotopic to the field of that order. In Proposition B of [45], Rocco and Rocha prove the following:

Theorem 20 *Let G be the Heisenberg group of degree a for the prime p. If F is a proper semifield of order p^a, then $G(F)$ is not isomorphic to G.*

Note that Theorem 20 implies that for every prime p and integer a satisfying either p is odd and $a \geqslant 3$ or $p = 2$ and $a \geqslant 4$, then there exists a semifield group of order p^{3a} that is not isomorphic to the Heisenberg group.

If F is a (pre)-semifield, then we define F^{op} by $a *^{\text{op}} b = b * a$. Obviously, if F is commutative, then $F^{\text{op}} = F$. It is possible to have F isotopic to F^{op} when F is not isotopic to a commutative semifield. Both Hiranime (in Theorem 5.1 of [20]) and Knarr and Stroppel (Proposition 3.2 (b) of [27]) have proved the following:

Theorem 21 *If F is a semifield, then $G(F) \cong G(F^{\text{op}})$.*

We say that $(F_1, +, *_1)$ and $(F_2, +, *_2)$ are *anti-isotopic* if there exist (additive) isomorphisms $\alpha, \beta, \gamma : F_1 \to F_2$ such that

$$\gamma(a *_1 b) = \beta(b) *_2 \alpha(a), \quad \forall a, b \in F_1.$$

It is not difficult to see that F_1 and F_2 are anti-isotopic if and only if F_1 and F_2^{op} are isotopic. We have seen that if F_1 and F_2 are either isotopic or anti-isotopic, then $G(F_1)$ and $G(F_2)$ are isomorphic. In fact, these are the only times that these groups can be isomorphic. This is the content of the next theorem. This is proved by Hiranime as Theorem 5.1 of [20] and Knarr and Stroppel as Theorem 6.6 of [27].

Theorem 22 *If F_1 and F_2 are (pre-)semifields, then $G(F_1)$ and $G(F_2)$ are isomorphic if and only if F_1 and F_2 are isotopic or anti-isotopic.*

We can define an equivalence relation on semifields by saying that two semifields are equivalent if they are isotopic or anti-isotopic. Theorem 22 shows that the isomorphism classes of the semifield groups are in bijection to the equivalence classes of this equivalence relation. To emphasize this point, the number of semifield groups of order p^{3a} equals the number of unique semifields up to isotopism or anti-isotopism.

As we have stated above, the field of order 8 is the only semifield of order 8, so the Heisenberg group is the unique semifield group of order $2^9 = 512$. Kleinfeld in [26] and Knuth in [28] computed the semifields of order 16. There are two isotopism classes of proper semifields of order 16 and the two classes are not anti-isotopic. Thus, there exist three semifield groups of order 2^{12} up to isomorphism. Knuth [28] and Walker [51] determined the semifields of order 32. There are six isotopism classes of semifields of order 32, and of these, two pairs are anti-isotopic. Thus, there exist four semifield groups of order 2^{15}. Finally, using a computer search, Rúa et al. [46] found all the semifields of order 64. They have determined that there are 332 isotopism classes of semifields of order 64. The number of equivalence classes under isotopsim and anti-isotopism is 184, so the number of semifield groups of order 2^{18} is 184.

The semifields of order 27 were initially computed by Dickson [13]. It is known that all of the proper semifields of order 27 are isotopic, so there are two semifield groups of order 3^9. The semifields of order 81 were computed by Dempwolff [12] using the computer. There are 27 isotopism classes of semifields of order 81 and once anti-isotopisms are included, there are 19 equivalence classes under isotopism and anti-isotopism. It follows that there are 19 semifield groups of order 3^{12}. Rúa et al. [48] have also used the computer to find the semifields of order 243. There are 23 isotopism classes of semifields of order 243 and accounting for both isotopism and anti-isotopism, there are 15 equivalence classes; so the number of semifield groups of order 3^{15} is 15.

The semifields of order 5^3 were also computed by Dickson [13] with mistakes corrected by Long [37]. There are four isotopism classes of semifields of order 5^3. It is possible that there is one pair of anti-isotopic semifields that are not isotopic. Thus, we have not determined whether there are three or four semifield groups of order 5^9. In [7], they give enough information to compute the semifields of order 5^4; however, they have not computed the isotopism classes of the semifields of order 5^4. The general case of semifields of order p^3 was initially studied by Albert in [1]. Menchetti proved a conjecture of Kaplansky regarding the full classification of these semifields in [42]. Using this classification, we should be able to determine the semifield groups of order p^9. Finally, we mention that the semifields of order 7^4 have also been classified [8], so we should also be able to determine the semifield groups of order 7^{12}. We summarize these results in Table 1.

Table 1 Number of semifield groups

$\|F\|$	# isotopism classes	$\|G(F)\|$	# $G(F)$	# commutative semifields
2^3	1	2^9	1	1
2^4	3	2^{12}	3	1
2^5	6	2^{15}	4	2
2^6	332	2^{18}	184	2
2^7	?	2^{21}	?	2
3^3	2	3^9	2	2
3^4	27	3^{12}	19	2
3^5	23	3^{15}	15	7
5^3	4	5^9	?	2
5^4	?	5^{12}	?	?
7^3	?	7^9	?	?
7^4	356	7^{12}	227	2

In this table, we consider possible values of $|F|$. For each value of $|F|$, we list the number of isotopism classes of semifields of order $|F|$. We list $|G(F)|$ (which is $|F|^3$). We list the number of semifield groups of order $|G(F)|$ which equals the number of equivalence classes under isotopism, anti-isotopism, and the number of isotopism classes that contain a commutative semifields which equals the number of semifield groups having more than two abelian subgroups of order $|F|^2$.

10 Commutative Semifields and Seminuclei

We now see that if $G(F)$ is a semifield group, then the number of abelian subgroups of order $|F|^2$ can be determined by the structure of F. This next result is proved by Verardi (Theorem 3.14 of [50]), Hiranime (Proposition 4.2 (i) of [20]), and Knarr and Stroppel (Lemma 4.3 of [27]).

Theorem 23 *Let F be a semifield. Then, $G(F)$ has more than two abelian subgroups of order $|F|^2$ if and only if F is isotopic to a commutative semifield.*

In particular, F is not isotopic to a commutative semifield if and only if $G(F)$ has exactly two abelian subgroups of order $|F|^2$. We summarize the number of isotopism classes of commutative semifields in Table 1. For the most part, the references for the results in that table are the same as the references for the last couple paragraphs of the previous section, although we also need to refer to [47].

When F is a commutative semifield, we can determine the number of abelian subgroups of $G(F)$ of order $|F|^2$. We define

$$\mathrm{Mid}(F) = \{z \in F \mid x * (z * y) = (x * z) * y, \forall x, y \in F\},$$

which is called the *middle seminucleus* of F. When F is finite, $\mathrm{Mid}(F)$ is a field. The following theorem is proved by Verardi (Corollary 5.9 of [50]), Hiranime (Proposition 4.3 (ii) of [20]), and Knarr and Stroppel (Lemma 4.3 of [27]).

Theorem 24 *If F is a commutative semifield, then $|\mathrm{Mid}(F)| = p^h$, where $1 + p^h$ is the number of abelian subgroups of order $|F|^2$ in $G(F)$.*

It is not difficult to see that F is a vector space over $\mathrm{Mid}(F)$. This implies when $|F| = p^a$ that h divides a where $|\mathrm{Mid}(F)| = p^h$. Recall that we knew from before that $h \leqslant a$. We note that the Heisenberg groups are the only semifield groups of order 2^9 and 2^{12} that have more than two abelian subgroups of the maximal possible order. For 2^{15}, 2^{18}, 2^{21}, 3^9, 3^{12}, 5^9, and 7^{12} there is for each order, one semifield group other than the Heisenberg group that has more than two abelian subgroups maximal possible order. For 2^{15} and 2^{21}, we see from Theorem 12 that the semifield group other than the Heisenberg group with more than two abelian subgroups of maximal order will have exactly three abelian subgroups of maximal order. The non-Heisenberg group of order 2^{18} having more than two abelian subgroups of maximal order will have five abelian subgroups of maximal order.

In a similar vein, the non-Heisenberg group of order 3^9 and 3^{15} having more than two abelian subgroups of maximal order have four abelian subgroups of maximal order. The non-Heisenberg group of order 3^{12} with more than two abelian subgroups of maximal order has ten abelian subgroups of maximal order. Finally, the non-Heisenberg group of 5^9 with more than two abelian subgroups of maximal order has six abelian subgroups of maximal order.

Define

$$\mathrm{R}(F) = \{z \in F \mid (x * y) * z = x * (y * z), \forall x, y \in F\}.$$

This is the *right seminucleus* of F. When F is finite, $R(F)$ is also a field, and F is a vector space over $\mathrm{R}(F)$. The following result is due to Verardi (Theorem 5.12 of [50]).

Theorem 25 *Suppose F is a commutative semifield of order p^a, p is an odd prime, and $|\mathrm{R}(F)| = p^r$. For every element $v \in G(F) \backslash G(F)'$, set $Z(v) = Z(C_G(v))$.*

(a) *If $|Z(v)| = p^{a+k}$, then r divides k.*
(b) *If $v_1, \ldots, v_m$ in $G(F) \backslash G(F)'$ are chosen so that*

$$G(F) \backslash G(F)' = \bigcup_{i=1}^{m} (Z(v_i) \backslash G(F)'),$$

then $m \equiv 1 \pmod{p^r}$.

Note that the left seminucleus can be defined analogously, and that a similar theorem will hold with the left seminucleus replacing the right seminucleus. This shows that some information regarding the group structure of $G(F)$ can be obtained from the semifield structure of F when F is commutative. It seems likely to us that there are other results regarding the structure of semifields that can be applied to obtain group structure of $G(F)$. We note that there has been much more work studying commutative semifield than general semifields in the literature, and we expect that many of the results regarding commutative semifields can be applied to semifield groups. Finally, we would like to see results relating the semifield structure of F with the group structure of $G(F)$ when F is not commutative.

References

1. Albert, A.A.: On nonassociative division algebras. Trans. Am. Math. Soc. **72**, 296–309 (1952)
2. Albert, A.A.: Finite noncommutative division algebras. Proc. Am. Math. Soc. **9**, 928–932 (1958)
3. Albert, A.A.: Isotopy for generalized twisted fields. An. Acad. Brasil. Ci. **33**, 265–275 (1961)
4. Beisiegel, B.: Semi-extraspezielle p-Gruppen. Math. Z. **156**, 247–254 (1977)
5. Bosma, W., Cannon, J., Playoust, C.: The Magma algebra system I: the user language. J. Symbolic Comput. **24**, 235–265 (1997)
6. Chillag, D., MacDonald, I.D.: Generalized Frobenius groups. Israel J. Math. **47**, 111–122 (1984)
7. Combarro, E.F., Rúa, I.F., Ranilla, J.: New advances in the computational exploration of semi-fields. Int. J. Comput. Math. **88**, 1990–2000 (2011)
8. Combarro, E.F., Rúa, I.F., Ranilla, J.: Finite semifields with 7^4 elements. Int. J. Comput. Math. **89**, 1865–1878 (2012)
9. Cordero, M., Wene, G.P.: A survey of finite semifields. Discrete Math. **208**(209), 125–137 (1999)
10. Cronheim, A.: T-groups and their geometry. Illinois J. Math. **9**, 1–30 (1965)
11. Dark, R., Scoppola, C.M.: On Camina groups of prime power order. J. Algebra **181**, 787–802 (1996)
12. Dempwolff, U.: Semifield planes of order 81. J. Geom. **89**, 1–16 (2008)
13. Dickson, L.E.: Linear algebras in which division is always uniquely possible. Trans. Am. Math. Soc. **7**, 370–390 (1906)
14. Dickson, L.E.: On commutative linear algebras in which division is always uniquely possible. Trans. Am. Math. Soc. **7**, 514–522 (1906)
15. Dolfi, S., Moretó, A., Navarro, G.: The groups with exactly one class of size a multiple of p. J. Group Theory **12**, 219–234 (2009)
16. Fernández-Alcober, G.A., Moretó, A.: Groups with two extreme character degrees and their normal subgroups. Trans. Am. Math. Soc. **353**, 2171–2192 (2001)
17. Goldstein, D., Guralnick, R.M., Lewis, M.L., Moretó, A., Navarro, G., Tiep, P.H.: Groups with exactly one irreducible character of degree divisible by p. Algebra Number Theory **8**, 397–428 (2014)
18. Hall, P.: The classification of prime-power groups. J. Reine Angew. Math. **182**, 130–141 (1940)
19. Heineken, H.: Nilpotente Gruppen, deren sämtliche Normalteiler charakteristisch sind, Arch. Math. (Basel) **33**(1979/80), 497–503
20. Hiramine, Y.: Automorphisms of p-groups of semifield type. Osaka J. Math. **20**, 735–746 (1983)
21. Huppert, B.: Endliche Gruppen I. Springer, Berlin, New York (1967)
22. Huppert, B.: Character Theory of Finite Groups. Walter de Gruyter & Co., Berlin (1998)
23. Isaacs, I.M.: Finite Group Theory. American Mathematical Society, Providence, RI (2008)

24. Isaacs, I.M., Lewis, M.L.: Camina p-groups that are generalized Frobenius complements. Arch. Math. (Basel) **104**, 401–405 (2015)
25. Kantor, W.M.: Finite semifields in finite geometries, groups, and computation. Walter de Gruyter GmbH & Co. KG, Berlin, pp. 103–114 (2006)
26. Kleinfeld, E.: Techniques for enumerating Veblen-Wedderburn systems. J. Assoc. Comput. Mach. **7**, 330–337 (1960)
27. Knarr, N., Stroppel, M.J.: Heisenberg groups, semifields, and translation planes. Beitr. Algebra Geom. **56**, 115–127 (2015)
28. Knuth, D.E.: Finite semifields and projective planes. J. Algebra **2**, 182–217 (1965)
29. Knuth, D.E.: A class of projective planes. Trans. Am. Math. Soc. **115**, 541–549 (1965)
30. Lewis, M.L.: Character tables of groups where all nonlinear irreducible characters vanish off the center in Ischia group theory 2008, 174–182. World Scientific Publishing, Hackensack, NJ (2009)
31. Lewis, M.L.: Brauer pairs of Camina p-groups of nilpotence class 2. Arch. Math. (Basel) **92**, 95–98 (2009)
32. Lewis, M.L.: Classifying Camina groups: a theorem of Dark and Scoppola. Rocky Mountain J. Math. **44**, 591–597 (2014). Erratum on "Classifying Camina groups: a theorem of Dark and Scoppola" [MR3240515], Rocky Mountain J. Math. **45**, 273 (2015)
33. Lewis, M.L.: Centralizers of Camina groups with nilpotence class 3. J. Group Theory
34. Lewis, M.L., Moretó, A., Wolf, T.R.: Non-divisibility among character degrees. J. Group Theory **8**, 561–588 (2005)
35. Lewis, M.L., Wilson, J.B.: Isomorphism in expanding families of indistinguishable groups. Groups Complex. Cryptol. **4**, 73–110 (2012)
36. Lewis, M.L., Wynn, C.: Supercharacter theories of semiextraspecial p-groups and Frobenius groups (Submitted)
37. Long, F.W.: Corrections to Dicksons table of three dimensional division algebras over F_5. Math. Comp. **31**, 1031–1033 (1977)
38. Macdonald, I.D.: Some p-groups of Frobenius and extraspecial type. Israel J. Math. **40**, 350–364 (1981)
39. Macdonald, I.D.: More on p-groups of Frobenius type. Israel J. Math. **56**, 335–344 (1986)
40. Mann, A.: Some finite groups with large conjugacy classes. Israel J. Math. **71**, 55–63 (1990)
41. Mann, A., Scoppola, C.M.: On p-groups of Frobenius type. Arch. der Math. **56**, 320–332 (1991)
42. Menichetti, G.: On a Kaplansky conjecture concerning three-dimensional division algebras over a finite field. J. Algebra **47**, 400–410 (1977)
43. Nenciu, A.: Brauer pairs of VZ-groups. J. Algebra Appl. **7**, 663–670 (2008)
44. Newman, M.F., O'Brien, E.A., Vaughan-Lee, M.R.: Groups and nilpotent Lie rings whose order is the sixth power of a prime. J. Algebra **278**, 383–401 (2004)
45. Rocco, N.R., Rocha, J.S.: A note on finite semifields and certain p-groups of class 2. Discrete Math. **275**, 355–362 (2004)
46. Rúa, I.F., Combarro, E.F., Ranilla, J.: Classification of semifields of order 64. J. Algebra **322**, 4011–4029 (2009)
47. Rúa, I.F., Combarro, E.F.: Commutative semifields of order 3^5. Commun. Algebra **40**, 988–996 (2012)
48. Rúa, I.F., Combarro, E.F., Ranilla, J.: Determination of division algebras with 243 elements. Finite Fields Appl. **18**, 1148–1155 (2012)
49. van der Waall, R.W., Kuisch, E.B.: Homogeneous character induction II. J. Algebra **170**, 584–595 (1994)
50. Verardi, L.: Gruppi semiextraseciali di esponente p. Ann. Mat. Pura Appl. **148**, 131–171 (1987)
51. Walker, R.J.: Determination of division algebras with 32 elements. Proc. Sympos. Appl. Math. **15**, 83–85 (1963)
52. Warfield Jr. R.B.: Nilpotent groups. In: Lecture Notes in Mathematics, vol. 513. Springer, Berlin, New York (1976)
53. Wynn, C.W.: Supercharacter theories of Camina pairs. Ph.D. Dissertation, Kent State University (2017)

The 3-Modular Character Table of the Automorphism Group of the Sporadic Simple O'Nan Group

Klaus Lux and Alexander Ryba

Abstract We compute the 3-modular character table of the group O′N.2. Much of the table is deduced character theoretically from the known 3-modular character table of the sporadic simple O'Nan group O′N. We finish the remaining questions module theoretically with an application of condensation.

Keywords Sporadic simple group · Automorphism group · Modular character · ATLAS project

Math subject classification 20-04 · 20C20 · 20C34 · 20C40

1 Introduction

This paper is a contribution to the modular ATLAS project, whose goal is to compute the modular character tables of all the groups given in the Atlas of finite groups, see [23]. The 3-modular character table of O′N.2, the automorphism group of the O′N group, is the second remaining open problem, see http://www.math.rwth-aachen.de/~MOC/work.html.

The O'Nan group O′N of order $2^9 \cdot 3^4 \cdot 5 \cdot 7^3 \cdot 11 \cdot 19 \cdot 31$ was discovered by M. O'Nan in 1976, see [14]. It was constructed by C. Sims as a rank five permutation group acting on 122760 points with point stabilizer $L_3(7).2$. The existence of the outer automorphism was proved by Andrilli in [1]. Other constructions of the simple group and its automorphism group are available in [18, 21].

K. Lux (✉)
Department of Mathematics, University of Arizona, 617 Santa Rita Rd., Tucson, AZ 85721, USA
e-mail: klux@math.arizona.edu

A. Ryba
Department of Computer Science, Queens College/CUNY, 65-30 Kissena Boulevard, Queens, NY 11367, USA
e-mail: Alexander.Ryba@qc.cuny.edu

J. Feldvoss et al. (eds.), *Advances in Algebra*, Springer Proceedings in Mathematics & Statistics 277, https://doi.org/10.1007/978-3-030-11521-0_12

A summary of the known modular character tables: We begin with tables for the triple cover $3 \cdot O'N$ of the simple group O′N. For the primes 5, 11, 19, and 31, [5] gave modular tables in form of the Brauer trees. However, some of these trees are only determined up to the positions of certain ordinary, algebraically conjugate characters. These ambiguities were resolved in the diploma thesis of M. Ottensmann, see [15]. The 7-modular table of $3 \cdot O'N$ was determined by Hiss, Henke, and Müller in [4]. Finally, the 2-modular and 3-modular character tables for $3 \cdot O'N$ were computed by Wilson and Jansen in [9].

For the group $3 \cdot O'N.2$, the modular character tables for the primes not equal to 3 are subdivided into the modular characters inflated from O′N.2 and the faithful modular characters. Since the faithful characters are all induced irreducible modular characters from $3 \cdot O'N$, they can be computed directly from the information given in the Atlas, [3]. So, we can focus on the modular tables of O′N.2. The Brauer trees for the factor group O′N.2 for the prime 31 were determined by Müller [13], and those for the primes 11 and 19 only require straightforward extensions of the principal block of O′N, so are unproblematic. In an appendix, we give an argument that determines the Brauer trees for O′N.2 for the prime 5.

The 7-modular character table of O′N.2 was computed in [4], and the 2-modular table is easy to derive from the 2-modular character table of O′N. However, as remarked in the paper by Wilson and Jansen, see [9], the 3-modular table of O′N.2 was left to be determined. The statement in [4] that this table is already known was made in error. We note that the Modular Atlas [23] does give (without reference) a partial modular character table in which many of the values of the irreducible Brauer characters of the two largest degrees on the outer conjugacy classes are left open. We reduce to four possibilities for these Brauer characters with purely character theoretic arguments and choose between the four options module theoretically.

We will make considerable use of the two irreducible representations of O′N.2 of dimension 154 over $\mathbb{F}_9$, the field with nine elements, which were constructed by Jansen and Wilson, see [8]. One of them, which we call 154^+, is stored in the Atlas of finite group representations, [25]. For our calculations that apply the computer algebra system GAP, [22], we access 154^+ from the database of the GAP package `AtlasRep`, [24].

2 The Characters and 3-Modular Blocks of O′N and O′N.2

The 3-modular character table of O′N was first computed in [9] and is available in the GAP character table library, see [2, 22]. The numberings we attach to blocks, ordinary, and modular irreducibles of O′N are those provided by GAP. Throughout the paper, we use χ (resp. φ) to denote irreducible ordinary (resp. modular) characters of O′N. We attach a subscript to indicate a position within the GAP character table. Thus, χ_2 (resp. φ_2) denotes the second ordinary (resp. modular) irreducible in the GAP table for O′N.

2.1 *The Blocks of* O′N

There are two 3-modular blocks of positive defect for O′N: The principal block of defect 4 with 14 irreducible modular characters, and a block of defect 2 with 5 irreducible modular characters, whose defect group is described in [10].

There are six blocks of defect zero, for which we summarize the information in the following table.

Ordinary character	χ_{21}	χ_{22}	χ_{25}	χ_{26}	χ_{27}	χ_{28}
Modular character	φ_{20}	φ_{21}	φ_{22}	φ_{23}	φ_{24}	φ_{25}
	169290_1	169290_2	175770	207360_1	207360_2	207360_3

It is often convenient to refer to characters by degree, as in the third line of the table, in this case a subscript indicates position among several GAP characters with a given degree.

2.2 *The Blocks of* O′N.2

An irreducible (ordinary or modular) character ψ of O′N either extends to two irreducible characters that we write as $\psi^{\pm}$ or gives an induced irreducible, written $\widehat{\psi}$, of O′N.2. We need only determine all the modular irreducible characters of the form $\varphi^{\pm}$ because the induced irreducibles vanish at outer classes of O′N.2 and are therefore already completely known.

The ordinary, irreducible characters of O′N.2 can be assigned to blocks by computing their central characters and reducing mod 3, see for example [11, Theorem 4.4.8]. This shows that O′N.2 has two blocks with positive defect (covering the two such blocks of O′N).

The modular irreducible characters in blocks of defect 0 for O′N.2 are read from the ordinary character table of O′N.2 and are shown in the following Table 1.

We complete a description of the 3-modular character table of O′N.2 by obtaining the decomposition matrices for the two blocks of positive defect in the remaining sections of this paper.

Table 1 The blocks of O′N.2 with defect 0

$\varphi_{20}^{\pm}$	$\varphi_{21}^{\pm}$	$\varphi_{22}^{\pm}$	$\varphi_{23}^{\pm}$	$\varphi_{24}^{\pm}$	$\varphi_{25}^{\pm}$
$169290_1^{\pm}$	$169290_2^{\pm}$	$175770^{\pm}$	$207360_1^{\pm}$	$207360_2^{\pm}$	$207360_3^{\pm}$

2.3 The Ordinary Irreducible Characters of O′N.2

We now specify the assignment of ± labels to an ordinary irreducible χ of O′N.2 that extends an irreducible of O′N. If χ takes a nonzero integral value at the outer class $2b$, then its sign is the sign of this value. This rule, covers all but 6 character extension pairs, one of which $\chi_{21}^{\pm}$ consists of characters of defect 0, which do not matter for our work. To settle the remaining five cases, we define χ_7^+, χ_{11}^+, χ_{20}^+, χ_{29}^+ and χ_{30}^+ to have values $8r2$ at class $8c$, $r7$ at class $56a$, $r5$ at class $10b$, $4r2$ at class $8c$, and $4r2$ at class $8c$, respectively. We remark that the choice of signs described here matches that used in the Atlas [3].

2.4 The Irreducible Modular Character $\varphi_2^+ = 154^+$

The character of the representation 154^+ is one of the two extensions of the irreducible modular character φ_2 of O′N. We choose this character as our definition of the extension φ_2^+. The Brauer character values of φ_2^+ on the inner 3-regular classes are as in [9]. On outer 3-regular classes, we compute them from the representation 154^+ with the GAP-function `BrauerCharacterValue`. This function implements the standard Brauer lifting as described in [7, Introduction, Sects. 2–4]. We obtain the following values on outer classes:

$2b$	$8b$	$8c$	$8d$	$10b$	$10c$	$14b$	$22a$
0	$-4r2$	$4r2$	0	$r5$	$-r5$	0	0

and

$38abc$	$56a$	$56b$	$56c$	$56d$
0	$k56*15$	$k56*5$	$k56$	$k56*3$

We remark that the names of outer conjugacy classes used here are compatible with those in GAP for the ordinary character table of O′N.2 and the 3-modular character table of O′N. There is a free choice of the classes $10b$ and $8b$ from the sets $\{10b, 10c\}$ and $\{8b, 8c\}$—this is because the ordinary character table of O′N.2 has automorphisms $(16a, 16b)(8b, 8c)(24a, 24b)(56a, 56b)(56c, 56d)$ and $(15a, 15b)(10a, 10b)(30a, 30b)$ whose restrictions to inner classes give automorphisms of the 3-modular character table of O′N. The class $56a$ is then determined since it is the only class that squares to an (inner) class whose character value on 154^+ is $-r7$ (rather than $r7$) and has class $8b$ for its 7th power. The classes $56b$, $56d$, and $56c$ are then obtained as the 3rd, 5th, and 15th powers of $56a$, respectively. The class names used above meet these requirements.

On 154^+ the traces of outer classes over $\mathbb{F}_9$, computed with the GAP function `FrobeniusCharacterValue`, are as follows:

2b	6b	8b	8c	8d	10b	10c	14b	22a	24a	24b
0	0	f^2	f^6	0	f^2	f^6	0	0	f^2	f^6

30a	30b	38abc	56a	56b	56c	56d
f^2	f^6	0	f^2	f^6	0	0

here f denotes the primitive element of $\mathbb{F}_9$ as given in GAP.

We observe that apart from the elements of order 38, the conjugacy class of an outer element of O′N.2 is determined by the order, the character value and the trace. However, the only ordinary characters not constant on the conjugacy classes of order 38 are the characters $\chi_{36}, \ldots, \chi_{41}$ (of degree 207360) of defect 0. In other words, the precise conjugacy class of an element of order 38 is irrelevant for the Brauer characters in the two blocks with positive defect and can, therefore, be ignored in our computations. For any other outer element of O′N.2, we may recover its class by evaluating it in the representation 154^+ and comparing the Brauer character and trace with entries in the tables given above.

3 The Block of Defect 2

The block of defect 2 for O′N involves five modular irreducibles: $\varphi_7 = 5643_1$, $\varphi_8 = 5643_2$, $\varphi_9 = 5643_3$, $\varphi_{13} = 10944$ and $\varphi_{18} = 41724$. The decomposition matrix for the block, as given in [9] and also in [23] is:

Block 2:	φ_7	φ_8	φ_9	φ_{13}	φ_{18}
$10944_1 = \chi_2$	.	.	.	1	.
$52668_1 = \chi_{11}$	.	.	.	1	1
$58311_1 = \chi_{12}$	1	.	.	1	1
$58311_2 = \chi_{13}$	.	.	1	1	1
$58311_3 = \chi_{14}$	.	1	.	1	1
$58653_1 = \chi_{15}$	1	1	1	.	1

In its action on characters in the block of defect 2 for O′N, the outer automorphism fixes $\varphi_7, \varphi_{13}, \varphi_{18}$ and fuses φ_8 and φ_9. Therefore, Clifford's theorem shows that there is one irreducible character that is induced and three pairs of extensions in the covering block of O′N.2, as summarized in the following table.

$\varphi_7^{\pm}$	$\widehat{\varphi_8}$	$\varphi_{13}^{\pm}$	$\varphi_{18}^{\pm}$
$(5643_1)^{\pm}$	11286	$10944^{\pm}$	$41724^{\pm}$

We shall show that the covering block for O′N.2 is described by Table 2 that contains a collapsed form of the decomposition matrix. Later we give an analogous collapsed decomposition matrix for the principal block. Each pair of extensions of an irreducible character of O′N belong to a single row (or column) of the table, according as the character is ordinary (or modular). Other rows (resp. columns) correspond to induced ordinary (resp. modular) irreducibles. The notation used for an entry in the table depends on the status (as induced or extended) for both its row and column.

If an irreducible modular character $\widehat{\varphi}$ of O′N.2 is induced from a pair of irreducibles, φ_a and φ_b of O′N, then its column contains decomposition numbers of O′N.2 that are immediately available from the decomposition matrix of O′N. In case a row corresponds to an ordinary irreducible character $\widehat{\chi}$ induced from χ, the entry $d_{\widehat{\chi},\widehat{\varphi}}$ may be calculated as $d_{\chi,\varphi_a} + d_{\chi,\varphi_b}$. Alternatively, in a row of the collapsed decomposition matrix that corresponds to a pair of extensions $\chi^{\pm}$ of χ, the entry gives the decomposition numbers $d_{\chi^+,\widehat{\varphi}} = d_{\chi^-,\widehat{\varphi}}$ that are calculated as $d_{\chi,\varphi_a} = d_{\chi,\varphi_b}$.

We now consider decomposition numbers that involve two irreducible modular characters $\varphi^{\pm}$ of O′N.2 that extend a character φ of O′N. For an induced ordinary irreducible character $\widehat{\chi}$, we have $d_{\widehat{\chi},\varphi^+} = d_{\widehat{\chi},\varphi^-} = d_{\chi,\varphi}$. Again the value is obtained directly from the decomposition matrix of O′N and we write it as the corresponding entry in our collapsed table.

Finally, in the case of an entry that corresponds to a pair of modular irreducibles $\varphi^{\pm}$ and a pair of ordinary irreducibles $\chi^{\pm}$, we must consider two decomposition numbers $n = d_{\chi^+,\varphi^+} = d_{\chi^-,\varphi^-}$ and $m = d_{\chi^+,\varphi^-} = d_{\chi^-,\varphi^+}$. The corresponding entry of the collapsed matrix is written as n/m. We abbreviate entries 1/0, 0/1, and n/n as $+$, $-$, and $n\pm$. Although n and m are not immediately available from the decomposition matrix of O′N, they do satisfy the relation $n + m = d_{\chi,\varphi}$.

Table 2 Collapsed decomposition matrix for the block with defect 2

Block Defect 2	$\varphi_7^{\pm}$	$\widehat{\varphi_8}$	$\varphi_{13}^{\pm}$	$\varphi_{18}^{\pm}$
	5643	11286	10944	41724
$10944 = \chi_2^{\pm}$	.	.	+	.
$52668 = \chi_{11}^{\pm}$	.	.	−	+
$58311 = \chi_{12}^{\pm}$	+	.	+	+
$116622 = \widehat{\chi_{13}}$	.	1	1	1
$58653 = \chi_{15}^{\pm}$	+	1	.	+

As usual, an entry marked by a dot indicates that no associated modular irreducible occurs in the decomposition of an associated ordinary irreducible. We remark that any integer or dot in our collapsed decomposition matrices is obtained straightforwardly from the decomposition matrix of O′N. We need only supply an argument for other entries.

The table contains information sufficient to write all character values of the 3-modular irreducibles in the block. We may, therefore, view it as including a specification of the signs of those modular irreducibles of O′N.2 that extend irreducibles of O′N.

We shall justify the unknown entries of the table (which in this case are marked as $+$ or $-$) by showing how to compute all of the irreducible modular characters in the block. Of course, knowledge of an irreducible modular character does not pin down any entries in its column, but the complete set allows us to fill in the table, row by row.

Characters $\widehat{\varphi_8}, \varphi_{13}^{\pm}$: As we have already noted, the irreducible $\widehat{\varphi_8}$ presents no problem: its values on outer classes of O′N.2 are all 0. The characters φ_{13}^{+} and φ_{13}^{-} are also straightforward to write down. The decomposition matrix for O′N just has a single entry of 1 in the row for χ_2, which shows that $\varphi_{13}^{\pm}$ are restrictions of the ordinary characters $\chi_2^{\pm}$, to the 3-regular classes. The $+$ in row $\chi_2^{\pm}$ of the collapsed matrix defines which of $\varphi_{13}^{\pm}$ is φ_{13}^{+}.

Characters $\varphi_7^{\pm}$: The antisymmetric square of the character φ_2 of O′N decomposes as $\varphi_{10} + \varphi_7$, a sum of irreducibles from different blocks. It follows that one of the extensions of φ_7 to O′N.2 gives the component of the antisymmetric square of φ_2^{+} that lies in the block of defect 2. However, the character φ_2^{+} is available from Sect. 2 and therefore the required component of its antisymmetric square can be calculated. This allows us to compute the characters $\varphi_7^{\pm}$. Note that the component of the antisymmetric square in the principal block provides the characters $\varphi_{10}^{\pm}$. Our choices of signs for $\varphi_7^{\pm}$ and $\varphi_{10}^{\pm}$ give the decomposition $\Lambda^2\varphi_2^{+} = \varphi_7^{-} + \varphi_{10}^{-}$.

Characters $\varphi_{18}^{\pm}$: The following equation links known character values:

$$(\chi_{12}^{+})' - (\chi_{11}^{+})' = \varphi_{13}^{+} - \varphi_{13}^{-} + \varphi_7^{+}.$$

The terms on the left are restrictions of ordinary irreducibles to 3-regular classes and the terms on the right are modular irreducibles. It follows that φ_{13}^{-} is contained in $(\chi_{11}^{+})'$, and the decomposition matrix for O′N shows that their difference is one of the extensions of φ_{18}. The final $+$ in row $\chi_{11}^{\pm}$ of the collapsed matrix defines which of $\varphi_{18}^{\pm}$ is φ_{18}^{+}. This completes our construction of the modular irreducibles in the block, their values provide the decompositions that furnish all entries of our collapsed decomposition matrix given in Table 2.

4 Character Theoretic Analysis of the Principal Block

The decomposition matrix for the principal block of O′N [9, 23] is:

Block 1:	φ_1	φ_2	φ_3	φ_4	φ_5	φ_6	φ_{10}	φ_{11}	φ_{12}	φ_{14}	φ_{15}	φ_{16}	φ_{17}	φ_{19}
$1 = \chi_1$	1	.	.	.	.	.	.	.	.	.	.	.	.	.
$13376_1 = \chi_3$	.	.	.	.	.	.	1	1	.	.	.	.	.	.
$13376_2 = \chi_4$	.	.	.	.	.	.	1	.	1	.	.	.	.	.
$25916_1 = \chi_5$	1	1	.	.	.	.	.	.	.	.	1	.	.	.
$25916_2 = \chi_6$	1	1	.	.	.	.	.	.	.	.	.	1	.	.
$26752 = \chi_7$	2	1	.	.	1	1	.	.	.	1	.	.	.	.
$32395_1 = \chi_8$	2	1	.	.	.	1	1	.	.	1	.	.	.	.
$32395_2 = \chi_9$	2	1	.	.	1	.	1	.	.	1	.	.	.	.
$37696 = \chi_{10}$	1	2	.	.	.	.	1	.	.	.	.	.	1	.
$64790_1 = \chi_{16}$	3	3	1	.	1	1	1	.	.	1	.	.	1	.
$64790_2 = \chi_{17}$	3	3	.	1	1	1	1	.	.	1	.	.	1	.
$85064 = \chi_{18}$	3	4	1	1	1	1	.	.	.	.	1	1	1	.
$116963 = \chi_{19}$	6	4	.	.	1	1	3	1	1	2	.	.	1	.
$143374 = \chi_{20}$	2	1	.	.	1	1	2	.	.	1	.	.	.	1
$175616_1 = \chi_{23}$	3	3	1	1	1	1	2	.	.	1	.	.	1	1
$175616_2 = \chi_{24}$	3	3	1	1	1	1	2	.	.	1	.	.	1	1
$234080_1 = \chi_{29}$	4	5	1	1	2	2	2	.	.	.	1	1	2	1
$234080_2 = \chi_{30}$	4	6	1	1	1	1	4	1	1	1	.	.	2	1

The irreducible modular characters in the block are: $\varphi_1 = 1$, $\varphi_2 = 154$, $\varphi_3 = 342_1$, $\varphi_4 = 342_2$, $\varphi_5 = 495_1$, $\varphi_6 = 495_2$, $\varphi_{10} = 6138$, $\varphi_{11} = 7238_1$, $\varphi_{12} = 7238_2$, $\varphi_{14} = 25606$, $\varphi_{15} = 25761_1$, $\varphi_{16} = 25761_2$, $\varphi_{17} = 31249$, and $\varphi_{19} = 104346$.

The outer automorphism fixes $\varphi_1, \varphi_2, \varphi_{10}, \varphi_{14}, \varphi_{17}, \varphi_{19}$ and fuses the other 3-modular irreducibles in the block in pairs. Therefore, Clifford's theorem shows that there are 16 irreducible modular characters in the principal block of O′N.2, which we summarize in Table 3.

We now show that the principal block for O′N.2 has the collapsed decomposition matrix given in Table 4.

Again, the justification requires a construction for all irreducible Brauer characters in the block.

Characters $\widehat{\varphi_3}, \widehat{\varphi_5}, \widehat{\varphi_{11}}, \widehat{\varphi_{15}}, \varphi_1^{\pm}, \varphi_2^{\pm}, \varphi_{10}^{\pm}$: These characters are all already known. The induced characters vanish on outer elements of O′N.2, while the known

Table 3 Modular irreducibles in the principal 3-block of O′N.2

$\varphi_1^{\pm}$	$\varphi_2^{\pm}$	$\widehat{\varphi_3}$	$\widehat{\varphi_5}$	$\varphi_{10}^{\pm}$	$\widehat{\varphi_{11}}$	$\varphi_{14}^{\pm}$	$\widehat{\varphi_{15}}$	$\varphi_{17}^{\pm}$	$\varphi_{19}^{\pm}$
$1^{\pm}$	$154^{\pm}$	684	990	$6138^{\pm}$	14768	$25606^{\pm}$	51552	$31249^{\pm}$	$104346^{\pm}$

Table 4 Collapsed decomposition matrix for the principal block

Block 1	$\varphi_1^{\pm}$	$\varphi_2^{\pm}$	$\widehat{\varphi_3}$	$\widehat{\varphi_5}$	$\varphi_{10}^{\pm}$	$\widehat{\varphi_{11}}$	$\varphi_{14}^{\pm}$	$\widehat{\varphi_{15}}$	$\varphi_{17}^{\pm}$	$\varphi_{19}^{\pm}$
	1	154	684	990	6138	14476	25606	51522	31249	104346
$1 = \chi_1^{\pm}$	+	.	.	.	.	.	.	.	.	.
$26752 = \widehat{\chi_3}$	.	.	.	.	1	1	.	.	.	.
$51832 = \widehat{\chi_5}$	1	1	.	.	.	.	.	1	.	.
$26752 = \chi_7^{\pm}$	1±	-	.	1	.	.	+	.	.	.
$64790 = \widehat{\chi_8}$	2	1	.	1	1	.	1	.	.	.
$37696 = \chi_{10}^{\pm}$	+	1±	.	.	+	.	.	.	+	.
$129580 = \widehat{\chi_{16}}$	3	3	1	2	1	.	1	.	1	.
$85064 = \chi_{18}^{\pm}$	2/1	2±	1	1	.	.	.	1	+	.
$116963 = \chi_{19}^{\pm}$	4/2	2±	.	1	2/1	1	1±	.	+	.
$143374 = \chi_{20}^{\pm}$	1±	-	.	1	1±	.	+	.	.	+
$175616 = \chi_{23}^{\pm}$	2/1	2/1	1	1	1±	.	-	.	+	-
$175616 = \chi_{24}^{\pm}$	2/1	1/2	1	1	1±	.	+	.	+	+
$234080 = \chi_{29}^{\pm}$	2±	3/2	1	2	1±	.	.	1	1±	-
$234080 = \chi_{30}^{\pm}$	2±	2/4	1	1	2±	1	+	.	1±	+

matrix representations of degrees 1 and 154 give $\varphi_1^{\pm}$ and $\varphi_2^{\pm}$. Finally, the characters $\varphi_{10}^{\pm}$ are identified as a byproduct of the identification of $\varphi_7^{\pm}$ in Sect. 3. Only the following 3 pairs of modular irreducibles remain to be determined: $\varphi_{14}^{\pm}, \varphi_{17}^{\pm}, \varphi_{19}^{\pm}$. We observe from the collapsed matrix that their values follow from knowledge of all of the entries corresponding to other irreducibles in the rows for $\chi_7^{\pm}, \chi_{10}^{\pm}, \chi_{20}^{\pm}$, respectively. Our remaining effort is concentrated on these rows.

We apply the following general facts about a group with structure $G.2$. Let ψ be a Brauer character of $G.2$, denote by ψ^{fr} the Frobenius conjugate character and define the Brauer character ψ^{frt} as the tensor product of ψ^{fr} with the sign character. Then, the multiplicity of an irreducible Brauer character φ of $G.2$ in a (reducible) Brauer character ψ of $G.2$ is equal to the multiplicity of the irreducible Brauer character φ^{fr} (resp. φ^{frt}) in ψ^{fr} (resp. ψ^{frt}). We shall apply this observation in cases where ψ is a restriction χ' of an ordinary irreducible character χ to the 3-regular classes. The actions of fr and frt on some already known Brauer characters of O′N.2 are as follows:

ψ	$\varphi_1^{\pm}$	$\varphi_2^{\pm}$	$\varphi_{10}^{\pm}$	$(\chi_7^{\pm})'$	$(\chi_{10}^{\pm})'$	$(\chi_{20}^{\pm})'$
ψ^{fr}	$\varphi_1^{\pm}$	$\varphi_2^{\mp}$	$\varphi_{10}^{\pm}$	$(\chi_7^{\mp})'$	$(\chi_{10}^{\pm})'$	$(\chi_{20}^{\mp})'$
ψ^{frt}	$\varphi_1^{\mp}$	$\varphi_2^{\pm}$	$\varphi_{10}^{\mp}$	$(\chi_7^{\pm})'$	$(\chi_{10}^{\mp})'$	$(\chi_{20}^{\pm})'$

Note that $\varphi_1^{\pm}$ and $\varphi_{10}^{\pm}$ are moved by frt but $(\chi_7^{\pm})'$ and $(\chi_{20}^{\pm})'$ are not. This accounts for the three entries of 1± in the rows of the table marked by these ordinary irreducibles. Similarly, the entry of 1± in the row for $\chi_{10}^{\pm}$ follows since the restriction of the character is fixed by fr but $\varphi_2^{\pm}$ is not.

Characters $\varphi_{14}^{\pm}$: We now show that $\varphi_{14}^{+} = (\chi_7^{+})' - \varphi_1^{+} - \varphi_1^{-} - \widehat{\varphi_5} - \varphi_2^{-}$. In other words, we need to show that the second entry in the row for $\chi_7^{\pm}$ of the collapsed decomposition matrix is $-$ rather than $+$. The final $+$ in the row defines which of $\varphi_{14}^{\pm}$ is φ_{14}^{+}, and all other entries are known. We argue that if the second entry was a $+$, then $\psi := (\chi_7^{+})' - \varphi_1^{+} - \varphi_1^{-} - \widehat{\varphi_5} - \varphi_2^{+}$ would be a genuine Brauer character. Now, the scalar product of the tensor product of φ_2^{+} and ψ with the projective indecomposable character $\Phi_7^{+} = \chi_{12}^{+} + \chi_{15}^{+}$ (corresponding to the irreducible 5643-dimensional Brauer character φ_7^{+} in the second block) is -1, a contradiction.

The ambiguity in characters $\varphi_{17}^{\pm}, \varphi_{19}^{\pm}$: We can place one more character theoretic restriction on each of these characters, the argument is similar to that used to determine $\varphi_{18}^{\pm}$ in Sect. 3. We observe the following relations between known characters:

$$(\chi_{18}^{+})' - (\chi_{10}^{+})' = \varphi_1^{+} + \varphi_1^{-} + \varphi_2^{+} + \varphi_2^{-} + \widehat{\varphi_3} + \widehat{\varphi_5} - \varphi_{10}^{+} + \widehat{\varphi_{15}}$$

$$(\chi_{29}^{+})' - (\chi_{20}^{-})' - (\chi_{10}^{+})' - (\chi_{10}^{-})' = \widehat{\varphi_3} + \widehat{\varphi_5} - \varphi_{10}^{+} - \varphi_{10}^{-} - \varphi_{14}^{-} + \widehat{\varphi_{15}}.$$

The first relation shows that φ_{10}^{+} is a constituent of χ_{10}^{+}, which confirms the $+$ in the corresponding entry of our collapsed table. The second relation shows that φ_{14}^{-} is a constituent of at least one of χ_{20}^{-}, χ_{10}^{+} and χ_{10}^{-}. However, for the group O′N, φ_{14} is not a constituent of χ_{10}. This shows that φ_{14}^{-} is a constituent of χ_{20}^{-} and confirms the $+$ in the corresponding entry of the collapsed table. Apart from their final $+$ entries, which merely define which of $\varphi_{17}^{\pm}$ and $\varphi_{19}^{\pm}$ are φ_{17}^{+} and φ_{19}^{+}, the rows for $\chi_{10}^{\pm}$ and $\chi_{20}^{\pm}$ each have only one further entry that is still unknown. This leaves two possibilities for each of the character pairs $\varphi_{17}^{\pm}, \varphi_{19}^{\pm}$. Hence, there remain four possibilities for the decomposition matrix, according as we obtain φ_{17}^{+} and φ_{19}^{+} by subtracting $(\varphi_1^{+}, \varphi_2^{+})$, $(\varphi_1^{+}, \varphi_2^{-})$, $(\varphi_1^{-}, \varphi_2^{+})$, or $(\varphi_1^{-}, \varphi_2^{-})$ from the reducible character pair $((\chi_{10}^{+})' - \varphi_2^{+} - \varphi_2^{-} - \varphi_{10}^{+}, (\chi_{20}^{+})' - \varphi_1^{+} - \varphi_1^{-} - \widehat{\varphi_5} - \varphi_{10}^{+} - \varphi_{10}^{-} - \varphi_{14}^{+})$. We now decide between these cases module theoretically.

5 Module Theoretic Computations

We note that for the group O′N, the irreducible characters φ_{17} and φ_{19} appear with multiplicities 8 and 2 in the symmetric cube of φ_2. Moreover, in characteristic 3, tensors of the form $v \otimes v \otimes v$ span a submodule in the symmetric cube of any module V. The *deleted symmetric cube* of V is defined as the quotient of the symmetric cube by this submodule (see [6]). If V has dimension n, then the deleted symmetric cube, which we write as $\bar{S}^3(V)$, has dimension $n(n+1)(n+2)/6 - n$. We deduce that there are extensions of both φ_{17} and φ_{19} that have nonzero multiplicity in $\bar{S}^3(\varphi_2^{+})$. Modules affording these extension characters are constituents in the deleted symmetric cube of 154^{+}. If we could compute these constituents, we could read off their characters. Unfortunately, the deleted symmetric cube has dimension 620466 and is too big to chop directly with M. Ringe's C-Meataxe, [17], [16]. However, we can read off character theoretic information about its constituents by applying condensation.

Condensation [11, 19] is a powerful and general procedure that applies in the following context. Let G be a group, F a field of characteristic p and K a subgroup of G of order prime to p. Write e for the average of the elements of K, computed in the group algebra FG. For a given FG-module M, a *condensation*, or *condensed module* Me, is just the K-fixed space M^K. It is a module for the *condensation subalgebra* $eFGe \subset FG$. The composition factors of the condensed $eFGe$-module Me are in one-to-one correspondence with the composition factors S of M that have $Se \neq 0$ (see [11, 19]).

Here, we take G to be O′N.2, the prime p to be 3, and the field F to be $\mathbb{F}_9$. The FG-module 154^+ is represented by a pair of matrices giving the action of standard generators [26] for O′N.2 as given in [25]. The standard generators are elements c, d with c in class $2b$ and d in class $4a$ such that cd has order 22. Standard generators a and b for O′N are obtained from c and d by using the words $a = (cdd)^{-2}dd(cdd)^2$, $b = d$ (see [25]). Generators x and y for a subgroup $4^3.L_3(2)$ in O′N are given by the words $x = (ab)^{-4}a(ab)^4$, $y = (abb)^{-4}(abababbbababb)^4(abb)^4$ (see [25]). In terms of x and y, the element $u = (yxyxy)^4$ lies in the normal 4^3 subgroup and generates this group together with its conjugates $v = u^x$, $w = u^y$. As generators for a condensation subgroup K of type $4^2 \times 2$, we take u, w and v^2.

By construction the condensation subgroup K lies inside the simple group O′N. The fusion of the elements of K into O′N.2 is found by listing elements of K and using the Brauer character values of 154^+ on the inner classes as given in [9]. The fusion is as follows:

1a	2a	4a	4b
1	7	6	18

It determines the condensed dimensions of the irreducible modules for O′N character theoretically. Condensed dimensions of irreducibles for O′N.2 follow by first restricting to O′N. (This relies on the fact that K is contained in the simple group O′N.) The following table gives the dimensions of the simple modules for O′N.2 in its first row and the corresponding condensed dimensions in the second. (It turns out that two modular irreducibles have the same degree if and only if their condensed modules have the same degree.)

1	154	684	990	5643	11286	6138	14476	10944	25606	51522
1	2	24	42	183	330	194	444	368	818	1610

31249	41724	169290	175770	207360	104346
953	1304	5292	5526	6480	3254

We apply [20] to obtain actions of the elements $ece, ede, ecde$ and $edce$ (where c, d are standard generators for O′N.2) of the condensation algebra on the 19202-dimensional condensed module Me obtained from the O′N.2-module $M = \bar{S}^3(154^+)$. As in any application of condensation, we need to consider the technical question of whether the selected elements generate the full condensation algebra. In general, this is difficult to answer, but in this case, we can show that they generate a subalgebra $C \leq eFGe$ that is big enough for our purposes. What we prove is that the composition factors for the $eFGe$-module Me remain irreducible on restriction to C.

We can count the number of composition factors of the $eFGe$-module Me character theoretically. Observe that no irreducible FG-module condenses to 0, which means that there is a bijection between the composition factors of the $eFGe$-module Me and the O′N.2-module M. Moreover, the composition factors for the restriction of M to O′N are readily available (using the known 3-modular character table of O′N). Hence, the known lists of irreducibles for O′N that fuse and extend to give irreducibles of O′N.2, allow us to count the number (but not necessarily the identities) of composition factors of the O′N.2-module M. We find exactly 62 composition factors.

Now, an application of M. Ringe's C-Meataxe [17] shows that when viewed as a C-module, the module Me also has exactly 62 composition factors. In particular, every $eFGe$-composition factor of Me is irreducible on restriction to C.

We choose composition factors S_1e and S_2e of the C-module Me with degrees 953 and 3254—there are two non-isomorphic candidates for each, the condensations of irreducible O′N.2-modules with characters in $\varphi_{17}^{\pm}$ and $\varphi_{19}^{\pm}$, respectively. We finish the identification of these characters by computing traces of selected elements of the condensation algebra on S_1e and S_2e. The key observation that we apply is the following trace formula from [11]: Suppose that $g \in G$ acts on an FG-module M, then the trace of ege on the condensed module Me can be rewritten as follows:

$$\mathrm{Tr}_{Me}(ege) = \mathrm{Tr}_M(ege) = \mathrm{Tr}_M(eeg) = \mathrm{Tr}_M(eg) = \frac{1}{|K|} \sum_{x \in Kg} \mathrm{Tr}_M(x).$$

We compute traces on S_1e and S_2e for ece and $edce$ (two of our generating elements for C) and compare the results to the traces predicted using the trace formula for each of the four possibilities for the Brauer character table. In order to apply the formula from [11], we require the following Tables 5 and 6 that list classes represented in the cosets Kc and Kdc. In each table, the first line names a class (or set of classes in the case of 38abc) and the second line gives the number of elements in the class or set of classes. The entries in the second line are calculated by listing all elements of the coset under consideration and obtaining their classes by the method described in 2.4.

Only the second of the candidate tables listed at the end of Sect. 4 gives correct predictions for the traces, which shows that the decomposition matrix for O′N.2 is described by the collapsed decomposition matrices given in Tables 2 and 4 together with the defect 0 blocks given in Table 1.

Table 5 Conjugacy classes in the coset Kc

2b	6b	8c	8d	10c	22a	24a	24b	30b	38abc	56a	56d
1	1	1	4	4	4	2	3	4	4	2	2

Table 6 Conjugacy classes in the coset Kdc

8d	10c	14b	22a	24a	24b	30b	38abc	56a	56c	56d
1	2	7	4	2	4	1	6	2	2	1

Appendix: Brauer Trees for O′N.2 in Characteristic 5

There are six blocks with Brauer trees for O′N.2 in characteristic 5 and only the 4th block in the GAP labeling of all blocks is not straightforward. It contains the following ordinary irreducibles:

Nr.	GAP-Nr.	Degree
1	6	51832
2	7	26752
3	8	26752
4	12	52668
5	13	52668

Its tree is a stem, more precisely, an unfolding of a tree for O′N. The middle node is the ordinary irreducible of degree 51832 and its neighbors are the two irreducibles of degree 52688. Consider the following permutation of conjugacy classes of O′N.2: $\pi = 15(ab)16(ab)8(bc)10(bc)24(ab)30(ab)56(ab)56(cd)$. We note that π is an automorphism of the ordinary character table that switches the 7th and 8th ordinary irreducibles of degree 26752 while fixing the other nodes of the Brauer tree under consideration. Accordingly, we may suppose that the Brauer tree has the following form.

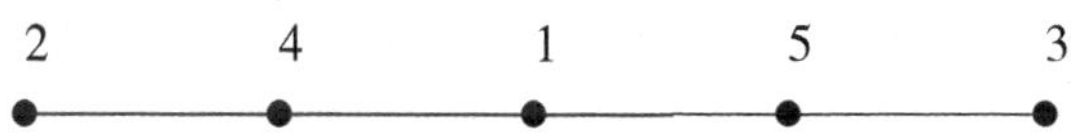

We further observe that the restriction of π to the 3-regular classes is an automorphism of the 3-modular character table of O′N.2, which means that any relabeling of conjugacy classes that are necessary to ensure that the 5-modular Brauer tree is as given above does not result in a change to the 3-modular character table computed in earlier sections. In other words, our descriptions of the 3-modular and 5-modular character tables of O′N.2 are compatible. A detailed discussion of character table automorphisms as label permutations and compatibility of character tables is available in Sect. 10 of [12].

References

1. Andrilli, S.F.: On the uniqueness of O'Nan's sporadic simple group. ProQuest LLC, Ann Arbor, MI (1979). http://gateway.proquest.com/openurl?url_ver=Z39.88-2004&rft_val_fmt=info:ofi/fmt:kev:mtx:dissertation&res_dat=xri:pqdiss&rft_dat=xri:pqdiss:8008854. Thesis (Ph.D.)–Rutgers The State University of New Jersey - New Brunswick
2. Breuer, T.: The GAP Character Table Library, Version 1.2.1 (2012). GAP package
3. Conway, J., Curtis, R., Norton, S., Parker, R., Wilson, R.: Atlas of Finite Groups. Clarendon Press, Oxford, England (1985)
4. Henke, A., Hiss, G., Müller, J.: The 7-modular decomposition matrices of the sporadic O'Nan group. J. London Math. Soc. (2) **60**(1), 58–70 (1999). https://doi.org/10.1112/S0024610799007735
5. Hiss, G., Lux, K.: Brauer Trees of Sporadic Groups. Oxford Science Publications. The Clarendon Press, Oxford University Press, New York (1989)
6. James, G.D.: The decomposition of tensors over fields of prime characteristic. Math. Z. **172**(2), 161–178 (1980). https://doi.org/10.1007/BF01182401
7. Jansen, C., Lux, K., Parker, R., Wilson, R.: An atlas of Brauer characters. London Mathematical Society Monographs. New Series, vol. 11. The Clarendon Press, Oxford University Press, New York. Appendix 2 by T. Breuer and S. Norton, Oxford Science Publications (1995)
8. Jansen, C., Wilson, R.A.: Two new constructions of the O'Nan group. J. London Math. Soc. (2)**56**(3), 579–583 (1997). https://doi.org/10.1112/S0024610798005742
9. Jansen, C., Wilson, R.A.: The 2-modular and 3-modular decomposition numbers for the sporadic simple O'Nan group and its triple cover. J. London Math. Soc. (2) **57**(1), 71–90 (1998). https://doi.org/10.1112/S0024610798005730
10. Koshitani, S., Kunugi, N., Waki, K.: Broué's conjecture for non-principal 3-blocks of finite groups. J. Pure Appl. Algebra **173**(2), 177–211 (2002). https://doi.org/10.1016/S0022-4049(01)00170-0
11. Lux, K., Pahlings, H.: Representations of groups. In: A Computational Approach. Cambridge Studies in Advanced Mathematics, vol. 124. Cambridge University Press, Cambridge (2010). https://doi.org/10.1017/CBO9780511750915
12. Lux, K., Ryba, A.J.E.: The 13-modular character table of 2.Suz.2. Comm. Algebra (to appear)
13. Müller, J.: Private communication (2017)
14. O'Nan, M.E.: Some evidence for the existence of a new simple group. Proc. London Math. Soc. (3) **32**(3), 421–479 (1976). https://doi.org/10.1112/plms/s3-32.3.421
15. Ottensmann, M.: Vervollständigung der Brauerbäume von 3.ON in Charakteristik 11, 19 und 31 mit Methoden der Kondensation. Diploma thesis, RWTH Aachen (2000)
16. Parker, R.A.: The computer calculation of modular characters (the meat-axe). In: Computational Group Theory, pp. 267–274. Durham (1982); Academic Press, London (1984)
17. Ringe, M.: The C–Meataxe. Manual RWTH Aachen (1994)
18. Ryba, A.J.E.: A new construction of the O'Nan simple group. J. Algebra **112**(1), 173–197 (1988). https://doi.org/10.1016/0021-8693(88)90141-X
19. Ryba, A.J.E.: Computer condensation of modular representations. J. Symbolic Comput. **9**(5–6), 591–600 (1990). https://doi.org/10.1016/S0747-7171(08)80076-4. Computational Group Theory, Part 1
20. Ryba, A.J.E.: Condensation of symmetrized tensor powers. J. Symbolic Comput. **32**(3), 273–289 (2001). https://doi.org/10.1006/jsco.2001.0459
21. Soicher, L.H.: A new existence and uniqueness proof for the O'Nan group. Bull. London Math. Soc. **22**(2), 148–152 (1990). https://doi.org/10.1112/blms/22.2.148
22. The GAP Group: GAP – Groups, Algorithms, and Programming, Version 4.7.5 (2014). http://www.gap-system.org
23. The Modular Atlas homepage. http://www.math.rwth-aachen.de/~MOC/
24. Wilson, R., Parker, R., Nickerson, S., Bray, J., Breuer, T.: AtlasRep, a GAP interface to the atlas of group representations, Version 1.5 (2011). Refereed GAP package

25. Wilson, R., Walsh, P., Tripp, J., Suleiman, I., Parker, R., Norton, S., Nickerson, S., Linton, S., Bray, J., Abbott, R.: ATLAS of Finite Group Representations. http://brauer.maths.qmul.ac.uk/Atlas/
26. Wilson, R.A.: Standard generators for sporadic simple groups. J. Algebra **184**(2), 505–515 (1996). https://doi.org/10.1006/jabr.1996.0271

Low Degree Cohomology of Frobenius Kernels

Nham V. Ngo

Abstract Let G be a simple algebraic group defined over an algebraically closed field of characteristic $p > 0$. For a positive integer r, let G_r be the r-th Frobenius kernel of G. We determine in this paper a number m such that the cohomology $\mathrm{H}^n(G_r, k)$ is isomorphic to $\mathrm{H}^n(G_1, k)$ for all $n \leq m$ where m depends on p and the type of G.

Keywords Group schemes · Frobenius kernels · Cohomology

1 Introduction

Let G be a reduced group scheme defined over an algebraically closed field k of prime characteristic p. For each r, the r-th Frobenius kernel G_r of G is defined as the kernel of the morphism $F_r : G \to G$ induced from the ring endomorphism $f \mapsto f^{p^r}$ on $k[G]$. In modular representation theory, the cohomology of Frobenius kernels G_r is closely related to that of G [4, II.4.12]. While the first Frobenius kernels have been intensively studied (see [4, II.12] for a summary), higher Frobenius kernels are more challenging to work with. Explicit calculations have been made for G_r cohomology in the case of low degree (see for example [1, 5]) and in the case G of rank 1 (cf. [6]). We now prove that in low degrees (depending on the prime p) the G_r cohomology is the same as the G_1 cohomology. In particular,

Main Theorem *Let G be a simple algebraic group and p a very good prime for G. Then for each $n < \frac{p}{c}$ where the values of c are determined from the following table*

Type	A_n	B_n	C_n	D_n	E_6	E_7	E_8	F_4	G_2
c	1	2	2	2	3	4	6	4	3

N. V. Ngo (✉)
University of North Georgia–Gainesville, Oakwood, GA 30566, USA
e-mail: nvngo@ung.edu

J. Feldvoss et al. (eds.), *Advances in Algebra*, Springer Proceedings
in Mathematics & Statistics 277, https://doi.org/10.1007/978-3-030-11521-0_13

there are $B^{(r)}$-module isomorphisms

$$H^n(B_r,k)\cong H^n(B_1,k)^{(r-1)}\cong\begin{cases}S^{\frac{n}{2}}(\mathfrak{u}^*)^{(r)} & \textit{if } n \textit{ is even},\\ 0 & \textit{otherwise}.\end{cases}$$

Furthermore, there are G-module isomorphisms

$$H^n(G_r,k)^{(-r)}\cong ind_B^G H^n(B_r,k)^{(-r)}\cong\begin{cases}ind_B^G S^{\frac{n}{2}}(\mathfrak{u}^*) & \textit{if } n \textit{ is even},\\ 0 & \textit{otherwise}.\end{cases}$$

Note that our result generalizes computations for type A in [2, Theorem 1.8] and [5, Corollary 3.4].

2 Notation

Let k be an algebraically closed field of characteristic $p > 2$. Throughout the paper, we fix a simple, simply connected algebraic group G over k, that is defined and split over the prime field $\mathbb{F}_p$. Denote by h the Coxeter number of G. Fix a maximal torus T of G, and denote by Φ the root system of T in G. Fix a set $\Pi = \{\alpha_1, \ldots, \alpha_n\}$ of simple roots in Φ, and let Φ^+ be the corresponding set of positive roots. Let $B \subset G$ be the Borel subgroup of G containing T and corresponding to Φ^+, the set of positive roots, and let $U \subset B$ be the unipotent radical of B.

For each $r \geq 1$, the r-th Frobenius kernel of a closed subgroup H of G is defined as the scheme-theoretic intersection of H with G_r. Given a rational H-module M, write $M^{(r)}$ for the module obtained by twisting the structure map for M by F_r.

The prime p is good for G if

- p is any prime for G of type A_n;
- $p \geq 3$ for G of type B_n, C_n, D_n;
- $p \geq 5$ for G of type E_6, E_7, F_4, G_2;
- $p \geq 7$ for G of type E_8.

Additionally, a good prime p is called *very good* for G provided that if G is of type A_n, then p is coprime to $n + 1$. Every root $\beta \in \Phi^+$ can be written as a sum of simple roots with nonnegative coefficients. Let c be the largest coefficient in all the expressions of positive roots in terms of simple roots. By studying Table 2 in [3, 12.2], one can obtain the values of c, depending on the type of G, exactly as shown in the table in the Main Theorem.

Let X be the weight lattice of Φ. Let $\mathfrak{g} = \mathrm{Lie}(G)$ be the Lie algebra of G, similarly $\mathfrak{b} = \mathrm{Lie}(B)$ and $\mathfrak{u} = \mathrm{Lie}(U)$. Denote by $S^\bullet(\mathfrak{u}^*)$ and $\Lambda^\bullet(\mathfrak{u}^*)$ respectively the symmetric algebra and exterior algebra over $\mathfrak{u}^* := \mathrm{Hom}_k(\mathfrak{u}, k)$. For simplicity, we write S^i and Λ^j instead of $S^i(\mathfrak{u}^*)$ and $\Lambda^j(\mathfrak{u}^*)$. Throughout this paper, the symbol $\otimes$ means the tensor product over the field k. Suppose H is an algebraic group over

k and M is a (rational) module of H. Denote by M^H the submodule consisting of all the fixed points of M under the H-action. If M is a B-module, then the induced G-module can be defined as $\text{ind}_B^G M = (k[G] \otimes M)^B$. The higher derived functor of $\text{ind}_B^G(-)$ is denoted by $R^i\text{ind}_B^G(-)$.

3 Cohomology of Frobenius Kernels

We assume G is a simple algebraic group and the characteristic p is a very good prime for G. We first prove the following.

Lemma 1 *Let a, b be nonnegative integers such that $a + b < \frac{p}{c}$. Then we have*

$$(S^a \otimes \Lambda^b)^{T_1} \cong \begin{cases} k & \text{if } a = b = 0, \\ 0 & \text{otherwise.} \end{cases}$$

Proof Suppose $(S^a \otimes \Lambda^b)^{T_1} \neq 0$. Then, there is a weight $\tau \in X$ such that $p\tau = \sigma + \lambda$ where σ, λ are respectively weights of S^a and Λ^b. Since $\sigma + \lambda \in \mathbb{N}\Phi$, so is $p\tau$. As p is a very good prime, it does not divide the order of the finite group $X/(\mathbb{N}\Phi)$; hence τ must be in the root lattice $\mathbb{N}\Phi$. Note that $\sigma + \lambda$ is the sum of $a + b$ positive roots (not necessarily distinct), we can write $\sigma + \lambda = \Sigma_{i=1}^m d_i\alpha_i$ with $m = |\Pi|$, simple roots α_i and nonnegative integers d_i for $1 \leq i \leq m$. Now observe that, for each $1 \leq i \leq m$, we have the coefficient

$$d_i \leq (a + b)c < \frac{p}{c}c = p.$$

In other words, $\sigma + \lambda$ cannot be a p-multiple of any nonzero root. Therefore, $\tau = 0$ and so $a = b = 0$. □

Theorem 1 *For each $n < \frac{p}{c}$, there is a $B^{(r)}$-module isomorphism*

$$H^n(B_r, k) \cong H^n(B_1, k)^{(r-1)} \cong \begin{cases} (S^{\frac{n}{2}})^{(r)} & \text{if } n \text{ is even,} \\ 0 & \text{otherwise.} \end{cases}$$

Proof We first prove the theorem for the case $r = 1$. Note that this has been verified for $p > h$ in all degrees ([4, Prop. II.12.11]). Taking the T_1-invariant functor on the spectral sequence for U_1-cohomology [2, Prop. 1.1], we obtain a spectral sequence

$$E_2^{2i,j} = (S^i)^{(1)} \otimes (\mathrm{H}^j(\mathfrak{u}, k))^{T_1} \Rightarrow \mathrm{H}^{2i+j}(B_1, k). \tag{1}$$

Now we consider all indices i, j such that $2i + j < \frac{p}{c}$. Since we can identify $\mathrm{H}^j(\mathfrak{u}, k)$ with a T-submodule of Λ^j, and by Lemma 1

$$(\Lambda^j)^{T_1} \cong \begin{cases} k & \text{if } j = 0, \\ 0 & \text{otherwise,} \end{cases}$$

it follows that $\mathrm{H}^j(\mathfrak{u}, k)^{T_1} = k$ if $j = 0$, and 0 otherwise. We then have

$$E_2^{2i,j} \cong \begin{cases} (S^i)^{(1)} & \text{if } j = 0, \\ 0 & \text{otherwise.} \end{cases}$$

This implies that the differentials of the spectral sequence $d_t^{2i,j} : E_t^{2i,j} \to E_t^{2i+t,j+1-t}$ are 0 for all $0 \le j < \frac{p}{c}$ since either $E_t^{2i,j} = 0$ or $E_t^{2i+t,j+1-t} = 0$ with $t \ge 2$. Therefore, the spectral sequence (1) has $E_\infty^{2i,j} = E_2^{2i,j}$ for all $2i + j < \frac{p}{c}$. This implies that $\mathrm{H}^n(B_1, k) \cong (S^{\frac{n}{2}})^{(1)}$ for even degree $n < \frac{p}{c}$.

Now assume that $r > 1$ and consider the spectral sequence in [6, Theorem 3.1.2]

$$\bigoplus S^{a_r(r)} \otimes \left[\left[(\Lambda^{b_1})^{T_1(-1)} \otimes S^{a_1} \otimes \Lambda^{b_2} \right]^{T_1(-1)} \otimes \cdots \otimes S^{a_{r-1}} \otimes \Lambda^{b_r} \right]^{T_1^{(r-1)}} \Rightarrow \mathrm{H}^n(B_r, k)$$

where the direct sum is taken over all a_s, b_t with $n = 2(a_1 + \cdots + a_r) + b_1 + \cdots + b_r$. Since $n < \frac{p}{c}$, we have $b_s, a_s + b_{s+1} < \frac{p}{c}$ for each $1 \le s \le r - 1$. Lemma 1 shows that

$$\left[[(\Lambda^{b_1})^{T_1(-1)} \otimes S^{a_1} \otimes \Lambda^{b_2}]^{T_1(-1)} \otimes \cdots \otimes S^{a_{r-1}} \otimes \Lambda^{b_r} \right]^{T_1^{(r-1)}} \cong \begin{cases} k, & a_s = b_s = b_r = 0 \\ & \text{for } 1 \le s \le r-1, \\ 0 & \text{otherwise.} \end{cases}$$

Thus, looking at the first page of the spectral sequence with $i + j < \frac{p}{c}$, we see that $E_1^{i,j} = 0$ except $E_1^{p^r(\frac{i+j}{2}),(2-p^r)(\frac{i+j}{2})} = (S^{\frac{i+j}{2}})^{(r)}$. It follows that $\mathrm{H}^n(B_r, k)$ is a subquotient of $(S^{\frac{n}{2}})^{(r)}$ for $n < \frac{p}{c}$. On the other hand, [4, II.10.14.5] gives us that as a vector space

$$(S^{\frac{n}{2}})^{(r)} = (S^{\frac{n}{2}})^{(1)} \cong \mathrm{H}^n(B_1, k) \hookrightarrow \mathrm{H}^n(B_r, k).$$

It then follows that $\mathrm{H}^n(B_r, k)$ is isomorphic to $(S^{\frac{n}{2}})^{(r)}$ as a B-module when n is even. □

Remark 1 Note that for each $n < \frac{p}{c}$, by [4, Lemma II.12.12] we have

$$R^m \mathrm{ind}_B^G(\mathrm{H}^n(B_r, k)^{(-r)}) \cong R^m \mathrm{ind}_B^G(S^{\frac{n}{2}}) = 0$$

for each $m > 0$. So the same technique can be applied to the spectral sequence

$$R^m \mathrm{ind}_B^G(\mathrm{H}^n(B_r, k)^{(-r)}) \Rightarrow \mathrm{H}^{n+m}(G_r, k)^{(-r)}$$

to get G_r-cohomology for degree up to $\frac{p}{c}$. Hence, we obtain the second part of the main theorem as follows.

Theorem 2 *For each $n < \frac{p}{c}$, there are isomorphisms of G-modules*

$$H^n(G_r, k)^{(-r)} \cong ind_B^G(H^n(B_r, k)^{(-r)}) \cong \begin{cases} ind_B^G(S^{\frac{n}{2}}) & \text{if } n \text{ is even,} \\ 0 & \text{otherwise.} \end{cases}$$

Consequently, for each $n < \frac{p}{c}$, we have

$$H^n(G_r, k) \cong H^n(G_1, k)^{(r-1)}.$$

Remark 2 We notice that Friedlander and Parshall already studied this problem in [2, Theorem 1.8]. They showed that if G is of type A with $p > h$, then for each $r > 1$, the inflation map $\mathrm{H}^n(G_r/G_{r-1}, k) \to \mathrm{H}^n(G_r, k)$ is an isomorphism for all $n < 2p - 1$. In a similar setting, Kaneda et al. were able to increase the upper bound of n in [5, 3.3, 3.4]. Our calculations slightly improve these results as we do not require $p > h$ and G does not have to be of type A.

Acknowledgements This paper is a part of the author's dissertation at the University of Georgia. He is grateful for the guidedance of his Ph.D. advisor Daniel K. Nakano and secondary advisor Christopher M. Drupieski.

References

1. Bendel, C.P., Nakano, D.K., Pillen, C.: Third cohomology for Frobenius kernels and related structures. In: Lie Algebras, Lie Superalgebras, Vertex Algebras and Related Topics. Proceedings of Symposia in Pure Mathematics, pp. 81-118. American Mathematical Society, Providence (2016)
2. Friedlander, E.M., Parshall, B.J.: Cohomology of infinitesimal and discrete groups. Math. Ann. **273**, 353–374 (1986)
3. Humphreys, J.E.: Introduction to Lie Algebras and Representation Theory. Springer-Verlag, New York-Berlin (1978)
4. Jantzen, J.C.: Representations of Algebraic Groups. American Mathematical Society, Providence (2003)
5. Kaneda, M., Shimada, N., Tezuka, M., Yagita, N.: Cohomology of infinitesimal algebraic groups. Math. Z. **205**, 61–96 (1990)
6. Ngo, N.V.: Cohomology for Frobenius kernels of SL_2. J. Algebra **396**, 39–60 (2013)

On the Finite W-Algebra for the Queer Lie Superalgebra

Elena Poletaeva

Abstract In this paper, we study the finite W-algebra W_χ for the queer Lie superalgebra $\mathfrak{g} = Q(N)$ associated with an arbitrary even nilpotent element χ in the coadjoint representation. We describe the annihilator $\mathfrak{g}^\chi$ and construct a set of elements in the generalized Whittaker module $U(\mathfrak{g})/I_\chi$ which under certain map project onto a homogeneous basis in $\mathfrak{g}^\chi$. In the case when the corresponding nilpotent element has Jordan blocks of equal size, these elements form a set of generators for W_χ.

Keywords Finite W-algebra · Queer Lie superalgebra · Kazhdan filtration

1 Introduction

Finite W-algebras are finitely generated associative algebras constructed from complex semi-simple Lie algebras and their nilpotent elements. A. Premet gave the general definition of these algebras and developed their theory in [20] (see also [10]). Let $\mathfrak{g}$ be a semi-simple Lie algebra and $e \in \mathfrak{g}$ be a nilpotent element. In the extreme case when $e = 0$, the finite W-algebra simply coincides with the universal enveloping algebra $U(\mathfrak{g})$. The special case when e is a *regular* nilpotent element was considered by B. Kostant, who was motivated by the study of Whittaker vectors and of Whittaker models. He proved that in this case the finite W-algebra coincides with the center of $U(\mathfrak{g})$ [9]. It turns out that a finite W-algebra is a quantization of the Poisson structure on the so-called Slodowy slice, i.e. a transversal slice to the orbit of e in the adjoint representation. Note that constructing a finite W-algebra, one can also associate it with a nilpotent element χ in the coadjoint representation, which is identified with e via a nondegenerate invariant symmetric bilinear form on $\mathfrak{g} : \chi(x) = (e|x)$. The finite W-algebra associated with χ is denoted by W_χ.

Premet's definition of a finite W-algebra makes sense for Lie superalgebras, and some of his results can be generalized for classical Lie superalgebras. However,

E. Poletaeva (✉)
School of Mathematical and Statistical Sciences, University of Texas Rio Grande Valley, Edinburg, TX 78539, USA
e-mail: elena.poletaeva@utrgv.edu

J. Feldvoss et al. (eds.), *Advances in Algebra*, Springer Proceedings in Mathematics & Statistics 277, https://doi.org/10.1007/978-3-030-11521-0_14

the result of B. Kostant does not hold in this case. The finite W-algebras for Lie superalgebras have been extensively studied by mathematicians and physicists in [1–3, 5, 13–19, 22–24].

For a Lie superalgebra, whose even part is a reductive Lie algebra we define W_χ associated with an even nilpotent element χ in the coadjoint representation. In [18] we studied in detail the queer Lie superalgebra $Q(N)$ in the regular case (i.e., the corresponding nilpotent element has a single Jordan block). In particular, we proved that W_χ for $Q(N)$ associated with a regular nilpotent χ is isomorphic to a quotient of the super-Yangian of $Q(1)$ (Theorem 6.2).

In [19] we considered the case when $N = nl$, and the corresponding nilpotent element has Jordan blocks each of size l. We proved that W_χ is isomorphic to a quotient of the super-Yangian of $Q(n)$ (Theorem 4.2).

A. Premet has proved that if $\mathfrak{g}$ is a semi-simple Lie algebra, then the associated graded algebra $Gr_K W_\chi$ with respect to the Kazhdan filtration is isomorphic to $S(\mathfrak{g}^\chi)$ (the symmetric algebra of the annihilator $\mathfrak{g}^\chi$ of χ in $\mathfrak{g}$). He introduced the map $P : W_\chi \to S(\mathfrak{g}^\chi)$, namely, he proved that for $X \in W_\chi$ the term of the highest Kazhdan degree and of the highest degree with respect to the good $\mathbb{Z}$-grading of $\mathfrak{g}$ belongs to $S(\mathfrak{g}^\chi)$ (see [20]).

In this work we study the finite W-algebra for $\mathfrak{g} = Q(N)$ associated with an *arbitrary* even nilpotent χ. We describe the annihilator $\mathfrak{g}^\chi$ and construct a set of elements in the generalized Whittaker module $U(\mathfrak{g})/I_\chi$ which project under the map P onto a homogeneous basis in $\mathfrak{g}^\chi$. In the case when the corresponding nilpotent element has Jordan blocks of equal size, these elements form a set of generators for W_χ.

In [19] (Corollary 5.16) we have realized the finite W-algebra for $Q(N)$, where $N = nl$ and the corresponding nilpotent element has Jordan blocks each of size l, inside $(U(Q(n)))^{\otimes l}$ as $W_\chi \cong ev^{\otimes l} \circ \Delta_l(Y(Q(n)))$, where Δ is the comultiplication and ev is the evaluation homomorphism (see [11]). This realization is in the spirit of [2, 4], where the authors used (shifted) truncated (super)-Yangians. In [13] Peng proved the similar result for $\mathfrak{gl}(ml|nl)$.

An interesting problem is to describe the finite W-algebra for $Q(N)$ associated with an arbitrary even nilpotent χ in terms of (shifted) super-Yangians.

For the benefit of the reader we include all the necessary definitions and facts on finite W-algebras and Lie superalgebras in Sects. 2 and 3, as they appeared in [19].

2 Finite W-Algebras for Lie Superalgebras

Let $\mathfrak{g} = \mathfrak{g}_{\bar{0}} \oplus \mathfrak{g}_{\bar{1}}$ be a Lie superalgebra with reductive even part $\mathfrak{g}_{\bar{0}}$. Let $\chi \in \mathfrak{g}_{\bar{0}}^* \subset \mathfrak{g}^*$ be an even nilpotent element in the coadjoint representation, i.e., the closure of the $G_{\bar{0}}$-orbit of χ in $\mathfrak{g}_{\bar{0}}^*$ (where $G_{\bar{0}}$ is the algebraic reductive group of $\mathfrak{g}_{\bar{0}}$) contains zero.

Definition 1 The *annihilator* of χ in $\mathfrak{g}$ is

$$\mathfrak{g}^{\chi} = \{x \in \mathfrak{g} \text{ s.t. } \chi([x, \mathfrak{g}]) = 0\}.$$

Definition 2 *A good $\mathbb{Z}$-grading for χ* is a $\mathbb{Z}$-grading $\mathfrak{g} = \oplus_{j \in \mathbb{Z}} \mathfrak{g}_j$ satisfying the following two conditions:

(1) $\chi(\mathfrak{g}_j) = 0$ if $j \neq -2$;
(2) $\mathfrak{g}^{\chi}$ belongs to $\bigoplus_{j \geq 0} \mathfrak{g}_j$.

Note that $\chi([\cdot, \cdot])$ defines a nondegenerate super skew-symmetric even bilinear form on $\mathfrak{g}_{-1}$. Let $\mathfrak{l}$ be a maximal isotropic (i.e., Lagrangian) subspace with respect to this form. We consider a nilpotent subalgebra $\mathfrak{m} = (\oplus_{j \leq -2} \mathfrak{g}_j) \bigoplus \mathfrak{l}$ of $\mathfrak{g}$. The restriction of χ to $\mathfrak{m}$, $\chi : \mathfrak{m} \to \mathbb{C}$, defines a one-dimensional representation $\mathbb{C}_{\chi} =< v >$ of $\mathfrak{m}$. Let I_{χ} be the left ideal of $U(\mathfrak{g})$ generated by $a - \chi(a)$ for all $a \in \mathfrak{m}$.

Definition 3 The induced $\mathfrak{g}$-module

$$Q_{\chi} := U(\mathfrak{g}) \otimes_{U(\mathfrak{m})} \mathbb{C}_{\chi} \cong U(\mathfrak{g})/I_{\chi}$$

is called *the generalized Whittaker module.*

Definition 4 *The finite W-algebra* associated with the nilpotent element χ is

$$W_{\chi} := \mathrm{End}_{U(\mathfrak{g})}(Q_{\chi})^{op}.$$

As in the Lie algebra case, the superalgebras W_{χ} are all isomorphic for different choices of good $\mathbb{Z}$-gradings and maximal isotropic subspaces $\mathfrak{l}$ [24].

If $\mathfrak{g}$ admits an even nondegenerate $\mathfrak{g}$-invariant supersymmetric bilinear form, then $\mathfrak{g} \cong \mathfrak{g}^*$ and $\chi(x) = (e|x)$ for some nilpotent $e \in \mathfrak{g}_{\bar{0}}$ (i.e. ad$e$ is a nilpotent endomorphism of $\mathfrak{g}$). By the Jacobson–Morozov theorem e can be included in $\mathfrak{sl}(2) =< e_+, h, e_- >$, where $e_+ = e$. As in the Lie algebra case, the linear operator ad$h$ defines a Dynkin $\mathbb{Z}$-grading $\mathfrak{g} = \bigoplus_{j \in \mathbb{Z}} \mathfrak{g}_j$, where

$$\mathfrak{g}_j = \{x \in \mathfrak{g} \text{ s.t. } \mathrm{ad}h(x) = jx\}.$$

As follows from the representation theory of $\mathfrak{sl}(2)$, the Dynkin $\mathbb{Z}$-grading is good for χ. Let $\mathfrak{g}^e := \mathrm{Ker}(\mathrm{ad}e)$. Clearly, $\mathfrak{g}^e = \mathfrak{g}^{\chi}$. For a Lie superalgebra $\mathfrak{g} = \mathfrak{g}_{\bar{0}} \oplus \mathfrak{g}_{\bar{1}}$ we will denote $\underline{\dim \mathfrak{g}} = (\dim \mathfrak{g}_{\bar{0}} | \dim \mathfrak{g}_{\bar{1}})$. Note that as in the Lie algebra case, $\underline{\dim \mathfrak{g}^e} = \underline{\dim \mathfrak{g}_0} + \underline{\dim \mathfrak{g}_1}$.

Note that by Frobenius reciprocity

$$\mathrm{End}_{U(\mathfrak{g})}(Q_{\chi}) = \mathrm{Hom}_{U(\mathfrak{m})}(\mathbb{C}_{\chi}, Q_{\chi}).$$

That defines an identification of W_{χ} with the subspace

$$Q_{\chi}^{\mathfrak{m}} = \{u \in Q_{\chi} \text{ s.t. } au = \chi(a)u \text{ for all } a \in \mathfrak{m}\}.$$

In what follows we denote by $\pi : U(\mathfrak{g}) \to U(\mathfrak{g})/I_\chi$ the natural projection. By the above

$$W_\chi = \{\pi(y) \in U(\mathfrak{g})/I_\chi \text{ s.t. } (a - \chi(a))y \in I_\chi \text{ for all } a \in \mathfrak{m}\},$$

or, equivalently,

$$W_\chi = \{\pi(y) \in U(\mathfrak{g})/I_\chi \text{ s.t. } \operatorname{ad}(a)y \in I_\chi \text{ for all } a \in \mathfrak{m}\}. \tag{1}$$

One can easily check that the multiplication in W_χ can be also defined by

$$\pi(y_1)\pi(y_2) = \pi(y_1 y_2)$$

for $y_i \in U(\mathfrak{g})$ such that $\operatorname{ad}(a)y_i \in I_\chi$ for all $a \in \mathfrak{m}$ and $i = 1, 2$.

Definition 5 A nilpotent $\chi \in \mathfrak{g}_{\bar{0}}^*$ is called *regular* nilpotent if the $G_{\bar{0}}$-orbit of χ has maximal dimension, i.e., the dimension of $\mathfrak{g}_{\bar{0}}^\chi$ is minimal. (Equivalently, a nilpotent $e \in \mathfrak{g}_{\bar{0}}$ is *regular* nilpotent, if $\mathfrak{g}_{\bar{0}}^e$ attains the minimal dimension, which is equal to $\operatorname{rank}\mathfrak{g}_{\bar{0}}$.)

Theorem 1 (Kostant [9]) *For a reductive Lie algebra $\mathfrak{g}$ and a regular nilpotent element $e \in \mathfrak{g}$, the finite W-algebra W_χ is isomorphic to the center of $U(\mathfrak{g})$.*

This theorem does not hold for Lie superalgebras. In fact, if $\mathfrak{g}$ is a basic Lie superalgebra (i.e., $\mathfrak{g}$ is a simple Lie superalgebra such that the representation of the Lie algebra $\mathfrak{g}_{\bar{0}}$ on $\mathfrak{g}_{\bar{1}}$ is completely reducible and $\mathfrak{g}$ admits an even non-degenerate $\mathfrak{g}$-invariant bilinear form), then the center of $U(\mathfrak{g})$ is even according to [6, 8], while W_χ has a nontrivial odd part. In this work we study the case when $\mathfrak{g}$ is the queer Lie superalgebra $Q(N)$. A. Sergeev proved that the center of $U(Q(N))$ is even in [21]. More precisely, he proved that this center is generated by the even elements $\sum_{i=1}^N e_{i,i}^{(2m+1)}$ (see (3) in Sect. 3 for notations). However, W_χ has a nontrivial odd part (see Corollary 1 in Sect. 4).

Let $T(\mathfrak{g})$ be the tensor algebra of $\mathfrak{g}$.

Definition 6 Define a $\mathbb{Z}$-grading on $T(\mathfrak{g})$ by setting the degree of $g \in \mathfrak{g}_j$ to be $j + 2$. This induces a filtration on $U(\mathfrak{g})$ and therefore on $U(\mathfrak{g})/I_\chi$ which is called the *Kazhdan filtration*. We will denote by Gr_K the corresponding graded algebras. Since by (1) $W_\chi \subset U(\mathfrak{g})/I_\chi$, we have an induced filtration on W_χ.

Theorem 2 (Premet [20]) *Let $\mathfrak{g}$ be a semi-simple Lie algebra. Then the associated graded algebra $Gr_K W_\chi$ is isomorphic to $S(\mathfrak{g}^\chi)$.*

To generalize this result to the super case, we assume that $\mathfrak{l}'$ is some subspace in $\mathfrak{g}_{-1}$ satisfying the following two properties:

(1) $\mathfrak{g}_{-1} = \mathfrak{l} \oplus \mathfrak{l}'$,
(2) $\mathfrak{l}'$ contains a maximal isotropic subspace with respect to the form defined by $\chi([\cdot, \cdot])$ on $\mathfrak{g}_{-1}$.

If $\dim(\mathfrak{g}_{-1})_{\bar{1}}$ is even, then $\mathfrak{l}'$ is a maximal isotropic subspace. Let $\mathfrak{l}^{\perp}$ be the orthogonal complement to $\mathfrak{l}$ with respect to the above bilinear form on $\mathfrak{g}_{-1}$. If $\dim(\mathfrak{g}_{-1})_{\bar{1}}$ is odd, then $\mathfrak{l}^{\perp} \cap \mathfrak{l}'$ is one-dimensional and we fix $\theta \in \mathfrak{l}^{\perp} \cap \mathfrak{l}'$ such that $\chi([\theta, \theta]) = 2$. It is clear that $\pi(\theta) \in W_\chi$ and $\pi(\theta)^2 = 1$.

Let $\mathfrak{p} = \bigoplus_{j \geq 0} \mathfrak{g}_j$. By the Poincaré-Birkhoff–Witt theorem, $U(\mathfrak{g})/I_\chi \cong S(\mathfrak{p} \oplus \mathfrak{l}')$ as a vector space. Therefore $Gr_K(U(\mathfrak{g})/I_\chi)$ is isomorphic to $S(\mathfrak{p} \oplus \mathfrak{l}')$ as a vector space. The good $\mathbb{Z}$-grading of $\mathfrak{g}$ induces the grading on $S(\mathfrak{p} \oplus \mathfrak{l}')$. For any $X \in U(\mathfrak{g})/I_\chi$ let $Gr_K(X)$ denote the corresponding element in $Gr_K(U(\mathfrak{g})/I_\chi)$, and $P(X)$ denote the highest weight component of $Gr_K(X)$ in this $\mathbb{Z}$-grading.

Theorem 3 ([18], Proposition 2.7) *Let $y_1, \ldots, y_p$ be a basis in $\mathfrak{g}^\chi$ homogeneous in the good $\mathbb{Z}$-grading. Assume that there exist $Y_1, \ldots, Y_p \in W_\chi$ such that $P(Y_i) = y_i$ for all $i = 1, \ldots, p$. Then*

(a) if $\dim(\mathfrak{g}_{-1})_{\bar{1}}$ is even, then $Y_1, \ldots, Y_p$ generate W_χ, and if $\dim(\mathfrak{g}_{-1})_{\bar{1}}$ is odd, then $Y_1, \ldots, Y_p$ and $\pi(\theta)$ generate W_χ;

(b) if $\dim(\mathfrak{g}_{-1})_{\bar{1}}$ is even, then $Gr_K W_\chi \cong S(\mathfrak{g}^\chi)$, and if $\dim(\mathfrak{g}_{-1})_{\bar{1}}$ is odd, then $Gr_K W_\chi \cong S(\mathfrak{g}^\chi) \otimes \mathbb{C}[\xi]$, where $\mathbb{C}[\xi]$ is the exterior algebra generated by one element ξ.

3 The Queer Lie Superalgebra $Q(N)$

Recall that the *queer* Lie superalgebra is defined as follows:

$$Q(N) := \left\{ \left(\begin{array}{c|c} A & B \\ \hline B & A \end{array} \right) \text{ s.t. } A, B \text{ are } N \times N \text{ matrices} \right\},$$

where matrices $\left(\begin{array}{c|c} A & 0 \\ \hline 0 & A \end{array} \right)$ are even, while $\left(\begin{array}{c|c} 0 & B \\ \hline B & 0 \end{array} \right)$ are odd.

Let $\operatorname{otr} \left(\begin{array}{c|c} A & B \\ \hline B & A \end{array} \right) = \operatorname{tr} B$.

Remark 1 $Q(N)$ has one-dimensional center $< z >$, where $z = 1_{2N}$. Let

$$SQ(N) = \{X \in Q(N) \text{ s.t. } \operatorname{otr} X = 0\}.$$

The Lie superalgebra $\tilde{Q}(N) := SQ(N)/ < z >$ is simple for $N \geq 3$, see [7].

Note that $\mathfrak{g} = Q(N)$ admits an *odd* non-degenerate $\mathfrak{g}$-invariant supersymmetric bilinear form

$$(x|y) := \operatorname{otr}(xy) \text{ for } x, y \in \mathfrak{g}.$$

Therefore, we identify the coadjoint module $\mathfrak{g}^*$ with $\Pi(\mathfrak{g})$, where Π is the change of parity functor. Let $e_{i,j}$ and $f_{i,j}$ be standard bases in $\mathfrak{g}_{\bar{0}}$ and $\mathfrak{g}_{\bar{1}}$ respectively:

$$e_{i,j} = \left(\begin{array}{c|c} E_{ij} & 0 \\ \hline 0 & E_{ij} \end{array}\right), \quad f_{i,j} = \left(\begin{array}{c|c} 0 & E_{ij} \\ \hline E_{ij} & 0 \end{array}\right), \tag{2}$$

where E_{ij} are elementary $N \times N$ matrices.

In [21] Sergeev recursively defined the elements $e_{i,j}^{(m)}$ and $f_{i,j}^{(m)}$ of $U(Q(N))$:

$$\begin{aligned} e_{i,j}^{(m)} &= \sum_{k=1}^{N} e_{i,k} e_{k,j}^{(m-1)} + (-1)^{m+1} \sum_{k=1}^{N} f_{i,k} f_{k,j}^{(m-1)}, \\ f_{i,j}^{(m)} &= \sum_{k=1}^{N} e_{i,k} f_{k,j}^{(m-1)} + (-1)^{m+1} \sum_{k=1}^{N} f_{i,k} e_{k,j}^{(m-1)}, \end{aligned} \tag{3}$$

where $e_{i,j}^{(0)} = \delta_{i,j}$ and $f_{i,j}^{(0)} = 0$. Then

$$\begin{aligned} &[e_{i,j}, e_{k,l}^{(m)}] = \delta_{j,k} e_{i,l}^{(m)} - \delta_{i,l} e_{k,j}^{(m)}, \quad [e_{i,j}, f_{k,l}^{(m)}] = \delta_{j,k} f_{i,l}^{(m)} - \delta_{i,l} f_{kj}^{(m)}, \\ &[f_{i,j}, e_{k,l}^{(m)}] = (-1)^{m+1} \delta_{j,k} f_{i,l}^{(m)} - \delta_{i,l} f_{k,j}^{(m)}, \\ &[f_{i,j}, f_{k,l}^{(m)}] = (-1)^{m+1} \delta_{j,k} e_{i,l}^{(m)} + \delta_{i,l} e_{k,j}^{(m)}. \end{aligned} \tag{4}$$

4 The Finite W-Algebra for $Q(N)$

Let e be an arbitrary nilpotent element in $\mathfrak{g} = Q(N)$, which is represented by a nilpotent $N \times N$-matrix consisting of n Jordan blocks of size l_i for $i = 1, \ldots, n$, where $l_1 \leq l_2 \leq \cdots \leq l_n$. Then $l_1 + l_2 + \cdots + l_n = N$. Let $s_p = l_1 + l_2 + \cdots + l_{p-1}$ for $2 \leq p \leq n$ and $s_1 = 0$. We include e in $\mathfrak{sl}(2) =< e_+, h, e_- >$, where $e_+ = e$:

$$\begin{aligned} e_+ &= \sum_{p=1}^{n} \sum_{i=1}^{l_p - 1} e_{s_p+i, s_p+i+1}, \\ e_- &= \sum_{p=1}^{n} \sum_{i=1}^{l_p - 1} i(l_p - i) e_{s_p+i+1, s_p+i}, \\ h &= \sum_{p=1}^{n} \sum_{i=1}^{l_p} (l_p - 2i + 1) e_{s_p+i, s_p+i}. \end{aligned}$$

Note that ad$h$ defines a Dynkin $\mathbb{Z}$-grading of $\mathfrak{g}$:

$$\mathfrak{g} = \bigoplus_{j \in \mathbb{Z}} \mathfrak{g}_j.$$

This induces a $\mathbb{Z}$-grading of $U(\mathfrak{g}) : U(\mathfrak{g}) = \bigoplus_{j \in \mathbb{Z}} U(\mathfrak{g})_j$.

Let wt$(g) = j$ for $g \in \mathfrak{g}_j$. Correspondingly, wt$(g) = j$ for $g \in U(\mathfrak{g})_j$. Let $E = \Pi(e)$. Since we have an isomorphism $\mathfrak{g}^* \cong \Pi(\mathfrak{g})$, an even nilpotent $\chi \in \mathfrak{g}^*$ can be defined by $\chi(x) := (x|E)$ for $x \in \mathfrak{g}$. Note that the Dynkin $\mathbb{Z}$-grading is good for χ. In particular, $\chi(\mathfrak{g}_j) = 0$ if $j \neq -2$. Let $g \in \mathfrak{g}_{-2}$ be an element from the standard basis of $Q(N)$ (see (2)). Then

$$\chi(e_{s_p+i+1, s_p+i}) = 1 \text{ for } i = 1, \ldots, l_p - 1, \text{ and } \chi(g) = 0 \text{ otherwise.} \tag{5}$$

We have that

$$\mathfrak{g}^{\chi} = \mathfrak{g}^{E} =< e^k_{p,q} \mid f^k_{p,q} >,$$

for $1 \le p, q \le n$ and $k = 0, 1, \dots, \min\{l_p, l_q\} - 1$, where

$$e^k_{p,q} = \sum_{r=1}^{l_p-k} e_{s_p+r, s_{q+1}-l_p+r+k}, \quad f^k_{p,q} = \sum_{r=1}^{l_p-k} (-1)^{r+k-1} f_{s_p+r, s_{q+1}-l_p+r+k}, \text{ if } p \le q,$$
$$e^k_{p,q} = \sum_{r=1}^{l_q-k} e_{s_p+r, s_q+r+k}, \quad f^k_{p,q} = \sum_{r=1}^{l_q-k} (-1)^{r+k-1} f_{s_p+r, s_q+r+k} \text{ if } p > q.$$

Note that

$$\underline{\dim}(\mathfrak{g}^{\chi}) = \left(\sum_{p=1}^{n} l_p(2n-2p+1) \mid \sum_{p=1}^{n} l_p(2n-2p+1) \right).$$

Note also that

$$\mathrm{wt}(e^k_{p,q}) = \mathrm{wt}(f^k_{p,q}) = l_q - l_p + 2k, \text{ if } p \le q, \tag{6}$$

$$\mathrm{wt}(e^k_{p,q}) = \mathrm{wt}(f^k_{p,q}) = l_p - l_q + 2k, \text{ if } p > q. \tag{7}$$

Note that the $\mathbb{Z}$-grading of $\mathfrak{g}$ is not necessarily even. In fact, it is not even when there exist $1 \le p < q \le n$ such that $l_p \not\equiv l_q \pmod 2$. In this case

$$(\mathfrak{g}_{-1})_{\bar{0}} =< e_{s_p+r, s_q+\frac{l_q-l_p-1}{2}+r}, \quad e_{s_q+\frac{l_q-l_p+1}{2}+r, s_p+r} >,$$
$$(\mathfrak{g}_{-1})_{\bar{1}} =< f_{s_p+r, s_q+\frac{l_q-l_p-1}{2}+r}, \quad f_{s_q+\frac{l_q-l_p+1}{2}+r, s_p+r} >,$$

where $1 \le p < q \le n$ are such that $l_p \not\equiv l_q \pmod 2$ and $1 \le r \le l_p$.

Let n_p be the number of q such that $1 \le p < q \le n$ and $l_p \not\equiv l_q \pmod 2$. Then

$$\underline{\dim}(\mathfrak{g}_{-1}) = \left(\sum_{p=1}^{n} 2l_p n_p \mid \sum_{p=1}^{n} 2l_p n_p \right).$$

In particular, $\dim(\mathfrak{g}_{-1})_{\bar{1}}$ is even. On can choose a Lagrangian subspace $\mathfrak{l}$ in $\mathfrak{g}_{-1}$ with respect to the form $\chi([\cdot,\cdot])$ as follows:

$$\mathfrak{l} =< e_{s_p+r, s_q+\frac{l_q-l_p-1}{2}+r} \mid f_{s_p+r, s_q+\frac{l_q-l_p-1}{2}+r} >,$$

where $1 \le p < q \le n$ are such that $l_p \not\equiv l_q \pmod 2$ and $1 \le r \le l_p$. We consider a nilpotent subalgebra $\mathfrak{m} = (\oplus_{j \le -2} \mathfrak{g}_j) \bigoplus \mathfrak{l}$ of $\mathfrak{g}$. The left ideal I_χ and W_χ are defined now as usual. Recall that we denote by $\pi : U(\mathfrak{g}) \to U(\mathfrak{g})/I_\chi$ the natural projection.

Consider the following elements in $U(\mathfrak{g})/I_\chi$:

$$\pi(e^{(l_p+k)}_{s_{p+1},s_{q+1}-l_p+1}), \quad \pi(f^{(l_p+k)}_{s_{p+1},s_{q+1}-l_p+1}) \text{ if } p \le q, \tag{8}$$

$$\pi(e^{(l_q+k)}_{s_p+l_q,s_q+1}), \quad \pi(f^{(l_q+k)}_{s_p+l_q,s_q+1}) \text{ if } p > q, \tag{9}$$

where $1 \le p, q \le n$ and $k = 0, 1, \ldots, \min\{l_p, l_q\} - 1$.

Theorem 4 *For* $1 \le p, q \le n$ *and* $k = 0, 1, \ldots, \min\{l_p, l_q\} - 1$

$$P\Big(\pi(e^{(l_p+k)}_{s_{p+1},s_{q+1}-l_p+1})\Big) = e^k_{p,q}, \quad P\Big(\pi(f^{(l_p+k)}_{s_{p+1},s_{q+1}-l_p+1})\Big) = f^k_{p,q} \text{ if } p \le q, \tag{10}$$

$$P\Big(\pi(e^{(l_q+k)}_{s_p+l_q,s_q+1})\Big) = e^k_{p,q}, \quad P\Big(\pi(f^{(l_q+k)}_{s_p+l_q,s_q+1})\Big) = f^k_{p,q} \text{ if } p > q. \tag{11}$$

Proof We will use the following realization of $Q(N)$ given by Nazarov and Sergeev in [12]. Let the indices i, j run through $-N, \ldots, -1, 1, \ldots, N$. Put $p(i) = 0$ if $i > 0$ and $p(i) = 1$ if $i < 0$. As a vector space $Q(N)$ is spanned by the elements

$$F_{ij} = E_{ij} + E_{-i,-j}$$

Note that $F_{-i,-j} = F_{ij}$. The elements F_{ij} with $i > 0$ form a basis of $Q(N)$.

For any indices $N \ge 1$ and $i, j = \pm 1, \ldots, \pm N$, we denote by $F^{(m)}_{ij}$ the following element of $U(Q(N))$:

$$F^{(m)}_{ij} = \sum_{k_1,\ldots,k_{m-1}} (-1)^{p(k_1)+\cdots+p(k_{m-1})} F_{ik_1} F_{k_1k_2} \ldots F_{k_{m-2}k_{m-1}} F_{k_{m-1}j}. \tag{12}$$

Note that

$$F^{(m)}_{ij} = (-1)^{m-1} F^{(m)}_{-i,-j},$$

$$F^{(m)}_{ij} = e^{(m)}_{i,j}, \quad \text{for } i, j > 0, \tag{13}$$

$$F^{(m)}_{ij} = (-1)^{m+1} f^{(m)}_{-i,j}, \quad \text{for } i < 0, j > 0.$$

Then

$$F^{(m)}_{ij} = f^{(m)}_{i,-j}, \quad \text{for } i > 0, j < 0. \tag{14}$$

Recall that $U(\mathfrak{g})/I_\chi \simeq S(\mathfrak{p} \oplus \mathfrak{l}')$ as a vector space. The good $\mathbb{Z}$-grading of $\mathfrak{g}$ induces a grading on $S(\mathfrak{p} \oplus \mathfrak{l}')$.

Definition 7 For any $X \in U(\mathfrak{g})/I_\chi$, let $\text{hwc}(X)$ be the highest weight component of X in this $\mathbb{Z}$-grading, and let $\text{hwt}(X)$ be its weight. Also, for $X \in U(\mathfrak{g})/I_\chi$ we denote by $\deg(X)$ the Kazhdan degree of $Gr_K(X)$.

Note that if $x = F_{i_1 j_1} F_{i_2 j_2} \dots F_{i_m j_m}$, and $X = \pi(x)$, then

$$\mathrm{hwt}(X) \le \mathrm{wt}(x) + 2m, \tag{15}$$

$$\deg(X) \le \mathrm{wt}(x) + 2m. \tag{16}$$

One can prove (15) and (16) by induction on m. If $\mathrm{hwt}(X) = \mathrm{wt}(x) + 2m$, then $\mathrm{hwc}(X) \in \mathbb{C}_\chi$ and $\mathrm{hwt}(X) = 0$. Note also that

$$\mathrm{wt}(F_{\pm s_p \pm a, \pm s_q \pm b}) = l_p - l_q - 2a + 2b \tag{17}$$

for $1 \le p, q \le n$, $1 \le a \le l_p$, $1 \le b \le l_q$. Thus

$$\mathrm{wt}(F_{ik_1} F_{k_1 k_2} \dots F_{k_{m-2} k_{m-1}} F_{k_{m-1} j}) = \mathrm{wt}(F_{ij}).$$

Set $w = \pi(F_{ik_1} F_{k_1 k_2} \dots F_{k_{m-2} k_{m-1}} F_{k_{m-1} j})$. Then

$$\mathrm{hwt}(w) \le \mathrm{wt}(F_{ij}) + 2m, \tag{18}$$

$$\deg(w) \le \mathrm{wt}(F_{ij}) + 2m. \tag{19}$$

Let $p \le q$. Let $k = 0, 1, \dots l_p - 1$. Let $i = s_{p+1}$, $j = \pm(s_{q+1} - l_p + 1)$ and $m = l_p + k$.

Assume that $\mathrm{hwt}(w) = \mathrm{wt}(F_{ij}) + 2m$. Then $\mathrm{wt}(F_{ij}) + 2m = 0$. On the other hand, by (17) $\mathrm{wt}(F_{ij}) = l_q - 3l_p + 2$. Then $l_q - 3l_p + 2 = -2m = -2(l_p + k)$, hence $l_q - l_p + 2 = -2k$. A contradiction, since $l_q \ge l_p$, and $k \ge 0$. Thus

$$\mathrm{hwt}(w) \le \mathrm{wt}(F_{ij}) + 2(m-1) = l_q - l_p + 2k = \mathrm{wt}(e^k_{p,q}) = \mathrm{wt}(f^k_{p,q}),$$

see (6).

(1) Let $j = s_{q+1} - l_p + 1$ (> 0). It follows from (5) that

$$\mathrm{hwt}(w) = \mathrm{wt}(F_{ij}) + 2(m-1)$$

if and only if

$$w = \pi(F_{i,i-1} F_{i-1,i-2} \dots F_{s_p+r, s_{q+1}-l_p+r+k} \dots F_{j+2,j+1} F_{j+1,j})$$

for some $1 \le r \le l_p - k$, and in this case

$$\mathrm{hwc}(w) = \begin{cases} F_{s_p+r, s_{q+1}-l_p+r+k}, & if \quad \mathrm{hwt}(w) \ne 0, \\ F_{s_p+r, s_{q+1}-l_p+r+k} + c,\ c \in \mathbb{C}_\chi, & if \quad \mathrm{hwt}(w) = 0. \end{cases} \tag{20}$$

Note that

$$\sum_{r=1}^{l_p-k} F_{s_p+r,s_{q+1}-l_p+r+k} = \sum_{r=1}^{l_p-k} e_{s_p+r,s_{q+1}-l_p+r+k} = e^k_{p,q}. \tag{21}$$

Then according to (12)

$$\mathrm{hwc}(\pi(F_{ij}^{(m)})) = \begin{cases} e^k_{p,q} & if \quad \mathrm{hwt}(\pi(F_{ij}^{(m)})) \neq 0, \\ e^k_{p,q} + d, d \in \mathbb{C}_\chi & if \quad \mathrm{hwt}(\pi(F_{ij}^{(m)})) = 0. \end{cases} \tag{22}$$

Note that if $d \neq 0$, then $\deg(d) = 0$ and

$$\deg(F_{s_p+r,s_{q+1}-l_p+r+k}) = \mathrm{wt}(F_{s_p+r,s_{q+1}-l_p+r+k}) + 2 =$$
$$l_q - l_p + 2k + 2 = \mathrm{wt}(F_{ij}) + 2m$$

for $r = 1, \dots, l_p - k$ (see (17)). On the other hand, by (19) we have that $\deg(\pi(F_{ij}^{(m)})) \leq \mathrm{wt}(F_{ij}) + 2m$. Hence

$$\deg(\pi(F_{ij}^{(m)})) = \mathrm{wt}(F_{ij}) + 2m \text{ and } P((\pi(F_{ij}^{(m)})) = e^k_{p,q}. \tag{23}$$

(2) Let $j = -s_{q+1} + l_p - 1$ (< 0). It follows from (5) that

$$\mathrm{hwt}(w) = \mathrm{wt}(F_{ij}) + 2(m-1)$$

if and only if

$$w = \pi(F_{i,i-1}F_{i-1,i-2}\dots F_{s_p+r,-s_{q+1}+l_p-r-k}\dots F_{j-2,j-1}F_{j-1,j})$$

for some $1 \leq r \leq l_p - k$, and in this case

$$\mathrm{hwc}(w) = \begin{cases} F_{s_p+r,-s_{q+1}+l_p-r-k}, & if \quad \mathrm{hwt}(w) \neq 0, \\ F_{s_p+r,-s_{q+1}+l_p-r-k} + c, c \in \mathbb{C}_\chi, & if \quad \mathrm{hwt}(w) = 0. \end{cases} \tag{24}$$

Note that

$$\sum_{r=1}^{l_p-k} (-1)^{r+k-1} F_{s_p+r,-s_{q+1}+l_p-r-k} = \sum_{r=1}^{l_p-k} (-1)^{r+k-1} f_{s_p+r,s_{q+1}-l_p+r+k} = f^k_{p,q}.$$

Then according to (12)

$$\mathrm{hwc}(\pi(F_{ij}^{(m)})) = \begin{cases} f^k_{p,q} & if \quad \mathrm{hwt}(\pi(F_{ij}^{(m)})) \neq 0, \\ f^k_{p,q} + d, d \in \mathbb{C}_\chi & if \quad \mathrm{hwt}(\pi(F_{ij}^{(m)})) = 0. \end{cases} \tag{25}$$

Note that

$$\deg(F_{s_p+r,-s_{q+1}+l_p-r-k}) = \mathrm{wt}(F_{s_p+r,-s_{q+1}+l_p-r-k}) + 2 =$$
$$l_q - l_p + 2k + 2 = \mathrm{wt}(F_{ij}) + 2m$$

for $r = 1, \ldots, l_p - k$ (see (17)). On the other hand, by (19) we have that $\deg(\pi(F_{ij}^{(m)}))$ $\leq \mathrm{wt}(F_{ij}) + 2m$. Hence

$$\deg(\pi(F_{ij}^{(m)})) = \mathrm{wt}(F_{ij}) + 2m \text{ and } P((\pi(F_{ij}^{(m)})) = f_{p,q}^k. \tag{26}$$

Then (10) follows from (23) and (26) according to (13) and (14).

Let $p > q$. Let $k = 0, 1, \ldots, l_q - 1$. Let $i = s_p + l_q$, $j = \pm(s_q + 1)$ and $m = l_q + k$.

Assume that $\mathrm{hwt}(w) = \mathrm{wt}(F_{ij}) + 2m$. Then $\mathrm{wt}(F_{ij}) + 2m = 0$. But by (17) $\mathrm{wt}(F_{ij}) = l_p - 3l_q + 2$. Then $l_p - 3l_q + 2 = -2m = -2(l_q + k)$, hence $l_p - l_q + 2 = -2k$. A contradiction, since $l_p \geq l_q$, and $k \geq 0$. Thus

$$\mathrm{hwt}(w) \leq \mathrm{wt}(F_{ij}) + 2(m-1) = l_p - l_q + 2k = \mathrm{wt}(e_{p,q}^k) = \mathrm{wt}(f_{p,q}^k),$$

see (7).

(1) Let $j = s_q + 1$ (> 0). By (5) we have that

$$\mathrm{hwt}(w) = \mathrm{wt}(F_{ij}) + 2(m-1)$$

if and only if

$$w = \pi(F_{i,i-1}F_{i-1,i-2}\ldots F_{s_p+r,s_q+r+k}\ldots F_{j+2,j+1}F_{j+1,j})$$

for some $1 \leq r \leq l_q - k$, and in this case

$$\mathrm{hwc}(w) = \begin{cases} F_{s_p+r,s_q+r+k}, & if \;\; \mathrm{hwt}(w) \neq 0, \\ F_{s_p+r,s_q+r+k} + c,\, c \in \mathbb{C}_\chi, & if \;\; \mathrm{hwt}(w) = 0. \end{cases} \tag{27}$$

Note that

$$\sum_{r=1}^{l_q-k} F_{s_p+r,s_q+r+k} = \sum_{r=1}^{l_q-k} e_{s_p+r,s_q+r+k} = e_{p,q}^k.$$

This implies (22) according to (12). Note that

$$\deg(F_{s_p+r,s_q+r+k}) = \mathrm{wt}(F_{s_p+r,s_q+r+k}) + 2 = l_p - l_q + 2k + 2 = \mathrm{wt}(F_{ij}) + 2m$$

for $r = 1, \ldots, l_q - k$ (see (17)). On the other hand, by (19) we have that $\deg(\pi(F_{ij}^{(m)}))$ $\leq \mathrm{wt}(F_{ij}) + 2m$. Hence

$$\deg(\pi(F_{ij}^{(m)})) = \mathrm{wt}(F_{ij}) + 2m \text{ and } P((\pi(F_{ij}^{(m)})) = e^k_{p,q}. \tag{28}$$

(2) Let $j = -s_q - 1$ (< 0). By (5) we have that

$$\mathrm{hwt}(w) = \mathrm{wt}(F_{ij}) + 2(m-1)$$

if and only if

$$w = \pi(F_{i,i-1}F_{i-1,i-2}\dots F_{s_p+r,-s_q-r-k}\dots F_{j-2,j-1}F_{j-1,j})$$

for some $1 \leq r \leq l_q - k$, and in this case

$$\mathrm{hwc}(w) = \begin{cases} F_{s_p+r,-s_q-r-k}, & if \quad \mathrm{hwt}(w) \neq 0, \\ F_{s_p+r,-s_q-r-k} + c, c \in \mathbb{C}_\chi, & if \quad \mathrm{hwt}(w) = 0. \end{cases} \tag{29}$$

Note that

$$\sum_{r=1}^{l_q-k}(-1)^{r+k-1}F_{s_p+r,-s_q-r-k} = \sum_{r=1}^{l_q-k}(-1)^{r+k-1}f_{s_p+r,s_q+r+k} = f^k_{p,q}.$$

This implies (25) according to (12). Note that

$$\deg(F_{s_p+r,-s_q-r-k}) = \mathrm{wt}(F_{s_p+r,-s_q-r-k}) + 2 = l_p - l_q + 2k + 2 = \mathrm{wt}(F_{ij}) + 2m$$

for $r = 1, \dots, l_q - k$ (see (17)). On the other hand, by (19) we have that $\deg(\pi(F_{ij}^{(m)})) \leq \mathrm{wt}(F_{ij}) + 2m$. Hence

$$\deg(\pi(F_{ij}^{(m)})) = \mathrm{wt}(F_{ij}) + 2m \text{ and } P((\pi(F_{ij}^{(m)})) = f^k_{p,q}. \tag{30}$$

Then (11) follows from (28) and (30) according to (13) and (14). □

Corollary 1 *If the nilpotent element e has Jordan blocks of equal size, then the elements given in (8) and (9) form a set of generators for* W_χ.

Proof Assume that $l_1 = l_2 = \cdots = l_n = l$. Then the elements in (8) and (9) can be written as follows:

$$\pi(e^{(l+k)}_{lp,l(q-1)+1}), \quad \pi(f^{(l+k)}_{lp,l(q-1)+1})$$

for $p, q = 1, \dots, n$ and $k = 0, \dots, l-1$. By (4) we have that

$$[e_{l(s-1)+i,l(t-1)+j}, e^{(m)}_{lp,l(q-1)+1}] = [f_{l(s-1)+i,l(t-1)+j}, e^{(m)}_{lp,l(q-1)+1}] = 0,$$

$$[e_{l(s-1)+i,l(t-1)+j}, f^{(m)}_{lp,l(q-1)+1}] = [f_{l(s-1)+i,l(t-1)+j}, f^{(m)}_{lp,l(q-1)+1}] = 0,$$

for all $l \geq i > j \geq 1$, $1 \leq p, q, s, t \leq n$ and $m \geq 1$. In other words,

$$e^{(m)}_{lp,l(q-1)+1}, \quad f^{(m)}_{lp,l(q-1)+1} \in U(\mathfrak{g})^{ad\,\mathfrak{m}}.$$

Hence $\pi(e^{(m)}_{lp,l(q-1)+1}), \pi(f^{(m)}_{lp,l(q-1)+1}) \in W_\chi$. The proof now follows from Theorems 4 and 3(a). □

Acknowledgements This work was supported by a grant from the Simons Foundation (#354874, Elena Poletaeva). I thank V. Serganova and V. Stukopin for very helpful discussions.

References

1. Briot, C., Ragoucy, E.: W-superalgebras as truncations of super-Yangians. J. Phys. A **36**(4), 1057–1081 (2003)
2. Brown, J., Brundan, J., Goodwin, S.: Principal W-algebras for $GL(m|n)$. Algebra Numb. Theory **7**, 1849–1882 (2013)
3. Brundan, J., Goodwin, S.: Whittaker coinvariants for $GL(m|n)$. arXiv:1612.08152v1
4. Brundan, J., Kleshchev, A.: Shifted Yangians and finite W-algebras. Adv. Math. **200**, 136–195 (2006)
5. De Sole, A., Kac, V.G.: Finite vs affine W-algebras. Japan. J. Math. **1**, 137–261 (2006)
6. Gorelik, M.: The Kac construction of the centre of $U(g)$ for Lie superalgebras. J. Nonlinear Math. Phys. **11**(3), 325–349 (2004)
7. Kac, V.G.: Lie superalgebras. Adv. Math. **26**, 8–96 (1977)
8. Kac, V.G.: Laplace operators of infinite-dimensional Lie algebras and theta functions. Proc. Natl. Acad. Sci. USA **81**, 645–647 (1984)
9. Kostant, B.: On Whittaker vectors and representation theory. Invent. Math. **48**, 101–184 (1978)
10. Losev, I.: Finite W-algebras. In: Proceedings of the International Congress of Mathematicians, vol. **III**, 1281–1307, Hindustan Book Agency, New Delhi (2010). arXiv:1003.5811v1
11. Nazarov, M.: Yangian of the queer Lie superalgebra. Comm. Math. Phys. **208**, 195–223 (1999)
12. Nazarov, M., Sergeev, A.: Centralizer construction of the Yangian of the queer Lie superalgebra. In: Studies in Lie Theory. Progress in Mathematics, vol. 243, pp. 417–441. Birkhäuser Boston, Boston, MA (2006)
13. Peng, Y.: Finite W-superalgebras and truncated super Yangians. Lett. Math. Phys. **104**, 89–102 (2014)
14. Peng, Y.: On shifted super Yangians and a class of finite W-superalgebras. J. Algebra **422**, 520–562 (2015)
15. Poletaeva, E.: On principal finite W-algebras for the Lie superalgebra $D(2,1;\alpha)$. J. Math. Phys. **57**(5), 051702 (2016)
16. Poletaeva, E.: On finite W-algebras for Lie superalgebras in non-regular case. In: Dobrev, V. (ed) Proceedings of the XI International Workshop "Lie Theory and Its Applications in Physics" (Varna, Bulgaria, 15–21 June 2015). Springer Proceedings in Mathematics and Statistics, vol. 191, pp. 477–488. Springer, Tokyo-Heidelberg (2016)
17. Poletaeva, E., Serganova, V.: On finite W-algebras for Lie superalgebras in the regular case. In: Dobrev, V. (ed) Proceedings of the IX International Workshop "Lie Theory and Its Applications in Physics" (Varna, Bulgaria, 20–26 June 2011). Springer Proceedings in Mathematics and Statistics, vol. 36, pp. 487–497 (2013)
18. Poletaeva, E., Serganova, V.: On Kostant's theorem for the Lie superalgebra $Q(n)$. Adv. Math. **300**, 320–359 (2016). arXiv:1403.3866v1
19. Poletaeva, E., Serganova, V.: On the finite W-algebra for the Lie superalgebra $Q(n)$ in the non-regular case. J. Math. Phys. **58**(11), 111701 (2017). arXiv:1705.10200
20. Premet, A.: Special transverse slices and their enveloping algebras. Adv. Math. **170**, 1–55 (2002)

21. Sergeev, A.: The centre of enveloping algebra for Lie superalgebra $Q(n, \mathbb{C})$. Lett. Math. Phys. **7**, 177–179 (1983)
22. Wang, W.: Nilpotent orbits and finite W-algebras. In: Geometric Representation Theory and Extended Affine Lie Algebras, pp. 71–105. Fields Institute Communications, vol. 59. American Mathematical Society, Providence, RI (2011). arXiv:0912.0689v2
23. Zeng, Y., Shu, B.: Finite W-superalgebras for basic Lie superalgebras. J. Algebra **438**, 188–234 (2015). arXiv:1404.1150v2
24. Zhao, L.: Finite W-superalgebras for queer Lie superalgebras. J. Pure Appl. Algebra **218**, 1184–1194 (2014)

An Alternate Proof to Derek Robinson's 1968 Local Characterization Theorem on $\mathcal{T}$-Groups

Matthew F. Ragland

Abstract Define $\mathcal{T}$ to be the class of groups in which normality is a transitive relation. Define $\mathscr{C}_p$ to be the class of finite groups G for which each subgroup of a Sylow p-subgroup of G is normal in the corresponding Sylow normalizer. In [5] Robinson proved that a finite group satisfies $\mathscr{C}_p$ for all primes p if and only if it is a finite solvable $\mathcal{T}$-group. Here a new proof of this classic and influential result is presented.

Keywords Solvable Group · T-Group · Supersolvable Group · Sylow Subgroup

1 Introduction

A group is said to be a $\mathcal{T}$-group if normality is a transitive relation. In [5], the first local characterization of finite solvable $\mathcal{T}$-groups was given. Many different local characterizations of $\mathcal{T}$-groups have appeared in the literature since Robinson's first, but Robinson's characterization has arguably been the most influential. According to Google Scholar, it has been cited 92 times in the current literature at the time of this writing. As in [5], define $\mathscr{C}_p$ (p is a fixed prime) to be the class of groups G where each subgroup of a Sylow p-subgroup P of G is normal in the normalizer, $N_G(P)$, of P. Robinson's result states that G is a finite $\mathscr{C}_p$-group for all primes p if and only if G is a finite solvable $\mathcal{T}$-group.

The aim of this short article is to provide a new proof of Robinson's well-known result which is much more elementary than Robinson's original. The theory of transfer, Grün's Second Theorem, and Frobenius's criterion for p-nilpotence are used in the original whereas the proof given here uses basic principles such as the subgroup and quotient closure properties of $\mathscr{C}_p$. Also, an attempt is made to keep the proof as self-contained as possible so that one not familiar with the area can read the proof with essentially no references save for a standard text on group theory. In particular, and fittingly, Robinson's text [6] is all that is needed.

M. F. Ragland (✉)
Auburn University at Montgomery, P.O. Box 244023, Montgomery 36124-4023, USA
e-mail: mragland@aum.edu

J. Feldvoss et al. (eds.), *Advances in Algebra*, Springer Proceedings in Mathematics & Statistics 277, https://doi.org/10.1007/978-3-030-11521-0_15

2 Preliminaries

Let us first record the definition of $\mathscr{C}_p$ and the theorem already mentioned in the introduction for the purposes of reference.

Definition 1 Define $\mathscr{C}_p$ to be the class of finite groups G for which each subgroup of a Sylow p-subgroup P of G is normal in the normalizer, $N_G(P)$, of P. That is, $G \in \mathscr{C}_p$ if $H \leq P \in \mathrm{Syl}_p(G) \Rightarrow H \trianglelefteq N_G(P)$.

Theorem 1 (Robinson [5]) *Let G be a finite group. Then G is a solvable $\mathcal{T}$-group if and only if G is a $\mathscr{C}_p$-group for all primes p.*

In [5], as a consequence of Theorem 1, Robinson deduced that subgroups and quotients of $\mathscr{C}_p$-groups are again $\mathscr{C}_p$-groups. In [1] Ballester-Bolinches and Esteban-Romero gave a quick proof showing $\mathscr{C}_p$ is a subgroup-closed class. Their proof is provided for completeness. Although full quotient closure without the aid of Theorem 1 appears difficult, it is not hard to show that G/N is a $\mathscr{C}_p$-group when G is a $\mathscr{C}_p$-group and N is either a p-group or a p'-group. Robinson's original proof of Theorem 1 is striking in the sense that it does not use these facts.

Lemma 1 (Robinson [5]) *$\mathscr{C}_p$ is a subgroup-closed class.*

Proof (**Ballester-Bolinches and Esteban-Romero** [1]) Let $K \leq H_p \leq H \leq G \in \mathscr{C}_p$ with $H_p \in \mathrm{Syl}_p(H)$. Since G is a $\mathscr{C}_p$-group, every Sylow normalizer of a Sylow p-subgroup of G containing H_p normalizes K. Let $G_p \in \mathrm{Syl}_p(G)$ with $H_p \leq G_p$ so that $K \trianglelefteq N_G(G_p)$. If $g \in N_H(H_p)$, then $H_p \leq {G_p}^g$ and so $K \trianglelefteq N_G({G_p}^g) = N_G(G_p)^g$. Hence $K \trianglelefteq \langle N_G(G_p), N_G(G_p)^g \rangle$. Sylow normalizers are abnormal (see 9.2.10 of [6]) and so we have $g \in \langle N_G(G_p), N_G(G_p)^g \rangle$. Thus g normalizes K so that $K \trianglelefteq N_H(H_p)$ showing that H is a $\mathscr{C}_p$-group. □

Lemma 2 *Let G be a finite $\mathscr{C}_p$-group. Then the following hold:*

1. *If N is a normal p-subgroup of G, then G/N is a $\mathscr{C}_p$-group.*
2. *If N is a normal p'-subgroup of G, then G/N is a $\mathscr{C}_p$-group.*

Proof Part (1) is clear from the definition of $\mathscr{C}_p$.

For part (2), let H/N be a subgroup of P/N where $P/N \in \mathrm{Syl}_p(G/N)$ with G/N a $\mathscr{C}_p$-group and N a normal p'-subgroup of G. Then there exist Sylow p-subgroups H_p and P_p of H and P, respectively, such that $H_p \leq P_p$, $H = H_pN$, and $P = P_pN$. Since $G \in \mathscr{C}_p$, we know $H_p \trianglelefteq N_G(P_p)$. Therefore

$$H/N = H_pN/N \trianglelefteq N_G(P_p)N/N = N_{G/N}(P/N).$$

Thus G/N is a $\mathscr{C}_p$-group. □

The following result is key in the new proof. We provide a proof for completeness.

Lemma 3 (Doerk [2]) *If G is a minimal non-supersolvable group, then G possesses a normal Sylow subgroup.*

Proof Let G be a minimal non-supersolvable group with p the smallest prime divisor of $|G|$. Let us suppose G is not p-nilpotent and let M be a maximal subgroup of G. Then M is supersolvable and by a theorem of Zappa (see 5.4.8 in [6]), $O_{p'}(M)$ is a normal Hall p'-subgroup of M. Hence M is p-nilpotent and we see that every maximal subgroup of G is p-nilpotent. By a result of Ito (see 10.3.3 in [6]), we have that G must contain a normal Sylow p-subgroup.

Now suppose G is p-nilpotent. Hence G has a normal Hall p'-subgroup, say H. Now H is supersolvable and hence contains a normal Sylow subgroup, say Q. But Q is characteristic in H and H is characteristic in G whence Q is normal in G. Since H is a Hall subgroup of G, we have Q is a Sylow subgroup of G as well. This completes the proof. □

In addition to the above results, we need the following three facts due to Gaschütz [3] concerning $\mathcal{T}$-groups. For proofs of Lemmas 4 and 5 see 13.4.5 and 13.4.7, respectively, in [6]. We provide a short proof of Lemma 6 for completeness.

Lemma 4 (Gaschütz [3]) *Let $N \trianglelefteq G$ with G a finite group satisfying the following:*

1. *G/N is a $\mathcal{T}$-group;*
2. *H subnormal in N implies $H \trianglelefteq G$;*
3. *N is a Hall subgroup of G.*

Then G is a $\mathcal{T}$-group.

Lemma 5 (Gaschütz [3]) *Subgroups of finite solvable $\mathcal{T}$-groups are $\mathcal{T}$-groups.*

Lemma 6 (Gaschütz [3]) *A finite solvable $\mathcal{T}$-group is supersolvable.*

Proof Let G be a finite solvable $\mathcal{T}$-group with N a minimal normal subgroup of G. N is an elementary abelian p-group for some prime p. Let $g \in N$ with g of order p. Then $\langle g\rangle \trianglelefteq N \trianglelefteq G$. Hence $\langle g\rangle \trianglelefteq G$ and we have that N is cyclic of order p. It is clear that G/N is also a $\mathcal{T}$-group and, by induction, must be supersolvable. Since N is cyclic, we must have G supersolvable. □

3 Proof of Theorem 1

Now we are in a position to provide a short proof of Theorem 1.

Proof If G is a finite solvable $\mathcal{T}$-group, then so are all of its subgroups by Lemma 5. Suppose $H \leq P \in \mathrm{Syl}_p(G)$. Then H subnormal in $N_G(P) \in \mathcal{T}$ gives rise to $H \trianglelefteq N_G(P)$. Thus G is a $\mathscr{C}_p$-group for all primes p.

Now suppose G is a finite $\mathscr{C}_p$-group for all primes p. We will induct on $|G|$. Suppose X is a proper subgroup of G. Then $X \in \mathscr{C}_p$ for all primes p by Lemma 1.

By induction, X is a solvable $\mathcal{T}$-group and hence, by Lemma 6, is supersolvable. So we can deduce that G is either supersolvable or G is a minimal non-supersolvable group. With the use of Lemma 3, we see that G contains a normal Sylow subgroup. Let P be a normal Sylow p-subgroup of G. By Lemma 2 part (1), we have that G/P is a $\mathscr{C}_p$-group. For primes $q \neq p$, Lemma 2 part (2) tells us that G/P is a $\mathscr{C}_q$-group. By induction we can deduce that G/P is a solvable $\mathcal{T}$-group. If H is subnormal in P, then by hypothesis, $H \trianglelefteq N_G(P) = G$. Since P is a Sylow subgroup, and hence a Hall subgroup, we can deduce from Lemma 4 that G is a $\mathcal{T}$-group. That G is solvable follows from the fact that P and G/P are both solvable. □

Acknowledgements Derek Robinson's 1968 paper [5] on $\mathcal{T}$-groups is the first mathematical article that the author read as a Ph.D. student under Professor James (Jim) Clark Beidleman's guidance. The author fondly remembers working through details of the paper and meeting regularly with Dr. Beidleman who used the result in his work often. He once suggested that the author's alternate proof presented here (arrived at many years after working through the original proof) be published. Sadly, Dr. Beidleman passed away recently [4]; this article is dedicated to his memory.

References

1. Ballester-Bolinches, A., Esteban-Romero, R.: Sylow permutable subnormal subgroups of finite groups. J. Algebra **251**, 727–738 (2002)
2. Doerk, K.: Minimal nicht überauflösbare, endliche Gruppen. Math. Z. **91**, 198–205 (1966)
3. Gaschütz, W.: Gruppen, in denen das Normalteilersein transitiv ist. J. Reine Angew. Math. **198**, 87–92 (1957)
4. Ragland, M.: James C. Beidleman (1937-2017). Adv. Group Theory Appl. **4**, 1–2 (2018)
5. Robinson, D.J.S.: A note on finite groups in which normality is transitive. Proc. Amer. Math. Soc. **19**, 933–937 (1968)
6. Robinson, D.J.S.: A course in the theory of groups. In: Graduate Texts in Mathematics, 2nd edn., vol. 80. Springer-Verlag, New York (1996)

Some Tables of Right Set Properties in Affine Weyl Groups of Type A

Leonard L. Scott and Ethan C. Zell

Abstract The tables of the title are a first attempt to understand empirically the sizes of certain distinguished sets, introduced by Hankyung Ko, of elements in affine Weyl groups. The distinguished sets themselves each have a largest element w, and all other elements are constructible combinatorially from that largest element. The combinatorics are given in the language of right sets, in the sense of Kazhdan–Lusztig. Collectively, the elements in a given distinguished set parameterize highest weights of possible modular composition factors of the "reduction modulo p" of a pth root of unity irreducible characteristic 0 quantum group module. Here, p is a prime, subject to conditions discussed below, in some cases known to be quite mild. Thus, the sizes of the distinguished sets in question are relevant to estimating how much time might be saved in any future direct approach to computing irreducible modular characters of algebraic groups from larger irreducible characters of quantum groups. Actually, Ko has described two methods for obtaining potentially effective systems of such sets. She has proved one method to work at least for all primes p as large as the Coxeter number h, in a context she indicates largely generalizes to smaller p. The other method, which produces smaller distinguished sets, is known for primes $p \geq h$ for which the Lusztig character formula holds, but is currently unknown to be valid without the latter condition. In the tables of this paper, we calculate, for all w indexing a (p-)regular highest weight in the (p-)restricted parallelotope, distinguished set sizes for both methods, for affine types A_3, A_4, and A_5. To keep the printed version of this paper sufficiently small, we only use those w indexing actual restricted weights in the A_5 case. The sizes corresponding to the two methods of Ko are listed in columns (6) and (5), respectively, of the tables. We also make calculations in column (7) for a third, more "obvious" system of distinguished sets (see part (1) of Proposition 1), to indicate how much of an improvement each of the first two systems provides. Finally,

This research was supported by Simons Foundation grant 359363 and the University of Virginia.

L. L. Scott (✉) · E. C. Zell
Department of Mathematics, University of Virginia, Charlottesville, VA 22903, USA
e-mail: lls2l@virginia.edu

E. C. Zell
e-mail: ecz4wu@virginia.edu

J. Feldvoss et al. (eds.), *Advances in Algebra*, Springer Proceedings in Mathematics & Statistics 277, https://doi.org/10.1007/978-3-030-11521-0_16

all calculations have been recently completed for affine type A_6, and the restricted cases are listed in this paper as a final table.

Keywords Right set · Sage · Modular characters · Quantum · Lusztig · Weyl

1 Introduction

In a recent paper, [9] Hankyung Ko has introduced two methods potentially useful for algorithmic calculations of irreducible modular characters of semisimple algebraic groups G in characteristic $p > 0$. It has been well known since the early '60s that the larger "Weyl" (or "standard") modules provide at least one setting for calculating such irreducible modular characters, using natural bilinear forms. See [12, 16][1] Such bilinear form approaches were largely forgotten when Lusztig proposed a modular character formula [8], shown to hold for primes p large relative to the root system in [3], with an explicit lower bound on p given in [6]. However, the large sizes of negative examples in [15], as well as the general need for results applying for smaller primes, led Williamson and collaborators to a recent series of papers [1, 2, 11]. These results are conceptually quite elegant, and already provide formulas in all types with lower bounds on p close to those originally proposed by Lusztig (and apply for all primes in type A). However, the cost in computational complexity is hard to estimate. Briefly, characters of irreducible modules are recovered in [11] from ingredients in new formulas for characters of tilting modules. The ingredients themselves are obtained from considerations and calculations using p-canonical bases, and calculation of the latter involves a bilinear form, see [14].

This background makes it reasonable to at least investigate what can be done to revive the original bilinear form approach, and the results of Ko we discuss here should be viewed in that context. Before going further, we introduce some of the notation used in [9]. Much of it also largely follows Jantzen [7].

2 Notation and Preliminaries

Let G be a semisimple simply connected algebraic group over an algebraically closed field k of characteristic $p > 0$. Fix a split maximal torus T with associated root system R and integral weight lattice $X = X(T)$. Fix also a choice R^+ for the positive roots in

[1]Steinberg had noticed the form used by Wong earlier. His context was different, but the form he found could be used to construct Wong's form. Steinberg worked with enveloping algebras, whereas Wong worked with Weyl modules, focusing on their irreducible heads. It is hard to know if Steinberg had modular irreducible representations in mind with his form, though such representations were, of course, one of his interests. He was later the AMS reviewer of Wong's paper and does not mention noticing such a potential application, only that he had previously observed the form in his (widely distributed) Yale lecture notes.

R, and a corresponding set $X^+ \subseteq X$ of dominant weights. The sum of all fundamental dominant weights is denoted ρ. The ordinary finite Weyl group associated to R is denoted W_f, and the affine Weyl group is W_p; it is the semidirect product of $p\mathbb{Z}R$ with W_f. (This semidirect product makes sense for p any positive integer, such as $p = 1$ or $p = h$, the Coxeter number. This will be useful later in our tables.) The "dot action" of an element $w \in W_p$ on an element $\gamma \in X$ is given by $w \cdot \gamma = w(\gamma + \rho) - \rho$. In this action, reflections in hyperplanes passing through $-\rho$ and orthogonal to fundamental roots in R form a fundamental set $s_1, s_2, \ldots$ of generators for W_f, in the sense of Coxeter groups. We choose an additional element of W_p to be the reflection s_0 in the hyperplane $\{x \mid (x + \rho, \alpha_0^\vee) = -p\}$ to complete the set of fundamental generators S for W_p. (Here α_0 is the maximum short root, and we do NOT follow Jantzen, who would use $+p$, instead of $-p$, to define the hyperplane. In the terminology of [9], our S consists of reflections in the walls of the top antidominant alcove C^-, not the bottom dominant alcove.) This is a good time to mention that the right set $\mathscr{R}(w)$ of an element $w \in W_p$ is defined to be $\{s \in S \mid ws < w\}$. Here the inequality is in the sense of Bruhat(-Chevalley), which, in this case, just means that ws is shorter than w. The terminology "right descent set" is used in place of "right set" in [9], and appears also in [14].

We will consider, for each dominant weight $\lambda \in X^+$, three finite-dimensional G-modules, $\Delta(\lambda)$, $\Delta^0(\lambda)$, and $L(\lambda)$. The third is the irreducible G-module of highest weight λ, whose character we wish to calculate. The first module is the Weyl (or "standard") module with highest weight λ. It may be obtained by "reduction mod p" (base change to k) from a module for a $\mathbb{Z}$-form of the enveloping algebra for the complex semisimple Lie algebra associated to R. This "integral" module is irreducible upon base change to $\mathbb{Q}$ or $\mathbb{C}$. It follows that the character of $\Delta(\lambda)$ is known, given by Weyl's character formula. With somewhat more care with coefficients, a similar process, using a quantum enveloping algebra at a root of unity, rather than a Lie algebra enveloping algebra, can be used to construct $\Delta^0(\lambda)$. See the discussion in [9], or the more available [5], which uses the notation $\Delta^{red}(\lambda)$. To summarize, $\Delta^0(\lambda)$ is the "reduction mod p" of an irreducible quantum enveloping algebra module at a root of unity (a pth root of unity, for p odd). For all but a few primes, its character is known to be given by Lusztig's character formula in simply laced types of all ranks (for all primes in type A), and in all types when $p > h$, see [13, p. 273]. The module $L(\lambda)$ is a homomorphic image (and the head) of $\Delta^0(\lambda)$, and the latter is a homomorphic image of $\Delta(\lambda)$.

It is sufficient, to motivate our tables, to work with the p-regular case. This also simplifies notation, since, when $p \geq h$, most questions about modules for G reduce to those for modules with highest weight in the orbit $W_p \cdot -2\rho$. Here $-2\rho \in C^-$ and, letting w_0 denote the longest word in the ordinary Weyl group W_f, we have $0 = w_0 \cdot -2\rho$. The weight 0 is the highest weight of the 1-dimensional trivial module $L(0)$. More generally, the dominant weights λ in the orbit $W_p \cdot -2\rho$ are precisely the elements $w_0 y \cdot -2\rho$ where $y \in W_p$ satisfies $\ell(w_0 y) = \ell(w_0) + \ell(y)$. That is, y is the shortest element in its right coset $W_f y$; also $w_0 y$ is the longest element in its (the same) right coset of W_f. Equivalently, y is the unique minimal element in its right coset (using the Bruhat order), and $w_0 y$ is the unique maximal element in its right coset. Ko denotes the set of all elements $w_0 y$ of W_p maximal in their right coset of W_f by W^+ (with no subscript p, perhaps to emphasize the independence of the definition on alcove geometry). Finally, if $w \in W^+$, then $w \cdot -2\rho$ is dominant, and we have already observed a converse. In this way, we have a natural 1-1 correspondence between dominant weights in $W_p \cdot -2\rho$ and elements of W^+.

3 The Results Motivating the Tables, and a Question

The following result in the p-regular case restates [9, Prop. 4.3] in conclusion (2), and combines it with a consequence (1) of [9, Prop. A.4]. An analog of [9, Prop. 4.3] in the p-singular case, presumably allowing smaller primes, is also sketched in [9, Rem. 4.4].

Proposition 1 *Suppose μ, λ are dominant weights with $L(\mu)$ a composition factor of $\Delta^0(\lambda)$ and with $\lambda \in W_p \cdot -2\rho$. Then also $\mu \in W_p \cdot -2\rho$* (well known), *so that there are unique elements $v, w \in W^+$ with $\mu = v \cdot -2\rho$ and $\lambda = w \cdot -2\rho$. Moreover, (1) $v \leq w$, and (2) $\mathscr{R}(w) \subseteq \mathscr{R}(v)$.*

Proof The assertion that λ and μ belong to the same "dot" orbit is the well-known linkage principle. Applying strong linkage in the case of Weyl modules, and using the fact that $\Delta^0(\lambda)$ is a homomorphic image of the Weyl module $\Delta(\lambda)$, shows μ is strongly linked to λ. This strong linkage implies conclusion (1) by using [9, Prop. A.4]. (Or use [10, Prop. 9.1].) Conclusion (2) of the proposition follows directly from [9, Prop. 4.3], which gives (2) after starting from a similar hypothesis in the more general case of regular dominant weights. (They are always parameterized as belonging to $W^+ \cdot \nu$ for some element ν in the top antidominant alcove C^-.)

The following is a variation, in the p-regular case, on a question raised by Ko at the beginning of [9, Sect. 5]. She also gives a version which applies in the p-singular case.

Question 1 *Suppose $p \geq h$, and let $v, w \in W^+$ be such that $L(v \cdot -2\rho)$ is a composition factor of $\Delta^0(w \cdot -2\rho)$ Is it then true that $\mathscr{R}(w) = \mathscr{R}(v)$?*

The arguments in the proof of [9, Prop. 5.8], in particular [9, Lem. 5.9], show that the question above reduces to the case of those $w \in W^+$ for which $w \cdot -2\rho$ is p-restricted. No form of the "Lusztig conjecture" is needed for this reduction. However, as Ko in effect observes, the question above has a positive answer whenever $\Delta^0(w \cdot -2\rho)$ is irreducible in all cases for which $w \in W^+$ and $w \cdot -2\rho$ is p-restricted. This is equivalent to the (Kato form) of the Lusztig conjecture, when $p > h$, or in affine type A (and others) when $p \geq h$. (Kato's version, for a given $p \geq h$, says that the Lusztig character formula holds for any irreducible module $L(w \cdot -2\rho)$ with $w \cdot -2\rho$ p-restricted. In the quantum case, the character formula is known to hold for all dominant $w \cdot -2\rho$, under the given, slightly different, requirements on p. See [13, p. 273].)

However, this Ko question potentially has positive answers even when the Lusztig conjecture does not. The tables we present suggest to us that computation time (for characters of irreducible G-modules) could be significantly reduced when the question has a positive answer. The weaker, but proved, assertion (2) in Proposition 1 provides a smaller reduction, but could also be useful.

4 The Tables

Assume $p \geq h$. The tables deal with those $w \in W^+$ for which $w \leq u$ in the Bruhat order for some $u \in W^+$ with the weight $u \cdot -2\rho$ p-restricted. We may rechoose $u \cdot -2\rho$ in the same p alcove as $(p-2)\rho$. Consequently, the w's and u's have a common unique maximal element $w_{\max}$. The $w \leq w_{\max}$ in W^+ may all be written uniquely as a product $w = w_0 y$ with $y \in W := W_p$ and lengths additive. The elements y are listed, as a reduced product of fundamental reflections, in column (1) of the tables below. The element w_0 is listed above each table, and the length of w recorded in column (4). Column (3) gives, $w \cdot -2\rho$ as a vector of coefficients at fundamental weights ("ω-coefficient"), in the generic case $p = h$. Column (2) gives, in all tables but that for affine A_6, the same element expressed as a vector ("ε-vector") in terms of the standard basis of the standard permutation module for W_f, a symmetric group for each W considered (affine type A_n for $n = 3, 4, 5, 6$). The main content lies in columns (5), (6), and (7), discussed informally in the abstract. Column (7) counts those $v \in W^+$ with $v \leq w$ in the Bruhat order; column (6) counts those among these v which also satisfy $\mathscr{R}(w) \subseteq \mathscr{R}(v)$. Finally, column (5) counts those among these v which satisfy $\mathscr{R}(w) = \mathscr{R}(v)$.

Recall that each $w = w_0 y$, where w_0 is $s_1 s_2 s_3 s_1 s_2 s_1$ for type A_3.

$y \in W$	$w \cdot (-2\rho)$	ω-coefficients	$l(w)$	(5)	(6)	(7)
$s_0 s_3 s_1 s_2$	(3, 1, −1, −3)	(2, 2, 2)	10	1	5	8
$s_0 s_1 s_2$	(3, −1, −1, −1)	(4, 0, 0)	9	1	1	4
$s_0 s_3 s_1$	(2, 1, −1, −2)	(1, 2, 1)	9	1	2	5
$s_0 s_3 s_2$	(1, 1, 1, −3)	(0, 0, 4)	9	1	1	4
$s_0 s_1$	(2, 0, −1, −1)	(2, 1, 0)	8	1	1	3
$s_0 s_3$	(1, 1, 0, −2)	(0, 1, 2)	8	1	1	3
s_0	(1, 0, 0, −1)	(1, 0, 1)	7	1	1	2
1	(0, 0, 0, 0)	(0, 0, 0)	6	1	1	1

For type A_4, w_0 is $s_1s_2s_3s_4s_1s_2s_3s_1s_2s_1$.

$y \in W$	$w \cdot (-2\rho)$	ω-coefficients	$l(w)$	(5)	(6)	(7)
$s_0s_4s_1s_2s_3s_0s_4s_2s_1s_0$	(6, 3, 0, −3, −6)	(3, 3, 3, 3)	20	1	31	52
$s_0s_1s_2s_3s_4s_3s_2s_1s_0$	(6, 0, 0, 0, −6)	(6, 0, 0, 6)	19	5	7	34
$s_0s_4s_1s_2s_3s_0s_4s_1s_0$	(6, 3, −3, −3, −3)	(3, 6, 0, 0)	19	3	4	26
$s_0s_4s_1s_2s_3s_0s_4s_2s_0$	(6, 1, 0, −3, −4)	(5, 1, 3, 1)	19	5	6	32
$s_0s_4s_1s_2s_3s_0s_4s_2s_1$	(5, 3, 0, −3, −5)	(2, 3, 3, 2)	19	2	15	38
$s_0s_4s_1s_2s_3s_2s_0s_1s_0$	(4, 3, 0, −1, −6)	(1, 3, 1, 5)	19	5	6	32
$s_0s_4s_3s_1s_0s_4s_2s_1s_0$	(3, 3, 3, −3, −6)	(0, 0, 6, 3)	19	3	4	26
$s_0s_1s_2s_3s_4s_3s_2s_0$	(6, 0, 0, −2, −4)	(6, 0, 2, 2)	18	3	5	26
$s_0s_1s_2s_3s_4s_3s_2s_1$	(5, 0, 0, 0, −5)	(5, 0, 0, 5)	18	2	2	21
$s_0s_4s_1s_2s_3s_0s_4s_0$	(6, 1, −1, −3, −3)	(5, 2, 2, 0)	18	2	3	22
$s_0s_4s_1s_2s_3s_0s_4s_1$	(5, 3, −2, −3, −3)	(2, 5, 1, 0)	18	3	5	22
$s_0s_4s_1s_2s_3s_0s_4s_2$	(5, 2, 0, −3, −4)	(3, 2, 3, 1)	18	2	10	27
$s_0s_4s_1s_2s_3s_2s_0s_1$	(4, 3, 0, −2, −5)	(1, 3, 2, 3)	18	2	10	27
$s_0s_4s_1s_2s_3s_2s_1s_0$	(4, 2, 0, 0, −6)	(2, 2, 0, 6)	18	3	5	26
$s_0s_4s_3s_1s_0s_4s_2s_1$	(3, 3, 2, −3, −5)	(0, 1, 5, 2)	18	3	5	22
$s_0s_4s_3s_1s_2s_0s_1s_0$	(3, 3, 1, −1, −6)	(0, 2, 2, 5)	18	2	3	22
$s_0s_1s_2s_3s_4s_3s_0$	(6, 0, −1, −2, −3)	(6, 1, 1, 1)	17	2	7	19
$s_0s_1s_2s_3s_4s_3s_2$	(5, 0, 0, −1, −4)	(5, 0, 1, 3)	17	1	3	17
$s_0s_4s_1s_2s_3s_0s_1$	(4, 3, −2, −2, −3)	(1, 5, 0, 1)	17	3	4	17
$s_0s_4s_1s_2s_3s_0s_4$	(5, 2, −1, −3, −3)	(3, 3, 2, 0)	17	1	6	19
$s_0s_4s_1s_2s_3s_2s_0$	(4, 2, 0, −2, −4)	(2, 2, 2, 2)	17	2	4	20
$s_0s_4s_1s_2s_3s_2s_1$	(4, 1, 0, 0, −5)	(3, 1, 0, 5)	17	1	3	17
$s_0s_4s_3s_1s_0s_4s_2$	(3, 2, 2, −3, −4)	(1, 0, 5, 1)	17	3	4	17
$s_0s_4s_3s_1s_2s_0s_1$	(3, 3, 1, −2, −5)	(0, 2, 3, 3)	17	1	6	19
$s_0s_4s_3s_1s_2s_1s_0$	(3, 2, 1, 0, −6)	(1, 1, 1, 6)	17	2	7	19
$s_0s_1s_2s_3s_4s_0$	(6, −1, −1, −2, −2)	(7, 0, 1, 0)	16	3	3	12
$s_0s_1s_2s_3s_4s_3$	(5, 0, −1, −1, −3)	(5, 1, 0, 2)	16	2	3	13
$s_0s_4s_1s_2s_0s_1$	(3, 3, −2, −2, −2)	(0, 5, 0, 0)	16	1	1	10
$s_0s_4s_1s_2s_3s_0$	(4, 2, −1, −2, −3)	(2, 3, 1, 1)	16	1	6	15
$s_0s_4s_1s_2s_3s_2$	(4, 1, 0, −1, −4)	(3, 1, 1, 3)	16	1	5	14
$s_0s_4s_3s_1s_0s_4$	(2, 2, 2, −3, −3)	(0, 0, 5, 0)	16	1	1	10
$s_0s_4s_3s_1s_2s_0$	(3, 2, 1, −2, −4)	(1, 1, 3, 2)	16	1	6	15
$s_0s_4s_3s_1s_2s_1$	(3, 1, 1, 0, −5)	(2, 0, 1, 5)	16	2	3	13
$s_0s_4s_3s_2s_1s_0$	(2, 2, 1, 1, −6)	(0, 1, 0, 7)	16	3	3	12
$s_0s_1s_2s_3s_4$	(5, −1, −1, −1, −2)	(6, 0, 0, 1)	15	2	3	9
$s_0s_4s_1s_2s_0$	(3, 2, −1, −2, −2)	(1, 3, 1, 0)	15	2	2	9
$s_0s_4s_1s_2s_3$	(4, 1, −1, −1, −3)	(3, 2, 0, 2)	15	1	5	11
$s_0s_4s_3s_1s_0$	(2, 2, 1, −2, −3)	(0, 1, 3, 1)	15	2	2	9
$s_0s_4s_3s_1s_2$	(3, 1, 1, −1, −4)	(2, 0, 2, 3)	15	1	5	11
$s_0s_4s_3s_2s_1$	(2, 1, 1, 1, −5)	(1, 0, 0, 6)	15	2	3	9
$s_0s_1s_2s_3$	(4, −1, −1, −1, −1)	(5, 0, 0, 0)	14	1	1	5
$s_0s_4s_1s_0$	(2, 2, 0, −2, −2)	(0, 2, 2, 0)	14	1	1	6
$s_0s_4s_1s_2$	(3, 1, −1, −1, −2)	(2, 2, 0, 1)	14	1	2	7
$s_0s_4s_3s_1$	(2, 1, 1, −1, −3)	(1, 0, 2, 2)	14	1	2	7
$s_0s_4s_3s_2$	(1, 1, 1, 1, −4)	(0, 0, 0, 5)	14	1	1	5
$s_0s_1s_2$	(3, 0, −1, −1, −1)	(3, 1, 0, 0)	13	1	1	4
$s_0s_4s_1$	(2, 1, 0, −1, −2)	(1, 1, 1, 1)	13	1	2	5
$s_0s_4s_3$	(1, 1, 1, 0, −3)	(0, 0, 1, 3)	13	1	1	4
s_0s_1	(2, 0, 0, −1, −1)	(2, 0, 1, 0)	12	1	1	3
s_0s_4	(1, 1, 0, 0, −2)	(0, 1, 0, 2)	12	1	1	3
s_0	(1, 0, 0, 0, −1)	(1, 0, 0, 1)	11	1	1	2
1	(0, 0, 0, 0, 0)	(0, 0, 0, 0)	10	1	1	1

For type A_5, w_0 is $s_1s_2s_3s_4s_5s_1s_2s_3s_4s_1s_2s_3s_1s_2s_1$. Below, the table includes only the entries with restricted weights:

$y \in W$	$w \cdot -2\rho$	ω-coefficients	$l(w)$	(5)	(6)	(7)
$s_0s_5s_4s_1s_0s_5s_2s_3s_4s_2s_1s_0s_5s_2s_3s_4s_0s_1s_2s_3$	(10, 6, 2, −2, −6, −10)	(4, 4, 4, 4, 4)	35	1	279	478
$s_0s_5s_4s_1s_0s_5s_2s_3s_4s_2s_1s_0s_5s_2s_3s_4s_0s_1s_2$	(9, 6, 2, −2, −6, −9)	(3, 4, 4, 4, 3)	34	2	138	381
$s_0s_5s_4s_1s_0s_5s_2s_3s_4s_2s_1s_0s_5s_2s_3s_0s_1s_2$	(9, 5, 2, −2, −6, −8)	(4, 3, 4, 4, 2)	33	5	108	301
$s_0s_5s_4s_1s_0s_5s_2s_3s_4s_2s_1s_0s_5s_2s_3s_4s_0s_1$	(8, 6, 2, −2, −5, −9)	(2, 4, 4, 3, 4)	33	5	108	301
$s_0s_5s_1s_2s_3s_4s_0s_5s_3s_4s_1s_2s_3s_0s_1s_2s_0$	(9, 5, 1, −2, −6, −7)	(4, 4, 3, 4, 1)	32	3	83	234
$s_0s_5s_4s_1s_0s_5s_2s_3s_4s_2s_1s_0s_5s_2s_3s_0s_1$	(8, 5, 2, −2, −5, −8)	(3, 3, 4, 3, 3)	32	20	55	244
$s_0s_5s_4s_1s_0s_5s_2s_3s_4s_2s_1s_0s_5s_2s_3s_4s_0$	(7, 6, 2, −1, −5, −9)	(1, 4, 3, 4, 4)	32	3	83	234
$s_0s_5s_1s_2s_3s_4s_0s_5s_3s_4s_1s_2s_3s_0s_1s_0$	(8, 5, 1, −2, −5, −7)	(3, 4, 3, 3, 2)	31	7	34	195
$s_0s_5s_1s_2s_3s_4s_0s_5s_3s_4s_1s_2s_3s_0s_1s_2$	(9, 5, 1, −3, −6, −6)	(4, 4, 4, 3, 0)	31	1	52	182
$s_0s_5s_4s_1s_0s_5s_2s_3s_4s_2s_1s_0s_5s_2s_0s_1$	(8, 4, 2, −2, −4, −8)	(4, 2, 4, 2, 4)	31	3	70	192
$s_0s_5s_4s_1s_0s_5s_2s_3s_4s_2s_1s_0s_5s_2s_3s_0$	(7, 5, 2, −1, −5, −8)	(2, 3, 3, 4, 3)	31	7	34	195
$s_0s_5s_4s_1s_0s_5s_2s_3s_4s_2s_1s_0s_5s_2s_3s_4$	(6, 6, 3, −1, −5, −9)	(0, 3, 4, 4, 4)	31	1	52	182
$s_0s_5s_1s_2s_3s_4s_0s_5s_3s_4s_1s_2s_0s_1s_0$	(8, 4, 1, −2, −4, −7)	(4, 3, 3, 2, 3)	30	6	28	158
$s_0s_5s_1s_2s_3s_4s_0s_5s_3s_4s_1s_2s_3s_0s_1$	(8, 5, 1, −3, −5, −6)	(3, 4, 4, 2, 1)	30	2	58	157
$s_0s_5s_1s_2s_3s_4s_0s_5s_3s_4s_1s_2s_3s_1s_0$	(7, 5, 1, −1, −5, −7)	(2, 4, 2, 4, 2)	30	3	61	162
$s_0s_5s_4s_1s_0s_5s_2s_3s_4s_2s_1s_0s_5s_2s_0$	(7, 4, 2, −1, −4, −8)	(3, 2, 3, 3, 4)	30	6	28	158
$s_0s_5s_4s_1s_0s_5s_2s_3s_4s_2s_1s_0s_5s_2s_3$	(6, 5, 3, −1, −5, −8)	(1, 2, 4, 4, 3)	30	2	58	157
$s_0s_5s_1s_2s_3s_4s_0s_5s_3s_1s_2s_0s_1s_0$	(8, 4, 0, −2, −3, −7)	(4, 4, 2, 1, 4)	29	1	37	124
$s_0s_5s_1s_2s_3s_4s_0s_5s_3s_4s_1s_2s_0s_1$	(8, 4, 1, −3, −4, −6)	(4, 3, 4, 1, 2)	29	2	49	131
$s_0s_5s_1s_2s_3s_4s_0s_5s_3s_4s_1s_2s_1s_0$	(7, 4, 1, −1, −4, −7)	(3, 3, 2, 3, 3)	29	11	31	134
$s_0s_5s_1s_2s_3s_4s_0s_5s_3s_4s_1s_2s_3s_0$	(7, 5, 1, −3, −5, −5)	(2, 4, 4, 2, 0)	29	3	22	108
$s_0s_5s_1s_2s_3s_4s_0s_5s_3s_4s_1s_2s_3s_1$	(6, 5, 1, −1, −5, −6)	(1, 4, 2, 4, 1)	29	9	26	110
$s_0s_5s_4s_1s_0s_5s_2s_3s_4s_1s_2s_3s_1s_0$	(5, 5, 3, −1, −5, −7)	(0, 2, 4, 4, 2)	29	3	22	108
$s_0s_5s_4s_1s_0s_5s_2s_3s_4s_2s_1s_0s_5s_0$	(7, 3, 2, 0, −4, −8)	(4, 1, 2, 4, 4)	29	1	37	124
$s_0s_5s_4s_1s_0s_5s_2s_3s_4s_2s_1s_0s_5s_2$	(6, 4, 3, −1, −4, −8)	(2, 1, 4, 3, 4)	29	2	49	131
$s_0s_5s_1s_2s_3s_4s_0s_5s_3s_1s_2s_0s_1$	(8, 4, 0, −3, −3, −6)	(4, 4, 3, 0, 3)	28	1	42	105
$s_0s_5s_1s_2s_3s_4s_0s_5s_3s_1s_2s_1s_0$	(7, 4, 0, −1, −3, −7)	(3, 4, 1, 2, 4)	28	1	42	108
$s_0s_5s_1s_2s_3s_4s_0s_5s_3s_4s_1s_2s_0$	(7, 4, 1, −3, −4, −5)	(3, 3, 4, 1, 1)	28	8	23	94
$s_0s_5s_1s_2s_3s_4s_0s_5s_3s_4s_1s_2s_1$	(6, 4, 1, −1, −4, −6)	(2, 3, 2, 3, 2)	28	7	9	94
$s_0s_5s_1s_2s_3s_4s_0s_5s_3s_4s_1s_2s_3$	(6, 5, 1, −2, −5, −5)	(1, 4, 3, 3, 0)	28	5	16	83
$s_0s_5s_1s_2s_3s_4s_0s_5s_3s_4s_2s_1s_0$	(7, 3, 1, 0, −4, −7)	(4, 2, 1, 4, 3)	28	1	42	108
$s_0s_5s_4s_1s_0s_5s_2s_3s_4s_1s_2s_1s_0$	(5, 4, 3, −1, −4, −7)	(1, 1, 4, 3, 3)	28	8	23	94
$s_0s_5s_4s_1s_0s_5s_2s_3s_4s_1s_2s_3s_1$	(5, 5, 2, −1, −5, −6)	(0, 3, 3, 4, 1)	28	5	16	83
$s_0s_5s_4s_1s_0s_5s_2s_3s_4s_2s_1s_0s_5$	(6, 3, 3, 0, −4, −8)	(3, 0, 3, 4, 4)	28	1	42	105
$s_0s_5s_1s_2s_3s_4s_0s_5s_3s_1s_2s_0$	(7, 4, 0, −3, −3, −5)	(3, 4, 3, 0, 2)	27	3	17	78
$s_0s_5s_1s_2s_3s_4s_0s_5s_3s_1s_2s_1$	(6, 4, 0, −1, −3, −6)	(2, 4, 1, 2, 3)	27	3	16	79
$s_0s_5s_1s_2s_3s_4s_0s_5s_3s_2s_1s_0$	(7, 3, 0, 0, −3, −7)	(4, 3, 0, 3, 4)	27	2	35	90
$s_0s_5s_1s_2s_3s_4s_0s_5s_3s_4s_1s_0$	(7, 3, 1, −3, −4, −4)	(4, 2, 4, 1, 0)	27	4	13	63
$s_0s_5s_1s_2s_3s_4s_0s_5s_3s_4s_1s_2$	(6, 4, 1, −2, −4, −5)	(2, 3, 3, 2, 1)	27	4	14	74
$s_0s_5s_1s_2s_3s_4s_0s_5s_3s_4s_2s_1$	(6, 3, 1, 0, −4, −6)	(3, 2, 1, 4, 2)	27	3	16	79
$s_0s_5s_4s_1s_0s_5s_2s_3s_1s_2s_1s_0$	(4, 4, 3, −1, −3, −7)	(0, 1, 4, 2, 4)	27	4	13	63
$s_0s_5s_4s_1s_0s_5s_2s_3s_4s_1s_2s_1$	(5, 4, 2, −1, −4, −6)	(1, 2, 3, 3, 2)	27	4	14	74
$s_0s_5s_4s_1s_0s_5s_2s_3s_4s_1s_2s_3$	(5, 5, 2, −2, −5, −5)	(0, 3, 4, 3, 0)	27	2	9	64
$s_0s_5s_4s_1s_0s_5s_2s_3s_4s_2s_1s_0$	(5, 3, 3, 0, −4, −7)	(2, 0, 3, 4, 3)	27	3	17	78
$s_0s_5s_1s_2s_3s_4s_0s_5s_3s_1s_0$	(7, 3, 0, −3, −3, −4)	(4, 3, 3, 0, 1)	26	3	11	55
$s_0s_5s_1s_2s_3s_4s_0s_5s_3s_1s_2$	(6, 4, 0, −2, −3, −5)	(2, 4, 2, 1, 2)	26	1	25	64
$s_0s_5s_1s_2s_3s_4s_0s_5s_3s_2s_1$	(6, 3, 0, 0, −3, −6)	(3, 3, 0, 3, 3)	26	6	17	68
$s_0s_5s_1s_2s_3s_4s_0s_5s_3s_4s_1$	(6, 3, 1, −2, −4, −4)	(3, 2, 3, 2, 0)	26	3	5	51
$s_0s_5s_1s_2s_3s_4s_0s_5s_3s_4s_2$	(6, 2, 1, 0, −4, −5)	(4, 1, 1, 4, 1)	26	3	11	53

$s_0s_5s_1s_2s_3s_4s_3s_0s_1s_2s_1$	(5, 4, 0, −1, −2, −6)	(1, 4, 1, 1, 4)	26	3	11	53
$s_0s_5s_4s_1s_0s_5s_2s_3s_1s_2s_1$	(4, 4, 2, −1, −3, −6)	(0, 2, 3, 2, 3)	26	3	5	51
$s_0s_5s_4s_1s_0s_5s_2s_3s_2s_1s_0$	(4, 3, 3, 0, −3, −7)	(1, 0, 3, 3, 4)	26	3	11	55
$s_0s_5s_4s_1s_0s_5s_2s_3s_4s_1s_2$	(5, 4, 2, −2, −4, −5)	(1, 2, 4, 2, 1)	26	1	22	59
$s_0s_5s_4s_1s_0s_5s_2s_3s_4s_2s_1$	(5, 3, 2, 0, −4, −6)	(2, 1, 2, 4, 2)	26	1	25	64
$s_0s_5s_1s_2s_3s_4s_0s_5s_1s_0$	(7, 3, −1, −3, −3, −3)	(4, 4, 2, 0, 0)	25	1	6	36
$s_0s_5s_1s_2s_3s_4s_0s_5s_3s_1$	(6, 3, 0, −2, −3, −4)	(3, 3, 2, 1, 1)	25	4	12	46
$s_0s_5s_1s_2s_3s_4s_0s_5s_3s_2$	(6, 2, 0, 0, −3, −5)	(4, 2, 0, 3, 2)	25	2	10	47
$s_0s_5s_1s_2s_3s_4s_0s_5s_3s_4$	(6, 2, 1, −1, −4, −4)	(4, 1, 2, 3, 0)	25	1	6	39
$s_0s_5s_1s_2s_3s_4s_3s_0s_1s_2$	(5, 4, 0, −2, −2, −5)	(1, 4, 2, 0, 3)	25	4	12	44
$s_0s_5s_1s_2s_3s_4s_3s_2s_0s_1$	(5, 3, 0, 0, −2, −6)	(2, 3, 0, 2, 4)	25	2	10	47
$s_0s_5s_4s_1s_0s_5s_2s_3s_1s_2$	(4, 4, 2, −2, −3, −5)	(0, 2, 4, 1, 2)	25	2	9	42
$s_0s_5s_4s_1s_0s_5s_2s_3s_2s_1$	(4, 3, 2, 0, −3, −6)	(1, 1, 2, 3, 3)	25	4	12	46
$s_0s_5s_4s_1s_0s_5s_2s_3s_4s_1$	(5, 3, 2, −2, −4, −4)	(2, 1, 4, 2, 0)	25	2	9	42
$s_0s_5s_4s_1s_0s_5s_2s_3s_4s_2$	(5, 2, 2, 0, −4, −5)	(3, 0, 2, 4, 1)	25	4	12	44
$s_0s_5s_4s_1s_0s_5s_3s_2s_1s_0$	(3, 3, 3, 1, −3, −7)	(0, 0, 2, 4, 4)	25	1	6	36
$s_0s_5s_4s_1s_2s_3s_0s_1s_2s_1$	(4, 4, 1, −1, −2, −6)	(0, 3, 2, 1, 4)	25	1	6	39
$s_0s_5s_1s_2s_3s_4s_0s_5s_1$	(6, 3, −1, −2, −3, −3)	(3, 4, 1, 1, 0)	24	2	7	31
$s_0s_5s_1s_2s_3s_4s_0s_5s_3$	(6, 2, 0, −1, −3, −4)	(4, 2, 1, 2, 1)	24	1	14	36
$s_0s_5s_1s_2s_3s_4s_3s_0s_1$	(5, 3, 0, −2, −2, −4)	(2, 3, 2, 0, 2)	24	3	4	33
$s_0s_5s_1s_2s_3s_4s_3s_2s_0$	(5, 2, 0, 0, −2, −5)	(3, 2, 0, 2, 3)	24	2	4	34
$s_0s_5s_4s_1s_0s_5s_2s_3s_1$	(4, 3, 2, −2, −3, −4)	(1, 1, 4, 1, 1)	24	3	9	32
$s_0s_5s_4s_1s_0s_5s_2s_3s_2$	(4, 2, 2, 0, −3, −5)	(2, 0, 2, 3, 2)	24	3	4	33
$s_0s_5s_4s_1s_0s_5s_2s_3s_4$	(5, 2, 2, −1, −4, −4)	(3, 0, 3, 3, 0)	24	2	8	33
$s_0s_5s_4s_1s_0s_5s_3s_2s_1$	(3, 3, 2, 1, −3, −6)	(0, 1, 1, 4, 3)	24	2	7	31
$s_0s_5s_4s_1s_2s_3s_0s_1s_2$	(4, 4, 1, −2, −2, −5)	(0, 3, 3, 0, 3)	24	2	8	33
$s_0s_5s_4s_1s_2s_3s_2s_0s_1$	(4, 3, 1, 0, −2, −6)	(1, 2, 1, 2, 4)	24	1	14	36
$s_0s_5s_1s_2s_3s_4s_0s_1$	(5, 3, −1, −2, −2, −3)	(2, 4, 1, 0, 1)	23	2	6	24
$s_0s_5s_1s_2s_3s_4s_0s_5$	(6, 2, −1, −1, −3, −3)	(4, 3, 0, 2, 0)	23	1	6	25
$s_0s_5s_1s_2s_3s_4s_3s_0$	(5, 2, 0, −1, −2, −4)	(3, 2, 1, 1, 2)	23	1	6	27
$s_0s_5s_1s_2s_3s_4s_3s_2$	(5, 1, 0, 0, −1, −5)	(4, 1, 0, 1, 4)	23	1	5	22
$s_0s_5s_4s_1s_0s_5s_2s_1$	(3, 3, 2, −2, −3, −3)	(0, 1, 4, 1, 0)	23	2	3	19
$s_0s_5s_4s_1s_0s_5s_2s_3$	(4, 2, 2, −1, −3, −4)	(2, 0, 3, 2, 1)	23	1	6	26
$s_0s_5s_4s_1s_0s_5s_3s_2$	(3, 2, 2, 1, −3, −5)	(1, 0, 1, 4, 2)	23	2	6	24
$s_0s_5s_4s_1s_2s_3s_0s_1$	(4, 3, 1, −2, −2, −4)	(1, 2, 3, 0, 2)	23	1	6	26
$s_0s_5s_4s_1s_2s_3s_2s_0$	(4, 2, 1, 0, −2, −5)	(2, 1, 1, 2, 3)	23	1	6	27
$s_0s_5s_4s_3s_1s_2s_0s_1$	(3, 3, 1, 1, −2, −6)	(0, 2, 0, 3, 4)	23	1	6	25
$s_0s_5s_1s_2s_3s_0s_1$	(4, 3, −1, −2, −2, −2)	(1, 4, 1, 0, 0)	22	2	2	14
$s_0s_5s_1s_2s_3s_4s_0$	(5, 2, −1, −1, −2, −3)	(3, 3, 0, 1, 1)	22	2	6	20
$s_0s_5s_1s_2s_3s_4s_3$	(5, 1, 0, −1, −1, −4)	(4, 1, 1, 0, 3)	22	1	5	18
$s_0s_5s_4s_1s_0s_5s_2$	(3, 2, 2, −1, −3, −3)	(1, 0, 3, 2, 0)	22	2	2	16
$s_0s_5s_4s_1s_0s_5s_3$	(2, 2, 2, 1, −3, −4)	(0, 0, 1, 4, 1)	22	2	2	14
$s_0s_5s_4s_1s_2s_0s_1$	(3, 3, 1, −2, −2, −3)	(0, 2, 3, 0, 1)	22	2	2	16
$s_0s_5s_4s_1s_2s_3s_0$	(4, 2, 1, −1, −2, −4)	(2, 1, 2, 1, 2)	22	1	9	22
$s_0s_5s_4s_1s_2s_3s_2$	(4, 1, 1, 0, −1, −5)	(3, 0, 1, 1, 4)	22	1	5	18
$s_0s_5s_4s_3s_1s_2s_0$	(3, 2, 1, 1, −2, −5)	(1, 1, 0, 3, 3)	22	2	6	20
$s_0s_5s_1s_2s_0s_1$	(3, 3, 0, −2, −2, −2)	(0, 3, 2, 0, 0)	21	1	1	10
$s_0s_5s_1s_2s_3s_0$	(4, 2, −1, −1, −2, −2)	(2, 3, 0, 1, 0)	21	2	2	12
$s_0s_5s_1s_2s_3s_4$	(5, 1, −1, −1, −1, −3)	(4, 2, 0, 0, 2)	21	1	5	14
$s_0s_5s_4s_1s_0s_5$	(2, 2, 2, 0, −3, −3)	(0, 0, 2, 3, 0)	21	1	1	10
$s_0s_5s_4s_1s_2s_0$	(3, 2, 1, −1, −2, −3)	(1, 1, 2, 1, 1)	21	1	4	14
$s_0s_5s_4s_1s_2s_3$	(4, 1, 1, −1, −1, −4)	(3, 0, 2, 0, 3)	21	2	5	15
$s_0s_5s_4s_3s_1s_0$	(2, 2, 1, 1, −2, −4)	(0, 1, 0, 3, 2)	21	2	2	12

$s_0s_5s_4s_3s_1s_2$	(3, 1, 1, 1, −1, −5)	(2, 0, 0, 2, 4)	21	1	5	14
$s_0s_5s_1s_2s_0$	(3, 2, 0, −1, −2, −2)	(1, 2, 1, 1, 0)	20	1	2	9
$s_0s_5s_1s_2s_3$	(4, 1, −1, −1, −1, −2)	(3, 2, 0, 0, 1)	20	1	2	9
$s_0s_5s_4s_1s_0$	(2, 2, 1, 0, −2, −3)	(0, 1, 1, 2, 1)	20	1	2	9
$s_0s_5s_4s_1s_2$	(3, 1, 1, −1, −1, −3)	(2, 0, 2, 0, 2)	20	1	2	10
$s_0s_5s_4s_3s_1$	(2, 1, 1, 1, −1, −4)	(1, 0, 0, 2, 3)	20	1	2	9
$s_0s_1s_2s_3$	(4, 0, −1, −1, −1, −1)	(4, 1, 0, 0, 0)	19	1	1	5
$s_0s_5s_1s_0$	(2, 2, 0, 0, −2, −2)	(0, 2, 0, 2, 0)	19	1	1	6
$s_0s_5s_1s_2$	(3, 1, 0, −1, −1, −2)	(2, 1, 1, 0, 1)	19	1	2	7
$s_0s_5s_4s_1$	(2, 1, 1, 0, −1, −3)	(1, 0, 1, 1, 2)	19	1	2	7
$s_0s_5s_4s_3$	(1, 1, 1, 1, 0, −4)	(0, 0, 0, 1, 4)	19	1	1	5
$s_0s_1s_2$	(3, 0, 0, −1, −1, −1)	(3, 0, 1, 0, 0)	18	1	1	4
$s_0s_5s_1$	(2, 1, 0, 0, −1, −2)	(1, 1, 0, 1, 1)	18	1	2	5
$s_0s_5s_4$	(1, 1, 1, 0, 0, −3)	(0, 0, 1, 0, 3)	18	1	1	4
s_0s_1	(2, 0, 0, 0, −1, −1)	(2, 0, 0, 1, 0)	17	1	1	3
s_0s_5	(1, 1, 0, 0, 0, −2)	(0, 1, 0, 0, 2)	17	1	1	3
s_0	(1, 0, 0, 0, 0, −1)	(1, 0, 0, 0, 1)	16	1	1	2
1	(0, 0, 0, 0, 0, 0)	(0, 0, 0, 0, 0)	15	1	1	1

5 Some Remarks on Construction of the Tables

In this section, we outline the algorithm for the construction of the previous tables. For all programs, we utilized Sage, a well-known programming language akin to Python, but with many more mathematical libraries. While each program is made up of many helper functions, we utilized two main programs: one to isolate the "maximal" element $w \in W_p$ (defined, when $p = h$, as the element w such that $w \cdot -2\rho = (p-2)\rho$)[2] and a second to analyze the properties of $\mathscr{R}(w)$, as well as the right sets of those elements in W^+ less than or equal to w in the Bruhat order.

Before we began actual program implementation, we constructed helper programs which serve the following purposes: create the standard ε-vectors (which are used in the calculation of the roots and ρ), create ρ, create the roots α, compute the dot action of a group element on a vector, and calculate the "height" of a weight, a metric used to tell the distance from the maximal element. Each of the roots α—with the exception of α_0—can be computed quickly using the formula $\alpha_i = \varepsilon_i - \varepsilon_{i+1}$. Since α_0 is the sum of all other α_i, we have that α_0 is the difference of the first and last ε−vector. Using these roots, we may compute ρ as half of the sum of all the positive roots. Finally, we define the height of a weight as in [4, p. 56]; that is, the height of a weight is defined in a way equivalent to the sum of its coefficients when written as a sum of fundamental roots. For a weight ν then, this is $(\nu, \rho^\vee)$ (which for type A is just (ν, ρ)). Notice that the functions all depend on an initial input concerning the size of the generating set of the affine Weyl group, e.g., type A_3, A_4, or larger. To construct w_0, the longest word in the ordinary Weyl group, there are finitely many elements, so a smaller program was written which computes the element of maximum length. This

[2]For $p \geq h$, the maximal element w may be defined by the condition that $w \cdot -2\rho$ belongs to the same p-alcove as $(p-2)\rho$. Exact equality need not hold when $p > h$. But p-alcoves contain only one integral weight when $p = h$.

calculation was carried out beforehand, and the results were recorded. For efficacy, these results were then hard coded into the two main programs.

In the first program's implementation, we begin by generating all of the above variables, including the element w_0 and the affine Weyl group itself (which Sage allows with the function WeylGroup()). The program then proceeds as a variation of the typical minimum function, where the comparator is the height metric; that is, we look for the element w such that $w \cdot -2\rho$ has height closest to $(p-2)\rho$. Say zw_0 is the maximal element w. Repurposing the language of [4, Lem. 3.12.5], we have that $l(z) = -l(w_0) + 2(\rho, \rho^\vee)$ (where the latter term is (ρ, ρ) for type A). Since the maximal element is dominant, we have that the maximal element has length $2(\rho, \rho)$. We then subtract $l(w_0)$, and generate all elements with length $l(z)$ using the Sage function W.elements_of_length(). We subtract this value and generate the smaller elements for three reasons. First, we desire our output in the form $w_0 y$. Second, there are substantially fewer elements of smaller length, which decreases computation time. Finally, we may restrict our search only to the dominant weights and in doing so, we must check whether $l(w_0 y) = l(w_0) + l(y)$. Specifically, we reapply w_0 on the left and check the aforementioned condition. We only continue with these elements. With the proper elements generated, we may now proceed as a typical minimum program, first calculating the height of each element, comparing it to $(p-2)\rho$, and keeping the closest element stored in a local variable. Once the list of elements is exhausted, the result yields the maximal element in simple reflection notation.

For the second program, we require additional helper functions, one which calculates the right set of a given element, and a second which converts ε-notation of vectors into ω-notation for the tables. Since tables in Sage are constructed from lists, we also require the relevant outputs be placed into ordered lists. On the other hand, the right sets are stored as the values of a dictionary with the corresponding element as the key. We begin the second program by generating the standard roots, the Weyl group, and the relevant vectors, in a similar fashion to the first program. Then, we generate all elements of length 1 less than that of the maximal element, collecting only those elements both smaller in the Bruhat order and associated with dominant weights. After exhausting all elements of this length, we decrease the length and repeat this process until we reach w_0. To calculate the right sets of these elements, we employ apply_simple_reflection_right() for each of the elements in the generating set given by W.generators(), and simply check if there is a reduction in length on the element in question. We then keep the output stored as a set object in the dictionary. At this stage, we have an ordered list of elements less than or equal to the maximal element in the Bruhat order, a list of corresponding vectors in ε-notation, a list of corresponding coefficients in ω-notation, and a dictionary of right sets.

For the final calculations, we iterate through the list of desired elements, comparing each element to those with higher indices. Through this comparison, we make three separate counts:

1. The number of elements $v \in W^+$ following the initial element w with $v \leq w$ in the Bruhat order (and which are already known to have dominant weights). This count is column (7) in the table.
2. The number of elements from the first count with $\mathscr{R}(w) \subseteq \mathscr{R}(v)$. This count is column (6) in the table.
3. The number of elements from the first count with $\mathscr{R}(w) = \mathscr{R}(v)$. This count is column (5) in the table.

Since each count is specific to each element, we create ordered lists corresponding to the order of the element list. We then store these lists as a table and call the latex() function on the table object to get Sage output suitable for LaTeX.

Of course, since the size of the ordinary Weyl group grows factorially, the number of elements of interest in the affine case grows quickly, increasing the computational burden. While we were able to run all programs up to and including type A_4 on a typical laptop, the calculations were usually interrupted for larger inputs. Therefore, we turned to Rivanna, an advanced computing cluster with significantly more computing power. Using Rivanna, we were able to acquire the remaining tables.

With the same approach, we have more recently been able to make all the calculations for affine A_6. The table below gives all the restricted weight cases (all restricted weights $w \cdot -2\rho$ with $w = w_0 y$, $l(w) = l(w_0) + l(y)$). Products of fundamental reflections have been abbreviated to the sequence of their subscripts, to save space. For similar space considerations, the column with ε-notation for $w \cdot -2\rho$ has been replaced with an (unrelated but useful) column giving the right set $\mathscr{R}(w)$ of w, as a set of indices of fundamental reflections. The full affine A_6 table, displaying similar rows for all 5260 dominant weights $w \cdot -2\rho$ in the restricted parallelotope, requires 93 pages to display, but is available upon request.

Recall that $w = w_0 y$ with $w_0 = s_1 s_2 s_3 s_4 s_5 s_6 s_1 s_2 s_3 s_4 s_5 s_1 s_2 s_3 s_4 s_1 s_2 s_3 s_1 s_2 s_1$.

$y \in W$	$\mathscr{R}(w)$	ω−notation	$l(w_0 y)$	(5)	(6)	(7)
06510623451062345210612345210643210	{0}	(5, 5, 5, 5, 5, 5)	56	1	3019	5260
0651062345106234521061234521064321	{6, 1}	(4, 5, 5, 5, 5, 4)	55	3	1524	4431
065106234510623452106123454231201	{1, 5}	(3, 5, 5, 5, 4, 5)	54	10	1277	3752
065106234510623452106123452106432	{6, 2}	(5, 4, 5, 5, 5, 3)	54	10	1277	3752
06510623451062345210651234231201	{1, 4}	(2, 5, 5, 4, 5, 5)	53	10	1078	3159
06510623451062345210612345423120	{2, 5, 0}	(4, 4, 5, 5, 4, 4)	53	75	658	3237
06510623451062345210612345210643	{6, 3}	(5, 5, 4, 5, 5, 2)	53	10	1078	3159
0651062345106234521065341231201	{1, 3}	(1, 5, 4, 5, 5, 5)	52	4	887	2648
0651062345106234521065123423120	{2, 4, 0}	(3, 4, 5, 4, 5, 4)	52	64	560	2776
0651062345106234521061234542312	{2, 5}	(5, 3, 5, 5, 3, 5)	52	8	943	2780
0651062345106234521061234542310	{3, 5, 0}	(4, 5, 4, 5, 4, 3)	52	64	560	2776
0651062345106234521061234521064	{6, 4}	(5, 5, 5, 4, 5, 1)	52	4	887	2648
065106234510623452106534231201	{1, 2}	(0, 4, 5, 5, 5, 5)	51	1	575	2222
065106234510623452106534123120	{2, 3, 0}	(2, 4, 4, 5, 5, 4)	51	16	362	2368
065106234510623452106512342312	{2, 4, 6}	(4, 3, 5, 4, 4, 5)	51	55	497	2434
065106234510623452106512342310	{3, 4, 0}	(3, 5, 4, 4, 5, 3)	51	29	376	2427
065106234510623452106123454231	{1, 3, 5}	(5, 4, 4, 5, 3, 4)	51	55	497	2434

065106234510623452106123454210	{4, 5, 0}	(4, 5, 5, 4, 4, 2)	51	16	362	2368
065106234510623452106123452106	{6, 5}	(5, 5, 5, 5, 4, 0)	51	1	575	2222
06510623453210623450123423012	{2, 4}	(5, 2, 5, 3, 5, 5)	50	4	715	2077
06510623451062345210653423120	{2, 0}	(1, 3, 5, 5, 5, 4)	50	2	695	2022
06510623451062345210653412312	{2, 3, 6}	(3, 3, 4, 5, 4, 5)	50	26	328	2109
06510623451062345210653412310	{3, 0}	(2, 5, 3, 5, 5, 3)	50	6	730	2109
06510623451062345210651234231	{1, 3, 4, 6}	(4, 4, 4, 4, 4, 4)	50	101	204	2159
06510623451062345210651234210	{4, 0}	(3, 5, 5, 3, 5, 2)	50	6	730	2109
06510623451062345210612345423	{3, 5}	(5, 5, 3, 5, 2, 5)	50	4	715	2077
06510623451062345210612345421	{1, 4, 5}	(5, 4, 5, 4, 3, 3)	50	26	328	2109
06510623451062345210612345210	{5, 0}	(4, 5, 5, 5, 3, 1)	50	2	695	2022
0651062345321062345340123012	{2, 3, 5}	(4, 2, 4, 4, 5, 5)	49	11	288	1825
0651062345321062345012342301	{1, 3, 4}	(5, 3, 4, 3, 5, 4)	49	11	296	1863
0651062345106234521065342312	{6, 2}	(2, 2, 5, 5, 4, 5)	49	6	635	1827
0651062345106234521065342310	{1, 3, 0}	(0, 3, 5, 5, 5, 3)	49	5	258	1534
0651062345106234521065341231	{1, 3, 6}	(3, 4, 3, 5, 4, 4)	49	44	393	1894
0651062345106234521065341210	{2, 4, 0}	(1, 5, 3, 5, 5, 2)	49	29	334	1578
0651062345106234521065123423	{3, 4, 6}	(4, 5, 3, 4, 3, 5)	49	11	296	1863
0651062345106234521065123421	{1, 4, 6}	(4, 4, 5, 3, 4, 3)	49	44	393	1894
0651062345106234521065123210	{3, 5, 0}	(2, 5, 5, 3, 5, 1)	49	29	334	1578
0651062345106234521061234542	{2, 4, 5}	(5, 5, 4, 4, 2, 4)	49	11	288	1825
0651062345106234521061234521	{1, 5}	(5, 4, 5, 5, 2, 2)	49	6	635	1827
0651062345106234521061234210	{6, 4, 0}	(3, 5, 5, 5, 3, 0)	49	5	258	1534
06510623453210645234123012	{2, 4}	(4, 0, 4, 5, 5, 5)	47	1	498	1366
06510623453210645234012301	{1, 3}	(5, 2, 2, 5, 5, 4)	47	2	507	1411
06510623453210623453412301	{1, 3, 5}	(2, 1, 5, 4, 5, 4)	47	25	279	1260
06510623453210623453401230	{3, 5, 0}	(4, 4, 2, 4, 4, 5)	47	35	312	1464
06510623453210623453401201	{1, 2, 4, 5}	(3, 3, 3, 4, 5, 3)	47	28	102	1277
06510623453210623450123423	{3, 4}	(5, 5, 2, 2, 5, 5)	47	1	381	1382
06510623453210623450123420	{2, 4, 0}	(5, 4, 4, 2, 4, 4)	47	35	312	1464
06510623453210623450123201	{1, 3, 5}	(4, 3, 5, 2, 5, 2)	47	27	282	1289
06510623451062345210653423	{6, 3, 0}	(0, 2, 5, 5, 3, 5)	47	12	184	1052
06510623451062345210653421	{1, 4, 6}	(1, 3, 4, 5, 4, 3)	47	35	255	1195
06510623451062345210653412	{2, 4, 6}	(2, 5, 2, 5, 3, 4)	47	27	282	1289
06510623451062345210653210	{1, 5, 0}	(0, 4, 5, 4, 5, 1)	47	12	177	1062
06510623451062345210653121	{1, 6, 2, 5}	(2, 4, 4, 4, 4, 2)	47	35	92	1196
06510623451062345210651234	{6, 4}	(4, 5, 5, 2, 2, 5)	47	2	507	1411
06510623451062345210651232	{2, 6, 3, 5}	(3, 5, 4, 3, 3, 3)	47	28	102	1277
06510623451062345210631210	{6, 2, 0}	(1, 5, 4, 5, 4, 0)	47	12	177	1062
06510623451062345210612345	{3, 5}	(5, 5, 5, 4, 0, 4)	47	1	498	1366
06510623451062345210612342	{2, 4, 6}	(4, 5, 4, 5, 1, 2)	47	25	279	1260
06510623451062345210612321	{1, 3, 6}	(3, 4, 5, 4, 3, 1)	47	35	255	1195
06510623451062345012341201	{1, 4, 0}	(5, 3, 5, 5, 2, 0)	47	12	184	1052
0651062345321064523412301	{1, 3, 4}	(3, 0, 4, 5, 5, 4)	46	4	199	1097
0651062345321064523401230	{3, 0}	(5, 3, 1, 5, 4, 5)	46	3	456	1274
0651062345321064523401201	{1, 2, 4}	(4, 2, 2, 5, 5, 3)	46	5	194	1121
0651062345321062345341230	{3, 5, 0}	(1, 1, 5, 4, 4, 5)	46	28	213	970
0651062345321062345341201	{1, 4, 5}	(2, 2, 4, 4, 5, 3)	46	14	183	1082
0651062345321062345340123	{3, 5}	(4, 5, 1, 3, 5, 5)	46	1	454	1250
0651062345321062345340120	{2, 4, 5, 0}	(3, 4, 2, 4, 4, 4)	46	57	119	1161
0651062345321062345301201	{1, 2, 5}	(3, 3, 4, 3, 5, 2)	46	14	184	1089

0651062345321062345012342	{2, 4}	(5, 5, 3, 1, 5, 4)	46	1	454	1250
0651062345321062345012340	{4, 0}	(5, 4, 5, 1, 3, 5)	46	3	456	1274
0651062345321062345012320	{2, 3, 5, 0}	(4, 4, 4, 2, 4, 3)	46	57	119	1161
0651062345321062340123201	{1, 3, 6}	(5, 2, 5, 2, 5, 1)	46	25	207	953
0651062345106234521065342	{2, 6, 4, 0}	(0, 3, 4, 5, 3, 4)	46	43	100	905
0651062345106234521065341	{1, 4, 6}	(1, 5, 2, 5, 2, 5)	46	25	207	953
0651062345106234521065321	{1, 5, 6}	(1, 3, 5, 4, 4, 2)	46	10	164	1001
0651062345106234521065312	{2, 5, 6}	(2, 5, 3, 4, 3, 3)	46	14	184	1089
0651062345106234521065123	{3, 5, 6}	(3, 5, 5, 2, 2, 4)	46	5	194	1121
0651062345106234521063210	{6, 1, 0}	(0, 4, 5, 5, 4, 0)	46	2	98	880
0651062345106234521063121	{1, 2, 6}	(2, 4, 4, 5, 3, 1)	46	10	164	1001
0651062345106234521061234	{3, 4, 6}	(4, 5, 5, 4, 0, 3)	46	4	199	1097
0651062345106234521061232	{2, 3, 6}	(3, 5, 4, 4, 2, 2)	46	14	183	1082
0651062345106234501234120	{2, 4, 0}	(5, 4, 4, 5, 1, 1)	46	28	213	970
0651062345106234501231201	{1, 3, 5, 0}	(4, 3, 5, 4, 3, 0)	46	43	100	905
065106234532106452341230	{3, 4, 0}	(2, 0, 4, 5, 4, 5)	45	12	155	866
065106234532106452341201	{1, 4}	(3, 1, 3, 5, 5, 3)	45	2	349	964
065106234532106452340123	{6, 3}	(5, 4, 0, 4, 5, 5)	45	3	405	1112
065106234532106452340120	{2, 4, 0}	(4, 3, 1, 5, 4, 4)	45	23	227	1031
065106234532106452301201	{1, 2, 5}	(5, 1, 2, 5, 5, 2)	45	9	149	840
065106234532106234534123	{3, 5, 6}	(0, 1, 5, 3, 5, 5)	45	8	128	710
065106234532106234534120	{2, 4, 5, 0}	(1, 2, 4, 4, 4, 4)	45	44	91	845
065106234532106234534012	{2, 4, 5}	(3, 5, 1, 3, 5, 4)	45	5	180	1014
065106234532106234534010	{1, 4, 5, 0}	(2, 4, 2, 4, 3, 5)	45	25	71	878
065106234532106234531201	{1, 5}	(2, 2, 5, 3, 5, 2)	45	3	331	927
065106234532106234530120	{2, 5, 0}	(3, 4, 3, 3, 4, 3)	45	27	216	1002
065106234532106234501234	{1, 4}	(5, 5, 4, 0, 4, 5)	45	3	405	1112
065106234532106234501232	{2, 3, 5}	(4, 5, 3, 1, 5, 3)	45	5	180	1014
065106234532106234501230	{3, 5, 0}	(4, 4, 5, 1, 3, 4)	45	23	227	1031
065106234532106234301201	{1, 2, 4, 6}	(4, 2, 4, 3, 5, 1)	45	40	87	821
065106234532106234012320	{2, 6, 3, 0}	(5, 3, 4, 2, 4, 2)	45	25	71	878
065106234510623452301201	{1, 2, 5, 0}	(3, 3, 4, 5, 3, 0)	45	13	56	773
065106234510623452106534	{6, 4, 0}	(0, 4, 3, 5, 2, 5)	45	9	133	761
065106234510623452106532	{2, 6, 5, 0}	(0, 3, 5, 4, 3, 3)	45	13	56	773
065106234510623452106531	{1, 6, 3, 5}	(1, 5, 3, 4, 2, 4)	45	40	87	821
065106234510623452106512	{2, 5, 6}	(2, 5, 5, 2, 1, 5)	45	9	149	840
065106234510623452106321	{6, 1}	(1, 3, 5, 5, 3, 1)	45	2	295	844
065106234510623452106312	{6, 2}	(2, 5, 3, 5, 2, 2)	45	3	331	927
065106234510623452106123	{6, 3}	(3, 5, 5, 3, 1, 3)	45	2	349	964
065106234510623450123412	{1, 2, 4}	(5, 5, 3, 5, 1, 0)	45	8	128	710
065106234510623450123410	{3, 4, 0}	(5, 4, 5, 4, 0, 2)	45	12	155	866
065106234510623450123120	{2, 3, 5, 0}	(4, 4, 4, 4, 2, 1)	45	44	91	845
065106234510623401231201	{1, 3, 0}	(5, 2, 5, 3, 4, 0)	45	9	133	761
06510623453210645234123	{3, 4, 6}	(1, 0, 4, 4, 5, 5)	44	7	116	658
06510623453210645234120	{2, 4, 0}	(2, 1, 3, 5, 4, 4)	44	16	178	769
06510623453210645234012	{2, 4, 6}	(4, 4, 0, 4, 5, 4)	44	20	206	916
06510623453210645234010	{1, 4, 0}	(3, 3, 1, 5, 3, 5)	44	9	140	797
06510623453210645231201	{1, 3, 5}	(4, 0, 3, 5, 5, 2)	44	14	173	732
06510623453210645230120	{2, 5, 0}	(5, 2, 1, 5, 4, 3)	44	19	177	787
06510623453210642301201	{1, 2, 6}	(5, 1, 3, 4, 5, 1)	44	7	119	682
06510623453210623453412	{2, 6, 4, 5}	(0, 2, 4, 3, 5, 4)	44	15	48	629

065106234532106234534 10	{4, 5, 0}	(1, 3, 3, 4, 3, 5)	44	8	123	720
06510623453210623453401	{1, 4, 5}	(2, 5, 1, 3, 4, 5)	44	10	139	795
06510623453210623453120	{2, 5, 0}	(1, 2, 5, 3, 4, 3)	44	15	166	734
06510623453210623453012	{2, 5}	(3, 5, 2, 2, 5, 3)	44	2	328	891
06510623453210623453010	{1, 3, 5, 0}	(2, 4, 3, 3, 3, 4)	44	36	82	769
06510623453210623450123	{1, 3, 5}	(4, 5, 4, 0, 4, 4)	44	20	206	916
06510623453210623450120	{2, 5, 0}	(3, 4, 5, 1, 2, 5)	44	19	177	787
06510623453210623431201	{1, 4, 6}	(3, 1, 5, 3, 5, 1)	44	15	158	711
06510623453210623430120	{2, 6, 4, 0}	(4, 3, 3, 3, 4, 2)	44	36	82	769
06510623453210623401232	{2, 3, 6}	(5, 4, 3, 1, 5, 2)	44	10	139	795
06510623453210623401230	{6, 3, 0}	(5, 3, 5, 1, 3, 3)	44	9	140	797
06510623451062345231201	{1, 5, 0}	(2, 2, 5, 5, 3, 0)	44	3	117	665
06510623451062345230120	{2, 5, 0}	(3, 4, 3, 5, 2, 1)	44	15	166	734
06510623451062345210653	{3, 6, 5, 0}	(0, 4, 4, 4, 2, 4)	44	14	49	663
06510623451062345210651	{1, 5, 6}	(1, 5, 4, 3, 1, 5)	44	7	119	682
06510623451062345210632	{6, 2, 0}	(0, 3, 5, 5, 2, 2)	44	3	117	665
06510623451062345210631	{1, 3, 6}	(1, 5, 3, 5, 1, 3)	44	15	158	711
06510623451062345210612	{2, 4, 6}	(2, 5, 5, 3, 0, 4)	44	14	173	732
06510623451062345012341	{1, 3, 4}	(5, 5, 4, 4, 0, 1)	44	7	116	658
06510623451062345012312	{1, 2, 3, 5}	(4, 5, 3, 4, 2, 0)	44	15	48	629
06510623451062345012310	{3, 5, 0}	(4, 4, 5, 3, 1, 2)	44	16	178	769
06510623451062342301201	{1, 2, 4, 0}	(4, 2, 4, 4, 4, 0)	44	14	49	663
06510623451062340123120	{2, 3, 0}	(5, 3, 4, 3, 3, 1)	44	8	123	720
0651064532106345234123	{3, 4, 5}	(0, 0, 3, 5, 5, 5)	43	1	63	476
0651062345321064523412	{2, 4, 6}	(1, 1, 3, 4, 5, 4)	43	16	135	591
0651062345321064523410	{4, 0}	(2, 2, 2, 5, 3, 5)	43	2	242	666
0651062345321064523401	{1, 4, 6}	(3, 4, 0, 4, 4, 5)	43	19	162	730
0651062345321064523120	{2, 3, 5, 0}	(3, 0, 3, 5, 4, 3)	43	32	72	599
0651062345321064523012	{2, 5, 6}	(5, 3, 0, 4, 5, 3)	43	8	130	716
0651062345321064523010	{1, 3, 5, 0}	(4, 2, 1, 5, 3, 4)	43	31	70	629
0651062345321064231201	{1, 3, 6}	(4, 0, 4, 4, 5, 1)	43	15	139	599
0651062345321064230120	{6, 2, 0}	(5, 2, 2, 4, 4, 2)	43	6	114	648
0651062345321062345341	{1, 6, 4, 5}	(0, 3, 3, 3, 4, 5)	43	10	42	547
0651062345321062345340	{4, 5, 0}	(1, 5, 1, 2, 5, 5)	43	5	104	583
0651062345321062345312	{2, 5, 6}	(0, 2, 5, 2, 5, 3)	43	7	102	557
0651062345321062345310	{3, 5, 0}	(1, 3, 4, 3, 3, 4)	43	17	143	635
0651062345321062345301	{1, 3, 5}	(2, 5, 2, 2, 4, 4)	43	15	162	706
0651062345321062345012	{1, 2, 5}	(3, 5, 4, 0, 3, 5)	43	8	130	716
0651062345321062345010	{1, 5, 0}	(2, 4, 4, 2, 2, 5)	43	6	114	648
0651062345321062343120	{2, 6, 4, 0}	(2, 1, 5, 3, 4, 2)	43	26	64	581
0651062345321062343012	{2, 4, 6}	(4, 4, 2, 2, 5, 2)	43	15	162	706
0651062345321062343010	{1, 3, 6, 4, 0}	(3, 3, 3, 3, 3, 3)	43	20	22	608
0651062345321062340123	{1, 3, 6}	(5, 4, 4, 0, 4, 3)	43	19	162	730
0651062345321062340120	{2, 6, 4, 0}	(4, 3, 5, 1, 2, 4)	43	31	70	629
0651062345106342301201	{1, 2, 0}	(5, 1, 3, 5, 4, 0)	43	2	67	558
0651062345106234523120	{2, 5, 0}	(1, 2, 5, 5, 2, 1)	43	7	129	546
0651062345106234523012	{1, 2, 5}	(3, 5, 2, 5, 2, 0)	43	7	102	557
0651062345106234523010	{1, 3, 5, 0}	(2, 4, 3, 5, 1, 2)	43	26	64	581
0651062345106234521065	{6, 5, 0}	(0, 4, 5, 3, 1, 5)	43	2	67	558
0651062345106234521063	{6, 3, 0}	(0, 4, 4, 5, 1, 3)	43	8	103	580
0651062345106234521061	{1, 4, 6}	(1, 5, 4, 4, 0, 4)	43	15	139	599

0651062345106234501234	{2, 3, 4}	(5, 5, 5, 3, 0, 0)	43	1	63	476
0651062345106234501231	{1, 3, 5}	(4, 5, 4, 3, 1, 1)	43	16	135	591
0651062345106234501210	{2, 4, 5, 0}	(3, 4, 5, 3, 0, 3)	43	32	72	599
0651062345106234231201	{1, 4, 0}	(3, 1, 5, 4, 4, 0)	43	8	103	580
0651062345106234230120	{2, 4, 0}	(4, 3, 3, 4, 3, 1)	43	17	143	635
0651062345106234012312	{1, 2, 3, 6}	(5, 4, 3, 3, 3, 0)	43	10	42	547
0651062345106234012310	{3, 0}	(5, 3, 5, 2, 2, 2)	43	2	242	666
0612345064512340123120	{2, 3, 0}	(5, 5, 2, 1, 5, 1)	43	5	104	583
065106453210634523412	{2, 4, 5}	(0, 1, 2, 5, 5, 4)	42	3	83	434
065106234532106452341	{1, 4, 6}	(1, 2, 2, 4, 4, 5)	42	10	120	520
065106234532106452340	{6, 4, 0}	(2, 4, 0, 3, 5, 5)	42	4	98	545
065106234532106452312	{2, 6, 3, 5}	(2, 0, 3, 4, 5, 3)	42	13	44	476
065106234532106452310	{3, 5, 0}	(3, 1, 2, 5, 3, 4)	42	13	122	535
065106234532106452301	{1, 6, 3, 5}	(4, 3, 0, 4, 4, 4)	42	30	66	586
065106234532106452010	{1, 5, 0}	(5, 1, 1, 5, 2, 5)	42	5	87	465
065106234532106423120	{2, 6, 3, 0}	(3, 0, 4, 4, 4, 2)	42	14	47	500
065106234532106423012	{6, 2}	(5, 3, 1, 3, 5, 2)	42	2	223	605
065106234532106423010	{1, 6, 3, 0}	(4, 2, 2, 4, 3, 3)	42	10	40	528
065106234532106234534	{4, 5, 6}	(0, 4, 2, 2, 5, 5)	42	1	58	459
065106234532106234531	{1, 6, 3, 5}	(0, 3, 4, 2, 4, 4)	42	24	57	490
065106234532106234530	{3, 5, 0}	(1, 5, 2, 1, 5, 4)	42	12	122	525
065106234532106234510	{5, 0}	(1, 3, 5, 2, 2, 5)	42	1	196	540
065106234532106234501	{1, 5}	(2, 5, 3, 1, 3, 5)	42	2	223	605
065106234532106234312	{2, 4, 6}	(1, 1, 5, 2, 5, 2)	42	12	110	463
065106234532106234310	{3, 6, 4, 0}	(2, 2, 4, 3, 3, 3)	42	14	45	514
065106234532106234301	{1, 3, 4, 6}	(3, 4, 2, 2, 4, 3)	42	30	65	572
065106234532106234012	{1, 2, 4, 6}	(4, 4, 4, 0, 3, 4)	42	30	66	586
065106234532106234010	{1, 6, 4, 0}	(3, 3, 4, 2, 2, 4)	42	10	40	528
065106234532106230120	{6, 2, 0}	(5, 2, 5, 1, 1, 5)	42	5	87	465
065106234510634231201	{1, 3, 0}	(4, 0, 4, 5, 4, 0)	42	4	91	494
065106234510634230120	{2, 0}	(5, 2, 2, 5, 3, 1)	42	1	196	540
065106234510623452312	{1, 6, 2, 5}	(0, 2, 5, 5, 2, 0)	42	7	39	368
065106234510623452310	{3, 5, 0}	(1, 3, 4, 5, 1, 2)	42	12	115	487
065106234510623452301	{1, 3, 5}	(2, 5, 2, 5, 1, 1)	42	12	110	463
065106234510623452106	{6, 4, 0}	(0, 4, 5, 4, 0, 4)	42	4	91	494
065106234510623452010	{1, 4, 5, 0}	(2, 4, 4, 4, 0, 3)	42	14	47	500
065106234510623450123	{2, 3, 5}	(4, 5, 5, 2, 1, 0)	42	3	83	434
065106234510623450121	{1, 2, 4, 5}	(3, 5, 4, 3, 0, 2)	42	13	44	476
065106234510623423120	{2, 4, 0}	(2, 1, 5, 4, 3, 1)	42	12	115	487
065106234510623423012	{1, 2, 4, 6}	(4, 4, 2, 4, 3, 0)	42	24	57	490
065106234510623423010	{1, 3, 4, 0}	(3, 3, 3, 4, 2, 2)	42	14	45	514
065106234510623401231	{1, 3, 6}	(5, 4, 4, 2, 2, 1)	42	10	120	520
065106234510623401210	{2, 4, 0}	(4, 3, 5, 2, 1, 3)	42	13	122	535
061234506451234230120	{2, 4, 0}	(4, 5, 1, 2, 5, 1)	42	12	122	525
061234506451234012312	{1, 2, 3}	(5, 5, 2, 2, 4, 0)	42	1	58	459
061234506451234012310	{1, 3, 0}	(5, 5, 3, 0, 4, 2)	42	4	98	545
06510645321063452341	{1, 4, 5}	(0, 2, 1, 5, 4, 5)	41	5	74	389
06510645321063452312	{2, 3, 5}	(1, 0, 2, 5, 5, 3)	41	3	72	368
06510623453210645234	{6, 4}	(1, 3, 1, 3, 5, 5)	41	1	161	440
06510623453210645231	{1, 6, 3, 5}	(2, 1, 2, 4, 4, 4)	41	21	49	430
06510623453210645230	{3, 6, 5, 0}	(3, 3, 0, 3, 5, 4)	41	8	36	450

06510623453210645210	{2, 5, 0}	(4, 0, 2, 5, 2, 5)	41	10	99	402
06510623453210645201	{1, 5, 6}	(5, 2, 0, 4, 3, 5)	41	4	83	442
06510623453210642312	{2, 3, 6}	(2, 0, 4, 3, 5, 2)	41	6	81	409
06510623453210642310	{6, 3, 0}	(3, 1, 3, 4, 3, 3)	41	6	82	454
06510623453210642301	{1, 3, 6}	(4, 3, 1, 3, 4, 3)	41	12	115	502
06510623453210642010	{1, 6, 4, 0}	(5, 1, 2, 4, 2, 4)	41	8	32	399
06510623453210623453	{3, 5, 6}	(0, 4, 3, 1, 5, 4)	41	4	78	417
06510623453210623451	{1, 5, 6}	(0, 3, 5, 1, 3, 5)	41	2	80	425
06510623453210623450	{2, 5, 0}	(1, 5, 3, 0, 4, 5)	41	9	108	460
06510623453210623431	{1, 3, 4, 6}	(1, 2, 4, 2, 4, 3)	41	23	49	414
06510623453210623430	{3, 6, 4, 0}	(2, 4, 2, 1, 5, 3)	41	12	40	439
06510623453210623410	{6, 4, 0}	(2, 2, 5, 2, 2, 4)	41	3	80	449
06510623453210623401	{1, 4, 6}	(3, 4, 3, 1, 3, 4)	41	12	115	502
06510623453210623012	{1, 2, 6}	(5, 3, 4, 0, 2, 5)	41	4	83	442
06510623453210623010	{1, 6, 3, 0}	(4, 2, 4, 2, 1, 5)	41	8	32	399
06510623451234230120	{2, 4, 5, 0}	(0, 1, 5, 2, 5, 1)	41	15	39	307
06510623451063423120	{2, 3, 0}	(3, 0, 4, 5, 3, 1)	41	3	82	422
06510623451063423012	{1, 2, 6}	(5, 3, 1, 5, 3, 0)	41	2	80	425
06510623451063423010	{1, 3, 0}	(4, 2, 2, 5, 2, 2)	41	3	80	449
06510623451062345231	{1, 6, 3, 5}	(0, 3, 4, 5, 1, 1)	41	17	43	337
06510623451062345230	{2, 3, 5, 0}	(1, 5, 2, 5, 1, 0)	41	15	39	307
06510623451062345210	{4, 5, 0}	(1, 3, 5, 4, 0, 3)	41	3	82	422
06510623451062345201	{1, 4, 5}	(2, 5, 3, 4, 0, 2)	41	6	81	409
06510623451062345012	{2, 4, 5}	(3, 5, 5, 2, 0, 1)	41	3	72	368
06510623451062342312	{1, 2, 4, 6}	(1, 1, 5, 4, 3, 0)	41	17	43	337
06510623451062342310	{3, 4, 0}	(2, 2, 4, 4, 2, 2)	41	6	82	440
06510623451062342301	{1, 3, 4, 6}	(3, 4, 2, 4, 2, 1)	41	23	49	414
06510623451062342010	{1, 4, 0}	(3, 3, 4, 3, 1, 3)	41	6	82	454
06510623451062340123	{2, 3, 6}	(5, 4, 5, 1, 2, 0)	41	5	74	389
06510623451062340121	{1, 2, 4, 6}	(4, 4, 4, 2, 1, 2)	41	21	49	430
06510623451062301210	{2, 5, 0}	(5, 2, 5, 2, 0, 4)	41	10	99	402
06123450645341230120	{2, 5, 0}	(5, 4, 0, 3, 5, 1)	41	9	108	460
06123450645123423012	{1, 2, 4}	(4, 5, 1, 3, 4, 0)	41	4	78	417
06123450645123423010	{1, 3, 4, 0}	(3, 5, 1, 2, 4, 2)	41	12	40	439
06123450645123401231	{1, 3}	(5, 5, 3, 1, 3, 1)	41	1	161	440
06123450645123401210	{1, 2, 4, 0}	(4, 5, 3, 0, 3, 3)	41	8	36	450
0651064532106345234	{4, 5, 0}	(0, 3, 0, 4, 5, 5)	40	1	67	335
0651064532106345231	{1, 3, 5}	(1, 1, 1, 5, 4, 4)	40	10	81	336
0651064532106342312	{2, 3, 4, 6}	(0, 0, 2, 5, 5, 2)	40	4	25	243
0651062345341230120	{2, 3, 5, 0}	(1, 0, 4, 3, 5, 1)	40	16	35	280
0651062345321064523	{3, 5, 6}	(2, 2, 1, 3, 5, 4)	40	3	68	372
0651062345321064521	{1, 6, 2, 5}	(3, 0, 2, 4, 3, 5)	40	10	34	332
0651062345321064520	{2, 6, 5, 0}	(4, 2, 0, 3, 4, 5)	40	7	29	353
0651062345321064231	{1, 3, 6}	(2, 1, 3, 3, 4, 3)	40	8	90	374
0651062345321064230	{6, 3, 0}	(3, 3, 1, 2, 5, 3)	40	4	73	394
0651062345321064210	{2, 6, 4, 0}	(4, 0, 3, 4, 2, 4)	40	18	42	348
0651062345321064201	{1, 4, 6}	(5, 2, 1, 3, 3, 4)	40	8	92	385
0651062345321062345	{2, 5, 6}	(0, 4, 4, 0, 4, 5)	40	5	70	370
0651062345321062343	{3, 4, 6}	(1, 3, 3, 1, 5, 3)	40	3	69	360
0651062345321062341	{1, 4, 6}	(1, 2, 5, 1, 3, 4)	40	6	88	366
0651062345321062340	{2, 6, 4, 0}	(2, 4, 3, 0, 4, 4)	40	20	45	392

0651062345321062310	{6, 3, 0}	(3, 1, 5, 2, 1, 5)	40	4	65	346
0651062345321062301	{1, 3, 6}	(4, 3, 3, 1, 2, 5)	40	8	92	385
0651062345321062010	{6, 1, 0}	(5, 1, 3, 3, 1, 5)	40	1	43	326
0651062345123423012	{1, 2, 4, 5}	(0, 1, 5, 3, 4, 0)	40	10	25	244
0651062345123423010	{1, 3, 5, 4, 0}	(0, 2, 4, 2, 4, 2)	40	10	13	280
0651062345106342312	{1, 2, 3, 6}	(2, 0, 4, 5, 3, 0)	40	5	28	299
0651062345106342310	{3, 0}	(3, 1, 3, 5, 2, 2)	40	1	146	390
0651062345106342301	{1, 3, 6}	(4, 3, 1, 5, 2, 1)	40	6	88	366
0651062345106342010	{1, 4, 0}	(5, 1, 2, 5, 1, 3)	40	4	65	346
0651062345106234523	{2, 6, 3, 5}	(0, 4, 3, 5, 1, 0)	40	10	25	244
0651062345106234521	{1, 6, 4, 5}	(0, 3, 5, 4, 0, 2)	40	5	28	299
0651062345106234520	{2, 4, 5, 0}	(1, 5, 3, 4, 0, 1)	40	16	35	280
0651062345106234501	{1, 3, 4, 5}	(2, 5, 5, 2, 0, 0)	40	4	25	243
0651062345106234231	{1, 3, 4, 6}	(1, 2, 4, 4, 2, 1)	40	18	38	312
0651062345106234230	{2, 3, 6, 4, 0}	(2, 4, 2, 4, 2, 0)	40	10	13	280
0651062345106234210	{4, 0}	(2, 2, 5, 3, 1, 3)	40	1	146	390
0651062345106234201	{1, 4, 6}	(3, 4, 3, 3, 1, 2)	40	8	90	374
0651062345106234012	{2, 4, 6}	(4, 4, 5, 1, 1, 1)	40	10	81	336
0651062345106232010	{1, 3, 5, 0}	(4, 2, 4, 3, 0, 4)	40	18	42	348
0651062345106230121	{1, 6, 2, 5}	(5, 3, 4, 2, 0, 3)	40	10	34	332
0612345064534123012	{1, 2, 5}	(5, 4, 0, 4, 4, 0)	40	5	70	370
0612345064534123010	{1, 3, 5, 0}	(4, 4, 0, 3, 4, 2)	40	20	45	392
0612345064512342310	{3, 4, 0}	(2, 5, 1, 1, 5, 2)	40	5	69	356
0612345064512342301	{1, 3, 4}	(3, 5, 1, 3, 3, 1)	40	3	69	360
0612345064512342010	{1, 4, 0}	(3, 5, 2, 1, 3, 3)	40	4	73	394
0612345064512340123	{2, 3, 0}	(5, 5, 4, 0, 3, 0)	40	1	67	335
0612345064512340121	{1, 2, 4}	(4, 5, 3, 1, 2, 2)	40	3	68	372
0612345064512301210	{1, 2, 5, 0}	(5, 4, 3, 0, 2, 4)	40	7	29	353
065106453210634523	{3, 5, 0}	(1, 2, 0, 4, 5, 4)	39	5	74	294
065106453210634521	{1, 2, 5}	(2, 0, 1, 5, 3, 5)	39	4	57	269
065106453210634231	{1, 3, 4, 6}	(0, 1, 1, 5, 4, 3)	39	12	31	226
065106452341230120	{2, 3, 4, 0}	(0, 0, 3, 4, 5, 1)	39	7	19	197
065106234534123012	{1, 2, 3, 5}	(1, 0, 4, 4, 4, 0)	39	8	20	225
065106234534123010	{1, 3, 5, 0}	(1, 1, 3, 3, 4, 2)	39	14	33	260
065106234532106452	{2, 5, 6}	(3, 1, 1, 3, 4, 5)	39	4	56	298
065106234532106450	{6, 5, 0}	(5, 1, 0, 2, 5, 5)	39	1	38	255
065106234532106423	{6, 3}	(2, 2, 2, 2, 5, 3)	39	1	126	330
065106234532106421	{1, 2, 4, 6}	(3, 0, 3, 3, 3, 4)	39	17	38	293
065106234532106420	{2, 6, 4, 0}	(4, 2, 1, 2, 4, 4)	39	17	37	313
065106234532106234	{2, 4, 6}	(1, 3, 4, 0, 4, 4)	39	9	78	324
065106234532106231	{1, 3, 6}	(2, 1, 5, 1, 2, 5)	39	6	73	291
065106234532106230	{2, 6, 3, 0}	(3, 3, 3, 0, 3, 5)	39	9	30	310
065106234532106210	{6, 2, 0}	(4, 0, 4, 3, 1, 5)	39	2	57	287
065106234532106201	{6, 1}	(5, 2, 2, 2, 2, 5)	39	1	122	318
065106234512342310	{3, 4, 5, 0}	(0, 3, 3, 1, 5, 2)	39	4	23	250
065106234512342301	{1, 3, 4, 5}	(0, 2, 4, 3, 3, 1)	39	7	20	229
065106234512342010	{1, 4, 5, 0}	(0, 2, 5, 1, 3, 3)	39	5	27	253
065106234510634231	{1, 3, 6}	(2, 1, 3, 5, 2, 1)	39	4	70	282
065106234510634230	{2, 6, 3, 0}	(3, 3, 1, 5, 2, 0)	39	5	27	253
065106234510634210	{2, 4, 0}	(4, 0, 3, 5, 1, 3)	39	7	75	304
065106234510634201	{1, 4, 6}	(5, 2, 1, 5, 1, 2)	39	6	73	291

065106234510632010	{1, 5, 0}	(5, 1, 3, 4, 0, 4)	39	2	57	287
065106234510623452	{2, 6, 4, 5}	(0, 4, 4, 4, 0, 1)	39	8	20	225
065106234510623450	{3, 4, 5, 0}	(1, 5, 4, 3, 0, 0)	39	7	19	197
065106234510623423	{2, 3, 4, 6}	(1, 3, 3, 4, 2, 0)	39	7	20	229
065106234510623421	{1, 4, 6}	(1, 2, 5, 3, 1, 2)	39	4	70	282
065106234510623420	{2, 6, 4, 0}	(2, 4, 3, 3, 1, 1)	39	14	33	260
065106234510623401	{1, 3, 4, 6}	(3, 4, 5, 1, 1, 0)	39	12	31	226
065106234510623210	{3, 5, 0}	(3, 1, 5, 3, 0, 4)	39	7	75	304
065106234510623201	{1, 6, 3, 5}	(4, 3, 3, 3, 0, 3)	39	17	38	293
065106234510623012	{2, 5, 6}	(5, 3, 5, 1, 0, 2)	39	4	57	269
061234506453412310	{3, 5, 0}	(3, 4, 0, 2, 5, 2)	39	8	78	323
061234506453412301	{1, 3, 5}	(4, 4, 0, 4, 3, 1)	39	9	78	324
061234506453412010	{1, 4, 5, 0}	(5, 3, 0, 3, 3, 3)	39	9	30	310
061234506451234231	{1, 3, 4, 6}	(1, 5, 1, 1, 5, 1)	39	13	30	234
061234506451234230	{2, 3, 4, 0}	(2, 5, 1, 3, 3, 0)	39	4	23	250
061234506451234210	{2, 4, 0}	(2, 5, 2, 0, 4, 3)	39	8	78	323
061234506451234201	{1, 4}	(3, 5, 2, 2, 2, 2)	39	1	126	330
061234506451234012	{2, 4, 0}	(4, 5, 4, 0, 2, 1)	39	5	74	294
061234506451232010	{1, 3, 5, 0}	(4, 4, 2, 1, 2, 4)	39	17	37	313
061234506451230121	{1, 2, 5}	(5, 4, 3, 1, 1, 3)	39	4	56	298
061234506412301210	{1, 2, 0}	(5, 5, 2, 0, 1, 5)	39	1	38	255
06510645321063452	{2, 5, 0}	(2, 1, 0, 4, 4, 5)	38	5	62	243
06510645321063423	{3, 6, 4, 0}	(0, 2, 0, 4, 5, 3)	38	4	24	202
06510645321063421	{1, 2, 4, 6}	(1, 0, 1, 5, 3, 4)	38	12	27	191
06510645234123012	{2, 1, 4, 3}	(0, 0, 3, 5, 4, 0)	38	2	10	161
06510645234123010	{1, 3, 4, 0}	(0, 1, 2, 4, 4, 2)	38	8	20	186
06510623453412310	{3, 5, 0}	(1, 2, 2, 2, 5, 2)	38	4	58	235
06510623453412301	{1, 3, 5}	(1, 1, 3, 4, 3, 1)	38	7	51	214
06510623453412010	{2, 1, 5, 4, 0}	(2, 0, 3, 3, 3, 3)	38	9	10	211
06510623453210645	{1, 5, 6}	(4, 0, 1, 2, 5, 5)	38	1	48	219
06510623453210642	{2, 4, 6}	(3, 1, 2, 2, 4, 4)	38	6	64	267
06510623453210640	{6, 4, 0}	(5, 1, 1, 1, 5, 4)	38	2	48	230
06510623453210623	{2, 3, 6}	(2, 2, 4, 0, 3, 5)	38	3	53	263
06510623453210621	{1, 2, 6}	(3, 0, 4, 2, 2, 5)	38	1	53	246
06510623453210620	{6, 2, 0}	(4, 2, 2, 1, 3, 5)	38	1	50	264
06510623451234231	{1, 3, 4, 5}	(0, 4, 2, 1, 5, 1)	38	6	17	181
06510623451234230	{2, 3, 5, 4, 0}	(0, 3, 3, 3, 3, 0)	38	5	7	172
06510623451234210	{2, 4, 5, 0}	(0, 3, 4, 0, 4, 3)	38	12	30	229
06510623451234201	{1, 4, 5}	(0, 2, 5, 2, 2, 2)	38	3	42	211
06510623451232010	{1, 3, 5, 0}	(1, 1, 5, 1, 2, 4)	38	12	28	211
06510623451063423	{2, 3, 6}	(2, 2, 2, 5, 2, 0)	38	3	42	211
06510623451063421	{1, 2, 4, 6}	(3, 0, 3, 5, 1, 2)	38	12	30	227
06510623451063420	{2, 6, 4, 0}	(4, 2, 1, 5, 1, 1)	38	12	28	211
06510623451063210	{2, 5, 0}	(4, 0, 4, 4, 0, 4)	38	6	66	254
06510623451063201	{1, 5, 6}	(5, 2, 2, 4, 0, 3)	38	1	53	246
06510623451062345	{3, 6, 4, 5}	(0, 4, 5, 3, 0, 0)	38	2	10	161
06510623451062342	{2, 4, 6}	(1, 3, 4, 3, 1, 1)	38	7	51	214
06510623451062340	{3, 6, 4, 0}	(2, 4, 4, 2, 1, 0)	38	8	20	186
06510623451062321	{1, 6, 3, 5}	(2, 1, 5, 3, 0, 3)	38	12	30	227
06510623451062320	{2, 3, 6, 5, 0}	(3, 3, 3, 3, 0, 2)	38	9	10	211
06510623451062301	{1, 6, 3, 5}	(4, 3, 5, 1, 0, 1)	38	12	27	191

06123450645341231	{1, 6, 3, 5}	(2, 4, 0, 2, 5, 1)	38	11	28	216
06123450645341230	{2, 3, 5, 0}	(3, 4, 0, 4, 3, 0)	38	12	30	229
06123450645341210	{2, 4, 5, 0}	(4, 3, 0, 2, 4, 3)	38	14	33	262
06123450645341201	{1, 4, 5}	(5, 3, 0, 4, 2, 2)	38	3	53	263
06123450645312010	{1, 5, 0}	(5, 3, 1, 2, 2, 4)	38	1	50	264
06123450645123423	{2, 3, 4, 6}	(1, 5, 1, 2, 4, 0)	38	6	17	181
06123450645123421	{1, 2, 4, 6}	(1, 5, 2, 0, 4, 2)	38	11	28	216
06123450645123420	{2, 4, 0}	(2, 5, 2, 2, 2, 1)	38	4	58	235
06123450645123401	{1, 3, 4, 0}	(3, 5, 4, 0, 2, 0)	38	4	24	202
06123450645123210	{2, 3, 5, 0}	(3, 4, 2, 0, 3, 4)	38	14	33	262
06123450645123201	{1, 3, 5}	(4, 4, 2, 2, 1, 3)	38	6	64	267
06123450645123012	{2, 5, 0}	(5, 4, 4, 0, 1, 2)	38	5	62	243
06123450641232010	{1, 3, 0}	(4, 5, 1, 1, 1, 5)	38	2	48	230
06123450641230121	{1, 2, 6}	(5, 5, 2, 1, 0, 4)	38	1	48	219
0651064532106345	{1, 5, 0}	(3, 0, 0, 3, 5, 5)	37	1	45	183
0651064532106342	{2, 6, 4, 0}	(1, 1, 0, 4, 4, 4)	37	11	25	175
0651064532106321	{1, 2, 3, 6}	(0, 0, 1, 5, 2, 5)	37	5	14	123
0651064523412310	{3, 4, 0}	(0, 2, 1, 3, 5, 2)	37	2	35	171
0651064523412301	{1, 3, 4}	(0, 1, 2, 5, 3, 1)	37	2	31	155
0651064523412010	{1, 2, 4, 0}	(1, 0, 2, 4, 3, 3)	37	5	15	160
0651062345341231	{1, 3, 5}	(1, 3, 1, 2, 5, 1)	37	5	42	172
0651062345341230	{2, 3, 5, 0}	(1, 2, 2, 4, 3, 0)	37	9	22	163
0651062345341210	{2, 4, 5, 0}	(2, 1, 2, 2, 4, 3)	37	10	25	196
0651062345341201	{1, 2, 4, 5}	(2, 0, 3, 4, 2, 2)	37	6	18	178
0651062345321064	{1, 4, 6}	(4, 0, 2, 1, 5, 4)	37	4	54	199
0651062345321062	{6, 2}	(3, 1, 3, 1, 3, 5)	37	1	88	228
0651062345321060	{6, 3, 0}	(5, 1, 2, 0, 4, 5)	37	2	41	199
0651062345312010	{1, 2, 5, 0}	(2, 0, 4, 2, 2, 4)	37	4	18	183
0651062345123423	{2, 3, 4, 5}	(0, 4, 2, 2, 4, 0)	37	2	9	142
0651062345123421	{1, 2, 4, 5}	(0, 4, 3, 0, 4, 2)	37	6	18	169
0651062345123420	{2, 4, 5, 0}	(0, 3, 4, 2, 2, 1)	37	9	22	163
0651062345123210	{2, 3, 5, 0}	(1, 2, 4, 0, 3, 4)	37	11	25	194
0651062345123201	{1, 3, 5}	(1, 1, 5, 2, 1, 3)	37	5	44	179
0651062345106342	{2, 4, 6}	(3, 1, 2, 5, 1, 1)	37	5	44	179
0651062345106340	{3, 6, 4, 0}	(5, 1, 1, 5, 1, 0)	37	5	16	138
0651062345106321	{1, 6, 2, 5}	(3, 0, 4, 4, 0, 3)	37	6	23	194
0651062345106320	{2, 6, 5, 0}	(4, 2, 2, 4, 0, 2)	37	4	18	183
0651062345106234	{3, 4, 6}	(1, 3, 5, 2, 1, 0)	37	2	31	155
0651062345106232	{2, 6, 3, 5}	(2, 2, 4, 3, 0, 2)	37	6	18	178
0651062345106230	{3, 6, 5, 0}	(3, 3, 4, 2, 0, 1)	37	5	15	160
0651062345106201	{1, 6, 4, 5}	(5, 2, 5, 1, 0, 0)	37	5	14	123
0651062341232010	{1, 3, 4, 0}	(0, 1, 5, 1, 1, 5)	37	5	16	138
0612345064534210	{4, 5, 0}	(5, 2, 0, 1, 5, 3)	37	2	44	199
0612345064534123	{2, 6, 3, 5}	(2, 4, 0, 3, 4, 0)	37	6	18	169
0612345064534121	{2, 1, 6, 4, 5}	(3, 3, 0, 2, 4, 2)	37	7	9	181
0612345064534120	{2, 4, 5, 0}	(4, 3, 0, 4, 2, 1)	37	11	25	194
0612345064531210	{2, 5, 0}	(4, 3, 1, 1, 3, 4)	37	5	55	226
0612345064531201	{1, 5}	(5, 3, 1, 3, 1, 3)	37	1	88	228
0612345064512342	{2, 4, 6}	(1, 5, 2, 1, 3, 1)	37	5	42	172
0612345064512340	{3, 4, 0}	(2, 5, 3, 1, 2, 0)	37	2	35	171
0612345064512321	{2, 1, 6, 3, 5}	(2, 4, 2, 0, 3, 3)	37	7	9	181

0612345064512320	{2, 3, 5, 0}	(3, 4, 2, 2, 1, 2)	37	10	25	196
0612345064512301	{1, 3, 5, 0}	(4, 4, 4, 0, 1, 1)	37	11	25	175
0612345064312010	{1, 4, 0}	(5, 4, 0, 2, 1, 5)	37	2	41	199
0612345064123210	{2, 3, 0}	(3, 5, 1, 0, 2, 5)	37	2	44	199
0612345064123201	{1, 3, 6}	(4, 5, 1, 2, 0, 4)	37	4	54	199
0612345064123012	{6, 2, 0}	(5, 5, 3, 0, 0, 3)	37	1	45	183
065106453210634	{1, 6, 4, 0}	(2, 0, 0, 3, 5, 4)	36	3	17	137
065106453210632	{2, 6, 3, 0}	(0, 1, 0, 4, 3, 5)	36	6	14	115
065106452341231	{1, 3, 4, 6}	(0, 3, 0, 3, 5, 1)	36	8	19	128
065106452341230	{2, 3, 4, 0}	(0, 2, 1, 5, 3, 0)	36	3	12	121
065106452341210	{2, 4, 0}	(1, 1, 1, 3, 4, 3)	36	5	37	150
065106452341201	{1, 2, 4}	(1, 0, 2, 5, 2, 2)	36	2	27	136
065106452312010	{2, 1, 5, 3, 0}	(0, 0, 2, 4, 2, 4)	36	4	5	105
065106234534210	{1, 4, 5, 0}	(3, 0, 2, 1, 5, 3)	36	4	18	151
065106234534123	{2, 3, 5}	(1, 3, 1, 3, 4, 0)	36	2	27	136
065106234534121	{1, 2, 4, 5}	(2, 2, 1, 2, 4, 2)	36	5	15	148
065106234534120	{2, 4, 5, 0}	(2, 1, 2, 4, 2, 1)	36	8	19	142
065106234532106	{1, 3, 6}	(4, 0, 3, 0, 4, 5)	36	4	47	174
065106234531210	{2, 5, 0}	(2, 1, 3, 1, 3, 4)	36	3	44	172
065106234531201	{1, 2, 5}	(2, 0, 4, 3, 1, 3)	36	2	33	157
065106234512342	{2, 4, 5}	(0, 4, 3, 1, 3, 1)	36	2	27	136
065106234512340	{3, 4, 5, 0}	(0, 3, 5, 1, 2, 0)	36	3	12	121
065106234512321	{1, 2, 3, 5}	(1, 3, 3, 0, 3, 3)	36	4	14	147
065106234512320	{2, 3, 5, 0}	(1, 2, 4, 2, 1, 2)	36	8	19	142
065106234510634	{1, 3, 4, 6}	(4, 0, 2, 5, 1, 0)	36	6	18	119
065106234510632	{2, 5, 6}	(3, 1, 3, 4, 0, 2)	36	2	33	157
065106234510630	{3, 6, 5, 0}	(5, 1, 2, 4, 0, 1)	36	4	12	124
065106234510623	{3, 5, 6}	(2, 2, 5, 2, 0, 1)	36	2	27	136
065106234510620	{2, 4, 6, 5, 0}	(4, 2, 4, 2, 0, 0)	36	4	5	105
065106234312010	{1, 2, 4, 0}	(1, 0, 4, 2, 1, 5)	36	4	12	124
065106234123210	{2, 3, 4, 0}	(0, 2, 4, 0, 2, 5)	36	4	13	129
065106234123201	{1, 3, 4, 6}	(0, 1, 5, 2, 0, 4)	36	6	18	119
061234506453421	{1, 6, 4, 5}	(4, 2, 0, 1, 5, 2)	36	3	16	145
061234506453412	{2, 6, 4, 5}	(3, 3, 0, 3, 3, 1)	36	4	14	147
061234506453410	{3, 4, 5, 0}	(5, 2, 0, 4, 2, 0)	36	4	13	129
061234506453210	{3, 5, 0}	(5, 2, 1, 0, 4, 4)	36	4	46	175
061234506453121	{1, 6, 2, 5}	(3, 3, 1, 1, 3, 3)	36	5	17	160
061234506453120	{2, 5, 0}	(4, 3, 1, 3, 1, 2)	36	3	44	172
061234506451234	{1, 3, 4, 6}	(1, 5, 3, 0, 3, 0)	36	8	19	128
061234506451232	{2, 6, 3, 5}	(2, 4, 2, 1, 2, 2)	36	5	15	148
061234506451230	{3, 5, 0}	(3, 4, 3, 1, 1, 1)	36	5	37	150
061234506451201	{1, 4, 5, 0}	(5, 3, 4, 0, 1, 0)	36	6	14	115
061234506431210	{2, 4, 0}	(4, 4, 0, 1, 2, 5)	36	4	46	175
061234506431201	{1, 4, 6}	(5, 4, 0, 3, 0, 4)	36	4	47	174
061234506412321	{1, 2, 3, 6}	(2, 5, 1, 0, 2, 4)	36	3	16	145
061234506412320	{2, 6, 3, 0}	(3, 5, 1, 2, 0, 3)	36	4	18	151
061234506412301	{1, 6, 3, 0}	(4, 5, 3, 0, 0, 2)	36	3	17	137
06510645321063	{1, 6, 3, 0}	(1, 0, 0, 3, 4, 5)	35	4	11	96
06510645234210	{1, 4, 0}	(2, 0, 1, 2, 5, 3)	35	2	27	120
06510645234123	{2, 3, 4, 6}	(0, 3, 0, 4, 4, 0)	35	3	11	103
06510645234121	{1, 2, 4, 6}	(1, 2, 0, 3, 4, 2)	35	7	16	115

06510645234120	{2, 4, 0}	(1, 1, 1, 5, 2, 1)	35	3	29	110
06510645231210	{2, 3, 5, 0}	(0, 1, 1, 3, 3, 4)	35	6	15	100
06510645231201	{1, 2, 3, 5}	(0, 0, 2, 5, 1, 3)	35	2	10	91
06510642312010	{1, 2, 3, 0}	(0, 0, 3, 3, 1, 5)	35	1	6	85
06510623453421	{1, 4, 5}	(3, 1, 1, 1, 5, 2)	35	2	26	121
06510623453412	{2, 4, 5}	(2, 2, 1, 3, 3, 1)	35	1	24	121
06510623453410	{1, 3, 5, 4, 0}	(3, 0, 2, 4, 2, 0)	35	4	6	97
06510623453210	{1, 3, 5, 0}	(3, 0, 3, 0, 4, 4)	35	9	19	135
06510623453121	{1, 2, 5}	(2, 2, 2, 1, 3, 3)	35	1	27	133
06510623453120	{2, 5, 0}	(2, 1, 3, 3, 1, 2)	35	2	35	128
06510623451234	{1, 3, 4, 5}	(0, 4, 4, 0, 3, 0)	35	3	11	103
06510623451232	{2, 3, 5}	(1, 3, 3, 1, 2, 2)	35	1	24	121
06510623451230	{3, 5, 0}	(1, 2, 5, 1, 1, 1)	35	3	29	110
06510623451063	{1, 6, 3, 5}	(4, 0, 3, 4, 0, 1)	35	7	16	108
06510623451062	{2, 6, 4, 5}	(3, 1, 5, 2, 0, 0)	35	2	10	91
06510623451060	{4, 6, 5, 0}	(5, 1, 3, 3, 0, 0)	35	1	6	85
06510623431210	{2, 4, 0}	(1, 1, 3, 1, 2, 5)	35	3	30	118
06510623431201	{1, 2, 4, 6}	(1, 0, 4, 3, 0, 4)	35	7	16	108
06510623412321	{2, 1, 4, 3, 6}	(0, 3, 3, 0, 2, 4)	35	3	5	101
06510623412320	{2, 3, 6, 4, 0}	(0, 2, 4, 2, 0, 3)	35	4	6	97
06123454012321	{1, 2, 3, 5}	(1, 5, 1, 0, 1, 5)	35	3	11	93
06123450645342	{2, 6, 4, 5}	(5, 1, 0, 1, 5, 1)	35	3	11	93
06123450645341	{1, 3, 6, 4, 5}	(4, 2, 0, 3, 3, 0)	35	3	5	101
06123450645321	{1, 6, 3, 5}	(4, 2, 1, 0, 4, 3)	35	8	18	130
06123450645312	{2, 5, 6}	(3, 3, 1, 2, 2, 2)	35	1	27	133
06123450645310	{3, 5, 0}	(5, 2, 1, 3, 1, 1)	35	3	30	118
06123450645123	{1, 6, 3, 5}	(2, 4, 3, 0, 2, 1)	35	7	16	115
06123450645120	{2, 4, 5, 0}	(4, 3, 3, 1, 1, 0)	35	6	15	100
06123450643210	{3, 4, 0}	(5, 3, 0, 0, 3, 5)	35	2	35	140
06123450643121	{1, 2, 4, 6}	(3, 4, 0, 1, 2, 4)	35	8	18	130
06123450643120	{2, 6, 4, 0}	(4, 4, 0, 3, 0, 3)	35	9	19	135
06123450641232	{2, 3, 6}	(2, 5, 1, 1, 1, 3)	35	2	26	121
06123450641230	{6, 3, 0}	(3, 5, 2, 1, 0, 2)	35	2	27	120
06123450641201	{1, 6, 4, 0}	(5, 4, 3, 0, 0, 1)	35	4	11	96
0651234012321	{2, 1, 4, 3}	(0, 4, 2, 0, 1, 5)	34	1	6	71
0651064532106	{1, 6, 2, 0}	(0, 0, 0, 2, 5, 5)	34	1	6	60
0651064523421	{1, 4, 6}	(2, 1, 0, 2, 5, 2)	34	2	28	97
0651064523412	{2, 4, 6}	(1, 2, 0, 4, 3, 1)	34	2	26	95
0651064523410	{1, 3, 4, 0}	(2, 0, 1, 5, 2, 0)	34	2	10	78
0651064523210	{1, 3, 5, 0}	(1, 0, 1, 2, 4, 4)	34	6	13	85
0651064523121	{2, 1, 6, 3, 5}	(0, 2, 0, 3, 3, 3)	34	4	5	79
0651064523120	{2, 3, 5, 0}	(0, 1, 1, 5, 1, 2)	34	4	12	76
0651064231210	{2, 3, 0}	(0, 1, 2, 2, 2, 5)	34	1	18	82
0651064231201	{1, 2, 3, 6}	(0, 0, 3, 4, 0, 4)	34	3	9	75
0651062345342	{2, 4, 5, 0}	(4, 0, 1, 1, 5, 1)	34	5	13	79
0651062345341	{1, 3, 4, 5}	(3, 1, 1, 3, 3, 0)	34	2	9	85
0651062345321	{1, 3, 5}	(3, 1, 2, 0, 4, 3)	34	3	28	110
0651062345312	{2, 5}	(2, 2, 2, 2, 2, 2)	34	1	44	111
0651062345310	{1, 3, 5, 0}	(3, 0, 3, 3, 1, 1)	34	6	14	90
0651062345123	{1, 3, 5}	(1, 3, 4, 0, 2, 1)	34	2	26	95
0651062345120	{2, 4, 5, 0}	(2, 1, 5, 1, 1, 0)	34	4	12	76

0651062345106	{1, 6, 4, 5}	(4, 0, 4, 3, 0, 0)	34	3	9	75
0651062343210	{1, 3, 4, 0}	(2, 0, 3, 0, 3, 5)	34	3	11	96
0651062343121	{1, 2, 4, 6}	(1, 2, 2, 1, 2, 4)	34	6	13	94
0651062343120	{2, 6, 4, 0}	(1, 1, 3, 3, 0, 3)	34	6	14	90
0651062341232	{2, 3, 4, 6}	(0, 3, 3, 1, 1, 3)	34	2	9	85
0651062341230	{3, 6, 4, 0}	(0, 2, 5, 1, 0, 2)	34	2	10	78
0612345430121	{1, 2, 4, 5}	(2, 4, 0, 1, 1, 5)	34	3	11	85
0612345401232	{2, 3, 5, 0}	(1, 5, 1, 1, 0, 4)	34	5	13	79
0612345064534	{3, 6, 4, 5}	(5, 1, 0, 2, 4, 0)	34	1	6	71
0612345064532	{2, 6, 3, 5}	(5, 1, 1, 0, 4, 2)	34	3	11	85
0612345064531	{1, 6, 3, 5}	(4, 2, 1, 2, 2, 1)	34	6	13	94
0612345064512	{2, 1, 6, 4, 5}	(3, 3, 3, 0, 2, 0)	34	4	5	79
0612345064510	{4, 5, 0}	(5, 2, 2, 2, 1, 0)	34	1	18	82
0612345064321	{1, 3, 4, 6}	(4, 3, 0, 0, 3, 4)	34	6	17	108
0612345064312	{2, 4, 6}	(3, 4, 0, 2, 1, 3)	34	3	28	110
0612345064310	{3, 6, 4, 0}	(5, 3, 0, 3, 0, 2)	34	3	11	96
0612345064123	{1, 3, 6}	(2, 5, 2, 0, 1, 2)	34	2	28	97
0612345064120	{2, 6, 4, 0}	(4, 4, 2, 1, 0, 1)	34	6	13	85
0612345061201	{1, 6, 5, 0}	(5, 5, 2, 0, 0, 0)	34	1	6	60
065123430121	{1, 2, 4}	(1, 3, 1, 1, 1, 5)	33	1	15	67
065123401232	{2, 3, 4, 0}	(0, 4, 2, 1, 0, 4)	33	2	8	61
065106453210	{1, 2, 5, 0}	(0, 0, 1, 1, 5, 4)	33	2	7	54
065106452342	{2, 6, 4, 0}	(3, 0, 0, 2, 5, 1)	33	5	12	65
065106452341	{1, 3, 4, 6}	(2, 1, 0, 4, 3, 0)	33	4	11	69
065106452321	{1, 6, 3, 5}	(1, 1, 0, 2, 4, 3)	33	5	12	70
065106452312	{2, 6, 3, 5}	(0, 2, 0, 4, 2, 2)	33	2	9	67
065106452310	{1, 3, 5, 0}	(1, 0, 1, 5, 1, 1)	33	4	10	58
065106423210	{1, 3, 0}	(1, 0, 2, 1, 3, 5)	33	1	15	71
065106423121	{1, 2, 3, 6}	(0, 2, 1, 2, 2, 4)	33	2	7	67
065106423120	{2, 6, 3, 0}	(0, 1, 2, 4, 0, 3)	33	3	9	64
065106234534	{3, 4, 5, 0}	(4, 0, 1, 2, 4, 0)	33	2	8	61
065106234532	{2, 3, 5, 0}	(4, 0, 2, 0, 4, 2)	33	4	12	73
065106234531	{1, 3, 5}	(3, 1, 2, 2, 2, 1)	33	1	21	80
065106234512	{1, 2, 4, 5}	(2, 2, 4, 0, 2, 0)	33	2	9	67
065106234510	{1, 4, 5, 0}	(3, 0, 4, 2, 1, 0)	33	3	9	64
065106234321	{1, 3, 4, 6}	(2, 1, 2, 0, 3, 4)	33	4	12	80
065106234312	{2, 4, 6}	(1, 2, 2, 2, 1, 3)	33	1	21	80
065106234310	{1, 3, 6, 4, 0}	(2, 0, 3, 3, 0, 2)	33	4	4	66
065106234123	{1, 3, 4, 6}	(0, 3, 4, 0, 1, 2)	33	4	11	69
065106234120	{2, 6, 4, 0}	(1, 1, 5, 1, 0, 1)	33	4	10	58
061234543201	{1, 3, 4, 5}	(3, 3, 0, 0, 2, 5)	33	2	10	73
061234543012	{2, 4, 5, 0}	(2, 4, 0, 2, 0, 4)	33	4	12	73
061234540123	{1, 3, 5, 0}	(1, 5, 2, 0, 0, 3)	33	5	12	65
061234506453	{3, 5, 6}	(5, 1, 1, 1, 3, 1)	33	1	15	67
061234506451	{1, 6, 4, 5}	(4, 2, 2, 1, 2, 0)	33	2	7	67
061234506432	{2, 3, 4, 6}	(5, 2, 0, 0, 3, 3)	33	2	10	73
061234506431	{1, 3, 4, 6}	(4, 3, 0, 2, 1, 2)	33	4	12	80
061234506412	{1, 2, 4, 6}	(3, 4, 2, 0, 1, 1)	33	5	12	70
061234506410	{6, 4, 0}	(5, 3, 1, 2, 0, 1)	33	1	15	71
061234506120	{2, 6, 5, 0}	(4, 5, 1, 1, 0, 0)	33	2	7	54
06541230121	{1, 2, 3, 5}	(0, 3, 0, 2, 1, 5)	32	1	6	49

06512343201	{1, 3, 4}	(2, 2, 1, 0, 2, 5)	32	1	14	59
06512343012	{2, 4, 0}	(1, 3, 1, 2, 0, 4)	32	2	17	58
06512340123	{1, 3, 4, 0}	(0, 4, 3, 0, 0, 3)	32	2	8	51
06510645321	{1, 6, 2, 5}	(0, 1, 0, 1, 5, 3)	32	3	7	46
06510645234	{3, 6, 4, 0}	(3, 0, 0, 3, 4, 0)	32	2	8	51
06510645232	{2, 3, 6, 5, 0}	(2, 0, 0, 2, 4, 2)	32	3	4	49
06510645231	{1, 6, 3, 5}	(1, 1, 0, 4, 2, 1)	32	4	9	52
06510645210	{2, 1, 5, 4, 0}	(0, 0, 1, 5, 1, 0)	32	3	3	34
06510643210	{1, 2, 4, 0}	(0, 0, 2, 0, 4, 5)	32	1	6	46
06510642321	{1, 3, 6}	(1, 1, 1, 1, 3, 4)	32	2	16	60
06510642312	{2, 3, 6}	(0, 2, 1, 3, 1, 3)	32	1	13	58
06510642310	{1, 6, 3, 0}	(1, 0, 2, 4, 0, 2)	32	2	6	50
06510623453	{3, 5, 0}	(4, 0, 2, 1, 3, 1)	32	2	17	58
06510623451	{1, 4, 5}	(3, 1, 3, 1, 2, 0)	32	1	13	58
06510623432	{2, 3, 6, 4, 0}	(3, 0, 2, 0, 3, 3)	32	3	4	55
06510623431	{1, 3, 4, 6}	(2, 1, 2, 2, 1, 2)	32	3	9	60
06510623412	{1, 2, 4, 6}	(1, 2, 4, 0, 1, 1)	32	4	9	52
06510623410	{1, 6, 4, 0}	(2, 0, 4, 2, 0, 1)	32	2	6	50
06510623120	{2, 3, 6, 5, 0}	(0, 1, 5, 1, 0, 0)	32	3	3	34
06123454320	{2, 3, 5, 4, 0}	(4, 2, 0, 0, 2, 4)	32	2	4	52
06123454301	{1, 3, 5, 4, 0}	(3, 3, 0, 2, 0, 3)	32	3	4	55
06123454012	{2, 1, 5, 4, 0}	(2, 4, 2, 0, 0, 2)	32	3	4	49
06123450645	{2, 6, 4, 5}	(5, 1, 2, 0, 3, 0)	32	1	6	49
06123450643	{3, 4, 6}	(5, 2, 0, 1, 2, 2)	32	1	14	59
06123450641	{1, 4, 6}	(4, 3, 1, 1, 1, 1)	32	2	16	60
06123450612	{1, 6, 2, 5}	(3, 5, 1, 0, 1, 0)	32	3	7	46
06123450610	{3, 6, 5, 0}	(5, 4, 0, 2, 0, 0)	32	1	6	46
0654123201	{1, 3, 5}	(1, 2, 0, 1, 2, 5)	31	1	14	45
0654123012	{2, 3, 5, 0}	(0, 3, 0, 3, 0, 4)	31	3	8	43
0651234320	{2, 3, 4, 0}	(3, 1, 1, 0, 2, 4)	31	1	6	43
0651234301	{1, 3, 4, 0}	(2, 2, 1, 2, 0, 3)	31	1	6	45
0651234012	{1, 2, 4, 0}	(1, 3, 3, 0, 0, 2)	31	1	6	40
0651064532	{2, 6, 5, 0}	(1, 0, 0, 1, 5, 2)	31	2	6	35
0651064523	{3, 6, 5, 0}	(2, 0, 0, 3, 3, 1)	31	1	6	40
0651064521	{2, 1, 6, 4, 5}	(0, 1, 0, 4, 2, 0)	31	2	3	31
0651064321	{1, 2, 4, 6}	(0, 1, 1, 0, 4, 4)	31	3	7	40
0651064232	{2, 6, 3, 0}	(2, 0, 1, 1, 3, 3)	31	2	6	43
0651064231	{1, 3, 6}	(1, 1, 1, 3, 1, 2)	31	1	13	46
0651064210	{2, 1, 6, 4, 0}	(0, 0, 2, 4, 0, 1)	31	2	2	30
0651062345	{2, 4, 5, 0}	(4, 0, 3, 0, 3, 0)	31	3	8	43
0651062343	{3, 6, 4, 0}	(3, 0, 2, 1, 2, 2)	31	1	6	45
0651062341	{1, 4, 6}	(2, 1, 3, 1, 1, 1)	31	1	13	46
0651062312	{2, 1, 6, 3, 5}	(0, 2, 4, 0, 1, 0)	31	2	3	31
0651062310	{1, 3, 6, 5, 0}	(1, 0, 4, 2, 0, 0)	31	2	2	30
0612345432	{2, 3, 4, 5}	(5, 1, 0, 0, 1, 5)	31	1	5	32
0612345430	{3, 4, 5, 0}	(4, 2, 0, 1, 1, 3)	31	1	6	43
0612345401	{1, 4, 5, 0}	(3, 3, 1, 1, 0, 2)	31	2	6	43
0612345064	{2, 4, 6}	(5, 2, 1, 0, 2, 1)	31	1	14	45
0612345061	{1, 6, 3, 5}	(4, 4, 0, 1, 1, 0)	31	3	7	40
0612345012	{1, 2, 5, 0}	(2, 5, 1, 0, 0, 1)	31	2	6	35
065431201	{1, 2, 4, 5}	(0, 2, 0, 0, 3, 5)	30	1	6	31

065412320	{2, 3, 5, 0}	(2, 1, 0, 1, 2, 4)	30	2	6	34
065412301	{1, 3, 5, 0}	(1, 2, 0, 3, 0, 3)	30	3	6	35
065123432	{2, 3, 4, 6}	(4, 0, 1, 0, 1, 5)	30	1	5	27
065123430	{3, 4, 0}	(3, 1, 1, 1, 1, 3)	30	1	9	36
065123401	{1, 4, 0}	(2, 2, 2, 1, 0, 2)	30	1	9	36
065123012	{2, 1, 5, 3, 0}	(0, 3, 3, 0, 0, 1)	30	2	2	25
065106453	{1, 3, 6, 5, 0}	(0, 0, 0, 1, 5, 1)	30	2	2	20
065106452	{2, 4, 6, 5, 0}	(1, 0, 0, 3, 3, 0)	30	2	2	25
065106432	{2, 6, 4, 0}	(1, 0, 1, 0, 4, 3)	30	3	6	31
065106423	{6, 3, 0}	(2, 0, 1, 2, 2, 2)	30	1	9	36
065106421	{1, 2, 4, 6}	(0, 1, 1, 3, 1, 1)	30	2	5	28
065106234	{2, 6, 4, 0}	(3, 0, 3, 0, 2, 1)	30	3	6	35
065106231	{1, 6, 3, 5}	(1, 1, 3, 1, 1, 0)	30	2	5	28
065106210	{2, 1, 6, 5, 0}	(0, 0, 3, 3, 0, 0)	30	1	1	20
061234543	{1, 3, 4, 5}	(5, 1, 0, 1, 0, 4)	30	1	5	27
061234540	{2, 4, 5, 0}	(4, 2, 1, 0, 1, 2)	30	2	6	34
061234506	{2, 6, 3, 5}	(5, 3, 0, 0, 2, 0)	30	1	6	31
061234501	{1, 3, 5, 0}	(3, 4, 0, 1, 0, 1)	30	3	6	31
061234012	{2, 1, 6, 4, 0}	(1, 5, 1, 0, 0, 0)	30	2	2	20
06543120	{2, 4, 5, 0}	(1, 1, 0, 0, 3, 4)	29	2	6	25
06541232	{2, 6, 3, 5}	(3, 0, 0, 1, 1, 5)	29	1	5	22
06541230	{3, 5, 0}	(2, 1, 0, 2, 1, 3)	29	1	9	29
06541201	{2, 1, 5, 4, 0}	(0, 2, 0, 3, 0, 2)	29	2	2	22
06512343	{1, 3, 4, 6}	(4, 0, 1, 1, 0, 4)	29	2	5	23
06512340	{2, 4, 0}	(3, 1, 2, 0, 1, 2)	29	1	9	29
06512301	{1, 3, 5, 0}	(1, 2, 2, 1, 0, 1)	29	2	4	23
06510645	{1, 4, 6, 5, 0}	(0, 0, 0, 2, 4, 0)	29	1	1	15
06510643	{1, 3, 6, 4, 0}	(0, 0, 1, 0, 4, 2)	29	2	2	18
06510642	{2, 6, 4, 0}	(1, 0, 1, 2, 2, 1)	29	2	4	23
06510623	{2, 3, 6, 5, 0}	(2, 0, 3, 0, 2, 0)	29	2	2	22
06510621	{1, 6, 2, 5}	(0, 1, 2, 2, 1, 0)	29	2	3	19
06123454	{1, 2, 4, 5}	(5, 1, 1, 0, 0, 3)	29	1	5	22
06123450	{2, 3, 5, 0}	(4, 3, 0, 0, 1, 1)	29	2	6	25
06123401	{1, 3, 6, 4, 0}	(2, 4, 0, 1, 0, 0)	29	2	2	18
06123012	{2, 1, 6, 3, 0}	(0, 4, 2, 0, 0, 0)	29	1	1	15
0654312	{2, 6, 4, 5}	(2, 0, 0, 0, 2, 5)	28	1	5	17
0654310	{1, 3, 5, 4, 0}	(0, 1, 0, 0, 3, 3)	28	2	2	15
0654123	{1, 6, 3, 5}	(3, 0, 0, 2, 0, 4)	28	2	5	19
0654120	{2, 4, 5, 0}	(1, 1, 0, 2, 1, 2)	28	1	4	19
0651234	{1, 2, 4, 6}	(4, 0, 2, 0, 0, 3)	28	2	5	19
0651230	{2, 3, 5, 0}	(2, 1, 2, 0, 1, 1)	28	1	4	19
0651201	{1, 2, 5, 0}	(0, 2, 1, 2, 0, 1)	28	1	2	16
0651064	{1, 6, 4, 0}	(0, 0, 1, 1, 3, 1)	28	1	2	14
0651062	{2, 6, 5, 0}	(1, 0, 2, 1, 2, 0)	28	1	2	16
0612345	{1, 2, 3, 5}	(5, 2, 0, 0, 0, 2)	28	1	5	17
0612340	{2, 3, 6, 4, 0}	(3, 3, 0, 0, 1, 0)	28	2	2	15
0612301	{1, 6, 3, 0}	(1, 3, 1, 1, 0, 0)	28	1	2	14
065431	{1, 3, 6, 4, 5}	(1, 0, 0, 0, 2, 4)	27	1	2	11
065412	{2, 1, 6, 4, 5}	(2, 0, 0, 2, 0, 3)	27	1	2	13
065410	{1, 4, 5, 0}	(0, 1, 0, 1, 2, 2)	27	1	2	12
065123	{2, 1, 6, 3, 5}	(3, 0, 2, 0, 0, 2)	27	1	2	13

065120	{2, 5, 0}	(1, 1, 1, 1, 1, 1)	27	1	4	14
065106	{1, 3, 6, 5, 0}	(0, 0, 2, 0, 3, 0)	27	1	1	10
061234	{2, 1, 4, 3, 6}	(4, 2, 0, 0, 0, 1)	27	1	2	11
061230	{2, 6, 3, 0}	(2, 2, 1, 0, 1, 0)	27	1	2	12
061201	{2, 1, 6, 4, 0}	(0, 3, 0, 2, 0, 0)	27	1	1	10
06543	{4, 3, 6, 5, 0}	(0, 0, 0, 0, 1, 5)	26	1	1	6
06541	{1, 6, 4, 5}	(1, 0, 0, 1, 1, 3)	26	1	2	9
06512	{1, 6, 2, 5}	(2, 0, 1, 1, 0, 2)	26	1	2	10
06510	{1, 3, 5, 0}	(0, 1, 1, 0, 2, 1)	26	1	2	9
06123	{1, 2, 3, 6}	(3, 1, 1, 0, 0, 1)	26	1	2	9
06120	{2, 6, 4, 0}	(1, 2, 0, 1, 1, 0)	26	1	2	9
01234	{2, 1, 4, 3, 0}	(5, 1, 0, 0, 0, 0)	26	1	1	6
0654	{2, 4, 6, 5, 0}	(0, 0, 0, 1, 0, 4)	25	1	1	5
0651	{1, 6, 3, 5}	(1, 0, 1, 0, 1, 2)	25	1	2	7
0612	{1, 2, 4, 6}	(2, 1, 0, 1, 0, 1)	25	1	2	7
0610	{1, 3, 6, 4, 0}	(0, 2, 0, 0, 2, 0)	25	1	1	6
0123	{2, 1, 5, 3, 0}	(4, 0, 1, 0, 0, 0)	25	1	1	5
065	{2, 3, 6, 5, 0}	(0, 0, 1, 0, 0, 3)	24	1	1	4
061	{1, 3, 4, 6}	(1, 1, 0, 0, 1, 1)	24	1	2	5
012	{2, 1, 5, 4, 0}	(3, 0, 0, 1, 0, 0)	24	1	1	4
06	{2, 3, 6, 4, 0}	(0, 1, 0, 0, 0, 2)	23	1	1	3
01	{1, 3, 5, 4, 0}	(2, 0, 0, 0, 1, 0)	23	1	1	3
0	{2, 3, 5, 4, 0}	(1, 0, 0, 0, 0, 1)	22	1	1	2
e	{5, 4, 2, 3, 6, 1}	(0, 0, 0, 0, 0, 0)	21	1	1	1

6 Concluding Remarks

Though not listed in the tables, all the sets whose size is given in columns (5) and (6) above can be computed by our programs. A next step, which we hope to carry out, is to use this information, guided by the tables, to compute characters of $L(w \cdot -2\rho)$ in new cases. We would like to take this opportunity to thank the University of Virginia Advanced Research Computing Services group for their help, with special thanks to Jacalyn Huband.

References

1. Achar, P., Makasumi, S., Riche, S., Williamson, G.: Free-monodromic mixed tilting sheaves on flag varieties, 132 pp. arXiv:1703.05843v1
2. Achar, P., Makasumi, S., Riche, S., Williamson, G.: Koszul duality for Kac-Moody groups and characters of tilting modules, 50 pp. (2017). arXiv:1706.00183v1
3. Andersen, H., Jantzen, J., Soergel, W.: Representations of quantum groups at a pth root of unity and of semisimple groups in characteristic p. Astérique **220** (1994)
4. Cline, E., Parshall, B., Scott, L.: Infinitesimal Kazhdan-Lusztig theories. Contemp. Math. **139**, 43–73 (1983)
5. Cline, E., Parshall, B., Scott, L.: Reduced standard modules and cohomology. Trans. Am. Math. Soc. **361**(10), 5223–5261 (2009)
6. Fiebig, P.: An upper bound on the exceptional characteristics for Lusztig's character formula. J. Reine Angew. Math. **673**, 1–31 (2012)

7. Jantzen, J.: Representations of Algebraic Groups, 2nd edn. American Mathematical Society (2003)
8. Lusztig, G.: Some problems in the representation theory of finite Chevalley groups. In: The Santa Cruz Conference on Finite Groups (University of California, Santa Cruz, California, 1979), Proceedings of Symposia in Pure Mathematics, vol. 37, pp. 313-317. American Mathematical Society, Providence, R.I. (1980)
9. Ko, H.: Grade zero part of forced graded algebras. Sci. China Math. (2017). https://doi.org/10.1007/s11425-017-9163-0
10. Parshall, B., Scott, L.: A semisimple series for q-Weyl and q-Specht modules. Proc. Symp. Pure Math. **86**, 277–310 (2012)
11. Riche, S., Williamson, G.: Tilting modules and the p-canonical basis (2017). arXiv:1512.08296v3
12. Steinberg, R.: Lectures on Chevalley groups. Notes prepared by John Faulkner and Robert Wilson. Revised and corrected edition of the 1968 original. With a foreword by Robert R. Snapp. University Lecture Series, Vol. 66, xi+160 pp. American Mathematical Society, Providence, RI (2016)
13. Tanisaki, T.: Character formulas of Kazhdan-Lusztig type. In: Representations of Finite Dimensional Algebras and Related Topics in Lie Theory and Geometry, pp. 261–276, Fields Institute Communications, vol. 40, American Mathematical Society, Providence, RI (2004)
14. Thorge, L., Williamson, G.: The p-canonical basis for Hecke algebras. In: Categorification and Higher Representation Theory, pp. 333–361, Contemporary Mathematics, vol. 683, American Mathematical Society, Providence, RI (2017)
15. Williamson, G.: Schubert calculus and torsion explosion. With a joint appendix with Alex Kontorovich and Peter J. McNamara. J. Amer. Math. Soc. **30**(4), 1023–1046 (2017)
16. Wong, W.: Irreducible representations of finite Chevalley groups. J. Algebra **20**, 355–367 (1972)

Hypergroups All Nonidentity Elements of Which Are Involutions

Paul-Hermann Zieschang

Abstract The notion of a hypergroup generalizes the notion of a group. We introduce a hypergroup-theoretic generalization of the group-theoretic notion of an involution and characterize the hypergroups all nonidentity elements of which are involutions. Our characterization sheds new light on previous investigations in which a corresponding condition was considered within the theory of association schemes [10, 12] and table algebras [2]. We also show in how far hypergroups all nonidentity elements of which are involutions play a role in the investigation [11] of constrained hypergroups and their relationship to Coxeter hypergroups.

Keywords Hypergroup · Involution · Projective hypergroup · Table algebra

AMS Subject Classification: 20N20 · 51E15 · 51E20 · 51E24

1 Introduction

Let S be a set.

A map from $S \times S$ to the set of all non-empty subsets of S is called a *hyperoperation on* S.

Let μ be a hyperoperation on S. For any two elements p and q in S, we write pq instead of $\mu(p, q)$.

For any two subsets P and Q of S, we define PQ to be the union of the sets pq with $p \in P$ and $q \in Q$ and call the subset PQ of S the *complex product* of P and Q. If one of the two factors in a complex product consists of a single element, say s, we write s instead of $\{s\}$ in that product.

Following (and slightly generalizing) Marty's terminology in [7], we call S a *hypergroup* (with respect to μ) if the following three conditions hold:

H1 For any three elements p, q, and r in S, we have $p(qr) = (pq)r$.

P.-H. Zieschang (✉)
School of Mathematical and Statistical Sciences, University of Texas Rio Grande Valley, Edinburg, TX 78539, USA
e-mail: zieschang@utrgv.edu

J. Feldvoss et al. (eds.), *Advances in Algebra*, Springer Proceedings in Mathematics & Statistics 277, https://doi.org/10.1007/978-3-030-11521-0_17

H2 The set S contains an element e such that $se = \{s\}$ for each element s in S.
H3 For each element s in S, there exists an element s^* in S such that $q \in p^*r$ and $p \in rq^*$ for any three elements p, q, and r in S satisfying $r \in pq$.

An element e of S which satisfies $se = \{s\}$ for each element s in S is called a *neutral element* of S. A map * from S to S is called an *inverse function* of S if one has $q \in p^*r$ and $p \in rq^*$ for any three elements p, q, and r in S satisfying $r \in pq$.

In the literature, the term hypergroup is not always used in the sense of the above definition. In [3], for instance, a hypergroup is defined to be the vector space of all bounded Radon measures on a locally compact Hausdorff space satisfying certain conditions aiming toward harmonic-analytic goals. In [8], and that is another example, hypergroups are defined via commutative rings, and in [9], hypergroups appear as distinguished bases of finite-dimensional algebras over the field of the complex numbers. For a more complete overview of the various different mathematical concepts which have been called hypergroups since Marty's original definition, we refer to [1]. The definition which we use in this article is almost Marty's original definition. The only difference is that we do not assume that $pq = qp$ for any two hypergroup elements p and q. In other words, we do not restrict ourself to commutative hypergroups.

It should also be noted that hypergroups in the sense of Marty arise (additively written) in the definition of hyperrings as given by Krasner; cf. [6]. Connections between hyperrings and questions in number theory have been considered in [4, 5], and in articles quoted in these two publications.

The present article is organized as follows.

In Sect. 2, we show that each hypergroup possesses only one neutral element and only one inverse function. In Sect. 3, we compile several observations about products of hypergroup elements, and we provide some computational rules referring to complex products of subsets of hypergroups. In Sect. 4, we introduce closed subsets. The notion of a closed subset generalizes the notion of a subgroup from group theory to the theory of hypergroups and gives rise to the notion of an involution. In Sect. 5, we collect basic information about involutions. Section 6 contains our main result, Theorem 6.9. This theorem relates hypergroups all nonidentity elements of which are involutions to projective spaces. In Sect. 6, we also prove that hypergroups all nonidentity elements of which are involutions possess bases; cf. Theorem 6.5. This result may be of independent interest. In Sect. 7, we show in how far hypergroups all nonidentity elements of which are involutions arise in the investigation of constrained hypergroups. In Sect. 8, we relate our main result to results on table algebras obtained by Blau and Chen in [2].

We conclude this introduction with a useful observation on the inverse function of a hypergroup.

Lemma 1.1 *Let S be a hypergroup, let * be an inverse function of S, and let p, q, and r be elements of S. Then the statements $r \in pq$, $q \in p^*r$, $p \in rq^*$, $r^* \in q^*p^*$, $q^* \in r^*p$, and $p^* \in qr^*$ are pairwise equivalent.*

Proof Applying Condition H3 we obtain $q \in p^*r$ from $r \in pq$. Similarly, we obtain $p^* \in qr^*$ from $q \in p^*r$, $r^* \in q^*p^*$ from $p^* \in qr^*$, $q^* \in r^*p$ from $r^* \in q^*p^*$, $p \in rq^*$ from $q^* \in r^*p$, and $r \in pq$ from $p \in rq^*$. □

2 Neutral Elements and Inverse Functions

In this section, we show that each hypergroup possesses only one neutral element and only one inverse function.

Lemma 2.1 *Let S be a hypergroup, let e be a neutral element of S, and let* * *be an inverse function of S. Then we have* $e \in s^*s$ *for each element s in S.*

Proof Let s be an element of S. Since e is a neutral element of S, we have $s \in se$; cf. Condition H2. Thus, by Condition H3, $e \in s^*s$. □

Lemma 2.2 *Let S be a hypergroup, and let* * *be an inverse function of S. Then we have* $s^{**} = s$ *for each element s in S.*

Proof Let s be an element of S, and let e be a neutral element of S. Then, by Lemma 2.1, $e \in s^*s$. Thus, by Condition H3, $s \in s^{**}e$. On the other hand, by Condition H2, $s^{**}e = \{s^{**}\}$. Thus, $s \in \{s^{**}\}$, and that implies that $s^{**} = s$. □

Lemma 2.3 *Let S be a hypergroup, and let e be a neutral element of S. Then we have* $es = \{s\}$ *for each element s in S.*

Proof Let s be an element of S, and let r be an element in es. We shall be done if we succeed in showing that $r = s$. Let * be an inverse function of S. From $r \in es$, we obtain that $s^* \in r^*e$; cf. Lemma 1.1. On the other hand, by Condition H2, $r^*e = \{r^*\}$. Thus, $s^* \in \{r^*\}$, and that implies that $r^* = s^*$. It follows that $r = r^{**} = s^{**} = s$; cf. Lemma 2.2. □

Lemma 2.4 *Each hypergroup possesses exactly one neutral element.*

Proof Let S be a hypergroup, and let c and d be neutral elements of S. Since c is a neutral element of S, we have $cd = \{d\}$; cf. Lemma 2.3. Since d is a neutral element of S, we have $cd = \{c\}$; cf. Condition H2. It follows that $c = d$. □

For the remainder of this article, the uniquely determined neutral element of a hypergroup will always be denoted by 1. Thus, we have $s \cdot 1 = \{s\} = 1 \cdot s$ for each element s of a hypergroup S.

Lemma 2.5 *For each inverse function* * *of a hypergroup, we have* $1^* = 1$.

Proof From Lemma 2.1, we know that $1 \in 1^*1$. On the other hand, by Condition H2, $1^*1 = \{1^*\}$. Thus, $1 \in \{1^*\}$, and that implies that $1^* = 1$. □

Lemma 2.6 *Let S be a hypergroup, let * be an inverse function of S, and let p and q be elements in S. Then we have $1 \in pq$ if and only if $p = q^*$.*

Proof Assume first that $1 \in pq$. Then, by Condition H3, $p \in 1 \cdot q^*$. On the other hand, by Lemma 2.3, $1 \cdot q^* = \{q^*\}$. Thus, $p \in \{q^*\}$, and that implies that $p = q^*$.

Conversely, assume that $p = q^*$. Then $pq = q^*q$. Now recall from Lemma 2.1 that $1 \in q^*q$. Thus, $1 \in pq$. □

Lemma 2.7 *Each hypergroup possesses exactly one inverse function.*

Proof Let S be a hypergroup, let $^\flat$ and $^\sharp$ be inverse functions of S, and let s be an element in S. We shall see that $s^\flat = s^\sharp$.

Since $^\flat$ is assumed to be an inverse function of S, we obtain from Lemma 2.1 that $1 \in s^\flat s$. Thus, as $^\sharp$ is assumed to be an inverse function of S, Lemma 2.6 yields $s^\flat = s^\sharp$. □

For the remainder of this article, the uniquely determined inverse function of a hypergroup will always be denoted by *.

Recall from Lemma 2.1 that $1 \in s^*s$ for each element s of a hypergroup. An element s of a hypergroup is called *thin* if $s^*s = \{1\}$.

From Lemma 2.5, we know that $1^* = 1$. Thus, by Condition H2, $1^*1 = \{1\}$, and this says that 1 is thin.

3 Selected Observations

In this section, we compile several observations about products of hypergroup elements, and we provide some computational rules for complex products of subsets of hypergroups.

For each subset R of a hypergroup, we define R^* to be the set of all elements r^* with $r \in R$.

Lemma 3.1 *Let S be a hypergroup, and let p and q be elements in S. Then $(pq)^* = q^*p^*$.*

Proof Let s be an element in S. By definition, we have $s \in (pq)^*$ if and only if $s^* \in pq$; cf. Lemma 2.2. By Lemma 1.1, we have $s^* \in pq$ if and only if $s \in q^*p^*$. □

An element s of a hypergroup is called *symmetric* if $s^* = s$. A subset of a hypergroup is called *symmetric* if each of its elements is symmetric. Hypergroups are called *commutative* if $pq = qp$ for any two of their elements p and q.

Lemma 3.2 *Symmetric hypergroups are commutative.*

Proof Let S be a symmetric hypergroup, and let p and q be elements in S. Then, by Lemma 3.1, $pq = p^*q^* = (qp)^* = qp$. □

Lemma 3.3 *For any three subsets P, Q, and R of S, we have $(PQ)R = P(QR)$.*

Proof Let P, Q, and R be subsets of S, and let s be an element in $(PQ)R$. Then there exist elements t in PQ and r in R such that $s \in tr$. Since $t \in PQ$, there exist elements p in P and q in Q such that $t \in pq$. It follows that $s \in (pq)r$, and then that $s \in p(qr)$.

Since $s \in p(qr)$, qr contains an element u such that $s \in pu$. Since $u \in qr$, $u \in QR$. Thus, as $p \in P$ and $s \in pu$, $s \in P(QR)$.

Since s has been chosen arbitrarily among the elements of $(PQ)R$, this shows that $(PQ)R \subseteq P(QR)$. The reverse containment follows similarly. □

Lemma 3.4 *Let S be a hypergroup, and let P and Q be subsets of S. Then the following hold:*

(i) *If $P \subseteq Q$, then $P^* \subseteq Q^*$.*
(ii) *We have $(PQ)^* = Q^*P^*$.*

Proof

(i) Let s be an element in P^*. Then, by definition, $s^* \in P$. Thus, as $P \subseteq Q$, $s^* \in Q$, so, by definition, $s \in Q^*$.
(ii) This follows from Lemma 3.1. □

Lemma 3.5 *Let S be a hypergroup, and let R be a subset of S. Then the following hold:*

(i) *We have $R^{**} = R$.*
(ii) *If $R^* \subseteq R$, then $R^* = R$.*

Proof

(i) This follows from Lemma 2.2.
(ii) We are assuming that $R^* \subseteq R$. Thus, by Lemma 3.4(i), $R^{**} \subseteq R^*$. On the other hand, by (i), $R^{**} = R$. Thus, $R \subseteq R^*$. □

4 Closed Subsets

In this section, the letter S stands for a hypergroup.

A non-empty subset R of S is called *closed* if $R^*R \subseteq R$.

Note that S is a closed subset of S. Also $\{1\}$ is closed, since, by Lemma 2.5, $1^* = 1$ and, by Condition H2, $1 \cdot 1 \subseteq \{1\}$.

Lemma 4.1 *A subset R of S is closed if and only if $1 \in R$, $R^* = R$, and $RR = R$.*

Proof Assume first that R is closed. Then R is not empty. Let r be an element in R. Then, by definition, $r^*r \subseteq R$. On the other hand, by Lemma 2.1, $1 \in r^*r$. Thus, $1 \in R$.

From $1 \in R$, we obtain that $r^* \in r^* \cdot 1 \subseteq R$ for each element r in R; cf. Condition H2. It follows that $R^* \subseteq R$. Thus, by Lemma 3.5(ii), $R^* = R$.

From $1 \in R$, we obtain that $R = R \cdot 1 \subseteq RR$. From $R^* = R$, we also obtain that $RR = R^*R \subseteq R$. Thus, $RR = R$.

Assume now, conversely, that $1 \in R$, $R^* = R$, and $RR = R$. From $1 \in R$, we obtain that R is not empty. From $R^* = R$ and $RR = R$, we obtain that $R^*R = R$. □

Lemma 4.2 *Let $\mathcal{T}$ be a set of closed subsets of S. Then the intersection of the closed subsets T with $T \in \mathcal{T}$ is closed.*

Proof Let R denote the intersection of the closed subsets of S which belong to $\mathcal{T}$. From Lemma 4.1, we know that $1 \in T$ for each element T in $\mathcal{T}$. Thus, $1 \in R$. This shows that R is not empty.

For each element T in $\mathcal{T}$, we have $R \subseteq T$. Thus, by Lemma 3.4(i), $R^*R \subseteq T^*T \subseteq T$ for each element T in $\mathcal{T}$. It follows that $R^*R \subseteq R$. □

Lemma 4.3 *Let T and U be closed subsets of S. Then TU is closed if and only if $TU = UT$.*

Proof Assume first that TU is closed. Then, by Lemma 4.1, $(TU)^* = TU$. Since T and U are assumed to be closed, we also have $T^* = T$ and $U^* = U$. Thus, by Lemma 3.4(ii),

$$TU = (TU)^* = U^*T^* = UT.$$

Conversely, assuming that $TU = UT$ we obtain from Lemma 3.4(ii) that

$$(TU)^*TU = U^*T^*TU \subseteq U^*TU = U^*UT \subseteq UT = TU.$$

Thus, as $1 \in TU$, TU is closed. □

Let R be a subset of S.

We define $\langle R \rangle$ to be the intersection of the closed subsets of S which contain R as a subset.

From Lemma 4.2, we know that $\langle R \rangle$ is a closed subset of S. The closed subset $\langle R \rangle$ is said to be *generated* by R. The set R is called a *generating set* of $\langle R \rangle$.

Lemma 4.4 *Let P and Q be subsets of S with $P \subseteq Q$. Then $\langle P \rangle \subseteq \langle Q \rangle$.*

Proof By definition, $\langle P \rangle$ is the intersection of the closed subsets of S which contain P as a subset. On the other hand, by Lemma 4.2, $\langle Q \rangle$ is closed. Thus, as $P \subseteq Q \subseteq \langle Q \rangle$, $\langle P \rangle \subseteq \langle Q \rangle$. □

Let R be a non-empty subset of S. We set $R^0 := \{1\}$. For each positive integer n, we define $R^n := R^{n-1}R$.

Note that $\langle R\rangle = \langle R^* \cup R\rangle$ for each subset R of S. Thus, it is no loss of generality to restrict the investigation of generating sets to *-*invariant* generating sets, that is, to sets R satisfying $R^* = R$.

Lemma 4.5 *Let R be a non-empty *-invariant subset of S. Then $\langle R\rangle$ is equal to the union of the sets R^n with n a nonnegative integer.*

Proof Define Q to be the union of the sets R^n with n a nonnegative integer. We have to show that $\langle R\rangle = Q$.

Since R is assumed to be *-invariant, so is R^n for each nonnegative integer n; cf. Lemma 3.4(ii). Thus, for any two nonnegative integers k and l,

$$(R^k)^* R^l = R^k R^l = R^{k+l} \subseteq Q.$$

It follows that Q is closed. Thus, as $R \subseteq Q$, $\langle R\rangle \subseteq Q$.

Conversely, for each nonnegative integer n, we have $R^n \subseteq \langle R\rangle^n \subseteq \langle R\rangle$. Thus, the definition of Q yields $Q \subseteq \langle R\rangle$. □

Lemma 4.6 *Let R be a subset of $S \setminus \{1\}$, and let r be an element in R. Then $r \in \langle R \setminus \{r\}\rangle$ if and only if $\langle R \setminus \{r\}\rangle = \langle R\rangle$.*

Proof Since $R \setminus \{r\} \subseteq \langle R \setminus \{r\}\rangle$, we have $r \in \langle R \setminus \{r\}\rangle$ if and only if $R \subseteq \langle R \setminus \{r\}\rangle$. By definition, $R \subseteq \langle R \setminus \{r\}\rangle$ if and only if $\langle R\rangle \subseteq \langle R \setminus \{r\}\rangle$. Now recall from Lemma 4.4 that $\langle R \setminus \{r\}\rangle \subseteq \langle R\rangle$. Thus, $\langle R\rangle \subseteq \langle R \setminus \{r\}\rangle$ if and only if $\langle R \setminus \{r\}\rangle = \langle R\rangle$. □

Lemma 4.7 *Let P and Q be subsets of S with $\langle P\rangle\langle Q\rangle = \langle Q\rangle\langle P\rangle$. Then $\langle P \cup Q\rangle = \langle P\rangle\langle Q\rangle$.*

Proof We are assuming that $\langle P\rangle\langle Q\rangle = \langle Q\rangle\langle P\rangle$. Thus, by Lemma 4.3, $\langle P\rangle\langle Q\rangle$ is closed. Thus, as $P \cup Q \subseteq \langle P\rangle\langle Q\rangle$, $\langle P \cup Q\rangle \subseteq \langle P\rangle\langle Q\rangle$.

From Lemma 4.4, we know that $\langle P\rangle \subseteq \langle P \cup Q\rangle$ and $\langle Q\rangle \subseteq \langle P \cup Q\rangle$. Thus, as $\langle P \cup Q\rangle$ is closed, $\langle P\rangle\langle Q\rangle \subseteq \langle P \cup Q\rangle$. □

We conclude this section by introducing a notational simplification which will be useful in the remainder of this article.

If a subset R of S contains an element r with $R = \{r\}$, we write $\langle r\rangle$ instead of $\langle R\rangle$. We will also write $\langle r_1, \ldots, r_n\rangle$ instead of $\langle R\rangle$ if $r_1, \ldots, r_n$ are elements of a subset R of S with $R = \{r_1, \ldots, r_n\}$.

5 Involutions

In this section, the letter S stands for a hypergroup.

An element s in $S \setminus \{1\}$ is called *involution* if $\{1, s\}$ is a closed subset of S.

Lemma 5.1 *Let l be an involution of S. Then the following hold:*

(i) *We have* $l^* = l$.
(ii) *We have* $ll \subseteq \{1, l\}$.
(iii) *If* l *is not thin,* $ll = \{1, l\}$.

Proof

(i) Since l is assumed to be an involution, $l \neq 1$. Thus, as $1^* = 1$, $l^* \neq 1$. From the hypothesis that l is an involution, we also obtain that $\{1, l\}$ is closed. Thus, by Lemma 4.1, $l^* \in \{1, l\}$. Since $l^* \neq 1$, this implies that $l^* = l$.
(ii) Since l is an involution, $\{1, l\}$ is closed. Thus, $l^*l \subseteq \{1, l\}$. Thus, by (i), $ll \subseteq \{1, l\}$.
(iii) Assume that l is not thin. Then, by definition, $l^*l \neq \{1\}$. On the other hand, as $\{1, l\}$ is a closed subset of S, $l^*l \subseteq \{1, l\}$, and, by Lemma 2.1, $1 \in l^*l$. Thus, $l^*l = \{1, l\}$. Now (i) yields $ll = \{1, l\}$. □

Lemma 5.2 *Let* l *be an involution of* S*, let* p *be a thin element of* S*, and let* q *be an element of* pl*. Then* $q^*q \subseteq \langle l \rangle$.

Proof Since $q \in pl$ we obtain from Lemma 1.1 and Lemma 5.1(i) that $q^* \in lp^*$. Thus, as p is assumed to be thin,

$$q^*q \subseteq lp^*pl \subseteq ll \subseteq \langle l \rangle;$$

cf. Lemma 5.1(ii). □

Lemma 5.3 *Let* h *and* k *be involutions of* S*, and assume that* h *is thin and that* k *is not thin. Then* hk *does not contain an involution.*

Proof Assume, by way of contradiction, that hk contains an involution. We choose one of the involutions in hk and denote it by l.

Since h is an involution, we have $h^* = h$; cf. Lemma 5.1(i). Thus, $k \in hl$. Applying Lemma 5.2 to h and k in place of p and q we now obtain that $kk \subseteq \langle l \rangle$.

Now recall that k is not thin. Thus, by Lemma 5.1(iii), $k \in kk$. It follows that $k \in \langle l \rangle$. Since $k \neq 1$, this implies that $k = l$. Thus, as $k \in hl$, $k \in hk$. Equivalently, $h \in kk$. Thus, as $h \neq 1$ and $kk \subseteq \{1, k\}$, $h = k$, contradicting the fact that h is thin while k is not thin. □

The following lemma generalizes the combinatorial part of [2, Theorem 4.1].

Lemma 5.4 *Let* h *and* k *be two distinct involutions of* S*, and assume that* hk *is* **-invariant. Then* $\langle h, k \rangle = \{1, h, k\} \cup hk$ *and* $hk \cap \{1, h, k\}$ *is empty.*

Proof We are assuming that hk is *-invariant. Since h and k are assumed to be involutions, we also have $h^* = h$ and $k^* = k$; cf. Lemma 5.1(i). Thus, by Lemma 3.1,

$$hk = (hk)^* = k^*h^* = kh,$$

and then

$$\langle h\rangle\langle k\rangle = \{1, h, k\} \cup hk = kh \cup \{1, h, k\} = \langle k\rangle\langle h\rangle.$$

Thus, by Lemma 4.7, $\langle h, k\rangle = \langle h\rangle\langle k\rangle = \{1, h, k\} \cup hk$.

That $hk \cap \{1, h, k\}$ is empty follows from the fact that h and k are two distinct involutions. □

Lemma 5.5 *Let T be a closed subset of S, and let l be an involution of S with $Tl = lT$. Then $T\langle s\rangle = T\langle l\rangle$ for each element s in $T\langle l\rangle \setminus T$.*

Proof Let s be an element in $T\langle l\rangle \setminus T$. From $Tl = lT$, we obtain that $T\langle l\rangle = \langle l\rangle T$. Thus, by Lemma 4.3, $T\langle l\rangle$ is closed. Thus, as $T \subseteq T\langle l\rangle$ and $s \in T\langle l\rangle$, $T\langle s\rangle \subseteq T\langle l\rangle$.

From $s \in T\langle l\rangle \setminus T$, we obtain an element t in T such that $s \in tl$. It follows that $l \in t^*s \subseteq T\langle s\rangle$. Thus, $Tl \subseteq T\langle s\rangle$. Thus, as $T \subseteq T\langle s\rangle$, $T\langle l\rangle \subseteq T\langle s\rangle$. □

6 Projective Hypergroups

A hypergroup S is called *projective* if each element of $S \setminus \{1\}$ is an involution.

Each elementary abelian 2-group (viewed as a hypergroup) is an example of a projective hypergroup.

Lemma 6.1 *The following hold:*

(i) *Projective hypergroups are commutative.*
(ii) *Closed subsets of projective hypergroups are projective.*
(iii) *A projective hypergroup which has a thin nonidentity element consists of thin elements only.*

Proof

(i) By definition, each nonidentity element of a projective hypergroup is an involution, and from Lemma 5.1(i), we know that involutions are symmetric. Thus, projective hypergroups are symmetric, so that the claim follows from Lemma 3.2.
(ii) This follows from the definition of projective hypergroups.
(iii) This is a consequence of Lemma 5.3. □

Let S be a projective hypergroup.

A subset R of $S \setminus \{1\}$ is called *independent* if $\langle R \setminus \{l\}\rangle \neq \langle R\rangle$ for each element l in R.

Notice that subsets of $S \setminus \{1\}$ are independent if they have at most two elements.

Lemma 6.2 *Subsets of independent subsets of projective hypergroups are independent.*

Proof Let S be a projective hypergroup, let P and Q be subsets of S with $P \subseteq Q$, and assume that Q is independent. We will see that P is independent.

Let l be an element in P. Then, as $P \subseteq Q$, $l \in Q$. Thus, as Q is assumed to be independent, $\langle Q \setminus \{l\}\rangle \neq \langle Q\rangle$. Thus, by Lemma 4.6, $l \notin \langle Q \setminus \{l\}\rangle$. From $P \setminus \{l\} \subseteq Q \setminus \{l\}$,we also obtain that $\langle P \setminus \{l\}\rangle \subseteq \langle Q \setminus \{l\}\rangle$; cf. Lemma 4.4. It follows that $l \notin \langle P \setminus \{l\}\rangle$. Thus, by Lemma 4.6, $\langle P \setminus \{l\}\rangle \neq \langle P\rangle$.

Since l has been chosen arbitrarily from the set P, we have shown that P is independent. □

Proposition 6.3 *Let S be a projective hypergroup, let Q and R be finite independent subsets of S, and assume that $\langle Q\rangle \subseteq \langle R\rangle$. Then R contains a subset P such that $|P| = |Q|$ and $\langle (R \setminus P) \cup Q\rangle = \langle R\rangle$.*

Proof If Q is empty, we set $P := \emptyset$. Then $|P| = |Q|$ and $\langle (R \setminus P) \cup Q\rangle = \langle R\rangle$.

Assume that Q is not empty, let l be an element in Q, and set $Q' := Q \setminus \{l\}$. By induction, R contains a subset P' such that $|P'| = |Q'|$ and

$$\langle (R \setminus P') \cup Q'\rangle = \langle R\rangle.$$

Since Q is assumed to be independent, $l \notin \langle Q'\rangle$; cf. Lemma 4.6. On the other hand, we have $l \in \langle R\rangle$. Thus, as $\langle (R \setminus P') \cup Q'\rangle = \langle R\rangle$, there exist a subset O of $(R \setminus P') \cup Q'$ and an element k in $R \setminus P'$ such that $Q' \subseteq O$, $l \notin \langle O\rangle$, and $l \in \langle O \cup \{k\}\rangle$.

Now recall from Lemma 6.1(i) that S is commutative. Thus, by Lemma 4.7, $\langle O \cup \{k\}\rangle = \langle O\rangle\langle k\rangle$. Thus, as $l \in \langle O \cup \{k\}\rangle$, $l \in \langle O\rangle\langle k\rangle$. Thus, as $l \notin \langle O\rangle$,

$$\langle O\rangle\langle l\rangle = \langle O\rangle\langle k\rangle;$$

cf. Lemma 5.5.

We set $P := P' \cup \{k\}$. Since $k \notin P'$, we have $|P| = |P'| + 1 = |Q'| + 1 = |Q|$.

From $Q' \subseteq O$ and $O \subseteq (R \setminus P') \cup Q'$, we obtain that

$$(R \setminus P') \cup Q' = (R \setminus P') \cup O.$$

Thus, as $R \setminus P = (R \setminus P') \setminus \{k\}$ and $k \notin O$,

$$(R \setminus P) \cup Q' = ((R \setminus P') \cup Q') \setminus \{k\} = ((R \setminus P') \cup O) \setminus \{k\} = (R \setminus P) \cup O.$$

It follows that

$$(R \setminus P) \cup Q = (R \setminus P) \cup Q' \cup \{l\} = (R \setminus P) \cup O \cup \{l\},$$

so that, by Lemma 4.7,

$$\langle (R \setminus P) \cup Q\rangle = \langle R \setminus P\rangle\langle O\rangle\langle l\rangle.$$

From $P \setminus P' = \{k\}$ and $(R \setminus P') \cup Q' = (R \setminus P') \cup O$, we also obtain that

$$(R \setminus P) \cup O \cup \{k\} = (R \setminus P') \cup Q',$$

so that, again by Lemma 4.7,

$$\langle R \setminus P\rangle\langle O\rangle\langle k\rangle = \langle (R \setminus P') \cup Q'\rangle.$$

Now, recall that $\langle O\rangle\langle l\rangle = \langle O\rangle\langle k\rangle$. Thus,

$$\langle (R \setminus P) \cup Q\rangle = \langle (R \setminus P') \cup Q'\rangle.$$

Since $\langle (R \setminus P') \cup Q'\rangle = \langle R\rangle$, this implies that $\langle (R \setminus P) \cup Q\rangle = \langle R\rangle$. □

An independent subset R of a projective hypergroup S is called a *basis* of S if $\langle R\rangle = S$.

Lemma 6.4 *Let S be a projective hypergroup. Then the following hold:*

(i) *Each maximal independent subset of S is a basis of S.*
(ii) *Each minimal generating set of S is a basis of S.*

Proof

(i) Let P be a maximal independent subset of S. Since $1 \in \langle P\rangle$, it suffices to show that $l \in \langle P\rangle$ for each element l in $S \setminus \{1\}$.
Assume, by way of contradiction, that S contains an element h with $h \notin \langle P\rangle$, and set $Q := P \cup \{h\}$. Since $h \notin P$, the maximal choice of P implies that Q is not independent. Thus, Q contains an element k such that $\langle Q \setminus \{k\}\rangle = \langle Q\rangle$.
Since $h \in Q$, $h \in \langle Q\rangle$. Thus, as $h \notin \langle P\rangle$, $\langle P\rangle \neq \langle Q\rangle$. Thus, as $P = Q \setminus \{h\}$, $\langle Q \setminus \{h\}\rangle \neq \langle Q\rangle$. Thus, as $\langle Q \setminus \{k\}\rangle = \langle Q\rangle$, $k \neq h$. Thus, as $k \in Q = P \cup \{h\}$, $k \in P$.
From $k \in Q$ and $\langle Q \setminus \{k\}\rangle = \langle Q\rangle$, we obtain that $k \in \langle Q \setminus \{k\}\rangle$. Thus, we obtain elements $l_1, \ldots, l_n$ in $Q \setminus \{k\}$ with $k \in l_1 \cdots l_n$; cf. Lemma 4.5. Now recall from Lemma 6.1(i) that S is commutative. Thus, as each of the elements $l_1, \ldots, l_n$ is an involution, we may assume that the elements $l_1, \ldots, l_n$ are pairwise distinct; cf. Lemma 5.1(ii).
Assume that $\{l_1, \ldots, l_n\} \subseteq P$. Then $k \in l_1 \cdots l_n \subseteq \langle P \setminus \{k\}\rangle$. Thus, as $k \in P$, $\langle P \setminus \{k\}\rangle = \langle P\rangle$; cf. Lemma 4.6. This shows that P is not independent, contradiction.
Assume that $\{l_1, \ldots, l_n\} \nsubseteq P$. Then $\{1, \ldots, n\}$ contains an element i such that $l_i = h$. Since S is commutative, we may assume that $i = 1$. Then $k \in hl_2 \cdots l_n$. Thus, by Lemma 3.1, $h \in kl_n \cdots l_2 \subseteq \langle P\rangle$, contradiction.
(ii) Let R be a minimal generating set of S. We have to show that R is independent. Since R is a minimal generating set of S, we have $\langle R \setminus \{l\}\rangle \neq S$ for each element l in R. Thus, as $\langle R\rangle = S$, we have $\langle R \setminus \{l\}\rangle \neq \langle R\rangle$ for each element l in R, and that means that R is independent. □

Theorem 6.5 *Each projective hypergroup possesses a basis.*

Proof Let S be a projective hypergroup, let $\mathscr{C}$ be a chain of independent subsets of S, and let R be the union of the subsets of S which belong to $\mathscr{C}$. We claim that R is independent.

Assume that R is not independent. Then R contains an element l with $\langle R \setminus \{l\}\rangle = \langle R\rangle$. Then, by Lemma 4.6, $l \in \langle R \setminus \{l\}\rangle$. Thus, by Lemma 4.5, $R \setminus \{l\}$ contains elements $l_1, \ldots, l_n$ such that $l \in l_1 \cdots l_n$. Since R is the union of the subsets of S belonging to $\mathscr{C}$, $\mathscr{C}$ contains an element C with $l \in C$ and $\{l_1, \ldots, l_n\} \subseteq C$.

From $l \in l_1 \cdots l_n$ and $\{l_1, \ldots, l_n\} \subseteq C \setminus \{l\}$, we obtain that $l \in \langle C \setminus \{l\}\rangle$. Thus, as $l \in C$, $\langle C \setminus \{l\}\rangle = \langle C\rangle$; cf. Lemma 4.6. Since $l \in C$, this implies that C is not independent, contradicting $C \in \mathscr{C}$.

What we have seen is that each chain of independent subsets of S possesses an upper bound in the set of all independent subsets of S. Thus, by Zorn's Lemma, S possesses an independent subset which is maximal among the independent subsets of S. By Lemma 6.4(i), this set is a basis of S. □

A projective hypergroup S is said to be *finitely generated* if it contains a finite subset R such that $\langle R\rangle = S$.

Theorem 6.6 *Let S be a finitely generated projective hypergroup. Then the following hold:*

(i) *The hypergroup S possesses a finite basis.*
(ii) *Let Q be a finite basis of S, and let P be an independent subset of S. Then $|P| \leq |Q|$.*
(iii) *Any two bases of S have the same cardinality.*

Proof

(i) Since S is assumed to be finitely generated, S possesses a finite subset Q such that $\langle Q\rangle = S$. Among the subsets of Q which generate S, we choose P minimal. Then, by Lemma 6.4(ii), P is a basis of S.
(ii) Assume, by way of contradiction, that $|Q| + 1 \leq |P|$. Then P contains a subset O with $|Q| + 1 = |O|$. Since Q is assumed to be finite, so is O. Moreover, by Lemma 6.2, O is independent. Thus, as $\langle O\rangle \subseteq \langle Q\rangle$, $|O| \leq |Q|$; cf. Proposition 6.3. This contradiction shows that $|P| \leq |Q|$.
(iii) This follows from (ii). □

Let S be a finitely generated projective hypergroup. In Theorem 6.6(i), we saw that S possesses a finite basis, and in Theorem 6.6(iii), we saw that any two bases of S have the same number of elements. This number is called the *dimension* of S and will be denoted by $\dim(S)$.

Theorem 6.7 *Let S be a finitely generated projective hypergroup, and let T be a closed subset of S. Then the following hold:*

(i) *The projective hypergroup T is finitely generated, and* $\dim(T) \leq \dim(S)$.

(ii) *Assume that* $\dim(T) = \dim(S)$. *Then* $T = S$.

Proof

(i) Let Q be a basis of S. Since S is assumed to be finitely generated, Q is finite; cf. Theorem 6.6(i), (iii).
Let P be a basis of T. Then P is independent. Thus, as Q is a finite basis of S, $|P| \leq |Q|$; cf. Theorem 6.6(ii). Thus, as Q is finite, so is P. Since $\langle P \rangle = T$, this implies that T is finitely generated.
Since P is a basis of T and Q is a basis of S, we obtain from $|P| \leq |Q|$ that $\dim(T) \leq \dim(S)$.

(ii) Let P be a basis of T, and let Q be a basis of S. Then $|P| = |Q|$, since we are assuming that $\dim(T) = \dim(S)$. Thus, by Proposition 6.3, $T = \langle P \rangle = \langle Q \rangle = S$. □

Lemma 6.8 *Let S be a projective hypergroup. Then the following hold:*

(i) *Let h and k be two distinct elements in* $S \setminus \{1\}$. *Then* $\langle h, k \rangle$ *is the only closed subset of S which contains h and k and has dimension* 2.

(ii) *Let h, k, and l be elements in* $S \setminus \{1\}$. *Assume that* $h \neq l$ *and that* $h \in \langle k, l \rangle$. *Then* $k \in \langle h, l \rangle$.

(iii) *Let i, j, h, and k be elements in* $S \setminus \{1\}$. *Assume that* $i \neq j$, *that* $h \neq k$, *that* $i \notin \langle h, k \rangle$, $j \notin \langle h, k \rangle$, $h \notin \langle i, j \rangle$, *and* $k \notin \langle i, j \rangle$. *Assume that* $\dim(\langle i, j \rangle \cap \langle h, k \rangle) = 1$. *Then* $\dim(\langle i, k \rangle \cap \langle j, h \rangle) = 1$.

Proof

(i) Since $h \neq k$, $\langle h \rangle \neq \langle h, k \rangle$ and $\langle k \rangle \neq \langle h, k \rangle$. Thus, $\{h, k\}$ is independent. Thus, as $\langle h, k \rangle$ is generated by $\{h, k\}$, $\{h, k\}$ is a basis of $\langle h, k \rangle$. Since $h \neq k$, this implies that $\dim(\langle h, k \rangle) = 2$.
Let T be a closed subset of S with $\{h, k\} \subseteq T$ and $\dim(T) = 2$. Since $\{h, k\} \subseteq T$, $\langle h, k \rangle \subseteq T$. Thus, as $\dim(\langle h, k \rangle) = 2 = \dim(T)$, $\langle h, k \rangle = T$; cf. Theorem 6.7(ii).

(ii) We are assuming that $h \neq l$. Thus, by (i), $\dim(\langle h, l \rangle) = 2$.
Assume that $k = l$. Then $\langle k, l \rangle = \langle l \rangle$, and that means that $\dim(\langle k, l \rangle) = 1$. On the other hand, as $h \in \langle k, l \rangle$, $\langle h, l \rangle \subseteq \langle k, l \rangle$. Thus, as $\dim(\langle h, l \rangle) = 2$, we obtain from Theorem 6.7(i) that $2 \leq \dim(\langle k, l \rangle)$, contradiction.
Assume that $k \neq l$. Then, by (i), $\dim(\langle k, l \rangle) = 2$. Thus, as $\dim(\langle h, l \rangle) = 2$ and $\langle h, l \rangle \subseteq \langle k, l \rangle$, $\langle h, l \rangle = \langle k, l \rangle$; cf. Theorem 6.7(ii). It follows that $k \in \langle h, l \rangle$.

(iii) We are assuming that $\dim(\langle i, j \rangle \cap \langle h, k \rangle) = 1$. Thus, $S \setminus \{1\}$ contains an element m such that $\langle m \rangle = \langle i, j \rangle \cap \langle h, k \rangle$.
Since $j \notin \langle h, k \rangle$, $m \neq j$. Thus, as $m \in \langle i, j \rangle$, $i \in \langle m, j \rangle$; cf. (ii). Thus, as $m \in \langle h, k \rangle$, $i \in \langle j, h, k \rangle$. On the other hand, we know from Lemma 4.7 that $\langle j, h, k \rangle = \langle j, h \rangle \langle k \rangle$. Thus, $i \in \langle j, h \rangle \langle k \rangle$. Thus, $\langle j, h \rangle$ contains an element n such that $i \in n \langle k \rangle$.
From $i \in n \langle k \rangle$, we obtain that $i \in \langle n, k \rangle$. Thus, as $i \neq k$, $n \in \langle i, k \rangle$; cf. (ii).
From $i \in n \langle k \rangle \setminus \langle k \rangle$, we obtain that $n \neq 1$. Thus, as $n \in \langle i, k \rangle$ and $n \in \langle j, h \rangle$,

we have $1 \leq \dim(\langle i, k\rangle \cap \langle j, h\rangle)$.
Assume that $\dim(\langle i, k\rangle \cap \langle j, h\rangle) = 2$. Then $h \in \langle i, k\rangle$. Since $h \neq k$, this implies that $i \in \langle h, k\rangle$; cf. (ii). This contradiction shows that $\dim(\langle i, k\rangle \cap \langle j, h\rangle) = 1$. □

We will now show that each projective hypergroup gives rise to a projective space. We begin by recalling the definition of a projective space.

Let P and L be two sets which do not have an element in common, let I be a subset of $P \times L$, and set $\mathscr{P} := (P, L, I)$.

The elements in P are called the *points* of $\mathscr{P}$, the elements in L are called the *lines* of $\mathscr{P}$.

Let p be a point of $\mathscr{P}$, and let l be a line of $\mathscr{P}$. We say that p and l are *incident* if $(p, l) \in I$. In this case, we also say that p is *incident with* l or that l is *incident with* p.

The triple $\mathscr{P}$ is called a *projective space* if the following three conditions hold:

P1 For any two distinct points q and r of $\mathscr{P}$, there exists exactly one line l of $\mathscr{P}$ which is incident with p and q.
P2 Each line of $\mathscr{P}$ is incident with at least three points of $\mathscr{P}$.
P3 Let l and m be two different lines of $\mathscr{P}$ intersecting at a point q. Let t and u be two distinct points of $\mathscr{P}$, incident with l and distinct from q, let v and w be two distinct points of $\mathscr{P}$, incident with m and distinct from q. Then the lines of $\mathscr{P}$, incident with t and w and u and v, respectively, intersect.

Let m and n be lines of a projective space $\mathscr{P}$. It is easy to see and well known that the number of points of $\mathscr{P}$ incident with m is equal to the number of points of $\mathscr{P}$ incident with n. This number minus 1 is called the *order* of $\mathscr{P}$.

Theorem 6.9 *Let S be a projective hypergroup. Define*
$P := \{\langle l\rangle \mid l \in S \setminus \{1\}\}$,
$L := \{\langle h, k\rangle \mid h, k \in S \setminus \{1\},\ h \neq k\}$,
$I := \{(\langle h\rangle, \langle h, k\rangle) \mid h, k \in S \setminus \{1\},\ h \neq k\}$,
and $\mathscr{P} := (P, L, I)$. Then the following hold:

(i) *The triple $\mathscr{P}$ is a projective space.*
(ii) *Let h and k be two distinct elements in $S \setminus \{1\}$. Then $\mathscr{P}$ has order $|hk| + 1$.*

Proof

(i) Condition P1 follows from Lemma 6.8(i). Condition P2 holds, since we have $\langle h, k\rangle = \{1, h, k\} \cup hk$ for any two distinct elements h and k in $S \setminus \{1\}$; cf. Lemma 5.4. Condition P3 follows from Lemma 6.8(iii).
(ii) From Lemma 5.4, one obtains that $\langle h, k\rangle = \{1, h, k\} \cup hk$. This shows that the closed subset $\langle h, k\rangle$ contains $|hk| + 2$ involutions. □

From Theorem 6.9, one obtains serious restrictions on the order of finite projective hypergroups. In fact, let S be a finite projective hypergroup, and set $d := \dim(S)$.

If $d = 0$, we have $|S| = 1$, and if $d = 1$, we have $|S| = 2$.

Assume that $2 \leq d$, let h and k be two distinct elements in $S \setminus \{1\}$, and set $n := |hk| + 1$. Then the projective space associated to S (via Theorem 6.9) has order n. A standard counting argument in the theory of projective spaces shows that

$$|S \setminus \{1\}| = n^{d-1} + \cdots + n + 1.$$

Moreover, from the well-known identification of finite projective spaces of (projective) dimension at least 3 with vector spaces over finite fields, one obtains that n is a prime power if $4 \leq d$.

7 Constrained Hypergroups

In this section, we show in how far projective hypergroups arise in the investigation of constrained hypergroups.

Let S be a hypergroup, let L be a set of involutions of S, and assume that $S = \langle L \rangle$.

From Lemma 4.5, we obtain that, for each element s in S, there exists a nonnegative integer n satisfying $s \in L^n$. The smallest nonnegative integer n with $s \in L^n$ is called the *length* of s (with respect to L) and will be denoted by $\ell(s)$.

Since L consists of involutions only, L is symmetric. In particular, L is *-invariant. Thus, we have $\ell(s^*) = \ell(s)$ for each element s in S; cf. Lemma 3.4(ii).

From the definition of ℓ, one obtains that

$$\ell(r) \leq \ell(p) + \ell(q)$$

for any three elements p, q, and r in S with $r \in pq$. This observation advocates, of course, the interest in elements p, q, and r in S satisfying $\ell(r) = \ell(p) + \ell(q)$.

For each element q in S, we define $L_1(q)$ to be the set of all elements p in S such that pq contains an element r with $\ell(r) = \ell(p) + \ell(q)$.

For each element q in S, we define $L_{-1}(q)$ to be the set of all elements r in S such that there exists an element p in S satisfying $r \in pq$ and $\ell(r) = \ell(p) + \ell(q)$.

Notice that, for each element q in S, $1 \in L_1(q)$ and $q \in L_{-1}(q)$.

The set L is said to be *constrained* if $|sl| = 1$ for any two elements l in L and s in $L_1(l)$.[1] It is said to satisfy the *exchange condition* if $hs \subseteq sk \cup L_1(k)$ for any three elements k in L, s in $L_1(k)$, and h in L with $h \in L_1(s)$.[2]

The following result clarifies the relationship between constrained sets of involutions of a hypergroup and the exchange condition. It is due to Wang; cf. [11].

[1]Note that a set L of involutions is constrained if $|s\langle l\rangle| = 2$ for any two elements l in L and s in $L_1(l)$.

[2]It is easy to see that a set of thin involutions satisfying the exchange condition satisfies the group-theoretic exchange condition.

Theorem 7.1 *Let S be a hypergroup, let L be a constrained set of involutions of S, and assume that none of the elements of L is thin. Then L satisfies the exchange condition if and only if $L \cup \{1\}$ contains no projective hypergroup of dimension at least* 2.

Wang's proof of Theorem 7.1 is based on a delicate analysis of the sets $L_{-1}(s)$ and $L_1(s)$ with $s \in S$.

Let S be a hypergroup, let L be a non-empty set of involutions of S with $S = \langle L \rangle$. The hypergroup S is called *Coxeter hypergroup* with respect to L if L is constrained and satisfies the exchange condition. It is the idea of a Coxeter hypergroup to provide a framework for the theory of buildings, since each building gives rise to a Coxeter hypergroup.

Theorem 7.1 says that the difference between hypergroups which are Coxeter hypergroups with respect to a set of non-thin involutions and hypergroups generated by a constrained set of non-thin involutions is decided by projective hypergroups. Referring to the relationship between buildings and Coxeter hypergroups the theorem implies that thick buildings can be investigated just with the help of projective hypergroups and constrained hypergroups. It suggests that the investigation of thick buildings does not depend on the exchange condition.

8 An Application to Table Algebras

In this section, we relate Theorem 6.9 to a result of Blau and Chen on table algebras; cf. [2]. We first recall the definition of a table algebra. The definition is taken from [2].

Let A be a finite-dimensional associative algebra (with 1) over the field of the complex numbers, let B be a basis of A with $1 \in B$. The pair (A, B) is called a *table algebra* if the following three conditions hold:

T1 The structure constants of A with respect to B are nonnegative real numbers.
T2 The algebra A possesses an involutory algebra anti-automorphism * with $B^* = B$.
T3 For any two elements c and d in B, 1 is in the support of cd if and only if $d = c^*$.

Let (A, B) be a table algebra.

Let c, d, and e be elements in B. The structure constant of (A, B) defined by c, d, and e will be denoted by a_{cde}.

From [2, Remark 2.1], we know that there exists a unique algebra homomorphism δ from A to the field of the complex numbers such that $\delta(b^*) = \delta(b)$ for each element b in B. The elements $\delta(b)$ with $b \in B$ are called the *degrees* of (A, B).

The table algebra (A, B) is called *standard* if $a_{bb^*1} = \delta(b)$ for each element b in B. It is called *integral* if all its structure constants and degrees are rational integers.

For any two elements c and d in B, we denote by $\mathrm{supp}_A(c, d)$ the support of the product cd in A.

Lemma 8.1 *Let* (A, B) *be a table algebra. Then* B *is a hypergroup with respect to* supp_A.

Proof That supp_A is a hyperoperation on B follows from the definition of supp_A. Condition H1 follows from the associativity of A. Condition H2 follows from the definition of 1. Condition H3 follows from [1, Proposition 2.1]. □

Let (A, B) be a table algebra. An element b in $B \setminus \{1\}$ is called an *involution* of B if $\text{supp}_A(b, b) \subseteq \{1, b\}$. For any two elements c and d in B, we set $\sigma(c, d) := \{c, d\} \cup \text{supp}_A(c, d)$.

The following theorem is due to Blau and Chen; cf. [2, Theorem 1.1].

Theorem 8.2 *Let* (A, B) *be a standard integral table algebra, and assume that each element in* $B \setminus \{1\}$ *is an involution. Define*

$P := B \setminus \{1\}$,
$L := \{\sigma(c, d) \mid c, d \in B \setminus \{1\},\ c \neq d\}$,
$I := \{(c, \sigma(c, d)) \mid c, d \in B \setminus \{1\},\ c \neq d\}$,

and $\mathscr{P} := (P, L, I)$. *Then* $\mathscr{P}$ *is a projective space.*

Proof Considering Lemma 8.1 this follows from Theorem 6.9. □

In [2, Theorem 1.1(ii)], Blau and Chen have computed the order of the projective space $\mathscr{P}$ arising in Theorem 8.2.

Acknowledgements The author thanks an anonymous referee for several clarifying comments on the first submission of this article.

References

1. Blau, H.: Table algebras. Europ. J. Combin. **30**, 1426–1455 (2009)
2. Blau, H.I., Chen, G.: All-involution table algebras and finite projective spaces. Arch. Math. **105**, 313–322 (2015)
3. Bloom, W., Heyer, H.: Harmonic Analysis of Probability Measures on Hypergroups. de Gruyter, Berlin (1995)
4. Connes, A., Consani, C.: The hyperring of adèle classes. J. Number Theory **131**, 159–194 (2011)
5. Jun, J.: Hyperstructures of affine algebraic group schemes. J. Number Theory **167**, 336–352 (2016)
6. Krasner, M.: Approximation des corps valués complets de caractéristique p, $p{>}0$, par ceux de caractéristique zero, *Colloque d'Algèbre Supérieure* (Bruxelles, Décembre 1956) CBRM, Bruxelles (1957)
7. Marty, F.: Sur une généralisation de la notion de groupe, in Huitième Congres des Mathématiciens. Stockholm 45–59 (1934)
8. McMullen, J.R.: An algebraic theory of hypergroups. Bull. Austral. Math. Soc. **20**, 35–55 (1979)
9. Sunders, V.S., Wildberger, N.J.: Actions of finite hypergroups. J. Algebraic Combin. **18**, 135–151 (2003)

10. van Dam, E.R.: A characterization of association schemes from affine spaces. Des. Codes Cryptogr. **21**, 83–86 (2000)
11. Wang, A.: Constrained and Coxeter Table Algebras, Doctoral Dissertation, Department of Mathematical Sciences, Northern Illinois University, DeKalb, Illinois (2016)
12. Zieschang, P.-H.: Theory of Association Schemes. Springer Monographs in Mathematics, Berlin, Heidelberg, New York (2005)